TRACE ELEMENTS
IN
NATURAL WATERS

TRACE ELEMENTS IN NATURAL WATERS

Edited by
Brit Salbu, Dr. Philos.
Laboratory for Analytical Chemistry
Agricultural University of Norway
Ås, Norway

Eiliv Steinnes, Dr. Philos.
Department of Chemistry
University of Trondheim
Trondheim, Norway

CRC Press
Boca Raton Ann Arbor London Tokyo

Library of Congress Cataloging-in-Publication Data

Trace elements in natural waters / edited by Brit Salbu, Eiliv Steinnes.
p. cm.
Includes bibliographical references and index.
ISBN 0-8493-6304-7
1. Water chemistry. 2. Trace elements. I. Steinnes, Eiliv, 1938– .
GB855.T72 1994
628.1'61—dc20

94-13929
CIP

International Standard Book Number 0-8493-6304-7
Library of Congress Card Number 94-13929
Printed in the United States of America 3 4 5 6 7 8 9 0
Printed on acid-free paper

PREFACE

Natural waters are dispersed multielemental and multispecies systems that derive their composition from natural processes and that are affected by or the result of human activity. A few of the chemical elements are present in natural waters at concentrations of 10^{-5} M and above, and are termed macroelements. Macroelements are essential for characterizing the physico-chemical properties of the system (pH, ionic strength, color, etc.), and for a given system, their concentration levels may remain relatively stable.

Most elements in natural waters, however, are present at concentrations below 10^{-5} M and are termed trace elements. At such low concentration levels, microchemical phenomena, including interactions on phase boundaries and colloid chemistry, may become very significant and must be taken into account. Chemical reactions with trace elements are therefore much more difficult to predict than those involving the macroelements, and their concentration levels often show large temporal and spatial variations even within specific aquatic systems. It should be noted that most natural and artificially produced radionuclides appear in natural waters at trace concentrations and are influenced by microchemical processes like stable elements. Radionuclides are therefore not treated separately in this book.

Trace elements may be present in natural waters in a variety of physico-chemical forms (species) varying in size (nominal molecular weight), charge, redox and density properties influencing their mobility, and biological uptake. Information on the distribution of different species, the transformation processes, and the kinetics involved in these processes is therefore essential when assessing short- and long-term consequences of trace elements released to aquatic media from natural and anthropogenic sources.

The characterization of the different physico-chemical forms (species) constitutes a major challenge within analytical chemistry. Owing to the introduction of virtually contamination-free sampling and pre-analysis handling techniques and the development of increasingly sensitive analytical methods, considerable information about the speciation of aquatic trace elements has been achieved in recent years. By interfacing well-defined techniques with even more sensitive detection systems, major progress is expected in the future.

In order to present the state of the art with respect to trace elements in the various compartments of the hydrosphere, established scientists were invited to contribute with their specific competence. Chapters 1 through 4 deal with general aspects concerning sources and behavior of trace elements in aquatic systems, explain how to obtain reliable and meaningful analytical data, and discuss how to treat them statistically. In Chapters 5 through 11 the trace element occurrence and behavior in major parts of the hydrological cycle, respectively precipitation, lakes, rivers, interstitial waters, ground waters, estuaries, and oceans are reviewed. Chapter 12 is an attempt to summarize the state of the art regarding trace elements in natural waters and to point out some areas for future research.

In this book the liquid phase of H_2O is emphasized; thus there are no chapters on, for example, interactions of water vapor (volcanic) with mineral phases or on the occurrence and behavior of trace elements in ice and snow. The subject of trace elements in glacial ice is briefly mentioned in Chapter 12. Even though the present text is rather comprehensive, the reader may also find other important aspects that have been left out or only very briefly examined. For example, Chapter 6 on interstitial waters concentrates on sediment waters, while the water phase in the terrestrial unsaturated zone, including the surface soil where the plants derive their nutrition, is not discussed in depth. In spite of this, however, it is felt that the text covers most of the essential issues related to trace elements in natural waters.

The editors wish to express their sincere thanks to all the contributors for their positive response and enthusiasm which reflect their devotion to this field of science and for their cooperation and patience during the final stages of preparation.

Brit Salbu
Eiliv Steinnes

THE EDITORS

Brit Salbu, Cand. Real, Dr. Philos.
Director and Professor, Laboratory for Analytical Chemistry, Agricultural University of Norway, Ås, Norway

Dr. Salbu received the Cand. Real. degree in nuclear chemistry in 1974 and was awarded the Dr. Philos. degree in analytical nuclear chemistry in 1985, both at the Institute of Chemistry, University of Oslo, Oslo, Norway.

She served as assistant professor and senior scientist at the University of Oslo until 1987. Then she was appointed Director and Professor of the Isotope Laboratory, Agricultural University of Norway. After the reorganization of institutes at the Agricultural University of Norway, she was appointed Director of the Laboratory for Analytical Chemistry, which consists of four sections: organic analytical chemistry, inorganic analytical chemistry, isotope laboratory, and electron microscopy laboratory.

Dr. Salbu is a member of the Norwegian Academy of Science and Letters (Oslo). She is also a member of several national and international associations. She has served as a member of several scientific committees within the Norwegian Research Councils, as an advisory expert for the Ministry of Environment, and as a member of the governmental advisory group in connection with national preparedness associated with nuclear accidents.

Dr. Salbu is the author or co-author of more than 150 papers and book chapters and has edited one book. During the last 10 years, her research has concentrated on trace elements and radionuclides in the natural environments, and especially in waters with special emphasis on physico-chemical forms (speciation), transformation processes, mobility, and bioavailability.

In connection with bilateral and international research programs, she has been heavily involved in scientific expeditions, especially to the Ukraine, Belarus, and Russia, areas affected by the Chernobyl accident, and to the Kara Sea and Ural.

She is also a fellow of the Explorers Club.

Eiliv Steinnes, Cand. Real., Dr. Philos.
Professor of Environmental Science, Department of Chemistry, University of Trondheim, Trondheim, Norway

Dr. Steinnes was awarded the Cand. Real. degree in nuclear chemistry in 1963 and the Dr. Philos. degree in analytical chemistry in 1972, both by the University of Oslo. At the Norwegian Institute for Atomic Energy, he served as a research scientist from 1964 to 1968, as Head of the Radio-Analytical Section from 1968 to 1977, and as Head of the Analytical Division from 1978 to 1979. In 1976 he was Visiting Scientist at the U.S. Geological Survey, Reston, Virginia. Since 1980, he has held a chair in Environmental Science at the University of Trondheim where he served as Associate Dean of the Faculty of Science from 1981 to 1984 and as Rector of the College of Arts and Science from 1984 to 1990.

Dr. Steinnes is a member of the Norwegian Academy of Science and Letters (Oslo), the Royal Norwegian Society of Science and Letters (Trondheim) and the Norwegian Academy of Technical Sciences. He is also a member of several other national and international professional associations. Since 1971 he has served on the IUPAC Commission V.7 (Radiochemistry and Nuclear Methods) where he was Chairman from 1981 to 1983. He was President of the Norwegian Society of Soil Science from 1982 to 1986.

Dr. Steinnes is the author or co-author of over 300 papers and book chapters and has edited two books. During the last 20 years, his research has concentrated on trace elements in the natural environment, with particular emphasis on the importance of the atmospheric pathway for the supply of trace elements of natural and anthropogenic origin to terrestrial and aquatic ecosystems. He also contributed to the early studies on speciation of trace elements in natural waters.

CONTRIBUTORS

Bert Allard, Ph.D.
Department of Water and Environmental Studies
Linköping University
Linköping, Sweden

Sjur Andersen, Dr. Scient.
JORDFORSK
Ås, Norway

Petr Beneš, Ph.D., D.Sc.
Department of Nuclear Chemistry
Czech Technical University
Prague, Czech Republic

Hans Borg, Ph.D.
Institute of Applied Environmental Research
Laboratory for Aquatic Environmental Chemistry
Stockholm University
Solna, Sweden

Owen P. Bricker, Ph.D.
U.S. Geological Survey
Reston, Virginia

Kenneth W. Bruland, Ph.D.
Institute of Marine Sciences
University of California, Santa Cruz
Santa Cruz, California

Wolfgang Calmano, Ph.D.
Arbeitsbereich Umweltschutztechnik
Technische Universität Hamburg-Harburg
Hamburg, Germany

John R. Donat, Ph.D.
Department of Chemistry and Biochemistry
Old Dominion University
Norfolk, Virginia

Ulrich Förstner, Ph.D.
Arbeitsbereich Umweltschutztechnik
Technische Universität Hamburg-Harburg
Hamburg, Germany

Barry T. Hart, Ph.D.
Water Studies Centre & Department of Chemistry
Monash University
Melbourne, Australia

Tina Hines
Water Studies Centre & Department of Chemistry
Monash University
Melbourne, Australia

Jihua Hong
Arbeitsbereich Umweltschutztechnik
Technische Universität Hamburg-Harburg
Hamburg, Germany

Blair F. Jones, Ph.D.
U.S. Geological Survey
Reston, Virginia

Geoffrey E. Millward, Ph.D.
Department of Environmental Sciences
University of Plymouth
Plymouth, U.K.

Deborah H. Oughton, Ph.D.
Laboratory for Analytical Chemistry
Agricultural University of Norway
Ås, Norway

Howard B. Ross, Ph.D.
Department of Meteorology
Stockholm University
Stockholm, Sweden

Morten Schaanning, Dr. Scient.
JORDFORSK
Ås, Norway

Brit Salbu, Dr. Philos.
Laboratory for Analytical Chemistry
Agricultural University of Norway
Ås, Norway

Eiliv Steinnes, Dr. Philos.
Department of Chemistry
University of Trondheim
Trondheim, Norway

Andrew Turner, Ph.D.
Department of Environmental Sciences
University of Plymouth
Plymouth, U.K.

Stephen J. Vermette, Ph.D.
Department of Earth Sciences and Sciences Education
Buffalo State College
Buffalo, New York

Nils B. Vogt, Dr. Philos.
Nycomed Imaging AS
Oslo, Norway

Rolf D. Vogt, Cand. Scient.
Department of Chemistry
University of Oslo
Oslo, Norway

CONTENTS

Chapter 1

Main Factors Affecting the Composition of Natural Waters

Owen P. Bricker and Blair F. Jones

CONTENTS

I. INTRODUCTION

Natural waters acquire their chemical compositions from a variety of sources. They accumulate dissolved and suspended constituents through contact with the gases, liquids, and solids they encounter during their passage through the hydrologic cycle (Figure 1). The composition of the oceans has been constant on a time scale of millions of years;[1-3] however, the composition of surface waters and groundwaters continually evolves and changes on time scales of minutes to years, as these waters move along hydrologic flow paths that bring them into contact with a variety of geologic materials and biological systems.[4] Most terrestrial surface waters and groundwaters are supplied by rain and snow as a part of the hydrologic cycle. Atmospheric deposition is not pure water but contains dissolved substances that vary in amount and composition according to location and season. Under natural conditions, rain usually is slightly acidic because of reactions with atmospheric CO_2 and naturally occurring gaseous sulfur and nitrogen compounds. Naturally occurring organic acids may also contribute to the acidity of rain.[5] Anthropogenic emissions to the atmosphere in industrialized regions significantly increase rain acidity, creating the undesirable phenomenon referred to as "acid rain".[6,7] Close to coastal areas, sea salts comprise a large part of the dissolved substances in rain.[8] In western Australia, nearly 97% of the solutes in waters of the Yilgarn block are derived from marine aerosols.[9] Farther inland, the sea-salt component decreases, but other processes, such as dissolution of airborne dust, contribute to the dissolved-substances load.[10] Seasonal effects are observed in rain chemistry. For example, sulfate concentration in rain is commonly higher during summer than in winter because of increased photooxidation of SO_2 in the summer.[11]

In addition to material deposited in wet deposition, a substantial amount of particulate material may be entrained in the atmosphere by winds in arid regions and deposited far from the source. A good example is dust picked up by storms in the Sahara desert that is transported by winds across the Atlantic ocean to Bermuda and the U.S.[12-14] Volcanic eruptions eject particulates and gases into the atmosphere, and debris from large eruptions may be distributed over much of the globe.[15] These gases react with moisture in the atmosphere and affect the composition of rain. Particulate matter falls on the land surface and contributes to loads of suspended and dissolved substances in natural waters.

The composition of atmospheric deposition is modified by biological processes as the deposition passes through the vegetative canopy and upper organic soil horizons[16-18] and by mineral-weathering processes as the deposition passes through underlying mineral soil and bedrock.[19,20] Further changes in chemistry occur when the waters undergo evaporative concentration.[21] Eventually the waters reach the ocean, bearing their load of dissolved and suspended substances, and complete the hydrologic cycle. Thus, the main factors that affect the composition of natural waters are the interactions of the water with the gases, liquids, and solids that the waters contact during the hydrologic cycle. These interactions determine the chemical environments in which trace elements are found and place constraints on the

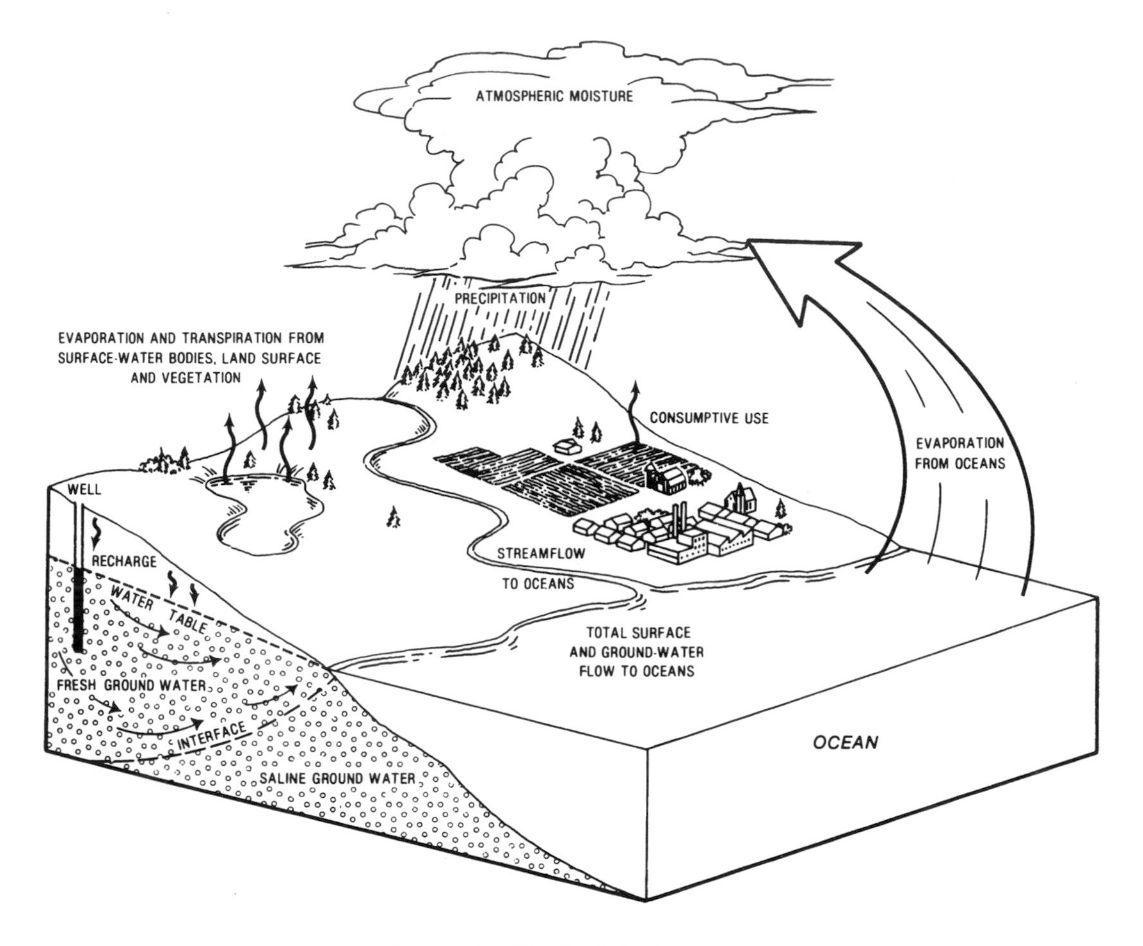

Figure 1 The hydrologic cycle.

transport and fate of trace elements in the environment behavior. In this chapter, we attempt (1) to define the major factors that govern the chemical milieu of natural waters and (2) set the stage for succeeding chapters that focus on trace element interactions in these waters.

II. PHYSICAL CONSIDERATIONS

Water (H_2O) is a unique compound. It has unusually high boiling and freezing points for a low molecular-weight compound and has a very high dielectric constant and surface tension.[22] When water is cooled, it reaches its maximum density at 4°C and then, with further cooling, decreases in density until 0°C, the temperature at which ice begins to form. Ice, the crystalline form of H_2O, is substantially less dense than liquid water.[23] These properties are of profound importance to living organisms and play a significant role in the weathering and erosional process by which H_2O shapes the earth's surface.

The properties of water are related to the structure of the water molecule.[24] The chemical bonds between the hydrogen ions and the oxygen ion are asymmetrical with a bond angle of 105°. The two hydrogen ions are on the same side of the oxygen ion. This arrangement leads to an unbalanced electrical charge that imparts a dipolar character to the molecule. In addition to electrostatic attraction due to the asymmetrical distribution of charge on the water molecule, the hydrogen ions have a capacity for specific interaction with other water molecules. This interaction, called hydrogen bonding, is largely responsible for the unique properties of water.

Because of the dipolar character of water molecules, they tend to be attracted by, and form "spheres" of, hydration around charged dissolved species. Generally, the higher the charge density of the dissolved ion, the larger the sphere of hydration.[25] Dissolved ions affect the physical properties and structure of water.[26,27] The dipolar water molecule also is strongly attracted to most mineral surfaces, thereby enhancing its effectiveness as a weathering agent.[28]

Water is not evenly distributed on the surface of the Earth; about 80% is in the world oceans; 19% is stored in pores and interstices in rocks beneath the earth's surface; 1% is in polar ice caps and glaciers; 0.002% is in streams, rivers, and lakes; and 0.0008% is in the atmosphere.[29] If only freshwater (excluding polar icecaps and glaciers) were considered, Lvovitch[30] estimates that 95% of freshwater is present as groundwater; 3.5% is in lakes, reservoirs, and rivers; and 1.5% is stored as soil moisture. The amount of freshwater discharged annually by rivers to the world's oceans is nearly equal to the total amount of water present in rivers and lakes.[29]

The average residence time of water in various reservoirs has been estimated by Nace.[31] Water in the oceans has a residence time of approximately 4000 years; in lakes and reservoirs, approximately 10 years; in rivers and streams, about 2 weeks; below the earth's surface as groundwater, 2 weeks to 10,000 years; in icecaps and glaciers 10 to 10,000 years; in soil, 2 weeks to 1 year; in the atmosphere, about 10 days.

Two physical factors, the residence time and the pathways or routes along which water moves through the system, are particularly important relative to the chemical composition of natural waters.[44] The residence time is important because the longer the residence time observation in a particular environment, the more opportunity there is for reactions between water and the materials with which it is in contact. Water pathways also play an important role in affecting the chemical composition of water. These pathways determine the materials that water contacts during its passage through the hydrologic system. In general, waters that follow shallow pathways contact more weathered and, consequently, less reactive materials than waters that move along deeper pathways. Therefore, the physical mechanisms that determine the travel time of water, and the pathways along which the water moves, can strongly affect the chemical evolution of the water.

III. SOURCES OF THE CHEMICAL CONSTITUENTS FOUND IN NATURAL WATERS

A number of reservoirs contribute material to natural waters. The major ion chemistry of fresh waters, however, is dominated by a relatively few reservoirs, primarily the atmosphere and the rocks of the earth's crust. Trace elements are also derived from these sources. The biosphere is a source of organic materials to natural waters, but biota also influence the distribution of particulate and dissolved materials through physical mechanisms, and the mobility and speciation of dissolved materials by biochemical processes. Some of the primary factors influencing the major element chemistry of natural waters are discussed below.

A. ATMOSPHERIC FACTORS

The atmospheric reservoir contributes significantly to the dissolved load of many natural waters. Atmospheric processes transport materials that are deposited at the earth's surface in the form of wet and dry deposition. In the past decade, much has been learned about the chemistry of wet deposition and its contribution to watershed systems.[32,33] Information about dry deposition is much less complete, and acceptable methods for its measurement have yet to be developed.[34,35] Estimates of the magnitude of dry deposition of sulfur and nitrogen are between 30 and 60% of the total deposition in the eastern U.S.[36] It appears that dry deposition may be a significant source of materials to watershed systems; however, until better methods are developed for its measurement, dry deposition cannot be quantified.

The chemistry of wet deposition from various localities is shown in Table 1. Rain from near coastal areas reflects a sea salt component (Table 1, columns a and b), but the Na to Cl ion ratios in rain from the midcontinent vary considerably from sea water ratios (Table 1, columns e and d). Rain in heavily industrialized areas contains higher concentrations of sulfate and nitrate and is lower in pH (Table 1, columns e and f) than rain in remote areas (Table 1, columns g and h). This difference reflects anthropogenic emissions of SO_2 and NO_x, primarily due to the combination from fossil fuels.

An assessment of the relative contribution of dissolved substances to surface waters by wet deposition compared to other sources can be made from hydrochemical budget studies of watersheds. In a region receiving wet deposition of similar composition, the atmospherically derived portion of dissolved substances in surface water will be greatest in watersheds underlain by rock types resistant to weathering. For this reason, the effects of acid deposition on surface waters are most apparent in regions underlain by resistant rock types. Geochemical budgets for some watersheds on different bedrock types are sho[illegible] in Table 2. These watersheds, located in the mid-Atlantic region of the U.S., all receive atmos[illegible] deposition of similar composition. Variations in major constituent fluxes into the watersheds [illegible] primarily to differences in rainfall amounts. In this region, atmospheric contributions to the an[illegible]

Table 1 **Precipitation-weighted mean annual concentrations of selected chemical constituents and pH, 1989 (concentration in milligrams per liter; pH in standard unit)**

	a American Samoa	b Olympic Park, WA	c Loch Vale, CO	d South Pass City, CO	e Leading Ridge, PA	f Bennett Bridge, NY	g Everglades Nat. Park, FL	h Manua Loa, HI
Ca	0.08	0.02	0.17	0.28	0.11	0.15	0.13	0.02
Mg	0.199	0.037	0.023	0.033	0.020	0.027	0.062	0.005
K	0.063	0.014	0.014	0.016	0.028	0.019	0.078	0.003
Na	1.677	0.320	0.087	0.23	0.068	0.077	0.487	0.038
NH_4	0.02	0.02	0.14	0.13	0.35	0.36	0.16	0.03
NO_3	0.04	0.08	0.70	0.61	2.14	2.33	0.65	0.09
Cl	3.05	0.58	0.08	0.13	0.17	0.18	0.85	0.06
SO_4	0.57	0.21	0.60	0.91	3.15	2.71	0.77	0.95
pH	5.52	5.45	5.31	5.42	4.16	4.23	5.17	4.82

loads range from about 30% in watersheds with reactive bedrock types to nearly 80% in watersheds underlain by bedrock types resistant to weathering. Some additional portion of the annual stream load is contributed to by atmospheric dry deposition to the watershed, and the remainder originates from watershed processes, primarily mineral weathering. Although atmospheric deposition inputs to watersheds may be substantial relative to the total budget, there are significant differences in the distribution of dissolved species in atmospheric deposition relative to streamflow. Major dissolved species in atmospheric deposition are hydrogen ion, sulfate, and nitrate with smaller amounts of alkali and alkaline-earth cations. Carbonate alkalinity and dissolved silica are usually below limits of detection. Watershed processes consume the hydrogen ion and nitrate and contribute alkali and alkaline-earth cations, silica, and carbonate alkalinity to the waters through mineral weathering reactions and biological processes. Chemical weathering also liberates trace elements that are contained in the minerals of the watershed.

Atmospheric deposition can make a significant contribution to the concentration of major elements in surface waters and groundwaters. It also may be an important source of trace metals to watersheds in industrialized regions. As early as the mid-19th century, Smith documented the deposition of metals by rain in and around the city of Manchester, England.[37] This work was extended to other industrialized regions in England and Europe with similar findings.[6] More recently, the importance of atmosphere transport and deposition of metals has been confirmed by other investigators in various parts of the world.[38-40] More detailed information on atmosphere contributions of trace metals to natural waters can be found in Chapter 6 of this volume.

B. GEOLOGIC FACTORS

The rocks that constitute the earth's crust are the primary reservoir and ultimate source of most major elements and trace elements in soils and in natural waters. Estimates of the composition of the earth were made by Clarke and Washington.[41] These estimates have been revised over the years as better information about the distribution and composition of the rocks of the earth's crust has been obtained.[42-46] The crustal abundance of the first 93 elements are shown in Figure 2.[46] Only a few elements — oxygen, silicon, aluminum, iron, calcium, sodium, potassium, and magnesium — exceed 1% by weight of the earth's crust. Horn and Adams[47] synthesized the crustal abundance data for 65 elements from previous investigations, and Hem (1985) added estimates for carbon and nitrogen based on the work of Parker.[46] The Horn and Adams compilation, as revised by Hem, is shown in Table 3. The average compositions of igneous rocks and the three major groups of sedimentary rocks, (sandstones, shales, and carbonates) are tabulated. ...lues in Table 3, especially for the rarer elements, are high because of the large ...crust inaccessible to direct sampling. Nevertheless, the estimates give a general ...on of elements among the major rock types. Carbonate rocks are enriched in minor ...de, strontium, and manganese. A number of trace metals and boron are strongly ...These differences are important with respect to natural water composition because ...types in the earth's crust is *not uniform.* Igneous rocks constitute nearly 95% of ...The average composition of the entire earth's crust, therefore, closely approxi-...position of igneous rocks. Sedimentary rocks, however, cover about 80% of the

Table 2 **Annual hydrochemical budgets for selected watersheds (mol/ka)**

	Hauver Br., MD Metabasalt		Soldiers Delight, MD Serpentine		Mill Run, VA Quartzite		Fishing Cr., MD Quartzite	
	Precipitation	Stream	Precipitation	Stream	Precipitation	Stream	Precipitation	Stream
H+	1093	0.6	930	0	449	82	535	3
Ca	82.5	928[a]	122	86	34	123	19	62
Mg	23.4	657	41	1445	8	187	6	99
Na	110	686[a]	132	62	39	106	17	174
K	14.2	35	78	9	17	101	14	101
Cl	230	758[a]	124	126	88	93	52	56
SO_4	356	680	451	331	225	448	233	118
NO_3	321	209	365	25	208	3	211	37
HCO_3	0	1680	0	2232	0	14	0	156
H_4SiO_4	0	1420	16	417	0	307	0	468
TDS (moles/ha)	2230	7060	2259	4733	1068	1444	1087	1274
I/O[b]		32%		48%		74%		85%

[a] Influenced by road salt; [b] Input/output.

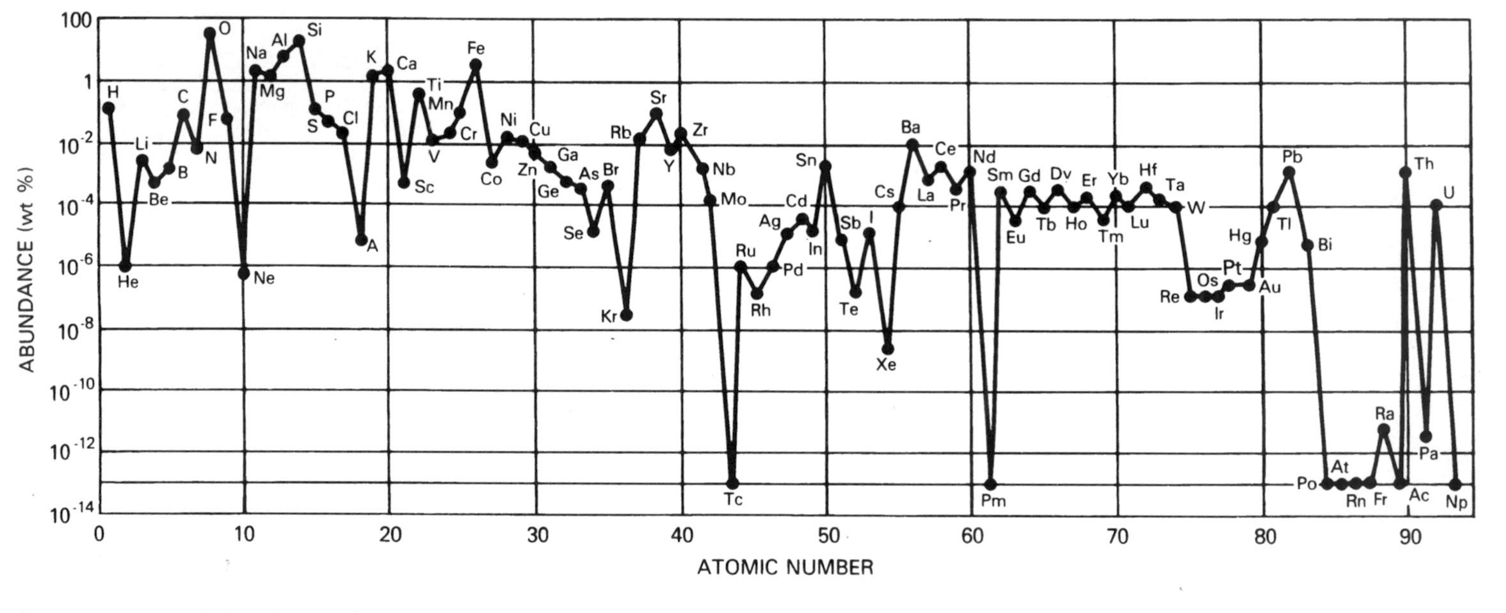

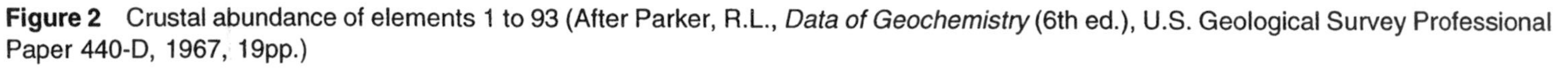
Figure 2 Crustal abundance of elements 1 to 93 (After Parker, R.L., *Data of Geochemistry* (6th ed.), U.S. Geological Survey Professional Paper 440-D, 1967, 19pp.)

Table 3 **Average composition, in parts per million, igneous rocks and some types of sedimentary rocks**

		Sedimentary Rocks		
Element	Igneous Rocks	Resistates (Sandstone)	Hydrolyzates (Shale)	Precipitates (Carbonates)
Si	285,000	359,000	260,000	34
Al	79,500	32,100	80,100	8,970
Fe	42,200	18,600	38,800	8,190
Ca	36,200	22,400	22,500	272,000
Na	28,100	3,870	4,850	393
K	25,700	13,200	24,900	2,390
Mg	17,600	8,100	16,400	45,300
Ti	4,830	1,950	4,440	377
P	1,100	539	733	281
Mn	937	392	575	842
F	715	220	560	112
Ba	595	193	250	30
S	410	945	1,850	4,550
Sr	368	28	290	617
C	320	13,800	15,300	113,500
Cl	305	15	170	305
Cr	198	120	423	7.1
Rb	166	197	243	46
Zr	160	204	142	18
V	149	20	101	13
Ce	130	55	45	11
Cu	97	15	45	4.4
Ni	94	2.6	29	13
Zn	80	16	130	16
Nd	56	24	18	8.0
La	48	19	28	9.4
N	46	—	600	—
Y	41	16	20	15
Li	32	15	46	5.2
Co	23	.33	8.1	.12
Nb	20	.096	20	.44
Ga	18	5.9	23	2.7
Pr	17	7.0	5.5	1.3
Pb	16	14	80	16
Sm	16	6.6	5.0	1.1
Sc	15	.73	10	.68
Th	11	3.9	13	.20
Gd	9.9	4.4	4.1	.77
Dy	9.8	3.1	4.2	.53
B	7.5	90	194	16
Yb	4.8	1.6	1.6	.20
Cs	4.3	2.2	6.2	.77
Hf	3.9	3.0	3.1	.23
Be	3.6	.26	2.1	.18
Er	3.6	.88	1.8	.45
U	2.8	1.0	4.5	2.2
Sn	2.5	.12	4.1	.17
Ho	2.4	1.1	.82	.18
Br	2.4	1.0	4.3	6.6
Eu	2.3	.94	1.1	.19

Table 3 (Continued) **Average composition, in parts per million, igneous rocks and some types of sedimentary rocks**

		Sedimentary Rocks		
Element	**Igneous Rocks**	**Resistates (Sandstone)**	**Hydrolyzates (Shale)**	**Precipitates (Carbonates)**
Ta	2.0	.10	3.5	.10
Tb	1.8	.74	.54	.14
As	1.8	1.0	9.0	1.8
W	1.4	1.6	1.9	.56
Ge	1.4	.88	1.3	.036
Mo	1.2	.50	4.2	.75
Lu	1.1	.30	.28	.11
Tl	1.1	1.5	1.6	.065
Tm	.94	.30	.29	.075
Sb	.51	.014	.81	.20
I	.45	4.4	3.8	1.6
Hg	.33	.057	.27	.046
Cd	.19	.020	.18	.048
In	.19	.13	.22	.068
Ag	.15	.12	.27	.19
Se	.050	.52	.60	.32
Au	.0036	.0046	.0034	.0018

Note: Values in parts per million. Dotted line indicates no data available.

Data from Wedepohl, K. H., *Geochemistry,* Holt, Rhinehart, and Winston, New York, 1971, 231.

surface of the continents.[29] Most natural waters occur at or near the surface of the earth and, consequently, have more contact with sedimentary rocks than with igneous rocks. Weathering reactions between natural waters and the minerals that constitute the rocks of the earths crust are the major geologic control on natural water chemistry. Physical weathering processes break rock into rock fragments, exposing fresh mineral surfaces to the atmosphere and hydrosphere. Chemical weathering processes, through the action of water, oxygen, carbon dioxide, and other acidic components, cause the chemical breakdown of primary bedrock minerals into secondary soil minerals, and release dissolved constituents to natural waters. As a result of these reactions, the chemical compositions of most surface waters and groundwaters show a strong relationships to the bedrock in the watershed where they occur.[19,20]

Literally thousands of minerals occur in the earth's crust and exhibit a wide range of composition and of reactivity with respect to the hydrosphere and atmosphere. Of all the possible minerals, less than a dozen constitute more than 98% of the rocks of the earth's crust (Table 4). The rest are much less abundant and usually are localized in unique occurrences, such as ore deposits, pegmatites, and evaporites. Natural water compositions, except in the vicinity of such anomalous deposits, will reflect weathering of the major rock-forming minerals. The importance of weathering of minerals is twofold: (1) the weathering reactions determine major dissolved species that occur in natural waters; and (2) trace elements commonly occur in solid-solution in rock-forming minerals and are released when minerals weather. Specific trace elements are associated with certain mineral groups and their aggregation into definite rock types; so, it is possible to make general predictions about both major and trace elements in natural waters from knowledge of the bedrock in the watershed where they occur.

Specific examples of the importance of mineral association in the supply of trace elements to natural waters can be seen in the frequency of occurrence of various minor solutes in different lithologic terrains. This association is usually based on chemical similarities of the trace elements with the major constituents, and resulting isomorphous solid solution, making up the primary minerals predominant in each rock type. Thus, transition metals are commonly associated with weathering of iron-bearing silicates or sulfides, but may have been lost in a previous weathering and depositional cycle to solution from secondary oxides or carbonates. The availability of trace elements to surficial weathering processes is not only a function of absolute crustal abundance, but also the distribution and weathering rate of primary mineral hosts. For example, strontium can be obtained as a secondary constituent from two of the most

Table 4 **Major mineral components of the earth's crust**

1. Feldspars (plagioclase, orthoclase)
2. Amphiboles and pyroxenes
3. Quartz
4. Micas (muscovite, biotite)
5. Clay minerals (illite, kaolinite, smectite)
6. Carbonates (calcite, aragonite, dolomite)
7. Evaporites (gypsum, halite)
8. Oxyhydroxides (goethite, ferrihydrite, birnessite)

common minerals in surface lithologies, calcite, and plagioclase, but relative dissolution rates render it definitely more available from the former than the latter. For comparison, zirconium is not only a rarer element, but is principally concentrated in a common, but not very abundant and highly resistant mineral, zircon.

It should be reemphasized that the principal lithologies of the earth's surface are largely composed of only a few major minerals, and this alone serves to limit associated trace element availability to natural waters. These minerals include quartz, alkali and plagioclase feldspar, muscovite and biotite mica, amphibole and pyroxene, calcite and dolomite, pyrite, gypsum and halite, clay minerals, and oxyhydroxides. Of these, quartz, alkali feldspar, and white mica are quite resistant to natural water dissolution at ambient temperatures. Important trace element associations, and their concentration levels with the common primary rock-forming minerals, are given in Table 5.

The stability of minerals in the earth's surface environment varies widely depending upon their composition and crystal structure. Goldich[48] developed a mineral weathering sequence for common rock-forming silicate minerals based on field studies of the weathering of granite gneiss, diabase, and amphibolite (Figure 3). Since the time of Goldich, numerous other researchers have attempted to predict reactivity or weathering sequences of minerals using a variety of different approaches. Reiche[49] devised an empirical "weathering potential index" based on the ratio of the sum of the alkali and alkaline earth elements minus water to the sum of the alkali and alkaline earth elements plus silica, aluminum, and iron in the mineral. Fairbairn[50] proposed a weathering index defined as the ratio of the volume of ions in the mineral unit cell to the total volume of the unit cell. Gruner[51] defined an "energy index" using Pauling's data for the electronegativities of the elements to explain silicate mineral stability. Curtis[52] and Kajiwora[53] attempted to use chemical thermodynamics to interpret weathering sequences. More recently, reactions between water and mineral surfaces at the atomic scale have been investigated with respect to mineral stability in the aqueous environment.[54,55] Despite the recent progress in this field, quantitative understanding of the fundamental mechanisms governing mineral stability at the atomic scale has not yet been fully achieved. Although a sequence of mineral stabilities or reactions cannot yet be precisely predicted from fundamental physico-chemical properties, mineral weathering and its effects on the chemistry of natural waters at the macroscopic scale is well defined.[29,56] Chemical reactions between natural waters and minerals produce dissolved constituents and solid residues. Many natural waters have compositions that reflect only mineral-water interactions.[19] Dissolved constituents released by weathering reactions enter groundwater-lake-stream systems, and the residual solid phases form the regolith and soil mantle. As discussed in considerable detail for lakes by Jones and Bowser[57], the understanding and quantification of processes controlling water and sediment chemistry require definition of the mineral phases as well as the bulk chemical composition of the earth materials in contact with natural fluids.

Atmospheric deposition in the form of rain or snow is the major source of water to the terrestrial environment. This water is usually mildly acidic due to reaction with atmospheric CO_2 and gaseous sulfur and nitrogen compounds. Soil waters are generally much more aggressive agents of mineral weathering than rain due to the high partial pressure of carbon dioxide P_{CO_2} of soil gases and, to a lesser extent, the presence of the other constituents such as organic acids (see discussion under biological factors). Weathering reactions between CO_2 charged soil waters and minerals are the major contributors of dissolved cations and carbonate alkalinity to natural waters. Mechanisms and rates of transfer of solutes between natural waters and soil, regolith, or sediment are a function of surface area, which, in turn, can be closely related to grain size. The finest grained, usually cryptocrystalline, materials commonl occurring as coatings on larger particles can exert an influence on water chemistry out of all propor to their total mass.[57,58]

Table 5 **Ranges of minor element contents of common rock-forming minerals**

Mineral	$\chi\%$	$0.\chi\%$	$0.0\chi\%$	$0.00\chi\%$	$0.000\chi\%$
Plagioclase	K	Sr	Ba, Rb, Ti, Mn	P, Ga, V, Zn, Ni	Pb, Cu, Li, Cr, Co, B
Potash feldspar	Na	Ca, Ba, Sr	Rb, Ti	P, Pb, Li, Ga, Mn	B, Zn, V, Cr, Ni, Co
Quartz				Al, Ti, Fe, Mg, Ca	Na, Ga, Li, Ni, B, Zn, Ge, Mn
Amphibole		Ti, F, K, Mn, Cl, Rb	Zn, Cr, V, Sr, Ni	Ba, Cu, P, Co, Ga, Pb	Li, B
Pyroxene	Al	Ti, Na, Mn, K	Cr, V, Ni, Cl, Sr	P, Cu, Co, Zn, Li, Rb	Ba, Pb, Ga, B
Biotite	Ti, F	Ca, Na, Ba, Mn, Rb	Cl, Zn, V, Cr, Li, Ni	Cu, Sr, Co, P, Pb, Ga	B
Magnetite	Ti, Al	Mg, Mn, V	Cr, Zn, Cu	Ni, Co	Pb, Mo
Olivine		Ni, Mn	Ca, Al, Cr, Ti, P, Co	Zn, V, Cu, Sc	Rb, B, Ge, Sr, As, Ga, Pb

Note: Within the columns, elements are listed in the general sequence of decreasing concentration.

Data from Wedepohl, K. H., *Geochemistry,* Holt, Rhinehart, and Winston, New York, 1971, 231.

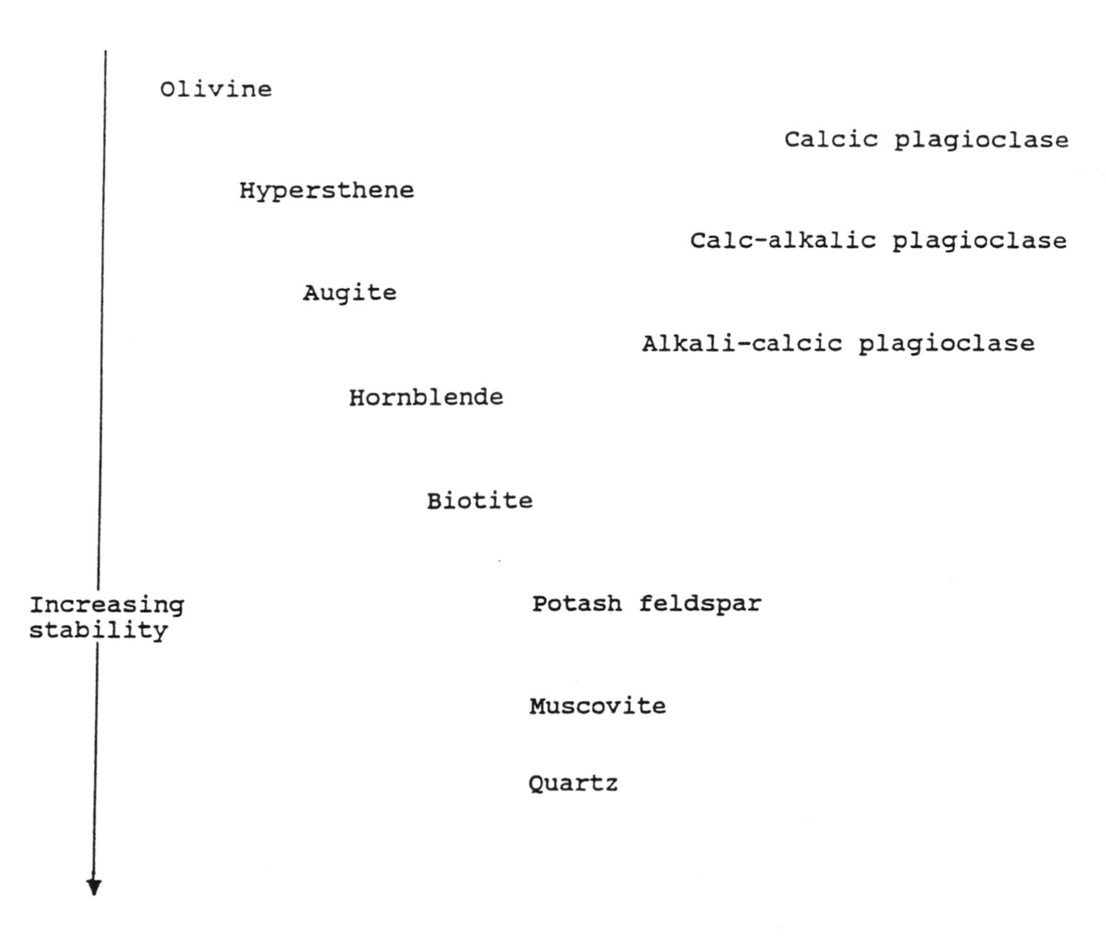

Figure 3 Stability of minerals in weathering (After Goldich, S.S., *J. Geol.*, 46, 17, 1938.)

Minerals exhibit two general types of weathering reactions: (1) congruent reactions in which the mineral dissolves completely, contributing all of its constituent elements to solution stoichiometrically; and (2) incongruent reaction in which some mineral constituents go into solution and others recombine to form a new solid phase.

The most important examples of congruent dissolution in natural waters are presented by the simple salts halite and gypsum. The most common examples of incongruent dissolution are provided by the primary rock-forming silicates, which typically weather by the formation of secondary oxides and/or clay minerals with the concomitant loss to solution of cations and silica. Different suites of trace elements in these minerals are associated with the major cations in solution, or the solid sesquioxide products of such reactions. Carbonate minerals can be involved in both types of reactions depending on their relative solubilities.

Good examples of mineral weathering reaction mechanisms can be drawn from typical trace elements associations. A ubiquitous minor constituent found in natural salt deposits is bromine, proxying for chlorine in halite solid solution. This element is released to meteoric waters on the congruent dissolution of halite:

$$\mathrm{Na(Cl,Br)} = \mathrm{Na^+} + \left(\mathrm{Cl^-} + \mathrm{Br^-}\right)$$

Similarly, strontium is a very common trace constituent in gypsum and will be released to dilute weathering waters by simple dissolution:

$$\mathrm{(Ca,Sr)SO_4 \cdot H_2O} = \left(\mathrm{Ca^{2+}} + \mathrm{Sr^{2+}}\right) + \mathrm{SO_4^{2-}} + \mathrm{H_2O}$$

In contrast, though strontium is also a very common constituent in aragonite, weathering often leads to incongruent recrystallization of the calcium carbonate with the Sr released to solution, as:

$$(Ca,Sr)CO_3 = CaCO_3 + Sr^{2+}$$

or with sufficient sulfate in the waters, another mineral is formed (insoluble celestite):

$$(Ca,Sr)CO_3 + SO_4^{=} = CaCO_3 + SrSO_4$$

The importance to trace element distribution of the incongruent dissolution of rock-forming silicates is in the associations of those elements with the relatively soluble major alkali and alkaline-earth constituents, as opposed to their association with the relatively insoluble silica and sesquioxides. Thus, the cobalt and nickel commonly associated with the predominantly ferrous iron and magnesium-bearing olivines and pyroxenes of mafic igneous rocks will be released on the mostly congruent dissolution of these silicates, only to be sorbed or coprecipitated with the formation of iron oxyhydroxides and, subsequently, perhaps even silica, as:

$$(Mg,Fe,Co,Ni)_2SiO_4 + 2H_2CO_3 = 2(Mg,Fe,Co,Ni)^{2+} + 2HCO_3^- + SiO_2 + 2OH^-$$

and

$$2Fe^{2+} + 2OH^- + O_2 = 2FeO(OH)$$

Similarly, the lithium or rubidium associated with alkali feldspars will be released to solution on weathering, but then can be involved in sorption on secondary clay such as kaolinite, as:

$$2(Na,K,Li,Rb)AlSi_3O_8 + 2H_2CO_3 + H_2O = 2(Na,K,Li,Rb)^+ + 2HCO_3^- + Al_2Si_2O_5(OH)_4 + 4SiO_2$$

Diagenetic processes in sediments can be responsible for the redistribution of elements between solution and solid phases. Thus, the development of a more anoxic, acidic, or complex-forming pore fluid environment resulting from the breakdown of organic matter can lead to either the variable mobilization or precipitation of different trace element groups, depending primarily on their redox properties and reactivity with dissolved oxygen or dissolved sulfide. Berner[59] has classified sedimentary environments geochemically based on these factors. He refers to the succession of processes in bacterial organic matter decomposition: oxygen consumption (respiration), nitrate reduction, sulfate reduction, and methane formation. Diagenetic environments corresponding to these processes are termed, respectively: oxic, postoxic (oxygen consumption without sulfate reduction — weakly reducing), sulfidic, and methanic (complete sulfate consumption and methane production — strongly reducing). Iron and manganese minerals, which also have distinctive trace element associations, are then taken to be characteristic of each environment. Hematite, geothite, ferrihydrite, and MnO_2-type minerals obviously indicate oxic conditions. Similarly, pyrite and marcasite reflect definitely sulfidic states. The most representative minerals of postoxic environments are the ferroso-ferric, 2:1 layer clay minerals, glauconite or nontronite ($Ca_{.25}Fe_2^{3+}Al_{.5}Si_{3.5}O_{10}(OH)$), which apparently require at least some iron reduction for their initial formation.[60] Except for rhodochrosite ($MnCO_3$), manganese is completely mobilized under postoxic conditions. Siderite ($FeCO_3$), vivianite $Fe_3(PO_4)_2{\cdot}8H_2O$, and rhodochrosite probably form as often under strongly [illegible] methanic conditions as in weakly reducing postoxic sediments; the main requirement is that all [illegible] from pore waters by bacterial reduction, which is readily achieved in many freshwater [illegible] ulfide minerals such as mackinawite (FeS) or greigite (Fe_3S_4) are indicative of [illegible] tions, insufficient to fully combine with available transition metal. In contrast, [illegible] quires strongly reducing environments to become stable over $MnCO_3$.

[illegible] genetic environments noted above particularly influence the trace elements associ- [illegible] ion metal group. The mobility of these constituents is constrained by sorption or [illegible] oxides, secondary carbonate, silicate, phosphate, or humic complexes. In contrast, trace element mobility is often enhanced by nonsulfidic reduction, but commonly prohibited by incorporation in stable resistant sulfide minerals like pyrite. Lacustrine examples of the added effects of pH and salinity, which control speciation, are given by Domagalski et al.[61]

C. BIOLOGICAL FACTORS

The physical and chemical characteristics of the environment are influenced by biological systems.[62-66] Many terrestrial aquatic organisms physically disturb the substrate (soil, sediment) on or in which they grow or reside. These organisms also take up chemical compounds from their surroundings and excrete other chemical compounds. In so doing, they affect the chemistry of natural waters. Microorganisms are ubiquitous in the environment. In their life processes they break down organic and inorganic materials into solid, liquid, and gaseous metabolic products. These compounds are excreted into the environment. Microbial reactions have a profound effect on the chemistry of natural waters. For instance, in soils, they elevate P_{CO_2} in the soil atmosphere, which makes soil waters much more effective as weathering agents.[67] In lacustrine, estuarine, and marine sediments, microorganism metabolism frequently controls the redox state of the sedimentary environment and, as a result, strongly affects the chemistry of the associated waters and minerals.[68-71] Finally, in groundwaters, bacteria commonly present to considerable depth in aquifers are capable of producing CO_2 as they metabolize organic material. Chapelle et al.[72] found *in situ* rates of bacterial CO_2 production in the range of 10^{-2} mmol l^{-1} yr^{-1} in the Hawthorn aquifer near Hilton Head Island, SC. Similar rates of microbial CO_2 production were found in a number of deep aerobic aquifers of the Atlantic Coastal Plain of South Carolina.[73,74] The geochemistry of dissolved organic carbon in coastal plain aquifers has been modeled.[74,75] Carbon dioxide enhances the dissolution of silicate and carbonate minerals along aquifer flow paths and may contribute to the development of secondary porosity in aquifer rocks.[76]

1. Physical Processes

In an abiotic world, sediment deposited in lakes, estuaries, and the oceans would remain in place unless disturbed by the physical actions of water currents or by gravitational forces. On the land surface, sediment formed by weathering processes would accumulate unless moved by the action of wind, running water, or by the force of gravity. Biota both enhance and retard the physical movement of sediment. For instance, terrestrial vegetation stabilizes sediment through its network of roots, inhibiting the erosive action of wind and running water. In a similar fashion, sediment in aquatic environments is stabilized by the growth of aquatic vegetation which inhibits the erosive actions of currents and gravitational slumping. Aquatic vegetation also may promote the accelerated deposition of sediment by impeding the free flow of sediment-laden waters, which decreases the fluid's competency to carry sediments in suspension.

The activities of biological systems may also physically disturb the environment. Terrestrial vegetation extends roots into the sediment. Trees uprooted by wind or other disturbances commonly turn over a large section of soil and regolith extending to the depth of the rooting zone (tree throws). As a result of tree throws, material from deep within the soil profile is brought to the surface where it is exposed to weathering. In forests, rock fragments and less weathered material from the depth is continually brought to the surface by such mechanisms. Burrowing organisms such as earthworms, insects, and animals serve as conveyer belts for exhuming fresh rock fragments and less weathered material and transporting it to the land surface. These mechanisms mix the soil column and continually expose material from depth to surface weathering.

In the aquatic environment, burrowing organisms transport buried sediment to the sediment-water interface and mix (bioturbation) the sediment column.[64,77-80] During bioturbation and transport of sediment from beneath the surface, interstitial waters are released to the water column.[81] Commonly, the bottom sediment becomes anoxic a short distance beneath the sediment-water interface. Interstitial waters in such reducing sediments can be rich in dissolved metals.[82,83] Mixing of these fluids with overlying waters by bioturbation releases the metals to the water column and makes them available for participation in chemical or biological processes. Models for bioturbation of sediments have been developed by Matisoff[84] and Matisoff and Robbins.[85] Minerals formed by diagenetic processes in the anoxic zone can be exhumed and transported into oxic conditions at the sediment surface.[86] These minerals, unstable in the oxic environment, react and alter the chemistry of bottom waters.

2. Biochemical Processes

In addition to physical activities that disturb and mix sediment and soils, organisms participate in biochemical processes that affect the environment. The organisms that probably have the greatest effect on the natural water chemistry are the microorganisms, primarily bacteria and fungi. Members of this large and diverse group are present in great numbers in virtually every environmental setting at and near the earth's surface. Microorganisms have been found in the atmosphere at altitudes up to 77 km above the surface, to depths of 2 km in the lithosphere, and 11 km in the hydrosphere.[87-89] They play a major role in the cycling of carbon, nitrogen, and sulfur in natural systems and are involved in the environmental chemistry of phosphorous, iron, manganese, and many other major and trace elements.[65,90] Microorganisms use inorganic and organic compounds as sources of energy and nutrients to form new cell structures and maintain old cells already formed.[91] They break down these compounds by redox processes, and the metabolic products are excreted into the environment.

Microbial processes and their products affect natural waters in a number of ways. They can enhance the dissolution of minerals through the production of acids or complexing agents, releasing the mineral constituents to solution. They can remove dissolved species from waters by incorporating them into their cells or by adsorption on cell walls, and they can cause precipitation of new mineral phases through alteration of redox state, changes in acid/base balances, or production of organic or inorganic ligands such as acetate or sulfide.[65,92]

The nitrifying bacteria *Nitrosomones* and *Nitrobacter* produce nitric acid by oxidation of ammonium:[65]

$$NH_4^+ + 2O_2 \rightarrow NO_3^- + 2H^+ + H_2O$$

The acid produced by this process enhances the weathering of minerals in soils, particularly carbonate minerals.[93,94] The Thiobacilli produce sulfuric acid by oxidation of reduced sulfur compounds:[89]

$$S_2^{\circ} + O_2 + OH^- \rightarrow S_2O_3^{2-} + H^+$$

$$S_2O_3^{2-} + 2O_2 + OH^- \rightarrow 2SO_4^{2-} + H^+$$

Sulfide minerals are oxidized by bacterially mediated reactions, releasing the associated metals and sulfate into the water. These processes may have a large effect on the acidity and chemical composition of natural waters.[95,96] The oxidation of pyrite (FeS_2), one of the most abundant sulfide minerals, is mediated by the bacteria *Thiobacillus* and *Ferrobacillus* ferroxidans. In this process the reduced sulfur is oxidized to sulfate, releasing ferrous ion and hydrogen ion to solution:

$$FeS_2 + 7/2\, O_2 + H_2O \rightarrow Fe^{2+} + 2SO_4^{2-} + 2H^+$$

The ferrous ion then hydrolyzes to form insoluble ferric hydroxide, releasing more hydrogen ion to the waters:

$$Fe^{2+} + 3H_2O \rightarrow Fe(OH)_3 + 3H^+$$

The oxidation of pyrite associated with disturbances caused by mining of coal and sulfide deposits has created a serious acidification problem in many mining districts.[97-99] These same types of microbially mediated reactions also occur with other metal sulfide minerals and are responsible for altering the acid/base balance and metal content of waters emanating from sulfide-rich rocks.[100-102]

Redox reactions, in turn, may affect trace elements. Iron released by oxidation of iron sulfide minerals and weathering of other iron-bearing minerals rapidly forms ferric hydroxide in oxic environments. The precipitation of ferric hydroxide may scavange trace elements from natural waters by coprecipitation or adsorption.[58,103,104] Oxyhydroxides form coatings on detrital sediment grains and are active in removing trace elements from solution.[105] The sediment particle with its oxyhydroxide coating and scavanged trace elements is transported by streams and rivers to lakes, estuaries, and the oceans. After deposition and burial the sediment commonly becomes anoxic because of microbial decomposition of organic matter. The oxyhydroxide coatings are unstable in this environment and dissolve, releasing iron and the associated trace metals to the interstitial waters in the sediment where they may be returned to the overlying waters by diffusion, advective, and bioturbation processes or combine with other dissolved species such as phosphate, carbonate, or sulfide to form new mineral phases in the sediments.[83,95,105]

Metals may also be solubilized by complexing agents formed during microbial biodegradation of organic matter. A large proportion of metal complexation in natural waters results from interactions with humic substances of poorly defined composition.[106,107] The physico-chemical nature of humate-metal interaction has been investigated by Lavigne et al.[108] for nickel (II) bound to fulvic acid and by Cacheris and Choppin[109] for thorium-humate complexes. Discrete ligand and continuous distribution models for metal-humate interactions have been discussed by Dzombak et al.[110]

Metals show a wide range of behavior with respect to complexation by organic ligands in natural waters. For instance, copper (II) is nearly 100% complexed in marine waters,[111-113] whereas cadmium and manganese are complexed to a negligible degree.[114,115] About 40% of the dissolved aluminum in freshwater lakes and streams commonly is present as an organic complex.[116]

Aquatic microorganisms can mediate specific reactions through extracellular enzymatic secretions either on the cell surface or free in solution.[117] These reactions enable the organisms to obtain necessary nutrients and trace elements by altering the redox state of chemical species or changing the nature of dissolved organic complexes. In these reactions, trace elements may be mobilized into, or taken up from, the waters.

Low molecular-weight organic acids are produced in soils by plants and microorganisms. Oxalic acid is commonly associated with fungal hyphae in soils.[118,119] Citrate, fumarate, malate, and oxalacetate are produced by a variety of celluolytic soil fungi.[120] Stable metal-organo chelates result from complexation with oxalic, citric, and gluconic acids.[121] Although a number of investigations have examined the effects of organic ligands on mineral dissolution, quantitative understanding of the role of these species in mineral weathering in natural systems is not well known.[122-126]

Biota affect the chemistry of natural waters through their physical activities and, more importantly, through the biochemical processes necessary for their life processes. Many reactions that occur in natural systems are biologically mediated and would not occur, or would take place at an infinitesimally slow rate, in the absence of biota. Weathering reactions, the most important contributors of dissolved metals and alkalinity to natural freshwaters, are primarily driven by microbially produced CO_2. Diagenetic processes responsible for the behavior of trace elements in sediments are closely tied to the microbial assimilation of carbon and reduction of sulfate. On a larger scale, the biochemical cycles of carbon and sulfur have exerted a major control on the level of atmospheric oxygen over Phanerozoic time.[127]

REFERENCES

1. **Conway, E. J.,** The chemical evolution of the oceans, *Irish Acad.,* 48B, 161, 1943.
2. **Garrels, R. M. and Mackenzie, F. T.,** Chemical history of the oceans deduced from postdepositional changes in sedimentary rocks, *Spec. Publ. Soc. Econ. Paleontol.,* 20, 193, 1974.
3. **Holland, H. D.,** *The Chemical Evolution of the Atmosphere and Oceans,* Princeton University Press, Princeton, NJ, 582, 1984.
4. **Bricker, O. P.,** Catchment flow systems, in *Acidification and Water Pathways,* Proceedings International Symposium on Acidification and Water Pathways, Bolkesjo, Norway, May 1987, 11.
5. **Keene, W. C. and Galloway, J. N.,** Organic acidity in precipitation of North America, *Atmos. Environ.,* 18, 2491, 1984.
6. **Smith, R. A.,** *Air and Rain: The Beginnings of a Chemical Climatology,* Longman, Green, London, 1872, 600.
7. **Cowling, E. B.,** Acid rain in historical perspective, *Environ. Sci. Technol.,* 16, 110, 1982.
8. **Jung, G. E. and Werby, R.,** The concentration of chloride, sodium, potassium, calcium, and sulfate in rain water over the United States, *J. Meteorol.,* 15, 417, 1958.
9. **McArthur, J. M., Turner, J., Lyons, W. B., and Thirlwal, M. F.,** Salt sources and water-rock interaction on the Yilgarn Block, Australia: isotopic and major element tracers, *Appl. Geochem.,* 4, 79, 1989.
10. **Fisher, D. W.,** 1968, Annual variations in chemical composition of atmospheric precipitation in eastern North Carolina and southeastern Virginia, U.S. Geological Survey Water-Supply Paper 1535-M.
11. **Calvert, J. G., Lazrus, A., Kok, G. L., Heikes, B. B., Walega, J. G., Lind, J., and Cantrell, C. A.,** Chemical mechanisms of acid generation in the troposphere, *Nature,* 317, 27, 1985.
12. **Sridhar, K., Jackson, M. L., Clayton, R. N., Gillette, D. A., and Hawley, J. W.,** Oxygen isotope ratios of quartz from wind-erosive soils of southwestern United States in relation to aerosol dust, *Soil Sci. Soc. Am. Proc.,* 42, 533, 1978.

13. **Le Roux, J., Clayton, R. N., and Jackson, M. L.,** Oxygen isotopic ratios in fine quality soil from sediments and soils of southern Africa, *Geochim. Cosmochim. Acta,* 44, 553, 1980.
14. **Jackson, M. L.,** Eolian additions to soils, in Proc Trans. 10th Intl. Congr. Soil Sci. (Moscow) 6, 1974, 458-465.
15. **Fisher, R. V. and Schmincke, H. U.,** *Pyroclastic Rocks,* Springer-Verlag, Heidelberg, 1984, 472.
16. **Lynch, J. A., Hanna, C. M., and Corbett, E. S.,** Predicting pH, alkalinity, and total acidity in stream water during episodic events, *Water Resour. Res.,* 22, 905, 1986.
17. **Puckett, L. J.,** The influence of forest canopies on the chemical quality of water and the hydrologic cycle, in *Chemical Quality of Water and the Hydrologic Cycle,* Averett, R. E. and D. M. McKnight, Eds., Lewis Publishers, Chelsea, MI, 1987, 3.
18. **Puckett, L. J.,** Estimates of ion sources in deciduous and coniferous throughfall, *Atmos. Environ.,* 24, 545, 1990.
19. **Garrels, R. M.,** Genesis of some groundwater from igneous rocks, in *Research in Geochemistry* 2, Ableson, P. H., Ed., John Wiley & Sons, New York, 1967, 405.
20. **Bricker, O. P. and Rice, K. C.,** Acidic deposition to streams: a geology-based method predicts their sensitivity, *Environ. Sci. Technol.,* 23, 379, 1989.
21. **Eugster, H. P. and Jones, B. F.,** Behavior of major solutes during closed-basin brine evolution, *Am. J. Sci.,* 279, 609, 1979.
22. **Horne, R. A.,** *Survey of Progress in Chemistry,* Vol. 4, Academic Press, 1968.
23. **Fletcher, W. H.,** *The Chemical Physics of Ice,* Cambridge University Press, 1970.
24. **Eisenberg, D. and Kautzman, W.,** *The Structure and Properties of Water,* Oxford University Press, 1968.
25. **Drost-Hansen, W.,** The structure of water and water-solute interactions, in *Equilibrium Concepts in Natural Water Systems,* Advances in Chemistry Series, No. 67, American Chemical Society, Washington, D.C., 1967, 70.
26. **Kay, R. L.,** The effect of water structure on the transport properties of electrolytes, in *Trace Inorganics in Water,* Advances in Chemistry Series, No. 73, American Chemical Society, Washington, D.C., 1968, 1.
27. **Stillinger, F. H.,** Water revisited, *Science,* 209, 451, 1980.
28. **Hem, J. D.,** *Study and Interpretation of the Chemical Characteristics of Natural Waters,* 3rd ed., U.S. Geological Survey Water-Supply Paper 2254, 263p, 1985.
29. **Garrels, R. M. and Mackenzie, F. T.,** *Evolution of Sedimentary Rocks,* W. W. Norton, New York, 1971, 397.
30. **Lvovitch, M. I.,** World water balance: general report, Proc. Symp. World Water Balance, *Int. Assoc. Sci. Hydrol.,* 2, 401-415, 1970.
31. **Nace, R. L., Ed.,** Scientific framework of world water balance, UNESCO Tech. Papers Hydrol. 7, 26 pp, 1971.
32. National Acid Precipitation Assessment Program, *Emissions, Atmospheric Processes and Deposition,* Vol. 1, U.S. Government Printing Office, Washington, D.C., 1990, reports 1-8.
33. National Acid Precipitation Assessment Program, *Aquatic Processes and Effects,* Vol. 2, U.S. Government Printing Office, Washington, D.C., 1990, reports 9-14.
34. **Hicks, B. B., Wesely, M. W., and Durham, J. L.,** Critique of methods to measure dry deposition, EPA-600/9-80-050, U.S. Environmental Protection Agency, Washington, D.C., 1980, 69 pp.
35. **Hicks, D. B., Hosker, R. P., Jr., Meyers, T. P., and Womack, J. D.,** Dry deposition inferential measurement techniques I. Design and tests of a prototype meteorological and chemical system for determining dry deposition, *Atmos. Environ.,* 25A, 2345-2359, 1991.
36. **Meyers, T. P., Hicks, B. B., Hosker, R. P., Jr., Womack, J. D., and Satterfield, L. C.,** Dry deposition inferential measurement techniques. II. Seasonal and annual deposition rates of sulfur and nitrate, *Atmos. Environ.,* 25A, 2361-2370, 1991.
37. **Smith, R. A.,** On the air and rain of Manchester, *Mem. Lit. Phil. Soc. Manchester,* Series 2, 10, 207, 1852.
38. **Cawse, P. A.,** Trace elements in the atmosphere of the U.K. ESNA Meeting, Aberdeen, AERE Harwell, 1981, 15 pp.
39. **Galloway, J. N., Thornton, J. D., Norton, S. A., Volchik, H. L., and McLean, R. A. N.,** Trace metals in atmospheric deposition: a review and assessment, *Atmos. Environ.,* 1, 1677, 1982.

40. **Steinnes, E.,** Impact of long-range atmospheric transport of heavy metals to the terrestrial environment in Norway, in *Lead, Mercury, Cadmium and Arsenic in the Environment,* SCOPE 1987, T. C. Hutchinson and K. M. Meema, Eds., Wiley & Sons, New York, 1987, 107.
41. **Clarke, F. W. and Washington, W. S.,** The composition of the earth's crust, U.S. Geological Survey Professional Paper 127, 1924, 117 pp.
42. **Fleischer, M.,** Recent estimates of the abundance of the elements in the earth's crust, U.S. Geological Survey Circular 285, 1953, 7 pp.
43. **Fleischer, M.,** The abundance and distribution of the chemical elements in the earth's crust, *J. Chem. Educ.,* 31, 446, 1954.
44. **Turekian, K. K. and Wedepohl, K. H.,** Distribution of the elements in some major units of the earth's crust, *Geol. Soc. Am. Bull.,* 72, 175, 1961.
45. **Taylor, S. R.,** Abundance of chemical elements in the continental crust — a new table, *Geochem. Cosmochim. Acta,* 28, 1273, 1964.
46. **Parker, R. L.,** Composition of the earth's crust, in *Data of Geochemistry* (6th ed.), U.S. Geological Survey Professional Paper 440-D, 1967, 19 pp.
47. **Horn, M. K. and Adams, J. A. S.,** Computer-derived geochemical balances and element abundance, *Geochim. Cosmochim. Acta,* 30, 279, 1966.
48. **Goldich, S. S.,** A study in rock weathering, *J. Geol.,* 46, 17, 1938.
49. **Reiche, P.,** Graphic representation of chemical weathering, *J. Sed. Petrol.,* 13, 58, 1943.
50. **Fairbairn, H. W.,** Packing in ionic minerals, *Bull. Geol. Soc. Am.,* 54, 1305, 1943.
51. **Gruner, J. W.,** An attempt to arrange silicates in the order of reaction energies at relatively low temperatures, *Am. Mineral.,* 35, 137, 1950.
52. **Curtis, C. D.,** Stability of minerals in surface weathering reactions, *Earth Surface Proc.,* 1, 63, 1976.
53. **Kajiwora, Y.,** A geochemical interpretation for the concept of Goldich's mineral stability series, Annual Report of the Institute of Geosciences, Univ. of Tsukuba, 6, 81, 1980.
54. **Lasaga, A. C.,** Atomic treatment of mineral-water surface reactions, in *Mineral-Water Interface Chemistry,* Hochella, M. F., Jr. and White, A. F., Eds., *Mineral Soc. Am. Rev. Mineral.,* 23, 1990, 87.
55. **Hochella, M. F., Jr.,** Atomic structure, microtopography, composition and reactivity of mineral surfaces, in *Mineral-Water Interface Chemistry,* Hochella, M. F., Jr. and White, A. F., Eds., *Mineral. Soc. Am. Rev. Mineral.,* 23, 1990, 87.
56. **Drever, J. I.,** *The Geochemistry of Natural Waters,* 2nd ed., Prentice Hall, Englewood Cliffs, NJ, 1982, 437.
57. **Jones, B. F. and Bowser, C. J.,** The mineralogy and related chemistry of lake sediments, in *Lakes: Chemistry, Geology, Physics,* Lerman, A., Ed., Spring-Verlag, New York, 1978, 179.
58. **Jenne, E. A.,** Controls on Mn, Fe, Ni, Cu, and Zn concentrations in soils and water: the significant role of hydrous Mn and Fe oxides, in *Trace Organisms in Water,* Advances in Chemistry Series, No. 73, Washington, D.C., Chemical Society, 1968, 337.
59. **Berner, R. A.,** A new geochemical classification of sedimentary environments, *J. Sediment. Petrol.,* 51, 359, 1981.
60. **Harder, H.,** Synthesis of iron layer-silicate minerals under natural conditions, *Clays Clay Minerals,* 26, 65, 1978.
61. **Domagalski, J. L., Eugster, H. P., and Jones, B. F.,** Trace metal geochemistry of Walker, Mono, and Great Salt Lakes, in *Fluid-Mineral Interactions: A tribute to H. P. Eugster,* Spencer, R. J. and Chou, I-Ming, Eds., The Geochemical Society Special Publication No. 2, 1990, 315.
62. **Redfield, A. C.,** The biological control of chemical factors in the environment, *Am. Sci.,* 46, 205, 1958.
63. **Redfield, A. C., Ketchum, B. J., and Richards, F. A.,** The influence of organisms on the composition of sea water, in *The Sea,* Vol. 2, Hill, M.M., Ed., Wiley-Interscience, New York, 1963, 26.
64. **Aller, R. C.,** The influence of macrobenthos on chemical diagenesis of marine sediments, Ph.D. Dissertation, Yale University, 1977, 600 pp.
65. **Berthelin, J.,** Microbial weathering processes in natural environments, in *Physical and Chemical Weathering in Geochemical Cycles,* Lerman, A. and Meybeck, M., Eds., NATO ASI Series, Vol. 251, 1985, 33.
66. **Krumbein, W. E. and Dyer, B. D.,** This planet is alive — weathering and biology — a multifaceted problem, in *The Chemistry of Weathering,* J.I. Drever, Ed., NATO ASI Series, Vol. 149, 1984, 143.
67. **Kempe, S.,** Freshwater carbon and the weathering cycle, in *Physical and Chemical Weathering in Geochemical Cycles,* Lerman, A. and Meybeck, M., Eds., NATO ASI Series, Vol. 251, 1985, 197.

68. **Emerson, S. and Widmer, G.,** Early diagenesis in anaerobic lake sediments. II. Thermodynamic and kinetic factors controlling the formation of iron phosphate, *Geochim. Cosmochim. Acta,* 42, 1307, 1978.
69. **Berner, R. A.,** Kinetic models for the early diagenesis of nitrogen, sulfur, phosphorous, and silicon in anoxic marine sediments, in *The Sea,* Vol. 5, Goldberg, E. D., Ed., Wiley & Sons, New York, 1974, 427.
70. **Berner, R. A.,** *Early Diagenesis: A Theoretical Approach,* Princeton University Press, Princeton, NJ, 1980, 250.
71. **Holdren, G. R., Bricker, O. P., and Matisoff, G.,** A model for the control of dissolved manganese in the interstitial waters of Chesapeake Bay, in *Marine Chemistry in the Coastal Environment,* Church, T.M., Ed., Washington, D.C., American Chemical Society Series No. 18, 1975, 364.
72. **Chapelle, F. H., Morris, J. T., McMahon, P. B., and Zelibor, J. L., Jr.,** Bacterial metabolism and the delta ^{13}C composition of ground water, Floridan Aquifer South Carolina, *Geology,* 16, 117, 1988.
73. **Chapelle, F. H. and Lovley, D. R.,** Rates of microbial metabolism, in deep coastal plain aquifers, *Appl. Environ. Microbiol.,* 53, 2636, 1990.
74. **McMahon, P. B. and Chapelle, F. H.,** Geochemistry of dissolved inorganic carbon in a coastal plain aquifer. II. Modeling carbon sources, sinks and ^{13}C evolution, *J. Hydrol.,* 127, 109, 1991.
75. **Chapelle, F. H. and McMahon, P. B.,** Geochemistry of dissolved inorganic carbon in a coastal plain aquifer. I. Sulfate from confining beds as an oxidant in microbial CO_2 production, *J. Hydrol.,* 127, 85, 1991.
76. **Lundegard, P. O. and Land, L. S.,** Problem of secondary porosity: Frio formation (Oligocene), Texas Gulf Coast, *Geology,* 12, 399, 1984.
77. **Goreau, T. J.,** Quantitative effects of sediment mixing on stratigraphy and biogeochemistry: a signal theory approach, *Nature,* 265, 525, 1977.
78. **Hanor, J. S. and Marshall, J. S.,** Mixing of sediment by organisms, in *Trace Fossils,* Perkins, B. F., Ed., Louisiana State University Misc. Pub 71-1, 1971, 127.
79. **Shink, D. R. and Guinasso, N. L.,** Modeling the influence of bioturbation and other processes on calcium carbonate dissolution at the sea floor, in *The Fate of Fossil Fuel CO_2 in the Ocean,* Anderson, N.R. and Malahoff, A., Eds., Plenum Press, New York, 1977, 375.
80. **Turekian, K. K., Cochran, J. K., and DeMaster, D. J.,** Bioturbation in deep sea deposits: rates and consequences, *Oceanus,* 21, 34, 1978.
81. **Matisoff, G., Fisher, J. B., and Matis, S.,** Effects of benthic macroinvertebrates on the exchange of solutes between sediments and freshwater, *Hydrobiologia,* 122, 19, 1985.
82. **Murray, J. W., Grundmanis, V., and Smethis, W. M.,** Interstitial water chemistry in the sediments of Saanich Inlet, *Geochim. Cosmochim. Acta,* 42, 1011, 1978.
83. **Bricker, O.P.,** Environmental factors in the inorganic chemistry of natural systems: the estuarine benthic sediment environment, in *Environmental Inorganic Chemistry,* Irgolic, K. J. and Martell, A. E., Eds., VCH Publishers, Deerfield Beach, FL, 1985, 135.
84. **Matisoff, G.,** Mathematical models of bioturbation, in *Animal-Sediment Relations: The Biotic Alterations of Sediments,* McCall, P. L. and Tevesz, M. J., Eds., Plenum Press, New York, 1982, 289-329.
85. **Matisoff, G. and Robbins, J.A.,** A model for biological mixing of sediments, *J. Geol. Educ.,* 35, 144, 1987.
86. **Benninger, L. K., Aller, R. C., Cochran, J. K., and Turekain, K. K.,** Effects of biological sediment mixing on the ^{210}Pb chronology and trace metal distribution in a Long Island Sound sediment core, *Earth Planet Sci. Lett.,* 43, 241, 1979.
87. **Kutznetsov, S. I., Ivanov, M. V., and Lyalikova, N. N.,** *Introduction to Geological Microbiology,* McGraw Hill, New York, 1963.
88. **Zajic, J. E.,** *Microbial Biogeochemistry,* Academic Press, New York, 1969.
89. **Ehrlich, H. L.,** *Geomicrobiology,* Marcel Dekker, 1981.
90. **Lovley, D. R.,** Dissimilatory Fe (II) and Mn (IV) reduction, *Microbiol. Rev.,* 55, 259, 1991.
91. **McCarty, P. L.,** Energetics of organic matter degradation, in *Water Pollution Microbiology,* Mitchell, R., Ed., John Wiley & Sons, New York, 1971, 91.
92. **Brock, T. D., Smith, D. W., and Madigan, M. T.,** *Biology of Microorganisms,* Prentice Hall, Englewood Cliffs, NJ, 1984.
93. **Faurie, G.,** Etude in vitro du role de la nitrification sur la lixiviation du calcium dans les sols calcaires, *Sci. Sol.,* 4, 207, 1977.
94. **Duchaufour, Ph.,** *Pedology* (transl.), Aller and Urwin, London, 1982.

95. **Berner, R. A.,** Migration of iron and sulfur within anaerobic sediments during early diagenesis, *Am. J. Sci.,* 267, 19, 42, 1969.
96. **Goldhaber, M. B., Aller, R. C., Cochran, J. K., Rosenfeld, J. K., Martens, C. S., and Berner, R. A.,** Sulfate reduction, diffusion, and bioturbation in Long Island Sound sediments, Report of the FOAM group, *Am. J. Sci.,* 277, 193, 1977.
97. **Nordstrom, D. K., Jenne, E. A., and Ball, J. W.,** Redox equilibria of iron in acid mine waters, in *Chemical Modeling in Aqueous Systems: Speciation, Sorption, Solubility and Kinetics,* Jenne, E. A., Ed., Am. Chem. Soc. Symp. Series., 93, 1979, 51.
98. **Nordstrom, D. K.,** Aqueous pyrite oxidation and the consequent formation of secondary iron minerals, in *Acid Sulfate Weathering,* Kittrick, J. A., Fanning, D. S., and Hossner, L. R., Eds., Soil Sci. Soc. Amer. Spec. Pub., 10, 1982, 37.
99. **Chapman, B. M., Jones, D. R., and Jung, R. F.,** Processes controlling metal ion attenuation in acid mine drainage systems, *Geochim. Cosmochim. Acta,* 47, 1957, 1983.
100. **Nordstrom, D. K.,** Hydrogeochemical and microbiological factors affecting the heavy metal chemistry of an acid mine drainage system, Ph.D. dissertation, Standford University, 1977, 210 pp.
101. **Alpers, C. N. and Brimhall, G. H.,** Paleohydrologic evolution and geochemical dynamics of cumulative supergene metal enrichment at La Escondida, Atacama Desert, Northern Chile, *Econ. Geol.,* 84, 229, 1989.
102. **Nielson, A. M. and Beck, J. V.,** Chalcocite oxidation and coupled carbon dioxide fixation by *Thiobacillus Ferrooxidans, Science,* 175, 1124, 1972.
103. **Hem, J. D.,** Reactions of metal ion at surfaces of hydrous iron oxide, *Geochim. Cosmochim. Acta,* 41, 527, 1977.
104. **Lion, L. W., Altman, R. S., and Leckie, J. O.,** Trace-metal adsorption characteristics of estuarine particulate matter: Evaluation of contribution of Fe/Mn oxide and organic surface coatings, *Environ. Sci. Technol.,* 16, 660, 1982.
105. **Callender, E. and Bowser, C. J.,** Manganese and copper geochemistry of interstitial fluids from manganese model-rich pelagic sediments of the northeastern equatorial Pacific Ocean, *Am. J. Sci.,* 280, 1063, 1980.

√ 105a. **Cerling, T. E. and Turner, R. R.,** Formation of freshwater Fe-Mn coatings on gravel and the behavior of ^{60}Co, ^{90}Sr, and ^{137}Cs in a small watershed, *Geochim. Cosmochim. Acta,* 46, 1333, 1982.

106. **Cabaniss, S. E. and Shuman, M. S.,** Copper bindings by dissolved organic matter. I. Suwannee River fulvic acid equilibria, *Geochim. Cosmochim. Acta,* 52, 185, 1988a.
107. **Cabaness, S. E. and Shuman, M. S.,** Copper binding by dissolved organic matter. II. Variation in type and source of organic matter, *Geochim. Cosmochim. Acta,* 52, 195, 1988b.
108. **Lavigne, J. A., Langford, C. H., and Mark, M. K. S.,** Kinetic study of the speciation of nickel (II) bound to a fulvic acid, *Anal. Chem.,* 59, 2616, 1987.
109. **Cacheris, W. P. and Choppin, G. R.,** Dissociation kinetics of thorium-humate complex, *Radiochim. Acta,* 42, 185, 1987.
110. **Dzombak, D. A., Fish, W., and Morel, F. M. M.,** Metal-humate interactions. I. Discrete ligand and continuous distribution models, *Environ. Sci. Technol.,* 20, 669, 1986.
111. **Coale, K. H. and Bruland, K. W.,** Copper complexation in the Northeast Pacific, *Limnol. Oceanogr.,* 33, 1084, 1988.
112. **Sunda, W. G. and Hanson, A. K.,** Measurement of free cupric ion concentration in sea water by a ligand competition technique involving copper sorption onto C_{18} SEP-PAK cartridges, *Limnol. Oceanogr.,* 32, 537, 1987.
113. **Hering, J. G., Sunda, W. G., Ferguson, R. L., and Morel, F. M. M.,** A field comparison of two methods for the determination of cooper complexation: bacterial bioassay and fixed-potential amperometry, *Marine Chem.,* 20, 299, 1987.
114. **Sunda, W. G.,** Measurement of manganese, zinc, and cadmium complexation in seawater using chelex ion exchange equilibria, *Marine Chem.,* 14, 365, 1984.

√ 115. **Duinker, J. C. and Kramer, C. J. M.,** An experimental study on the speciation of dissolved zinc, cadmium, lead and copper in river Rhine and North Sea water by differential pulsed anodic striping voltammetry, *Marine Chem.,* 5, 207, 1977.

116. **Driscoll, C. T., Baker, J. P., Bisogni, J. J., and Schofield, C. L.,** Aluminium speciation and equilibria in dilute acidic surface waters of the adirondack region of New York State, in *Acidic Precipitation: Geological Aspects,* Bricker, O.P., Ed., Butterworth, Boston, 1984, 55.

117. **Price, N. M. and Morel, F. M. M.,** Role of extracellular enzymatic reactions in natural waters, in *Aquatic Chemical Kinetics,* Ed., W. Stumm, Wiley-Interscience, New York, 1990, 235.
118. **Graustein, W. C., Cormack, K., Jr., and Sollins, P.,** Calcium oxalate: occurrence in soils and effect on nutrient and geochemical cycles, *Science,* 198, 1252, 1977.
119. **Cormack, K., Jr., Sollins, P., Todd, R. L., Fogal, R., Todd, A. W., Fender, W. M., Crossley, M. E., and Crossley, D. A., Jr.,** The role of oxalic and bicarbonate in calcium cycling by fungi and bacteria: some possible implications for soil animals, in *Soil Organisms as Components of Ecosystems,* Lohn, U. and Persson, T., Eds., Proc. VI International Soil Zoology Colloquium, Ecological Bull. (Stockholm), Vol. 25.
120. **Moore, S. and Stapelfeldt, E. E.,** in Proc 3 Int. Biodegr. Symp., Sharpley, J. M. and Kaplan, A. M., Eds., 1976, 711.
121. **Eckhardt, F. E. W.,** Solubilization, transport, and deposition of mineral cations by microorganisms-efficient rock weathering agents, in *The Chemistry of Weathering,* Drever J. I., Eds., NATO ASI Series, Vol. 149, 1985, 161.
122. **Stumm, W., Furrer, G., Wieland, E., and Zinder, B.,** The effects of complex-forming ligands on the dissolution of oxides and aluminosilicates, in *The Chemistry of Weathering,* Drever, J. I., Eds., NATO ASI Series, Vol. 149, 1985, 55.
123. **Mast, M. A. and Drever, J. I.,** The effect of oxalate on the dissolution rate of oligoclase and tremolite, *Geochim. Cosmochim. Acta,* 51, 2559, 1987.
124. **Grandstaff, D. E.,** The dissolution rate of foresteritic olivine from Hawaiian beach sand, in *Rates of Chemical Weathering of Rocks and Minerals,* Colman, S. M. and Dethier, D. P., Eds., Academic Press, Orlando, 1986, 41.
125. **Huang, W. H. and Kiang, W. E.,** Laboratory dissolution of plagioclase feldspars in water and organic acids at room temperature, *Am. Mineral,* 57, 1849, 1972.
126. **Huang, W. H. and Keller, W. D.,** Dissolution of rock-forming silicate minerals in organic acids, *Am. Mineral,* 55, 2076, 1970.
127. **Berner, R. A.,** Biogeochemical cycles of carbon and sulfur and their effect on atmospheric oxygen over Phanerozoic time, *Palaeogeogr. Palaeoclimatol. Palaeoecol. (Global and Planetary Change Section),* 75, 97, 1989.
128. National Atmospheric Deposition Program, NADP/NTN annual data summary precipitation chemistry in the United States, 1989, 340.
129. **Wedepohl, K. H.,** *Geochemistry,* Holt, Rhinehart, and Winston, New York, 1971, 231.

Chapter 2

Trace Chemistry Processes

Petr Beneš and Eiliv Steinnes

CONTENTS

I. INTRODUCTION

Trace elements in natural waters participate in many processes which change their physico-chemical forms (speciation) or distribution in space (migration) and affect their uptake by organisms. Some of the processes occur only at very low concentrations of the participating element or are at least strongly affected by the low concentration. We can call them trace chemistry processes as long as they are largely of chemical nature or have important chemical aspects. The processes involved are oxidation/reduction, association/dissociation in solution, adsorption/desorption, precipitation/dissolution, and aggregation/disaggregation.

In this chapter, a brief general discussion of trace chemistry processes is presented. The emphasis is placed on the significant peculiarities in the behavior of trace elements in these processes rather than on the discussion of their principles, which are treated in detail elsewhere.[1] Other processes important for the migration of trace elements such as advection, sedimentation, and resuspension are not treated here. They will be mentioned, together with the details on trace chemistry processes pertaining to individual types of natural waters, in other chapters of this book where appropriate.

In order to characterize important general features of trace chemistry processes, it is worthwhile to briefly mention basic differences in the behavior of substances at low concentrations as compared with that at ordinary (macro) concentrations and to clearly define the range of trace concentrations. In fact the physico-chemical concept of "trace concentration" is based on the following typical differences of great practical significance:[2]

1. At low concentrations substances tend to adsorb on the surface of solid particles of low adsorption capacity dispersed in aqueous system or on compact solid phases of similar capacity in contact with the system. The latter case includes adsorption on the surface of vessels and tools used for handling and analysing water samples. While the adsorption can be neglected at ordinary concentrations of the substance, it can considerably affect its behavior and speciation at low concentrations.
2. Precipitation of a substance present in a very low concentration, if it occurs at all, cannot lead to the formation of particles larger than colloidal size or to the formation of a visible (compact) solid phase. The nature and abundance of the particles formed are difficult to determine and the particles often behave irregularly or unpredictably.
3. The behavior of a substance at low concentrations can be affected by its interaction with other substances present in the system at low concentrations, often as impurities. As the nature, concentration, or even presence of the other substances in the system may not be known, it is difficult to predict their effect on the behavior. This leads to uncertainty and/or irreproducibility in the composition of the system and to concomitant irreproducibility in the behavior of the studied substance. The impurities only exceptionally affect the behavior of substances present in macroconcentrations (e.g., due to catalytic or heteronucleation processes).

0-8493-6304-7/95/$0.00+$.50

Experience shows that these differences are observed at concentrations lower than about 10^{-5} mol dm^{-3}, and therefore this level has been proposed as the upper limit of trace concentrations.[2,3] Substances or elements present in natural waters in lower concentrations can be considered as trace substances and trace elements. They represent a majority of the components of natural waters; only a few elements do not fall into this category. For instance, in typical oxic fresh waters the average concentrations of only Ca, Mg, K, Na, S(SO_4^{2-}), and Cl are clearly above this limit, while the average concentration of several other elements are close to the limit, i.e., Fe, Mn, F, N (NH_4^+), and P (phosphates). The practical lower limit of trace concentrations is determined by the detection limit of the analytical method employed and for natural waters also by the lowest significant concentration of a given trace element, e.g., for toxic elements, the lower limit can be derived from the toxicity limit for man, fish, etc., considering the concentration factors relevant for the critical food chain. Usually the limits are several orders of magnitude higher than the extremely low concentrations where difficulties with thermodynamic definition of equilibria begin.[2,4]

The field of trace concentrations was first opened for investigation with the discovery of short-lived natural radionuclides at the beginning of this century. At that time the objective was separation, isolation, and identification of radionuclides available only in unweighable amounts. Very soon the need for basic knowledge of laws governing the behavior of substances in trace concentration became obvious for achieving this objective. Later radiochemists also used artificial radionuclides as tracers in the study of trace chemistry processes in aqueous solutions. Investigations of the behavior and speciation of nonradioactive trace elements in natural waters began in the sixties, when analytical methods of sufficient sensitivity became available and when the need of the knowledge of these topics was recognized. Most of the reviews available on the topics and discussions in this monograph are based on data obtained by nonradiochemical methods.

However, useful information for this purpose can also be extracted from laboratory experiments with radionuclides and aqueous solutions, which were extensively reviewed by Starik,[5] Evans and Muramatsu,[6] and Beneš and Majer.[2] Radiochemical methods are still the most efficient tool for investigations of trace chemistry processes at extremely low concentrations.[2,4]

A critical general analysis of the knowledge gathered so far on trace chemistry processes reveals that widely ranging and valuable experimental data have been obtained and interpreted qualitatively or semiquantitatively, but a reliable quantitative description of many processes is still lacking. There are several reasons for this unsatisfactory state. The behavior of trace elements is complicated and depends on factors which cannot always be experimentally determined or followed. The data of many authors are difficult to interpret because they underestimate the importance of the strict maintenance of conditions or a detailed description of methods and conditions employed. Many of the methods used for the study of behavior and speciation of trace elements are imperfect and do not enable reproducible or well interpretable data to be obtained. It can be hoped that our understanding of trace chemistry processes will be improved by further development of methods and by more careful and sophisticated analyses carried out both in the laboratory and in the field.

II. OXIDATION AND REDUCTION

When the speciation, migration, biological uptake, and other aspects of the behavior of a trace element in natural waters are to be analyzed, the general aquatic chemistry of the element is to be considered first. For many elements the chemical reactions occurring depend strongly on the oxidation state of the element. Also, the biological uptake differs significantly between different oxidation forms of most elements. That is why redox reactions are put in the first place for discussion in this text. However, the behavior of trace elements in natural waters may also indirectly be affected by redox reactions. These reactions may change the nature and concentration of various ligating or precipitating compounds in the water and thus induce effects on association, precipitation, and other reactions of trace elements with these compounds.

The general aspects of oxidation-reduction processes in natural water have been treated by several authors.[1,7,8] Most of the reasonably well interpreted existing data concern substances present in macroconcentrations in natural waters. It appears that many natural waters are far from redox equilibrium due to rapid time and space changes in redox conditions and slow redox reactions. Still, the oxidation states are often calculated using equilibrium models neglecting the kinetics of the reactions, which is usually not very well known. Such calculations are valuable only to the extent that they define the equilibrium state towards which the system is proceeding.

Redox reactions of trace elements in natural waters differ from those of macroconstituents in at least two ways. First, they will not themselves significantly affect the redox potential, in contrast to reactions involving elements present in macroconcentrations (O, S, and sometimes C, N, Fe, Mn). Second, the reactions can depend heavily on the presence of other trace constituents in the water, the nature and concentration of which need not be known. This significantly increases the uncertainty connected with the interpretation of redox behavior of trace elements.

Foreign trace constituents can affect both the oxidation state of a trace element and the kinetics of its change, either as oxidants/reductants or as ligands forming ion pairs or complexes with the trace element. Their function as oxidants/reductants may be important only if their reaction with the trace element in question proceeds more easily or rapidly than oxidation/reduction of the trace element by a macrocomponent of the water determining the overall redox potential. Effects of complexing ligands on the oxidation/reduction equilibria or kinetics have been described for interactions between trace elements and ligands at macroconcentrations.[1] Such effects are to be expected also for ligands present at trace concentrations, if the ligands form sufficiently stable complexes with the trace element.

The last important effect to be mentioned in connection with trace elements is that the rate of oxidation/reduction is likely to decrease with decreasing concentrations of the substance to be oxidized/reduced. This would mean that very low reaction rates could be observed at very low concentrations of trace elements, resulting in negligible or slow changes of their oxidation state. Consequently, oxidation states of trace elements might depend on their concentration in natural water. However, in the case that oxidation/reduction of a trace element proceeds by a bimolecular reaction at a large excess of the oxidizing/reducing agent, the reaction rate may be independent of the concentration of trace element, and no such effects occur.

III. ASSOCIATION AND DISSOCIATION IN SOLUTION

Association of trace elements with ions and molecules in solution, as well as dissociation of complex forms of the elements, are processes that have a profound effect on their behavior in natural waters. These processes largely determine the speciation of the trace elements and their adsorption, precipitation, and biological uptake. There exists a great number of association and dissociation processes relevant to trace element behavior in natural waters, reflecting the large variety of trace elements and available ligands present. Starting with the solvent medium itself, the association with water molecules (hydration), although important to the aquatic chemistry of all elements, has normally not been considered when speciation of trace elements in natural waters is discussed. On the other hand, association with ions produced by dissociation of the water molecule (protonation, hydrolysis) is always paid great attention, as these processes change the charge and also other properties of trace element species. The same is the case for dissociation of trace element species present in the form of acids or bases.

Other association/dissociation reactions affecting trace elements can be roughly classified according to the nature of ligands involved: formation or dissociation either of ion pairs, or of complexes with inorganic ions, low molecular weight organic species, or high molecular weight polymeric organic substances. The latter ligands often reach colloidal dimensions. However, as far as they are hydrophilic and their solutions are formed by spontaneous dissolution of organic matter in water, they are classified as lyophilic colloids[9] and are to be considered as truly dissolved. This means that their solutions represent homogenous dispersion systems that are stable from a thermodynamic point of view. Therefore, interaction of trace elements with these ligands should not be regarded as adsorption or sorption[10,11] as these terms are usually reserved for heterogeneous phase systems.

The above-mentioned processes may be quantitatively described by standard methods of complex chemistry[1,12-14] using computer programs accounting for competition among different elements and ligands.[15-18] Although trace elements cannot affect significantly the concentrations of free ligands present in macroconcentrations, their competition for trace ligands has to be taken into account. Speciation calculations are generally based on equilibrium models, which may lead to errors as many processes in natural waters are slow, and natural waters are known to be nonequilibrium systems (see Reference 1, p. 112). Further problems stem from the great complexity of natural waters with respect to chemical constituents and the lack of quantitative data to characterize the interactions between some components that may be present. Trace concentrations of the solutes involved are also likely to bring other problems which can be summarized as follows:

1. The speciation of trace elements in natural waters may be strongly influenced by association with other trace compounds for which few data will normally exist concerning their concentrations, type of interaction with the trace element in question, or even their nature. Frequently, the presence of trace element species that cannot be readily identified may be observed in natural waters. Among the ligands most often mentioned in this connection are organic substances such as humic and fulvic acids. A number of review articles have been published on the complexation of trace metals with humic substances in natural waters.[11,19-24] Both experimental evidence and theoretical considerations indicate that humic/fulvic acids are the predominant components for complexing capacity towards trace metals in most natural fresh waters.[25] However, because of the complicated nature of these substances both with respect to their composition and their interaction with trace elements, it is difficult to determine the abundance and structure of these complexes either by calculation or experimentally. Conditional stability constants for a number of humic/fulvic complexes have been determined, but these constants have limited applicability in model calculations. Problems of complexing properties connected with characterization of humic substances toward trace elements and possible solutions of these problems have recently been reviewed.[26,124] Experimental techniques for identification and characterization of individual trace element-humic/fulvic complexes are still at an early stage of development.[20,21]
   Other organic compounds that have been considered in calculations of complex equilibria affecting the speciation of trace elements in natural waters are citrate, glycine, cysteine, NTA, etc.[27,28] Sometimes the combined effect of several complexing agents is calculated using a "mixture model" approach as defined by Sposito.[29]
2. A second problem connected with association reactions in water concerns the formation of polynuclear species. On the basis of published values for polynuclear complexes,[30-32] known from the chemistry of macroconcentrations, it may seem as if polynuclear complexes can be ignored at trace concentrations as compared to other species of the element concerned. At iron concentrations below 10^{-4} *M*, for example, the concentration of the $Fe_2(OH)_2^{4+}$ ion is negligible compared to those of monomeric species (see Reference 1, p. 336). This has led some researchers to assume that polynuclear species practically do not exist at trace concentrations.[33-36]
   It seems clear, however, that polymerization and condensation are the primary stage in the formation of true colloids of trace elements. It may, therefore, be assumed that polynuclear species of many metals exist even at trace concentrations in a pH region where the solubility of the respective metal hydroxides is very low. It is apparent that products of extensive hydrolysis polymerize more easily than less hydrolyzed species of the same metal. It has been shown that the concentration of Zr solutions at which hydrolysis polymers begin to form decreases with increasing pH.[37] Some multivalent metal species may be expected to be particularly prone to forming polynuclear species at trace concentrations, such as Pa^{V}, where the polymerization of $Pa(OH)_4^+$ ions is significant even at about 5×10^{-8} *M* Pa.[38]
   It may, therefore, be assumed that polynuclear forms of a trace element will be different from those known for macroconcentrations of the same elements and occur at higher pH values, usually in the same region as the formation of true colloids occurs.[2] Experimental verification of these assumptions is, however, extremely difficult, as the detection and analysis of the polynuclear forms are complicated by simultaneous adsorption and formation of trace colloids. Furthermore, the polynuclear species are sometimes formed only under conditions of supersaturation with respect to the metal hydroxide (see Reference 1, p. 334) and are thus thermodynamically unstable. Considerable achievements have recently been done in the study of polymerization reactions at macroconcentrations relevant for early stages of precipitation, e.g., Schneider and Schwyn.[39] It is to be hoped that the knowledge obtained in this way may also throw some light on polymerization at trace concentrations.
3. The last category of problems to be mentioned here is the sometimes very slow rate of association reactions at trace concentrations. As discussed before, the rate of homogeneous reactions may depend on the concentrations of reacting compounds. For a second-order bimolecular reaction

$$X + A \rightarrow XA$$

the rate of change of [X] is

$$-\frac{d[X]}{dt} = k[X][A] \quad (1)$$

It follows from this equation that on transition from macro- to microconcentrations, e.g., from [X] = [A] = 10^{-3} *M* to [X] = [A] = 10^{-9} *M*, i.e., by 6 orders of magnitude, the reaction rate decreases by 12 orders. On the other hand, if only one of the reacting components (X) is present in trace amounts while the other (A) is a macrocomponent of the system, so that its concentration remains practically constant during the reaction, the reaction rate is independent of the concentration of the trace component. An example of this is the formation of chloride complexes of mercury in sea water.

Considering the difficulties and problems mentioned above, calculations of association/dissociation equilibria in solution usually prove insufficient for characterization of the speciation and interpretation of the behavior of trace elements in natural waters. Then the presence and abundance of individual dissolved species must be determined experimentally. Methods suitable for this purpose have been critically reviewed by several authors.[2,40-44] They include electrochemical techniques, extraction, ion exchange, adsorption, measurement of electrophoretic mobility, diffusion, etc. and exploit quantitative or qualitative differences in the behavior of the individual forms of trace elements in physico-chemical processes. The evaluation of data obtained by these methods is, however, not simple, particularly if three or more forms of the analyzed trace element are present simultaneously in comparable amounts.

The applicability of a particular method often depends on the knowledge of quantities characterizing the behavior of individual forms of trace element in the process employed. For instance, the diffusion coefficients or electrophoretic mobilities of individual ionic species are necessary for quantitative evaluation of data obtained by studying diffusion or electrophoresis, respectively. These quantities need not be known. Most of the methods significantly perturbate the analyzed system, and the resulting shift of equilibria may lead to errors. Further complications may arise from adsorption of trace elements on the walls of instruments or vessels employed for the analysis, storage, or sampling of natural waters. The adsorption can change the ratio between individual species of the element and obscure the kinetics of processes on which the method is based (e.g., diffusion, electrophoresis). All these problems and complications lead to the result that most of the methods do not yield precise quantitative data on the abundance of well-defined species of trace elements. Instead, the species are frequently classified into operationally defined categories such as "free", "labile", and "strongly bound" (complexed) forms.

IV. ADSORPTION AND DESORPTION

The term "adsorption" is frequently used in scientific literature to describe attachment of ions or molecules from a liquid or gaseous phase to a solid phase by means of some specific mechanism, e.g., by some kind of specific forces. In this chapter, however, the term adsorption denotes any process leading to the concentration of a trace element from an aqueous phase onto the surface of a solid phase, except for coprecipitation and active (i.e., physiological) biological uptake. Desorption stands for release into the aqueous phase of previously adsorbed trace element.

Adsorption is one of the most important processes affecting the speciation, migration, and biological availability of trace elements in natural waters. It plays an important role in the regulation of concentrations of dissolved elements in sediments. It is also more or less specific to trace elements, as species present in macroconcentrations are normally little affected by the process. While a significant part (sometimes 90% or more) of many trace elements is bound to suspended particles in natural waters or bound to porous solid phases in contact with the waters (bottom sediments or solid phases in aquifers and soils), the interaction of major elements with the solid phase can usually be neglected or constitutes a minor problem in understanding the behavior of the elements in the system. Trace elements adsorbed on solid phases can be released again into the water phase; so, desorption is also an important process in trace element cycling in natural waters.

Adsorption and desorption are also of practical importance for the analysis of speciation and behavior of trace elements in natural waters. On one hand, methods based on adsorption and desorption are often used for concentrating trace elements prior to their determination in natural waters and in speciation procedures.[2,44-48] On the other hand, adsorption of trace elements on container walls or on the surface of instruments used for sampling and analysis of natural waters may seriously complicate the determination of the concentration and speciation of trace elements. While the use of adsorption for analytical purposes is outside the scope of this book, the latter problem is briefly treated in Chapter 4. More details can be found elsewhere.[2,49]

The discussion in this chapter is focused on the interaction of trace elements with natural solids in contact with natural waters. These solids can be roughly classified into (1) inorganic solids such as metal oxides and hydroxides, clays, carbonates, and other minerals, (2) organic nonliving solids, and (3) living cells. Trace elements interact with them by a variety of processes, sometimes very complicated and little known. Because of their importance, these interactions have been extensively studied by laboratory model experiments or by the analysis of field data on the distribution of trace elements (and radionuclides) in environmental systems. The first approach enables easier study of principal factors affecting the interaction and easier interpretation of the results, but may suffer from limited applicability of the results to natural systems.[50] The experiments are frequently carried out with model solids representing components of natural solids. The latter approach represents the most straightforward study of the interaction, but the field data usually depend on many variables and processes (including other ones than adsorption/desorption) and are therefore difficult to interpret.[51,52]

For comprehensive reviews of the experimental methods used in the study of adsorption/desorption of trace elements in natural systems, the results obtained, and conclusions drawn, the reader is referred to several monographs or articles (see Reference 2, pp. 163, 175 and References 53 to 58). Here we only outline the most important findings and conclusions. It has been found that adsorption and desorption are affected to a larger or smaller extent by the following variables: pH and composition (or ionic strength) of the water, concentration of the adsorbate (trace element), composition and properties of the solid phase, solid-to-solution ratio, and temperature. In many cases the adsorption or desorption is a slow process, or the properties of the solid phase change with time, so that the time of contact also has to be considered. Except for the latter factor, effects of these variables can be explained and, under suitable circumstances, quantitatively predicted using adsorption models, based on the knowledge of adsorption mechanisms.

Three basic mechanisms of trace element adsorption in aqueous systems can be distinguished (see Reference 2, p. 171). Chemisorption is characterized by the formation of a chemical bond (sharing of electrons) between the trace element and a specific surface site on the solid phase. Electrostatic adsorption is due to the action of attractive coulomb forces between the electrically charged surface and oppositely charged ions or colloid particles of the trace element. Physical adsorption occurs as a result of nonspecific forces of attraction between the solid and the trace element. The latter forces, called dispersion or van der Waals forces, are due to interaction between dipoles, either permanent or induced. They are generally weaker than other forces acting in adsorption from aqueous solutions and therefore the physical adsorption is sometimes considered unimportant in water systems. However, the dispersion forces may operate in concert with the so-called hydrophobic effect, which denotes the tendency of species with low hydration energy to be expelled from the aqueous phase due to mutual attraction between the molecules of water.[59]

Quite a few adsorption models have been devised to describe the adsorption of solutes onto solid surfaces. They differ in the number and nature of the adsorption processes (mechanisms) included, in mathematical treatment of the interaction, and in other aspects (see References 59 to 63 for detailed discussions). According to the above criteria, the models fall into three categories: surface bonding (chemical) models, electrostatic models, and combined models. Surface bonding models consider chemisorption as the main mechanism and assume that the trace element is directly bound to the solid surface. The physical phenomena are either neglected or considered as correction factors only. The most typical models of this category interpret adsorption as surface complex formation, which can be described using equilibrium formation constants analogously to the formation of soluble complexes.[64-67] Older ion exchange models[68,69] also belong to this category, as well as models earlier referred to as "specific adsorption" models. Electrostatic models interpret the adsorption mainly as coulombic interaction, which depends on the structure of the solid/solution interface.[70-73] Chemical interaction, if considered, is included in the calculation of electrostatic interactions. Combined models take into account both electrostatic and chemical energy interactions between the trace element and the surface.[61,74] James and Healy[75] in their model were the first to consider also the solvation energy, pointing out that ion-solvent interactions present a barrier to the adsorption.

The models were rather successfully used for explanation or quantitative description of the adsorption/desorption of trace elements on pure solid phases, mainly oxides, representing components of natural solids. Probably the most precise description was achieved for the effect of pH. Adsorption of cations is known to increase with pH from almost nil to a maximum (often complete adsorption is observed). The position of the increase on the pH scale (so-called adsorption edge) and steepness of the increase depend on the nature of the cation, the substrate, and on other factors (see, for example, Reference 2, p. 175). An inverse pH dependence is observed for anions. These effects are interpreted in electrostatic models by the change in surface charge of the substrate and by hydrolysis of the cation, changing its charge and ionic radius. In chemical models the effects are explained by direct involvement of hydrogen or hydroxyl ions in the chemical exchange with the trace element and/or the dissociation of surface groups participating at the adsorption.

A characteristic increase in adsorption of hydrolyzable cations takes place in the pH region where hydrolysis begins. Although some models explain this effect by predominant adsorption of hydrolytic products, it has been shown that the effect can be interpreted without hydrolysis, by surface complexing. Good correlation was found between equilibrium constants for surface complex formation and the first hydrolysis constants, and it has been proposed to use this correlation for calculation of unknown formation constants.[67] The adsorption of trace metals may decrease again at high pH values due to formation of anionic hydroxo complexes. Very complicated pH effects can be observed if the trace element forms colloids (see Reference 2, p. 182).

A notable exception is apparent for the adsorption of certain cations on some clays or clay-containing natural solids, where very moderate or no pH effect was found in the pH region typical for natural waters.[76,77] This is due to the predominant adsorption on negative surface charges formed by isostructural replacement of lattice ions of the clays. These charges do not depend on pH.

The effect of changing composition of the water is interpreted in adsorption models in terms of changes in the structure and charge of the solid/solution interface, of competition between the trace element and other solutes for surface sites, and of changes in hydration of the trace element. Although the adsorption of simple ions is usually suppressed at increasing ionic strength of the solution, the diversity of these effects means that the adsorption of hydrolyzed ions, neutral molecules, and colloidal forms of trace elements need not be affected or may even increase (see Reference 2, p. 192).

Soluble complexing ligands and surfactants in the water may play a very important role in the adsorption. The complexing ligands usually suppress the adsorption due to formation of nonadsorbable complexes in the aqueous phase. However, the opposite effect has also been described and explained by the existence of well adsorbable complexes or by the formation of ternary complexes on the surface.[78-80] Presence of surfactants may significantly affect the surface properties of the solid phase and cause either a decrease or increase in the adsorption.[81]

The effect of the concentration of a trace element on its adsorption can be most conveniently expressed in the form of an adsorption isotherm. The effect often conforms to Langmuir or Freundlich isotherms which were derived from adsorption models under different assumptions. Typically the relative adsorption, expressed as the fraction of the trace element bound to the solid phase, decreases with increasing total concentration of the element, and the adsorption edge shifts to higher pH values. This is commonly explained by consecutive saturation of sorption sites and by energetic heterogeneity of the sites. However, at very low concentrations of trace elements, the relative adsorption may be independent on their concentration and no shift of adsorption edge occurs. Then the linear (or Henry's) isotherm is valid.[77] The opposite effect, i.e., increase in the relative adsorption with increasing concentration, is very rare and indicates a change in the speciation of the trace element in solution, adsorption of polynuclear species, or surface precipitation (see Reference 53, p. 31).

The effect of the composition and properties of the solid phase is represented in adsorption models by means of the surface charge and parameters characterizing the properties of the solid/solution interface or the complexing ability of surface sites. It is very difficult to calculate such parameters theoretically; therefore, the necessary data are sought by experimental measurements of the adsorption even for rather simple and well-defined systems. A similar approach must be used for natural solids, which are to be considered in this respect as multicomponent, multiligand systems. Several authors tried to calculate adsorption parameters of well-defined mixtures of different solids from known adsorption parameters of the solids[82-84] using surface complexation models. Only one author subjected the calculated parameters to experimental verification and found that the composite adsorption behavior of a mixture of oxides and a clay was either enhanced or decreased relative to what should be expected. A probable reason for the discrepancy was a kind of solid-solid interaction.[85]

A related study was carried out with real samples of river and wastewater sediments.[86] Adsorption properties of the sediments towards radium were compared with analogous properties of model solids (minerals) representing components of the sediments. It was found that the adsorption affinity of the sediments could not be easily derived from their composition or other known properties. No simple correlation with surface area, organic matter, oxidic coatings, or other components of the sediments was observed.

It appears that a more promising method to assess the role of individual components of natural solids in the adsorption is direct analysis of the solids. One possibility is to study the correlation between the contents of trace elements and the composition of natural solids. Using this approach, Sigg[52] concluded that calcium carbonate does not represent an important component of suspended solids for the adsorption of heavy metals in lake water. Copper and zinc correlated mainly with organic matter in the solids, whereas lead was probably sorbed by ferric oxide.[52] A second approach relies on the analysis of trace element associations with components of natural solids by means of selective extractions (see, for example, References 44 and 87). It is also possible to compare adsorption properties of an unaltered solid with those of the same solid after certain components of the solid are removed by leaching.[77,88-90]

The effect of solid/solution ratio (m/V, mass of the solid vs. volume of solution) is expressed in adsorption models as the change in concentration of adsorption sites or in total adsorption capacity. The position of the adsorption edge shifts to lower pH values with increasing m/V, which is easily predicted by the models. The models invariably assume that the concentration increases linearly with m/V. However, experimental data do not always corroborate such an assumption. In practical measurements, the effect of m/V is usually examined by means of a distribution or partition coefficient K_d defined as

$$K_d = c_s/c_w \left(m^3 \cdot kg^{-1}\right) \quad (2)$$

where c_s is the concentration of trace element in the solid phase (mg kg^{-1}) and c_w is the concentration in water (mg·m^{-3}). K_d should be independent of m/V if the adsorption obeys a linear adsorption isotherm. At higher surface loading, or if a pronounced energetic heterogeneity of the adsorption sites exists, K_d decreases with decreasing m/V. This was corroborated for adsorption of many trace elements and radionuclides. However, a number of cases were reported when a pronounced decrease in K_d was observed with an increase in m/V above 0.1 to 100 kg·m^{-3} (see, for example, References 51, 90 to 92). Three different explanations were advanced for the effect: aggregation of solid phase particles at higher m/V ratios,[91] presence of colloidal particles in the system,[93] and complexation of trace elements by organic compounds dissolved from sediments.[90,94] However, no convincing evidence has yet been presented for any of these explanations.

The effect of temperature is implicitly included in most adsorption models by the temperature dependence of the interaction terms or of the surface complex formation constants. Theoretical considerations show that a decrease in equilibrium adsorption with increasing temperature should prevail. However, few experimental data on the effect have been published. A pronounced effect of temperature on the kinetics of the adsorption of trace elements was observed (see, for example, Reference 90).

As already mentioned, significant difficulties are encountered in application of adsorption models to natural waters. The reasons may be summarized as follows:

1. No existing model predicts accurately the adsorption/desorption behavior of trace elements over a wide range of important variables even in simple, well-defined systems. Our knowledge of the adsorption/desorption mechanism and laws is still very imperfect.
2. Natural solids are complicated mixtures of various components whose adsorption properties are unpredictably affected by shielding (coating, armoring), coprecipitation, and other phenomena. Calculation of adsorption parameters theoretically[52] or from similar parameters experimentally determined for components of natural solids will hardly find broader application.[85,94] Consequently, the adsorption parameters as distribution coefficients or surface complexation constants must be experimentally determined for the solid phase present in the system under investigation. Obtaining suitable data is complicated by the fact that considerable variability in adsorption properties of the solid phases may exist in time and space at a given locality.
3. Experimental determination of the adsorption parameters may be hampered by failure to simulate natural conditions in laboratory experiments. The exact composition of the aqueous phase and speciation of the trace element are often poorly known or may be affected by the sampling of natural phases so that their effect is difficult to assess. There are also some indications that adsorption/desorption equilibria in natural waters are affected by some colloids which may escape our attention.[93] This is further complicated by experimental difficulties in measuring the distribution at very low or high solid/solution ratios (see, for example, References 95 and 96).
4. The adsorption parameters may depend on solid-solid interactions as exemplified by the effect of the solid/solution ratio on distribution coefficients (see above). This effect may have important implications as the solid/solution ratio varies within very broad ranges in natural systems.
5. Very limited or controversial information exists on the reversibility and kinetics of adsorption/desorption processes in natural waters. Most adsorption models assume full reversibility of the adsorption, which means that the same equilibrium distribution should be achieved by adsorption and desorption

if the final conditions are the same. Experimental findings are, however, often in disagreement with such an assumption. The kinetics of adsorption and desorption may be very slow in natural waters. The limited reversibility and the slow kinetics cause large complications to calculations of the real distribution of trace elements between phases.

Apart from explaining the effect of important factors on the adsorption/desorption of trace elements in natural waters, the adsorption models have been used to predict the speciation and/or migration of trace elements. In the first case, adsorption is included in computation of equilibria in multiligand, multielement water systems together with complexation, redox, and precipitation processes. The solid surface is usually considered as one of the ligands in the system, and its competition with other dissolved or surface ligands for trace elements is described using suitable adsorption models.[16,73,89,97,98] Several authors compared the distribution of trace elements found in this way with analogous distributions of the same elements between dissolved and solid phases determined experimentally.[67,98,99] The comparisons showed variable success of the computations, indicating that complications mentioned above may have occurred. Adsorption models were also used for the prediction of the uptake of trace elements on algal surfaces.[52,100]

Suitable consideration of adsorption/desorption phenomena in modeling the migration of trace elements in natural waters represents a still more difficult problem. In many migration models the interaction with a solid phase is either neglected or described in a very simplified way, using an equilibrium distribution coefficient K_d. In the latter case one value of K_d is usually employed for one modeled system despite the time and space variability for K_d which may exist due to the variability of some basic factors affecting the adsorption and desorption. The main reason is that it is very difficult to express the dependence of K_d on these basic factors quantitatively and to include this dependence in the migration models. Furthermore, K_d values determined experimentally or obtained by calculations from field data may also reflect other processes like coprecipitation, diffusion into small pores, etc. In many systems the rate of migration is so high and the rate of adsorption/desorption so slow that adsorption/desorption equilibrium cannot be achieved during the migration, and the kinetics of the adsorption/desorption must be taken into consideration in the calculations.

For these reasons considerable criticism has been expressed against indiscriminative use of the equilibrium distribution coefficient K_d in migration models, and it has been recommended that new approaches for the inclusion of adsorption/desorption into migration models be sought (see, for example, References 101 to 103). The approaches may be different for different types of waters. For instance, it has been proposed for the modeling of migration in surface streams that the effect of the interaction of migrating species with suspended solids be included in the forms of suitable kinetic parameters and their quantified dependence on the m/V ratio and temperature.[104] The effects of composition of suspended solids and water can be properly considered in the form of variability of the kinetic parameters. In the modeling of migration in groundwaters, K_d can be replaced by a multiparametric equilibrium isotherm of the Freundlich or Langmuir type, possibly in combination with simple kinetic relations.[101,105-107]

V. PRECIPITATION, COPRECIPITATION, AND DISSOLUTION

Precipitation and dissolution belong to the most important processes in natural waters as they largely, directly or indirectly through adsorption, control the composition of natural waters in contact with lithosphere and atmosphere. They may also significantly affect the speciation and migration of trace elements. In most cases, however, it is the solid phase of substances present in macroconcentrations in natural systems that is precipitated or dissolved, and which serves as a nonisotopic carrier for a trace element. Trace elements are hence most frequently coprecipitated with or adsorbed on such solid phases or are dissolved with it. Nevertheless, more or less independent precipitation and dissolution of a solid phase of a sparingly soluble compound of a trace element is possible at certain circumstances. Because of the low concentration of the compound, the size of particles formed in this way can reach only colloidal dimensions.* These particles are conveniently called true colloids to distinguish them from particulate

* Although some authors use different size limits for definition of colloids in natural waters,[1] the generally accepted range of colloidal dimensions (the diameter or the largest dimension of particles or molecules) in colloid chemistry is 1 to 200 (500) nm. In view of the widely recognized use of filtration through 450-nm membrane filters for distinguishing between "particulate" and "dissolved" species in natural waters, we recommend using the (1-5) to 450 nm range as a definition of colloids in natural waters. The range given for the lower limit of colloidal size stems from experimental difficulties in determining colloid sizes in this range.

forms often called "pseudocolloids", where the trace element is incorporated by coprecipitation or held on the surface of foreign particles by adsorption. Both terms have long been used to describe the nature of colloidal forms of radionuclides in solutions, so-called "radiocolloids". However, they can be equally well employed to classify colloidal forms of nonradioactive trace elements. In fact, the radioactivity itself has hardly anything to do with the formation of colloidal forms of radionuclides at trace concentrations. Therefore, the term "radiocolloid" is a misnomer and should be replaced by a more general term, e.g., "trace colloid",* valid both for radioactive and nonradioactive trace elements (see Reference 2, p. 63).

The original definition of true colloids referred to colloidal forms of radionuclides composed exclusively or primarily of molecules of a sparingly soluble compound of the radionuclide in question. As explained in detail elsewhere (see Reference 2, p.64), this definition is hardly acceptable even for radioactive trace colloids. First, it is experimentally very difficult, if not impossible, to determine the proportion of molecules of a given composition to other molecules present in a trace colloid. Second, it is highly arbitrary to draw a limit for the abundance of a given substance in the colloid beyond which the colloid does not have the properties of that substance and therefore is not a true colloid. Third, it is conceivable that a colloid may be formed by precipitation of a trace compound on the surface of a foreign solid particle ("armoring"). If the surface coverage by the armoring compound is sufficient, the properties of such a colloid will be largely determined by the armoring compound, although its proportion to other substances in the colloid may be very small.

Therefore, a new definition for true colloids has been proposed (see Reference 2, p.66). According to this definition, true colloids are "trace colloids formed by condensation of slightly soluble forms of a trace element, whose properties and physico-chemical behavior are analogous to the properties and behavior of the solid phase corresponding to slightly soluble forms of this element in macroconcentrations". By this definition, colloids formed by precipitation of a trace compound on the surface of a foreign solid particle may qualify as a true colloid. Colloidal forms of trace elements whose properties do not correspond to a true colloid are pseudocolloids. It has been suggested to distinguish between "reversible" and "irreversible" pseudocolloids (see Reference 2, p. 66). The former are formed by reversible adsorption of a trace element on the surface of a foreign particle, from which the element can be easily desorbed. In irreversible pseudocolloids, the trace element is either incorporated into the interior of the foreign particle or is irreversibly sorbed on it, so that it can be released only by dissolution of the particle. Pseudocolloids most frequently fall in the category of "particulate" or "suspended" forms of trace elements in natural waters, as most carrying particles are larger than 450 nm. On the other hand, true colloids[2] are typically smaller than 30 to 50 nm, unless they are formed by armoring of larger particles.

Formation and dissolution of true colloids can be interpreted by general principles of precipitation and dissolution described in the chemical literature.[1,108-111] Application of the principles for trace concentrations was discussed in detail by Beneš and Majer,[2] who also presented a compilation of data characterizing formation of true colloids and pseudocolloids in aqueous solutions, based on experiments with radionuclides. Similar data were compiled by Starik[5] and Davydov.[112] They can be useful as sources of information also for natural waters, if specific features of different water bodies are properly considered.

It appears that the main factors affecting precipitation and dissolution of true colloids are pH and composition of the solution, concentration of the trace element, time, and temperature. The pH affects mainly the precipitation and dissolution of oxides or hydroxides, since OH^- ions act as the precipitating agent in this case. However, pH can also affect the existence of other solid phases due to hydrolysis of the trace element or due to its effect on association/dissociation of the precipitating agent (e.g., anion). Presence of complex-forming ligands can suppress the precipitation because of the formation of soluble complexes with the trace element or with the agent. The abundance of true colloids may also depend on the ionic strength of the water. This is because the ionic strength affects the activities of components taking part in the precipitation or in the competing reactions.

A factor of primary importance is the concentration of the trace element to be precipitated. The probability or extent of precipitation increases with increasing concentration of the trace element according to the well-known solubility product principle. Although the applicability of the principle to very low concentrations has been questioned by some authors, it appears that qualified conclusions can be drawn from computations based on solubility products even for trace concentrations (see Reference

---

* Trace colloids are defined as colloids formed by trace elements present in solution at concentrations below 10^{-5} mol dm^{-3}. Important ways in which trace colloids differ from "macrocolloids" include a certain uncertainty or difference in nature, a small number of particles in unit solution volume, and the impossibility of forming (e.g., through coagulation) a separate compact solid phase.

2, p.108). Solubility equilibria are commonly included in the advanced computer codes for calculation of speciation of trace elements in natural waters.[15-18] The calculations also take into consideration effects of pH, complexing ligands, and ionic strength. The effect of temperature is accounted for by the temperature dependence of the solubility product and other equilibrium constants.

Application of the general principles outlined above to conditions prevailing in natural waters leads to the conclusion that formation of pseudocolloids is more probable for trace elements in natural waters than formation of true colloids, particularly at concentrations lower than 10^{-8} mol dm^{-3}. One of the reasons is that the adsorption of a trace element on particles present in the waters occurs more easily than the precipitation of its sparingly soluble compound. Condensation of a substance to form a more or less pure solid phase (nucleation) requires considerable supersaturation due to a higher solubility of very small particles with strongly curved surface (nuclei).[109,110] It is therefore highly probable that adsorption will prevail or the precipitation of a new phase will take place on the surface of any available particle already present in the system (heteronucleation). Such particles are normally abundant in natural waters. Unless the particles are sufficiently covered with an armoring layer of the new phase, pseudocolloids will be formed even in the case of precipitation. Furthermore, most solid phases show a tendency to coprecipitate other compounds present in the system, particularly those that are chemically similar. At the conditions in natural waters, which represent complex mixtures of many substances, precipitated solids will therefore scarcely be pure compounds but rather solid solutions, mixtures, or layered structures of many components. The properties of such particles would hardly conform to the definition of true colloids.

Coprecipitation and armoring phenomena make exact thermodynamic calculations of the abundance and composition of solid phases present in natural waters difficult. Drawing quantitative conclusions from equilibrium calculations is also complicated by the sometimes very slow kinetics of the precipitation and by slow changes in the structure and/or composition of the precipitated solid phase ("ageing"). Many slightly soluble substances tend to precipitate as amorphous solids, which then slowly transform into crystalline precipitates. It should be stressed, however, that the crystallinity of precipitates depends on the rate of crystal growth, which can be significantly affected by the concentration of the precipitated components. The slower the rate, the more perfect crystals will be obtained. It is therefore probable that solid phases formed by trace elements in natural waters will have more perfect crystalline structures than those formed by the same elements at macroconcentrations. The characteristic crystalline forms of ferric oxohydroxide found in very dilute solutions of iron may serve as an example.[113]

Apart from its role in the formation of pseudocolloids, coprecipitation may significantly affect the behavior of trace elements in soil solutions and aquifers, where the elements can be deposited on porous solid phases by coprecipitation with various macrocomponents of ground waters. Coprecipitation has also been extensively used for preconcentration and separation of trace elements and radionuclides in the analysis of natural waters. Therefore, the basic principles and laws of coprecipitation are of great importance for the chemistry of trace elements in natural waters. The principles and laws were described in several monographs and reviews.[2,5,114-118] At least four different mechanisms of coprecipitation can be distinguished: formation of isomorphous mixed crystals, formation of anomalous mixed crystals, adsorption, and internal adsorption. Sometimes a fifth class is added, referred to as occlusion or mechanical inclusion.

In the first two cases, the coprecipitation proceeds by formation of a solid solution where either the trace element replaces isomorphously ions of a macrocomponent of the solid phase, or various kinds of nonisomorphous incorporation take place (anomalous mixed crystals). Both the mechanisms are characterized by incorporation of the trace element into the crystal lattice of a solid phase and by its continuous distribution in the host crystals. The equilibrium distribution of the trace element between the solid phase and the solution depends little on the external conditions during the precipitation, such as excess of precipitating ion, pH, the presence of moderate concentrations of foreign polyvalent ions, etc. The distribution can be quantitatively described using a thermodynamic or, in the special case of nonequilibrium distribution, a kinetic constant. However, this possibility apparently has not yet found adequate applications in computer calculations of the distribution of trace elements in natural water systems. The difference between the two mechanisms is manifested in the fact that the distribution of the trace element between solution and isomorphous mixed crystals does not depend on the concentration ratio of the trace element and the macrocomponent in the system, whereas coprecipitation by formation of anomalous mixed crystals can depend on this ratio due to the limited miscibility of the components. The effects of other factors are similar for both mechanisms; the distribution depends on the nature of the trace element, composition of the solid phase, temperature, and time. A very profound effect on the distribution is exerted by the ratio of the solubility of the macrocomponent compound to that of the solubility of the trace element compound. The higher the ratio, the more is the trace element concentrated in the precipitate.

However, this rule is only approximately valid, and other characteristics of the coprecitating components also play a role: the molecular weight of the macrocomponent compound, the size of the anion of the coprecipitated compounds, etc. The coprecipitation often decreases with increasing temperature. The time dependence of the distribution is influenced not only by the rate of precipitation but also by recrystallization of the precipitated solid. Due to the recrystallization, a rather slow redistribution of the coprecipitated trace element can take place.

The other mechanisms of coprecipitation are characterized by discontinuous distribution of the trace element in the host crystals and by a more pronounced dependence of external conditions on the distribution. The adsorption mechanism of coprecipitation is observed when the trace element is deposited only on the crystal surface. One of the main features of this mechanism is the rapid establishment of equilibrium distribution of the trace element. Other characteristic features are identical with those of adsorption on performed solid phases discussed earlier. Internal adsorption differs from ordinary adsorption coprecipitation mainly in that the trace element is incorporated into the solid phase. It is assumed that the adsorption proceeds on the inner surface of crystals, namely on crystal dislocations, crystal layers, intercrystalline borders, etc. The distribution of the trace element between the solid and aqueous phases is quite reproducible and can be characterized by a distribution coefficient, decreasing with increasing concentration of the trace element and of foreign multivalent ions in the system. Here again the equilibrium distribution is rapidly established. Occlusion and mechanical inclusion are the least known mechanisms of coprecipitation. They are characterized by irregular distribution of the trace element, strongly dependent on the rate and conditions of the precipitation.

Dissolution of trace elements from true colloids or from solid phases formed by coprecipitation depends on the same equilibria as the precipitation or coprecipitation. However, the kinetics of dissolution may be significantly different. The kinetics is often ruled by reactions occurring on the surface of solid phases.[119,120] Dissolution of true colloids can be very slow if a depolymerization reaction is included (see Reference 2, p. 113).

From the above discussion, it is apparent that characterization of the solid forms of trace elements in natural waters by computation methods is even less reliable than similar characterization of the dissolved forms. Therefore, most of the data on solid forms was obtained experimentally. Experimental methods used for this purpose can be roughly classified into two groups: methods characterizing the physico-chemical properties of colloidal or particulate forms such as the size, shape, surface charge, composition, etc. and methods characterizing the bond or site of the trace element in the solid phase. Methods of the first group employ either separation of particles using a suitable procedure (centrifugation, ultrafiltration, dialysis, etc.) or study of their behavior in a physico-chemical process (electrophoresis, diffusion, chromatography, etc.) Use of the methods is complicated by the same factors as discussed in Section III and was critically reviewed by a number of authors.[2,5,41,43,45,121-123] Detailed surveys of possibilities to distinguish between true colloids and pseudocolloids have also been presented.[2,5]

Binding forms of trace elements in solids can be determined either by nondestructive microanalysis techniques such as electron microprobe, secondary ion mass spectroscopy, etc. or by selective extraction methods. Both the approaches were recently evaluated by Kersten and Förstner;[87] the selective extraction methods were extensively discussed also by other authors (see, for example, References 125 to 127).

VI. AGGREGATION AND DISAGGREGATION

Field observations and model experiments have shown that colloidal and particulate forms of trace elements in natural waters differ significantly in behavior and toxicity from ionic and molecular forms of the same elements. This difference depends strongly on the size or size distribution of colloidal and particulate forms, which can be changed by aggregation (coagulation, sticking together) and disaggregation (peptization, separation) of the particles. These processes may influence the movement and distribution of trace elements in surface waters due to change in the rates of sedimentation and resuspension of particles. They can affect the accessibility of trace elements carried by such particles or the kinetics of transfer of trace elements over the particle/solution boundary, as the accessibility and transfer depend on the specific surface area of the particles. This may change the biological uptake of the trace elements and their participation in chemical reactions. Therefore, it is useful to briefly comment on the principles and reactions governing aggregation and disaggregation of natural particles. The basic principles are derived from colloid and surface chemistry,[9,128] where aggregate stability is defined as the ability of a dispersion system to maintain steady size distribution of its particles (its state of dispersion). Lyophilic colloids are considered homogenous systems and as such can be thermodynamically stable. Association or dissociation of their "particles" can be viewed as chemical reactions and described using equilibrium and kinetic approaches common for chemical reactions.

Lyophobic colloids and coarse dispersions (particulate forms of trace elements) are heterogenous systems characterized by high surface energy and therefore are thermodynamically unstable. Their aggregate stability can still be high enough that the dispersion state of lyophobic particles will change very slowly or negligibly during the period of observation. The aggregate stability depends on the surface charge of the particles or on the "wrapping" of the particles in a layer of protecting molecules or lyophilic colloids (electrostatic and steric stabilization). The layer will render the particles lyophilic character and thus higher stability. Without the stabilization by charge or by a protective layer, the particles tend to adhere to each other upon contact brought about by collisions in Brownian motion or due to different migration velocity of the particles imposed by velocity gradients in the water, or by different sedimentation rates. The surface charge will prevent contact between particles by coulombic repulsion forces; the protective layer will prevent adhesion. The stabilization may depend on composition, size and shape of the particles, on pH and chemical composition of water, on properties and concentration of stabilizing lyophilic colloids, etc. Insufficiently stabilized particles will coagulate by formation of aggregates, the rate of coagulation depending on the concentration and on other factors. The most frequent reason for destabilization is an increase in concentration of electrolytes (ionic strength) which brings about a decrease in surface charge (zeta potential) and lower electrostatic repulsion. This frequently happens, for example, during estuarine mixing of freshwater with saline water, or on discharge into natural fresh water of wastewaters containing high concentrations of dissolved salts. Freshly coagulated aggregates can be disaggregated (peptized) by chemical and physical effects which would remove the reason for the destabilization and restore the aggregate stability of the original particles. For example, mere dilution of saline wastewater with receiving fresh water can peptize aggregates contained in the wastewater.

Although considerable progress has been achieved in description of the coagulation and peptization processes in well-defined systems by theoretical models, the models are still not able to account for all the important phenomena occurring during these processes.[128-133] Aggregation and disaggregation in natural systems are much more difficult to interpret because of poor knowledge of the composition and properties of naturally occurring particles and due to the large variability of chemical composition, temperature, hydrodynamic conditions, etc. in natural waters.

It has been proven that the adsorption of natural organic compounds plays an important role in the stabilization of particles in some natural waters. Some of the compounds contain both hydrophobic components (hydrocarbon chains) and hydrophilic groups (carboxyls, hydroxyls, etc.) and therefore tend to be concentrated at the water/particle interface. Since they are negatively charged, they may give the particles a net negative charge. Both the negative charge and the hydrophilic character of the particles rendered by the adsorption enhance the stability of the particles. This was inferred either from laboratory experiments with natural organic compounds and model solids[134-138] or from measurements carried out in natural water samples.[139,140] It has also been found that the effect may at least partially be counterbalanced by the presence of divalent metal cations, especially Ca^{2+}.[138,139]

An interesting question concerns the role of trace elements in affecting the stability of natural particles. This role depends on the nature of the particles. It will be negligible for particles other than true colloids of the trace element in question, since the adsorption of a trace element cannot significantly affect the surface charge or adsorption properties of natural particles because of the low concentration of the element. The stability of true colloids will, of course, depend primarily on the surface chemical properties of a slightly soluble compound of the trace element, forming either the predominant component of the colloid or an armoring layer on the surface of a foreign particle. Then the trace element itself will decisively influence the surface charge and adsorption properties of the colloid. Therefore, coagulation can be expected, for example, at pH values close to the isoelectric point of the hydroxide or oxide prevailing in the colloid.[141]

REFERENCES

1. **Stumm, W. and Morgan, J. J.,** *Aquatic Chemistry,* 2nd ed., John Wiley & Sons, New York, 1981.
2. **Beneš, P. and Majer, V.,** *Trace Chemistry of Aqueous Solutions,* Elsevier, Amsterdam, 1980.
3. **Beneš, P.,** State of trace amounts of elements in aqueous solutions, *Chem. listy,* 66, 561, 1972 (in Czech).

4. **Guillaumont, R., Adloff, J. P., and Peneloux, A.,** Kinetic and thermodynamic aspects of tracer-scale and single atom chemistry, *Radiochim. Acta,* 46, 169, 1989.
5. **Starik, I. E.,** *Principles of Radiochemistry,* 2nd ed., Nauka, Leningrad, 1969 (in Russian), (English translation of the 1st ed.: AEC-tr-6314, U.S. Atomic Energy Commission, 1964).
6. **Evans, E. A. and Muramatsu, M., Eds.,** *Radiotracer Techniques and Application,* Marcel Dekker, New York, 1977.
7. **Hem, J. D.,** *Study and Interpretation of Chemical Characteristics of Natural Waters,* Geological Survey Water Supply Paper No. 1473, Washington, D.C., 1975.
8. **Marcus, R. A.,** Electron transfer in homogeneous and heterogeneous systems, in *The Nature of Seawater,* Goldberg, E. D., Ed., Dahlem Konferenzen, Berlin, 1975, 477.
9. **Hiemenz, P. C.,** *Principles of Colloid and Surface Chemistry,* Marcel Dekker, New York, 1977.
10. **De Wit, J. C. M., van Riemsdijk, W. M., and Koopal, L. K.,** Proton and metal ion binding on humic substances, in *Metal Speciation, Separation and Recovery,* Vol. 2, Patterson, J. W. and Passino, R., Eds., Lewis Publishers, Chelsea, MI, 1990, 329.
11. **Sposito, G.,** Sorption of trace metals by humic materials in soils and natural waters, *CRC Crit. Revs. Environ. Control,* 16, 193, 1986.
12. **Butler, J. N.,** *Ionic Equilibrium, A Mathematical Approach,* Addison-Wesley, Reading, MA, 1964.
13. **Baes, C. F. and Mesmer, R. E.,** *The Hydrolysis of Cations,* Wiley-Interscience, New York, 1976.
14. **Buffle, J.,** *Complexation Reactions in Aquatic Systems: An Analytical Approach,* Ellis Horwood, Chichester, 1988.
15. **Jenne, E. A., Ed.,** *Chemical Modeling in Aqueous Systems,* ACS Symposium Series 93, American Chemical Society, Washington, D.C., 1979.
16. **Nordstrom, D. K. and Ball, J. W.,** Chemical models, computer programs and metal complexation in natural waters, in *Complexation of Trace Metals in Natural Waters,* Kramer, C. J. M. and Duinker, J. C., Eds., Nijhoff & Junk Publ., The Hague, 1984, 149.
17. **Nordstrom, D. K. and Munoz, J. L.,** *Geochemical Thermodynamics,* Benjamin Cummings Publishing Co., Menlo Park, CA, 1985.
18. **Waite, T. D.,** Mathematical modeling of trace element speciation, in *Trace Element Speciation: Analytical Methods and Problems,* Batley, G. E., Ed., CRC Press, Boca Raton, FL, 1989, 117.
19. **Lund, W.,** The complexation of metal ions by humic substances in natural waters, in NATO ASI Ser., Ser. G, 23 (Metal Speciation in the Environment), 1990, 43.
20. **Christman, R. F. and Gjessing, E. T., Eds.,** *Aquatic and Terrestrial Humic Materials,* Ann Arbor Science, Ann Arbor, MI, 1983.
21. **Aiken, G. R., McKnight, R. L., Weshaw, R. L., and Mc Carthy, P., Eds.,** *Humic Substances in Soil, Sediment and Water,* John Wiley & Sons, New York, 1985.
22. **Fish, W., Dzombak, D. A., and Morel, F. M. M.,** Metal-humate interaction. II. Application and comparison of models, *Environ. Sci. Technol.,* 20, 676, 1986.
23. **Cabaniss, S. E., Shuman, M. S., and Collins, B. J.,** Metal-organic binding: a comparison of models, in *Complexation of Trace Metals in Natural Waters,* Kramer, C. J. M. and Duinker, J. C., Eds., Nijhoff & Junk Publ., The Hague, 1984, 165.
24. **Gamble, D. S.,** Interactions between natural organic polymers and metals in soil and freshwater systems: Equilibria, in *The Importance of Chemical "Speciation" in Environmental Processes,* Bernhard, M., Brinckman, F. E., and Sadler, P. Y., Eds., Springer-Verlag, Berlin, 1986, 217.
25. **Buffle, J.,** Natural organic matter and metal-organic interactions in aquatic systems, in *Metal Ions in Biological systems,* Vol. 18, *Circulations of Metals in the Environment,* Sigel, H., Ed., Marcel Dekker, New York, 1984, 165.
26. **Perdue, E. M.,** Measurements of binding site concentrations in humic substances, in *Metal Speciation: Theory, Analysis and Application,* Kramer, J. R. and Allen, H. E., Eds., Lewis Publishers, Chelsea, MI, 1988, 135.
27. **Vuceta, J. and Morgan, J. J.,** Chemical modeling of trace metals in fresh waters: role of complexation and adsorption, *Environ. Sci. Technol.,* 12, 1302, 1978.
28. **Campbell, P. G. C. and Stokes, P. M.,** Acidification and toxicity of metals to aquatic biota, *Can. J. Fish. Aquat. Sci.,* 42, 2034, 1985.
29. **Sposito, G.,** Trace metals in contaminated waters, *Environ. Sci. Technol.,* 15, 396, 1981.
30. **Sillén, L. G. and Martell, A. E.,** *Stability Constants of Metal-ion Complexes,* Spec. Publ. No. 17, Chemical Society, London, 1964.

31. **Sillén, L. G. and Martell, A. E.,** *Stability Constants of Metal-ion Complexes,* Spec. Publ. No. 17, Suppl. No. 1, Chemical Society, London 1971.
32. **Smith, R. M. and Martell, A. E.,** *Critical Stability Constants,* Vol. 4, Plenum Press, New York, 1976.
33. **Kraus, K. A. and Nelson, F.,** Anion exchange studies. X. Ion exchange in concentrated electrolytes. Gold(III) in hydrochloric acid solutions, *J. Am. Chem. Soc.,* 76, 984, 1954.
34. **Milburn, R. M. and Vosburgh, W. C.,** Spectrophotometric study of the hydrolysis of iron(III) ion. II. Polynuclear species, *J. Am. Chem. Soc.,* 77, 1352, 1955.
35. **Korotkin, Yu. S.,** Hydrolysis of transuranium elements. IV. Region of the sorption uniformity of trace amounts of americium(III), *Radiokhimiya,* 16, 217, 1974 (in Russian).
36. **Davydov, Yu. P., Jefremenkov, V. M., Gratchok, M. A., and Bondar, Yu. J.,** Investigation of the hydrolytic properties of U(VI) and Fe(III) in solutions, *J. Radioanal. Chem.,* 30, 173, 1976.
37. **Zielen, A. J. and Connick, R. E.,** The hydrolytic polymerization of zirconium in perchloric acid solution, *J. Am. Chem. Soc.,* 78, 5785, 1956.
38. **Guillamont, R.,** Protactinium ions in aqueous solutions. IV. Hydrolysis and polymerization of pentavalent protactinium, *Bull. Soc. Chim. France,* 2106, 1965 (7), (in French).
39. **Schneider, W. and Schwyn, B.,** The hydrolysis of iron in synthetic, biological and aquatic media, in *Aquatic Surface Chemistry,* Stumm, W., Ed., John Wiley & Sons, New York, 1987, 167.
40. **Buffle, J.,** Speciation of trace elements in natural waters, *Trends Anal. Chem.,* 1, 90, 1980.
41. **Florence, T. M. and Batley, G. E.,** Chemical speciation in natural waters, *CRC Crit. Rev. Anal. Chem.,* 9, 219, 1980.
42. **Batley, G. E., Ed.,** *Trace Element Speciation: Analytical Methods and Problems,* CRC Press, Boca Raton, FL, 1989.
43. **Betti, M. and Papoff, P.,** Trace elements: data and information in the characterization of an aqueous ecosystem, *CRC Crit. Rev. Anal. Chem.,* 19, 271, 1988.
44. **Beneš, P.,** Speciation procedures, in *The Environmental Behaviour of Radium,* International Atomic Energy Agency, Vienna, 1990, 273.
45. **Mizuike, A.,** Recent developments in trace metal speciation in fresh water, *Pure Appl. Chem.,* 59, 555, 1987.
46. **van den Berg, C. M. G. and Kramer, J. R.,** Determination of complexing capacities of ligands in natural waters and conditional stability constants of the copper complexes by means of manganese dioxide, *Anal. Chim. Acta,* 106, 113, 1979.
47. **Davey, E. W. and Soper, A. E.,** Apparatus for the *in situ* concentration of trace metals from seawater, *Limnol. Oceanogr.,* 20, 1019, 1975.
48. **Batley, G. E.,** Physicochemical separation methods for trace element speciation in aquatic samples, in *Trace Element Speciation: Analytical Methods and Problems,* Batley, G. E., Ed., CRC Press, Boca Raton, FL, 1989, 43.
49. **Batley, G. E.,** Collection, preparation and storage of samples for speciation analysis, in *Trace Element Speciation: Analytical Methods and Problems,* Batley, G. E., Ed., CRC Press, Boca Raton, FL, 1989, 1.
50. **Hamilton, E. I.,** K_d values: an assessment of field v. laboratory measurements, in *Application of Distribution Coefficients to Radiological Assessment Models,* Sibley, T. H. and Myttenaere, C., Eds., Elsevier, London, 1986, 35.
51. **Beneš, P.,** Radium in (continental) surface waters, in *The Environmental Behaviour of Radium,* International Atomic Energy Agency, Vienna, 1990, 373.
52. **Sigg, L.,** Surface chemical aspects of the distribution and fate of metal ions in lakes, in *Aquatic Surface Chemistry,* Stumm, W., Ed., John Wiley & Sons, New York, 1987, 319.
53. **Salomons, W. and Förstner, U.,** *Metals in the Hydrocycle,* Springer-Verlag, Berlin, 1984, 24.
54. **Jenne, E. A.,** Trace elements adsorption by sediments and soils — sites and processes, in *Molybdenum in the Environment,* Vol. 2, Marcel Dekker, New York, 1976, 425.
55. **Hart, B. T.,** Uptake of trace metals by sediments and suspended particulates: a review, in *Sediment/Freshwater Interaction,* Vol. 9, Sly, P. G., Ed., Nijhoff & Junk Publ., The Hague, 1982, 299.
56. **Onishi, Y., Serne, R. J., Arnold, E. M., Cowan, C. E., and Thompson, F. L.,** Critical review of radionuclide transport, sediment transport and water quality mathematical modelling and radionuclide adsorption/desorption mechanisms, Rep. NUREG/CR-1322 (PNL-2901), Batelle Pacific Northwest Lab., Richland, WA, 1981.
57. **Anderson, M. A. and Rubin, A. J., Eds.,** *Adsorption of Inorganics at the Solid Water Interface,* Ann Arbor Science, Ann Arbor, MI, 1981.

58. **Stumm, W., Ed.,** *Aquatic Surface Chemistry,* John Wiley & Sons, New York, 1987.
59. **Westall, J. C.,** Adsorption mechanisms in aquatic surface chemistry, in *Aquatic Surface Chemistry,* Stumm, W., Ed., John Wiley & Sons, New York, 1987, 1.
60. **Schindler, P. W. and Stumm, W.,** The surface chemistry of oxides, hydroxides and oxide minerals, in *Aquatic Surface Chemistry,* Stumm, W., Eds., John Wiley & Sons, New York, 1987, 83.
61. **Davis, J. A. and Leckie, J. O.,** Speciation of adsorbed ions at the oxide/water interface, in *Chemical Modeling in Aqueous Systems,* Jenne, E. A., Ed., American Chemical Society, Washington, D.C., 1979, 299.
62. **Morel, F. M. M., Westall, J. C., and Yeasted, J. G.,** Adsorption models. A mathematical analysis in the framework of general equilibrium calculations, in *Adsorption of Inorganics at the Solid Water Interface,* Anderson, M. A. and Rubin, A.J., Eds., Ann Arbor, Science, Ann Arbor, MI, 1981, 263.
63. **Leckie, J. O. and James, R. O.,** Control mechanisms for trace metals in natural waters, in *Aqueous Environmental Chemistry of Metals,* Rubin, A. J., Ed., Ann Arbor Science, Ann Arbor, MI, 1974, 1.
64. **Huang, C. P. and Stumm, W.,** Specific adsorption of cations on hydrous γ-Al_2O_3, *J. Colloid Interface Sci.,* 43, 409, 1973.
65. **Schindler, P. W., Fuerst, B., Dick, R., and Wolf, P. V.,** Ligand properties of surface silanol groups. I. Surface complex formation with Fe^{3+}, Cu^{2+}, Cd^{2+} and Pb^{2+}, *J. Colloid Interface Sci.,* 55, 469, 1976.
66. **Huang, C. P., Hsieh, Y. S., Park, S. W., Ozden Carapcioglu, M., Bowers, A. R., and Elliot, H. A.,** Chemical interactions between heavy metal ions and hydrous solids, in *Metals Speciation, Separation and Recovery,* Patterson, J. W. and Passino, R., Eds., Lewis Publishers, Chelsea, MI, 1987, 437.
67. **Bourg, A. C. M. and Mouvet, C.,** A heterogenous complexation model of the adsorption of trace metals on natural particulate matter, in *Complexation of Trace Metals in Natural Waters,* Kramer, C. J. M. and Duinker, J. C., Eds., Nijhoff & Junk Publ., The Hague, 1984, 267.
68. **Ahrland, S., Grenthe, I., and Norén, B.,** Ion exchange properties of silica gel. I. The sorption of Na^{+}, Ca^{2+}, Ba^{2+}, VO_2^{2+}, Gd^{3+}, Zr(IV), Nb(V), U(VI) and Pu(IV), *Acta Chem. Scand.,* 14, 1059, 1960.
69. **Dugger, D. L., Stanton, J. S., and Irby, B. N.,** The exchange of twenty metal ions with weakly acidic silanol group of silica gel, *J. Phys. Chem.,* 68, 757, 1964.
70. **Grahame, D. G.,** Electrical double layer, *J. Chem. Phys.,* 23, 1166, 1955.
71. **Yates, D. E., Levine, S., and Healy, T. W.,** Site-binding model of the electrical double layer at the oxide/water interface, *J. Chem. Soc. Faraday Trans. 1,* 70, 1807, 1974.
72. **Bowden, J. W., Posner, A. M., and Quirk, J. P.,** Ionic adsorption on variable charge mineral surfaces. Theoretical-charge development and titration curves, *Aust. J. Soil Res.,* 15, 121, 1977.
73. **Westall, J. and Hohl, H.,** A comparison of electrostatic models for the oxide/solution interface, *Adv. Colloid Interface Sci.,* 12, 265, 1980.
74. **Kinniburgh, D. G. and Jackson, M. L.,** Cation adsorption by hydrous metal oxides and clays, in *Adsorption of Inorganics at the Solid Water Interface,* Anderson, M. A. and Rubin, A. J., Eds., Ann Arbor Science, Ann Arbor, MI, 1981, 91.
75. **James, R. O. and Healy, T. W.,** Adsorption of hydrolyzable metal ions at the oxide-water interface. III. Thermodynamic model of adsorption, *J. Colloid Interface Sci.,* 40, 65, 1972.
76. **Bourg, A. C. M.,** Role of fresh water/sea water mixing on trace metal adsorption phenomena, in *Trace Metals in Sea Water,* Wong, C. S., Burton, J. D., Boyle, E., Bruland, K., and Goldberg, E. D., Eds., Plenum Press, New York, 1983, 195.
77. **Beneš, P., Lam Ramos, P., and Poliak, R.,** Factors affecting interaction of radiocesium with freshwater solids. I. pH, composition of water and the solids, *J. Radioanal. Nucl. Chem.,* 113, 359, 1989.
78. **Benjamin, M. M. and Leckie, J. O.,** Effects of complexation by Cl, SO_4 and S_2O_3 on adsorption behaviour of Cd on oxide surfaces, *Environ. Sci. Technol.,* 16, 162, 1982.
79. **Bourg, A. C. M. and Schindler, P. W.,** Ternary surface complexes. I. Complex formation in the system silica-Cu(II)-ethylendiamine, *Chimia,* 32, 166, 1978.
80. **Davis, J. A. and Leckie, J. O.,** Effect of adsorbed complexing ligands on trace metal uptake by hydrous oxides, *Environ. Sci. Technol.,* 12, 1309, 1978.
81. **Beveridge, A. and Pickering, W. F.,** The influence of surfactants on the adsorption of heavy metals by clays, *Water Res.,* 17, 215, 1983.
82. **Oakley, S. M., Nelson, P. O., and Williamson, K. J.,** Model of trace-metal partitioning on marine sediments, *Environ. Sci. Technol.,* 15, 474, 1981.
83. **Luoma, S. N. and Davis, J. A.,** Requirements for modeling trace metal partitioning to oxidized estuarine sediments, *Mar. Chem.,* 12, 159, 1983.

84. **Honeyman, B. D.,** Metal and metalloid adsorption at the oxide/water interface in systems containing mixtures of adsorbents, Ph.D. thesis, Stanford University, Stanford, CA, 1984.
85. **Leckie, J. O.,** Adsorption and transformation of trace element species at sediment/water interfaces, in *The Importance of Chemical "Speciation" in Environmental Processes,* Bernhard, M., Brinckman, F. E., and Sadler, P. J., Eds., Springer-Verlag, Berlin, 1986, 237.
86. **Beneš, P. and Strejc, P.,** Interaction of radium with freshwater sediments and their mineral components. IV. Waste water and riverbed sediments, *J. Radioanal. Nucl. Chem.,* 99, 407, 1986.
87. **Kersten, M. and Förstner, U.,** Speciation of trace elements in sediments, in *Trace Element Speciation: Analytical Methods and Problems,* Batley, G. E., Ed., CRC Press, Boca Raton, FL, 1989, 245.
88. **Lion, L. W., Altmann, R. S., and Leckie, J. O.,** Trace-metal adsorption characteristics of estuarine particulate matter: evaluation of contribution of Fe/Mn oxide and organic surface coatings, *Environ. Sci. Technol.,* 16, 660, 1982.
89. **Tang, Hong-Xiao and Xue, Han-Bin,** The model of aquatic suspended sediments as a multicomponent adsorbent to heavy metals, in *Heavy Metals in the Environment,* Proc. Int. Conf. Heidelberg, 1983, 884.
90. **Beneš, P., Jurak, M., and Cernik, M.,** Factors affecting interaction of radiocobalt with river sediments. II. Composition and concentration of sediment, temperature, *J. Radioanal. Nucl. Chem.,* 132, 225, 1989.
91. **O'Connor, D. J. and Connolly, J. P.,** The effect of concentration of adsorbing solids on the partition coefficient, *Water Res.,* 14, 1517, 1980.
92. **Meier, H., Zimmerhackl, E., Zeitler, G., Menge, P., and Hecker, W.,** Influence of liquid/solid ratios in radionuclide migration studies, *J. Radioanal. Nucl. Chem.,* 109, 139, 1987.
93. **Morel, F. M. M. and Gschwend, P. M.,** The role of colloids in the partitioning of solutes in natural waters, in *Aquatic Surface Chemistry,* Stumm, W., Ed., John Wiley & Sons, New York, 1987, 405.
94. **Cremers, A. and Maes, A.,** Radionuclide partitioning in environmental systems: a critical analysis, in *Application of Distribution Coefficients to Radiological Assessment Models,* Sibley, T. H. and Myttenaere, C., Eds., Elsevier, London, 1986, 4.
95. **Chang, C. C. Y., Davis, J. A., and Kuwabara, J. S.,** A study of metal ion adsorption at low suspended-solid concentrations, *Estuarine Coastal Shelf. Sci.,* 24, 419, 1987.
96. **Beneš, P., Kuncova, M., Slovak, J., and Lam Ramos, P.,** Analysis of the interaction of radionuclides with solid phase in surface waters using laboratory model experiments: methodical problems, *J. Radioanal. Nucl. Chem.,* 125, 295, 1988.
97. **Bourg, A. C. M.,** ADSORP, a chemical equilibria computer program accounting for processes in aquatic systems, *Environ. Technol. Lett.,* 3, 305, 1982.
98. **Osaki, S., Miyoshi, T., Sugihara, S., and Takashima, Y.,** Adsorption of Fe(III), Co(II) and Zn(II) onto particulates in fresh waters on the basis of the surface complexation model. II. Stabilites of metal species dissolved in fresh waters, *Sci. Tot. Environ.* 99, 115, 1990.
99. **Mouvet, C. and Bourg, A. C. M.,** Speciation (including adsorbed species) of copper, lead, nickel and zinc in the Meuse River. Observed results compared to values calculated with chemical equilibrium program, *Water Res.,* 17, 641, 1983.
100. **Xue, Han-Bin, Stumm, W., and Sigg, L.,** The binding of heavy metals to algal surfaces, *Water Res.,* 22, 917, 1988.
101. **Sibley, T. H. and Myttenaere, C., Eds.,** *Application of Distribution Coefficients to Radiological Assessment Models,* Elsevier, London, 1986, pp. 3 and 424.
102. **Schweich, D. and Sardin, M.,** Methodology for determining distribution coefficients and alternative description of the sorption process, in *Application of Distribution Coefficients to Radiological Assessment Models,* Sibley, T. M. and Myttenaere, C., Eds., Elsevier, London, 1986, 15.
103. **Beneš, P.,** Interaction of radionuclides with solid phase in the modelling of migration of radionuclides in surface waters, in *Impact des accidents d'origine nucléaire sur l'environnement,* Proceedings from the 4th International Symposium of Radioecology, Cadarache (France), Vol. 1, 1988, C60.
104. **Beneš, P., Cernik, M., and Lam Ramos, P.,** Factors affecting interaction of radiocesium with freshwater solids. II. Contact time, concentration of the solid and temperature, *J. Radioanal. Nucl. Chem.,* 159, 201, 1992.
105. **Couchat, Ph., Brissaud, F., and Gyraud, J. P.,** A study of strontium-90 movement in a sandy soil, *Soil Sci. Soc. Am. J.,* 44, 7, 1980.
106. **Southworth, G. R.,** Movement of radiotracer metal cations through a forest soil column, *Environ. Int.,* 13, 197, 1987.

107. **Coughtrey, P. J.,** Models for radionuclide transport in soils, *Soil Use Manage.,* 4, 84, 1988.
108. **Feitknecht, W. and Schindler, P.,** *Solubility Constants of Metal Oxides, Metal Hydroxides and Metal Salts in Aqueous Solutions,* Butterworths, London, 1963.
109. **Nielsen, A. E.,** *Kinetics of Precipitation,* Macmillan, New York, 1964.
110. **Schindler, P. W.,** Heterogenous equilibria involving oxides, hydroxides, carbonates and hydroxide carbonates, in *Equilibria Concepts in Natural Water Systems,* Stumm, W., Ed., Adv. in Chem. Ser., No. 67, American Chemical Society, Washington, D.C., 1967, 196.
111. **Walton, A. G.,** *The Formation and Properties of Precipitates,* Wiley Interscience, New York, 1979.
112. **Davydov, Yu. P.,** *State of Radionuclides in Solutions,* Nauka i Tekhnika, Minsk, 1978 (in Russian).
113. **Lengweiler, H., Buser, W., and Feitknecht, W.,** Die Ermittlung der Löslichkeit von Eisen(III)-hydroxiden mit ^{59}Fe(II). Der Zustand kleinster Mengen Eisen(III)-hydroxid in wässeriger Lösung, *Helv. Chim. Acta,* 44, 805, 1961.
114. **Rolia, E.,** Theory and practice of precipitation and coprecipitation, with particular reference to hydrometallurgical processing, *Can. Mines Branch Inf. Circ.,* IC 312, Ottawa, 1974.
115. **Melikhov, I. V. and Berdonosov, S. S.,** Classification of coprecipitation phenomena, Radiokhimiya, 16, 3, 1974 (in Russian).
116. **Zhukova, L. A.,** *Theory of Static and Dynamic Precipitation and Coprecipitation of Ions,* Energoizdat, Moscow, 1981 (in Russian).
117. **Solozhenkin, P. M., Ed.,** *Coprecipitation with Hydroxides,* No. 2, Tadshikskii Gosudarstvennyi Universitet, Dushanbe, 1977 (in Russian).
118. **Melikhov, I. V. and Merkulova, M. S.,** *Cocrystallization,* Khimiya, Moscow, 1975 (in Russian).
119. **Stumm, W. and Furrer, G.,** The dissolution of oxides and aluminium silicates; examples of surface-coordination-controlled kinetics, in *Aquatic Surface Chemistry,* Stumm, W., Ed., John Wiley & Sons, New York, 1987, 197.
120. **Schott, J. and Petit, J.-C.,** New evidence for the mechanisms of dissolution of silicate minerals, in *Aquatic Surface Chemistry,* Stumm, W., Ed., John Wiley & Sons, New York, 1987, 293.
121. **Steinnes, E.,** Physical separation techniques in trace element speciation studies, in *Trace Element Speciation in Surface Waters,* Leppard, G. G., Ed., Plenum Press, New York, 1981, 37.
122. **De Mora, S. J. and Harrison, R. M.,** The use of physical separation techniques in trace metal speciation studies, *Water Res.,* 17, 723, 1983.
123. **Batley, G. E.,** Physicochemical separation methods for trace element speciation in aquatic samples, in *Trace Element Speciation: Analytical Methods and Problems,* Batley, G. E., Ed., CRC Press, Boca Raton, FL, 1989, 43.
124. **Buffle, J. and Altmann, R. S.,** Interpretation of metal complexation by heterogeneous complexants, in *Aquatic Surface Chemistry,* Stumm, W., Ed., John Wiley & Sons, New York, 1987, 351.
125. **Jenne, E. A., Kennedy, V. C., Burchard, J. M., and Ball, J. W.,** Sediments collection and processing for selective extraction and for total trace element analysis, in *Contaminants and Sediments,* Baker, R. A., Ed., Ann Arbor Science, Ann Arbor, MI, 1980, 169.
126. **Luoma, S. N. and Bryan, G. W.,** A statistical assessment of the form of trace metals in oxidized estuarine sediments employing chemical extractants, *Sci. Total Environ.,* 17, 165, 1981.
127. **Pickering, W. F.,** Selective chemical extraction of soil components and bound metal species, *CRC Crit. Rev. Anal. Chem.,* 12, 233, 1981.
128. **van Olphen, H.,** *An Introduction to Clay Colloid Chemistry,* 2nd ed., Wiley-Interscience, New York, 1977.
129. **Gregory, Y.,** Effects of polymers on colloid stability, in *The Scientific Basis of Flocculation,* Ives, K. J., Ed., Sijthoff and Noordhoff, The Netherlands, 1978, 101.
130. **Verwey, E. J. W. and Overbeek, Th. G.,** *Theory of the Stability of Lyophobic Colloids,* Elsevier, Amsterdam, 1978.
131. **O'Melia, C. R.,** Particle-particle interactions, in *Aquatic Surface Chemistry,* Stumm, W., Ed., John Wiley & Sons, New York, 1987.
132. **Lyklema, J.,** Surface chemistry of colloids in connection with stability, in *The Scientific Basis of Flocculation,* Ives, K. J., Ed., Sijthoff and Noordhoff, The Netherlands, 1978, 3.
133. **Hirtzel, C. J. and Rajagopalan, R.,** *Colloidal Phenomena,* Noyes Publications, Park Ridge, NJ, 1985.
134. **Niehof, R. A. and Loeb, G. I.,** The surface charge of particulate matter in seawater, *Limnol. Oceanogr.,* 17, 7, 1972.

135. **Tipping, E.,** The adsorption of aquatic humic substances by ironoxides, *Geochim. Cosmochim. Acta,* 45, 191, 1981.
136. **Tipping, E. and Heaton, M. J.,** The adsorption of aquatic humic substances by two oxides of manganese, *Geochim. Cosmochim. Acta,* 47, 1393, 1983.
137. **Davis, J. A.,** Adsorption of natural dissolved organic matter at the oxide/water interface, *Geochim. Cosmochim. Acta,* 46, 2381, 1982.
138. **Ali, W., O'Melia, C. R., and Edzwald, J. K.,** Colloidal stability of particles in lakes: measurement and significance, *Water Sci. Technol.,* 17, 701, 1984.
139. **Hunter, K. A. and Liss, P. S.,** The surface charge of suspended particles in estuarine and coastal waters, *Nature,* 282, 823, 1979.
140. **Gibbs, R. J.,** Effect of natural organic coatings on the coagulation of particles, *Environ. Sci. Technol.,* 17, 237, 1983.
141. **Grebenshcsikova, V. I. and Davydov, Yu. P.,** State of Pu(IV) in the region of pH = 1.0-12.0 at the plutonium concentration of $2 \cdot 10^{5}$M, *Radiokhimiya,* 7, 191, 1965 (in Russian).

Chapter 3

Strategies of Sampling, Fractionation, and Analysis

Brit Salbu and Deborah H. Oughton

CONTENTS

0-8493-6304-7/95/$0.00+$.50

I. INTRODUCTION

Natural waters are dispersed multicomponent systems containing trace elements and radionuclides derived from both natural and anthropogenic sources, in concentrations varying according to:

1. The source term (i.e., spatial and tidal variations in the flux of matter from atmosphere, geo-, biosphere, and anthropogenic discharges)
2. Transport processes and interactions taking place within the water columns (e.g., microchemical processes influencing physico-chemical forms)
3. Removal processes (e.g., particle growth and sedimentation, biological uptake)

as described in more detail in several chapters of this book.

As the concentration of microcomponents in natural waters is strongly influenced by geological and biological cycles, large variations are seen when comparing either different water systems or similar systems situated in different geological settings. Within the same water system, concentrations may vary as a function of distance from the source, depth, or the time of sampling (e.g., season). In zones where waters of different qualities mix (e.g., estuaries), variable concentrations of trace elements are particularly pronounced.

Following the introduction of noncontaminating techniques for sampling, sample handling, and analysis, as well as developments within analytical techniques and instruments, the concentration levels of trace elements in unpolluted natural waters have been shown to be a factor of 10 to 1000 lower than previously accepted. Thus, the progress made in our understanding of trace element behavior in natural water systems is closely related to improvements within analytical chemistry.

Total concentrations of trace elements and radionuclides are usually determined for the following purposes:

1. *Survey investigations.* In order to estimate trace element fluxes, budgets, or annual discharges, or to follow up regulatory advices, intervention levels, etc., standardized routine methods of analysis, capable of determining a limited number of microcomponents, are usually applied.
2. *Establishing background values.* When reference or background values for trace elements and radionuclides are established, accurate element specific analytical methods of high precision, capable of determining the actual concentrations of the microcomponent of interest, should be utilized.
3. *Fractionation of samples for speciation purposes.* When samples are fractionated for speciation purposes, the total concentrations serve as reference values (100%) accounting for losses and/or contamination taking place during fractionation.

Information on macrocomponents in natural waters is required for characterizing the systems (e.g., pH, Eh, ionic strength), when data on microcomponents is interpreted and compared with literature data. When assessing mobility and bioavailability, however, information on total concentrations is of limited value, and knowledge of the physico-chemical forms in natural waters, transformation processes, and kinetics involved is essential. This chapter will, therefore, focus on the analytical strategy for the determination of the physico-chemical forms of trace elements and radionuclides in natural waters.

A. PHYSICO-CHEMICAL FORMS OF MICROCOMPONENTS

In dilute aquatic systems, microchemical phenomena which are negligible at macroconcentrations become significant. Reaction rates may decrease as the probability of effective collisions decreases. Furthermore, reaction directions and mechanisms may change as other processes, such as chemistry at phase boundaries or chemistry of colloidal systems, become predominant.[1] Dilute concentration phenomena observed when applying fractionation techniques are well known from radiochemistry and are similarly valid for stable elements at trace concentrations (Chapter 3).

Trace elements are defined as elements present in concentrations where microchemical phenomena and secondary reactions no longer can be neglected, i.e., concentrations lower than 10^{-5} M (1 mg/l for elements having atomic weight 100 Da).[1] The phrase "trace element" should, therefore, include most radionuclides in natural water systems.

In natural waters, most trace elements can be present in a variety of physico-chemical forms varying in size (molecular weight), charge, redox state, density, and magnetic properties.[2] The borderline between categories ranging from simple ions, molecules or complexes, through hydrolysis products and polymers, colloids and pseudocolloids, to species sorbed or incorporated in suspended inorganic or organic particles and bio-organisms is difficult to establish as there is continual transition between species (Figure 1). As

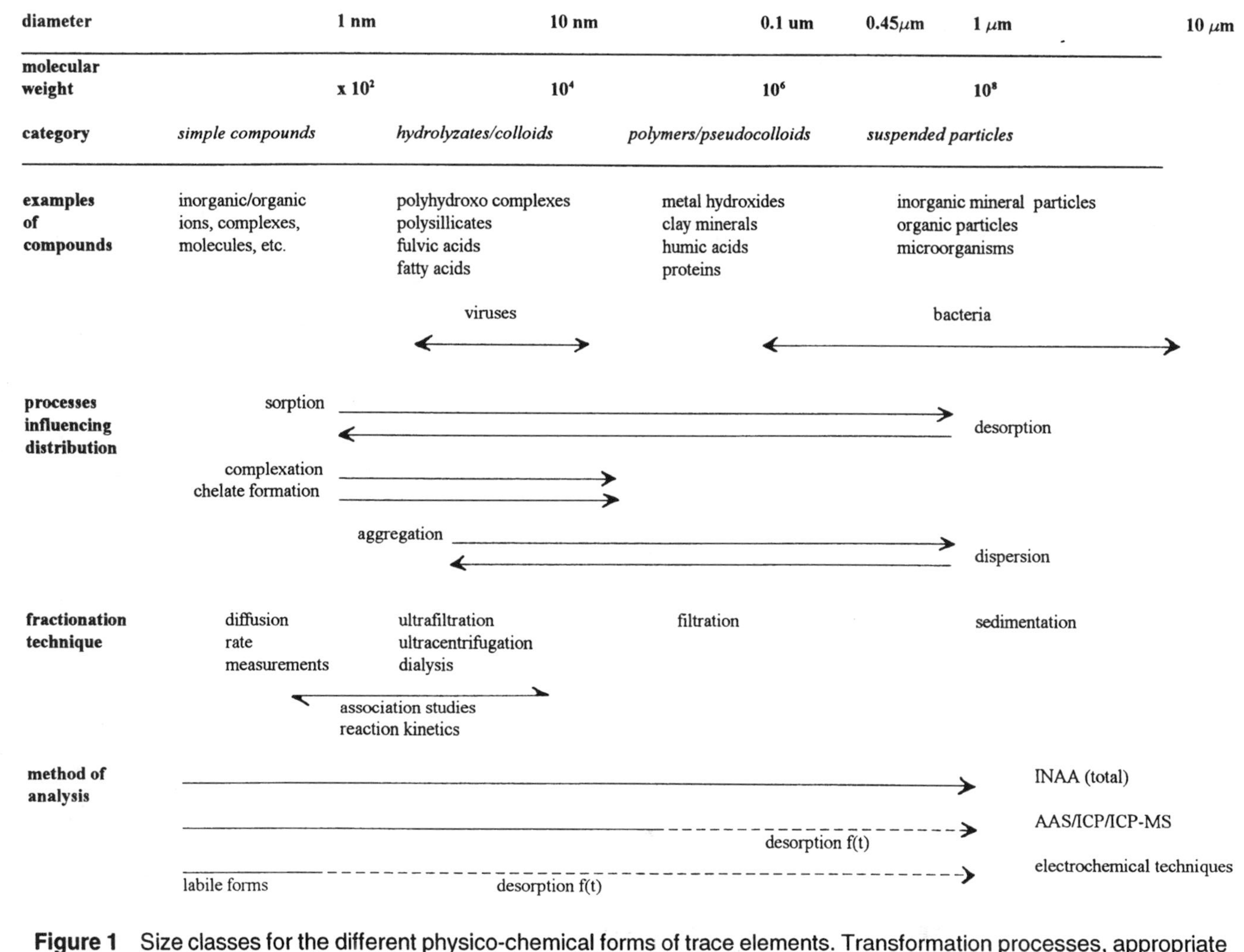

Figure 1 Size classes for the different physico-chemical forms of trace elements. Transformation processes, appropriate fractionation techniques, and methods of analysis are indicated.

Table 1 **Fraction of an element (%) present in low molecular weight forms, M_w < 10 kDa**

Element	Groundwater Bore-Hole Kise	Groundwater Spring Åstadalen	Lake Tyrifjorden	Lake Trehørningen	Lake Diplane	River Åsta	River Glomma
Ca	80	70	80	95	85	80	95
Al	10	70	80	75	n.d.	50	10
Sc	20	100	30	20	65	40	10
La	40	10	35	75	30	<5	n.d.
Ce	n.d.	<20	n.d.	30	50	15	<25
Cr	<5	<20	10	10	30	20	50
Mn	50	<40	50	25	n.d.	70	65
Fe	<1	<10	<5	35	<5	25	85
Co	<1	<10	50	75	90	30	20
Zn	20	10	5	75	100	20	55

Note: n.d. = no data. From Salbu, B. *Proc. Int. Conf. Modern Trends in Activation Analysis,* Vol. 1, Heydorn, K. Ed., Copenhagen, 1986, 135. With permission.

microchemical phenomena, e.g., physico-chemical forms of trace elements, as well as mobility and bioavailability cannot be derived directly from knowledge of macrochemistry, experimental data from studies on natural aquatic systems are needed. However, the lack of sufficiently sensitive noninfluencing, species-specific detectors necessitates the use of fractionation techniques applicable for the differentiation of species (e.g., size-classes). Information on species reported in the literature is therefore closely related to the fractionation techniques applied. Based on size fractionation of waters of different origin (Table 1),[3] the proportion of an element present as low molecular weight (LMW) forms varies substantially, depending on macrochemical conditions (pH, Eh, ionic strength) as well as the quality and quantity of interacting components (e.g., total organic C, Fe, Al, Mn). Information on transformation processes and kinetics involved is, however, still scarce in literature.

B. TRANSFORMATION PROCESSES

As natural waters are dynamic systems, transformation processes influencing the distribution pattern of species, and thereby mobility and bioavailability, can occur due to changing physical and chemical conditions. For example, temperature plays an important but somewhat overlooked role. The distribution of Al species at 25°C has been shown to be significantly different from that at 2°C (Figure 2), and colloidal species play a significant role in cold waters.[4] Contribution and removal of species occur frequently in the water body. Due to interactions with natural components, the distribution of species exhibits spatial and temporal variation,[5] and different equilibria are established under different environmental conditions, as well as during postcollection storage and treatment of water samples (Section IV).

At low pH, trace elements are usually present in LMW forms. If pH increases, polymerization may occur due to the rapid formation of mono- or polynuclear hydrolysis products. When acidified tributaries, containing high levels of potentially toxic LMW Al species,[6] mix with limed waters, the increase in pH gives rise to an instantaneous formation of Al polymers.[7] The polymerization of the LMW Al species is accompanied by a rapid precipitation of Al on surfaces of fish gills, which is thought to account for acute toxicity to fish observed in such mixing zones.[7,8]

Unless pH is low, loss of LMW forms due to sorption (Figure 1) may arise if sorption sites are offered from surfaces of foreign colloids or particles. Association with high molecular weight (HMW) naturally occurring organic complexing or chelating agents (e.g., humic substances) reduces the mobile fraction, and the size distribution of elements is shifted towards HMW species. Association with LMW ligands (e.g., fulvic acid) changes the polarity and reactivity of species which is believed to influence the bioavailability of trace elements.

If pH decreases, species are mobilized due to dissolution and displacement processes (e.g., desorption from surfaces of colloids and particles). The mobilization of trace metals from insoluble to soluble forms as a result of anthropogenic acidification of lakes and rivers is a well documented phenomenon.[9,10] However, the kinetics of these processes may be rather slow. Studies on synthetic, dilute solutions of Al showed that a rapid polymerization takes place when pH increases, whereas the dissolution process is rather slow when pH decreases.[11]

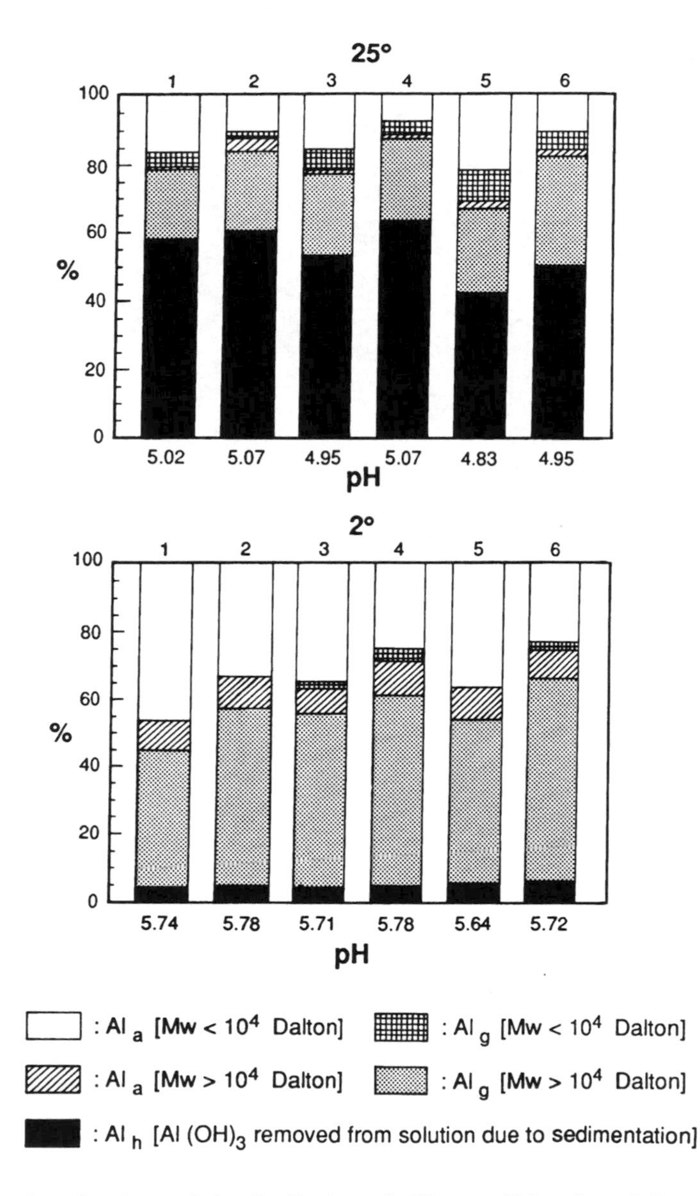

Figure 2 pH and molecular weight distribution of different Al-fractions (%) present in identical solutions stored for 1 month at 2°C and 25°C respectively. (From Lydersen, E., Salbu, B., Poleo, A. B. S., and Muniz, I. P., *Water Air Soil Pollut.,* 51, 203, 1990. With permission.)

Natural colloids (diameter 1 to 450 nm) are important transporting agents for trace elements and radionuclides in natural water systems (i.e., precipitation,[12,13] rivers,[5,14] lakes,[15,16] groundwater,[16,17] estuaries,[18,19] seawaters,[20,21] and wastewaters[22,23]). Studies on a mountain lake in the area of Norway heavily contaminated by fallout from the Chernobyl accident showed that the river inlet water to the lake during snowmelt is characterized by ^{137}Cs associated with colloidal material while ^{90}Sr is present in LMW forms.[24] Based on the size distribution pattern in inlet and outlet water, budget calculations show that ^{137}Cs retained in the lake (ca. 50%) is colloidal, while ^{90}Sr as LMW species is transported through the lake and downstream.

Processes influencing the stability of colloids are of major importance for the distribution of species in aquatic systems,[1,25] particularly in zones where waters of different qualities mix. The surfaces of colloids in natural waters are usually negatively charged, partly due to organic coatings,[26] and electrical double layers are created. Stability of colloids depends on the interaction between attractive (van der Waals) and repulsive (Coulomb) forces, which is influenced by chemical conditions. In mixing zones with increasing ionic strength (e.g., estuaries), the double layer of colloids and thereby Coulomb repulsion

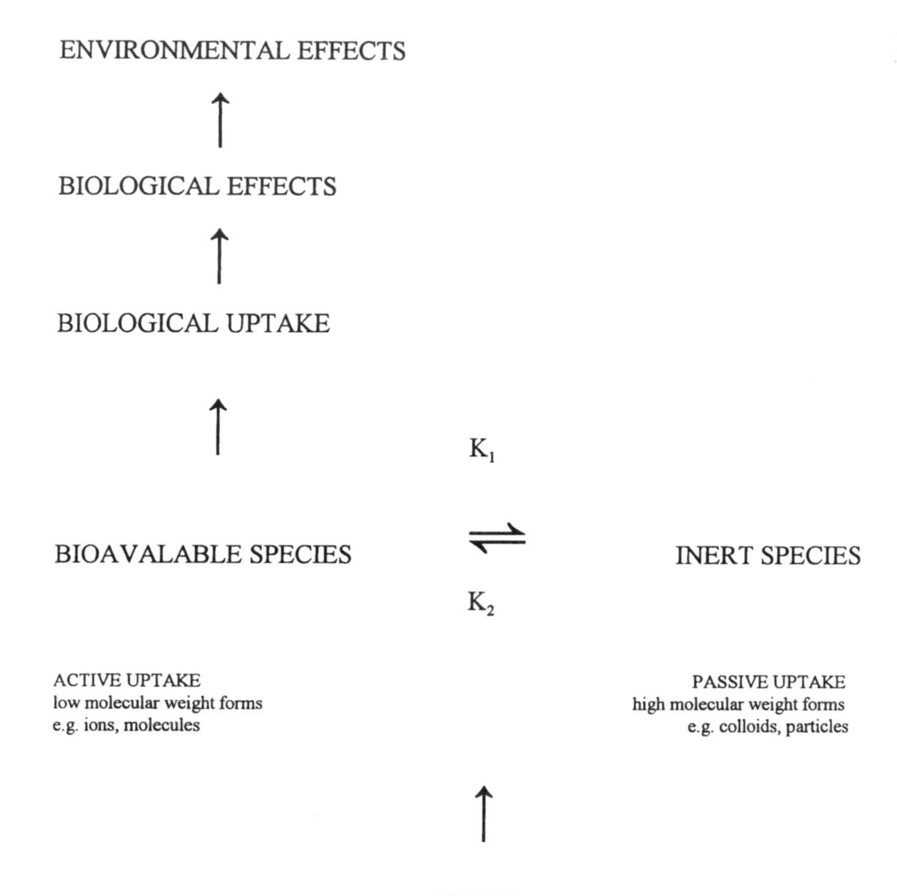

Figure 3 Differentiating between physico-chemical forms is essential for assessing biological uptake of trace elements and radionuclides.

is suppressed and, due to van der Waals attraction, particle growth occurs until sedimentation takes place. Thus, species sorbed to or incorporated in aggregates are removed from the water phase. If peptizers (e.g., hydrophobic cations) are absorbed onto the aggregate surfaces, the interparticle attractions weaken, and dispersion of aggregates and associated trace elements occurs.

Biological cycles influence the distribution of elements, as some species are bioavailable while others are biologically inert. LMW forms, often referred to as "free ions" or labile "species" in literature, are believed to be bioavailable (Figure 3),[27] while "ionic forms of trace metals" are considered the most toxic species for aquatic organisms.[29] Complexation with natural ligands may reduce the toxicity of metal ions, as has been demonstrated for Cu using bacterial bioassay,[30] for Cu, Cd, and Pb using algae[31] and for Al in fish.[32,33] However, lipid-soluble complexes of Cu and Hg have been reported to be more toxic to marine organisms (growth depression) than simple ions.[34,35] Furthermore, organometallic compounds of certain trace metals (Hg, Pb, Sm) are more toxic than simple inorganic ions,[28,35] although the opposite is valid for As, and there is evidence that LMW organic compounds are important for active biological uptake of Cd in fish.[36] Trace elements associated with colloids and particles are considered inert. However, for filtering organisms, passive uptake of colloids and particles may represent an important pathway.

As solid-water interactions, involving species penetrating diffuse double layers, are slow processes, complete equilibria among species may not have been established when samples are collected. For nonequilibrium systems (i.e., mixing zones), the time needed for fractionation and analysis to be performed is critical if relevant information on actual species is to be obtained. Thus, transformation processes occurring during storage or analysis[37] of samples should be avoided and samples should be fractionated *in situ* or shortly after sampling if physico-chemical forms are to be distinguished.

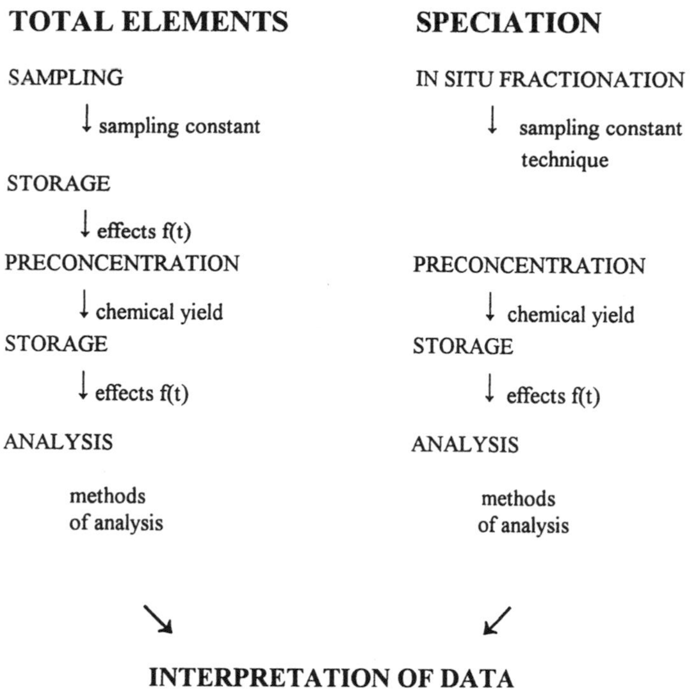

TOTAL ELEMENTS	SPECIATION
SAMPLING	IN SITU FRACTIONATION
↓ sampling constant	↓ sampling constant technique
STORAGE	
↓ effects f(t)	
PRECONCENTRATION	PRECONCENTRATION
↓ chemical yield	↓ chemical yield
STORAGE	STORAGE
↓ effects f(t)	↓ effects f(t)
ANALYSIS	ANALYSIS
methods of analysis	methods of analysis
↘	↙

INTERPRETATION OF DATA

Figure 4 Analytical strategy for the determination of trace elements and their physico-chemical forms.

C. ANALYTICAL STRATEGY

Analytical results and the interpretation of data are closely related to the analytical strategy (Figure 4), the procedures, and methods chosen and will be dependent on:

1. Representative noncontaminated samples and subsamples collected for analysis (Section II)
2. Rapid fractionation techniques applied *in situ* or shortly after sampling (Section III)
3. Preconcentration techniques and chemical yields (Section IV)
4. Method of analysis, determination limits, precision, and accuracy (Section V)

General aspects associated with the analytical strategy for the determination of trace elements in natural waters, including sampling,[39,40] speciation,[41-46] and analytical methodology,[43-48] have been discussed in several textbooks and review articles.

II. SAMPLING

The concentrations of trace elements and their various species in natural water samples are generally rather low and show spatial and temporal variations. Thus, the choice of sampling site, time, procedure, and equipment will influence the analytical results. The quality of a representative water sample can be defined as the degree to which the sample retains its composition and properties after the removal from the original environment. For trace elements, the main factors affecting the quality are contribution of elements due to contamination and loss of species due to sorption and volatilization. The development of noncontaminating techniques for representative sample collection and preanalysis handling has been essential for advances reached within trace element chemistry in natural waters. Speciation studies require that contamination control during analytical work is even more crucial. For radionuclides, however, loss of species represents the major source of error.

A. REPRESENTATIVE SAMPLING

A representative sample should reflect not only the concentration of elements in the water phase, but also the distribution of species equal to that within the bulk from which it is withdrawn. As HMW species such as colloids or particles are inhomogeneously distributed in the aquatic phase,[49] the probability of including these species decreases with decreasing volume collected. Thus, large volumes may be needed to avoid systematic discrimination of certain species. For many trace elements and radionuclides, the concentrations in the untreated sample may be too low to enable direct measurement, and preconcentration of large volumes of water is needed (Section IV). Volumes of up to 1000 l may be needed for the determination of transuranics, e.g., plutonium in ocean waters, while in lakes situated in Scandinavian areas affected by the Chernobyl accident, volumes of 200 to 400 l are needed. Close to the source, for instance in the Irish Sea contaminated by the effluent from the Sellafield nuclear installation or for water systems within the Chernobyl 30-km zone, volumes of 10 to 100 l should be sufficient.

Representative sampling presupposes that water samples are not contaminated from the surroundings (e.g., from surface or stratified layers enriched in certain trace elements[50] and radionuclides[51] or from sediments) or contaminated due to improper sampling devices or sampling procedures (e.g., for anoxic waters). The degree to which a sample can be considered to be representative may be derived from determination of the sampling constant.[52] Replicates can be withdrawn from one large volume collected from a single sampling. Alternatively, aliquots can be withdrawn from different volumes obtained by replicate sampling of unstratified water masses. By comparing results obtained by using different procedures or equipments, systematic sampling errors can be disclosed. The sampling and subsampling precision are usually improved if preconcentration from a large volume is performed (Section IV).

Sampling can be performed discretely ("point sources", individual containers) or integrated with respect to spatial or temporal variations (e.g., pump systems, composite volumes). It is important to stress that a sample may only be representative for the specific time and place from which it is withdrawn, thus care should be taken when data is interpreted. Running waters are particularly subject to variable flow conditions and episodic changes;[53] hence, the sampling procedure may not be adequate for describing the dynamic system, and trace element fluxes may be under(over)estimated.[49]

Interpretation of data is closely connected to sampling precision and the precision of the method of analysis. For low concentrations, the interpretation of data referring to the precision of the method is usually valid as the precision of the analytical methods is lowered as determination limits are approached. For higher concentrations, the sampling precision (i.e., the variation of elements or species between replicates) may be significantly lower than that of the method, and should therefore be taken into account when interpreting results.

B. CONTAMINATION

Contamination during sampling and preanalysis handling has been shown to be a major problem for determination of trace metals and their species in waters.[40,54] Contamination includes contribution of element species from surroundings, due to sampling conditions and procedures, and from sampling equipment. The contamination risk depends on the concentration levels involved and the ubiquity of the element in question. Contamination can arise from conditions at the sampling site or on board research vessels (e.g., atmosphere, laboratory facilities) and improper sampling devices[54] (e.g., containers, tubes or membrane material, cleaning procedures). When research vessels are used, sampling should take place from rafts rowed away from the research vessel, and nonmetallic hydrolines are required.[55] In general, metal, glass, and rubber equipment should be avoided. Equipment (sampling containers, tubes, or membranes) made from Teflon or polyethylene of low trace element content is preferred,[40,56] while quartz or borosilicate glass is recommended for the collection of samples for mercury determination.[57]

Cleaning procedures developed for containers used for storage of water samples are suitable for sampling devices. All equipment should be properly cleaned using high-purity acid leaching (nitric acid[58] or, for marine samples, hydrochloric acid[55]) and rinsed with double-distilled, deionized water. Acetone rinsing is frequently used prior to acid leaching in order to remove organic contaminants. Contamination from dust is minimized by using parafilm covers, and preanalysis handling in a clean room or with clean bench facilities (filtered laminar flow system, Class 100) is required to secure "contamination free" data.

C. LOSS OF SPECIES

The presence of active sorption sites on the interior surface of the sampling container, especially after acid leaching, can lead to removal of LMW species. To minimize sorption, sampling equipment should be conditioned with an aliquot of the sample.[59] This procedure is especially applicable for pump systems, where the initial sample volume pumped through the equipment serves a conditioning purpose. Special consideration must be taken when anoxic samples are collected. During emptying of sampling containers under oxic conditions, trace elements can coprecipitate with Fe compounds and deposit on the internal surface of the sampler. With pump systems connected to glove boxes, samples can easily be collected and fractionated under N_2 conditions.

D. WATER SAMPLE COLLECTORS

As stated in the previous section, metallic or rubber surfaces of equipment or hydrolines should be avoided when collecting samples for trace element analysis; polyethylene or Teflon is preferred. A variety of sampling equipment has been designed with respect to specific water systems and to the purpose of investigation (e.g., speciation), as discussed in Chapters 6 to 12.

1. Rainwater can be collected by means of funnels having a defined surface area, placed in connection with collecting containers. Alternatively, wet-only collectors can be utilized, which open during precipitation and close under dry conditions (Chapter 6).
2. Open cylinders are frequently used for sampling well-mixed lake or ocean waters. The lids are open until reaching the chosen sampling depth, then the lids are closed by a messenger (solid weight on hydrolines). However, contaminant waters, surface layers, or stratified water may seriously influence the quality of the sample.
3. Closed cylinders are the most suitable sample collectors, especially for deep or bottom water collection (e.g., anaerobic conditions). Initially the cylinder is closed, and then opens due to the hydrostatic pressure (e.g., 10 m) and remains open until the chosen depth is reached. Then the lid is closed by a messenger (Chapters 10, 12).
4. Pump systems are applicable for low turbidity waters as large volumes can be collected and are practical for collecting shallow waters or groundwater (borehole). A prewashed, polyethylene tube can be lowered to the chosen depth, and waters are collected directly from the tube by means of a peristaltic pump. When a molecular weight discriminator (e.g., hollow fiber) is connected to the tube, *in situ* size fractionation can be performed simultaneously.[59] Alternatively, continuous flow centrifugation can be utilized.[53,60] Pump systems which allow contact of water to metal surfaces should be avoided.
5. *In situ* dialysis using dialysis bags (e.g., 100 ml ionized water) has proved most suitable for collecting the LMW fraction of trace elements in running waters (rivers or lakes) under stable conditions.[14,61,62] However, clogging of membranes may occur in slow-running groundwater.[16] Dialysis bags containing chelating resins have also been used for sampling and concentrating LMW metal cations in rivers.[63]
6. Soil or sediment waters are traditionally collected by squeezing or centrifuging cores under *in situ* temperature or redox conditions. However, porous ceramic cup samplers (vacuum pump suction filtration) having a pore size of about 1 μm seem useful,[64] although initial sorption of species and time-dependent clogging of pores must be accounted for. Dialysis bags, fiber cloth or cages containing ion-exchange, or chelating resins have been utilized for the collection of "labile" metal ions in sediment-water systems[65] (Chapter 7).

There is a great need for standardization of sampling strategies, sampling equipments, and procedures, as the quality assurance associated with sampling required, for instance, within the context of accreditation is rather difficult to document.

III. FRACTIONATION TECHNIQUES

Due to transformation processes occurring during storage, information on species actually present in waters is obtained when the fractionation is performed *in situ* or takes place immediately after sampling. Requirements which should be met by fractionation techniques applicable for speciation purposes can be summarized as follows:[27]

1. Fractionation should take place *in situ* or at the site immediately after sampling in order to avoid storage effects.
2. The fractionation should be rapid in order to avoid establishment of equilibria between species retained and species to be separated during the fractionation.
3. Equipment surface area to sample volume ratio should be small in order to reduce sorption. However, conditioning with a sample aliquot minimizes this effect.
4. The method should not be sensitive to clogging.
5. Stability of colloids and aggregates should not be disrupted (e.g., aggregation or dispersion)
6. Chemical reagents influencing the distribution of species should be avoided.

Techniques yielding large fractionated volumes are favorable as determination limits can be lowered by using preconcentration techniques. A variety of size and charge fractionation techniques have been utilized for speciation studies[44-47] in the laboratory, while techniques applicable *in situ* or at site are limited. As all techniques suffer from disadvantages, the following section will focus on *in situ* or at site techniques which are less susceptible to methodological effects.

A. SIZE-FRACTIONATION TECHNIQUES

Among size fractionation techniques, continuous flow centrifugation/ultracentrifugation, filtration/ultrafiltration, and dialysis *in situ* meet the majority of requirements for speciation studies in natural waters.[38,66]

1. Centrifugation

By artificially increasing the force of gravity to counteract the viscosity, mutual repulsion, Brownian movement, and the force of shear, colloidal-sized particles can sediment during centrifugation or ultracentrifugation. As the separation of particles is based on differences in sedimentation coefficients, K_S (diameter × density), small dense particles are separated together with larger, less dense particles. For the separation of particles with a similar K_S, special consideration must be given to the particle shape (e.g., plate-shaped clay minerals vs. rod-shaped viruses),[67] the particle charge, and establishment of liquid junction potentials (e.g., due to organic coatings),[26,68,69] the degree of external or internal water (macroporosity),[70] and the viscosity, which is reduced if the temperature during centrifugation increases.

For particles differing widely in K_S, the most important factor contributing to deviation from theoretical predictions is the tendency of small particles (e.g., clay material) to cohere and form coarser particles. Furthermore, sorption during centrifugation/ultracentrifugation, due to transport of species to tube walls and subsequent adhesion to the wall, is difficult to quantify. This effect depends on the facilities used and is enhanced for fixed angle rotors and minimized when using sector-shaped centrifuge tubes. Despite these effects, ultracentrifugation, especially continuous flow systems,[53,60] has proved to be a useful online system for collecting colloids or particles, particularly from low turbidity waters. After separation, surface analysis techniques (e.g., electron microscopy) are particularly useful for characterizing retained colloids and particles.[17,71]

When density centrifugation is performed, particles of a given density will migrate in a density gradient solution until reaching equilibrium with the surroundings. Thus, colloids and particles in natural waters can be separated according to density. Appropriate density gradient materials are physiologically inert, water-soluble compounds of high density, low viscosity, and low osmotic pressure (nonelectrolyte). Sucrose is a suitable density medium, frequently used for separating microorganisms. The method seems useful for radionuclides associated with colloidal materials but needs careful standardization. For trace elements, the risk of contamination is substantial.

2. Filtration

Being a rapid and simple technique, filtration of samples through membranes of various types and pore diameters is most commonly used for separation of particles in natural waters.[66,72,73] Determination of the trace elements associated with retained particulate material on the filter can be performed directly[17,71] (EM, INAA) or after dissolution (ETAAS, ICPMS). The filtrate can be analyzed directly or after preconcentration. For filters having pore diameters of 0.1 μm and higher, conventional suction filtration, using a vacuum pump, is often employed. For high turbidity waters, pressure filtration (e.g., N_2 gas) is often used as the filtration rate can be significantly increased. In order to avoid rupture of algal or bacterial cells, low pressure filtration should be applied (lower than 20 mmHg).[74] By interfacing the filtration equipment with a peristaltic pump sampling system, *in situ* or on-line filtration is attained.[59]

During filtration, methodological effects, such as clogging of membrane, concentration polarization, sorption of species, and salt retention, may seriously affect the results.[40,66,72] When clogging occurs, the effective pore diameter of the filter decreases, the filtration rate decreases while the inverse phase velocity increases, and the species in the filtrate are no longer accurately defined according to pore size.[16] The introduction of stirred cells will minimize clogging and concentration polarization, but mechanical disruption or dispersion of aggregates (e.g., humic substances) may occur. Cellulose ester filters (Millipore, 0.45 μm) act as depth filters, and clogging is of less importance than for polycarbonate screen filters (Nuclepore, 0.4 μm).[16,73] The commonly used glass fiber membranes are, however, poorly defined with respect to pore sizes and should be avoided in trace element speciation studies.

For low turbidity waters and at trace concentrations, the most serious sources of error are usually intrinsic trace metal impurities and adsorption to equipment surfaces. Using polycarbonate membranes (Nuclepore), contamination from membranes and losses due to sorption are significantly lower than for cellulose acetate membranes.[72,73] For Fe, Al, and Sc in filtered river waters, a loss of about 10% was attributed to sorption to cellulose acetate membranes.[14] Similarly, absorption losses of 20 to 40% have been reported for portable water spiked with about 20 μg/l Pb(II).[75] As the adsorption loss depends on the physico-chemical forms of an element in question (e.g., $PbCO_3$ is more strongly sorbed[76] to negatively charged silica than $Pb^{2+}aq$) and the surface charge of equipment walls (dependent on material, cleaning, and conditioning procedures), loss may be difficult to account for and may result in erroneous interpretation of the results from speciation studies.

3. Ultrafiltration

Size fractionation by ultrafiltration has been frequently used in the determination of organometallic compounds,[5,77] trace elements,[15-17,20-22,59] and radionuclides[12,23,24,78-82] associated with colloidal material.

Using conventional ultrafiltration cells, a small volume is forced through the membrane using compressed inert gas (N_2 or Ar) or a centrifuge. Membranes having nominal cut-off in the range 500 to 100,000 Da or pore diameter in the range 1 to 100 nm are often used. The introduction of a magnetic stirrer favorably suppresses the concentration polarization and reduces the membrane clogging, while mechanical disruption of aggregates may influence the results. For high turbidity waters, however, clogging and sorption effects may be unavoidable as a long separation time is usually needed.[1]

Clogging and sorption effects are minimized when tangential or cross-flow ultrafiltration is performed using hollow fiber membranes or cross-flow cassettes.[20,59] The hollow fiber cartridge, consisting of uniform diameter cylindrical tubes made of inert non-ionic polymers, provides a large surface area per unit volume (Figure 5a). The rapid axial flow in the tube interior and large surface area minimize clogging, concentration polarization, and changes in ionic strength, as reported for conventional ultrafiltration. Based on standard solutions of well-defined carbohydrates and proteins, the nominal pore size distributions of membranes are shown to be relatively narrow.[16] After conditioning the system with a sample aliquot, i.e., after at least a 250-ml ultrafiltrate has passed through the membrane, sorption has proved insignificant.[59]

In the concentrating mode, the water sample is transferred from the sample container into the molecular discriminator using a peristaltic pump. Ultrafiltrate is collected while the waste returns to the container, and species having nominal molecular weight higher than the membrane are concentrated (Figure 5 top).

Samples of ultrafiltrate, or of concentrate, can then be analyzed using element-sensitive techniques or electron microscopy.[17,71] The system can also be used for studying processes in nonequilibrium systems such as mixing zones. When studying the interaction of LMW species with HMW colloids or polymers (Figure 5 top), the LMW species are circulated through the fiber while the HMW fraction is concentrated in the mixing chamber. The reduction in LMW forms reflects the association with colloids, and the kinetics involved can be evaluated.[83]

Ultrafiltration *in situ* using hollow fiber cartridges or cross-flow cassettes can be performed by lowering the polyethylene tube to the chosen depth in the lake, river, etc. (Figure 5 bottom).[20,59] The peristaltic pump transfers water directly into the molecular discriminator and, after conditioning, the ultrafiltrate is collected. This technique has proved especially useful for integrated sampling, investigations of episodic changes, and in studies of trace element chemistry in nonequilibrium mixing zones.[7] After mixing of acidic tributary waters with limed river water, LMW forms of Al polymerized, and the size distribution pattern of Al obtained by hollow fiber ultrafiltration changed instantaneously.[7,8]

4. Dialysis

During dialysis, diffusion of species smaller than the pore size of the membrane applied takes place until dialytic equilibrium is established.[1,84] The rate of dialysis depends on the diffusion constant, and the time needed to reach equilibrium depends on charge, shape, and size of the species concerned, as well as experiment conditions.[1] Conventional laboratory dialysis experiments are well known from the early days of radiochemistry. Dialysis bags containing the sample solution are immersed into a solution which does not contain the species to be dialyzed.[85] Alternatively, LMW species in solution can be dialyzed into a bag containing distilled water[86] or a dialysis cell with two equal compartments, one containing the sample and in the other, distilled water can be used.[15] Conventional dialysis techniques suffer from the same

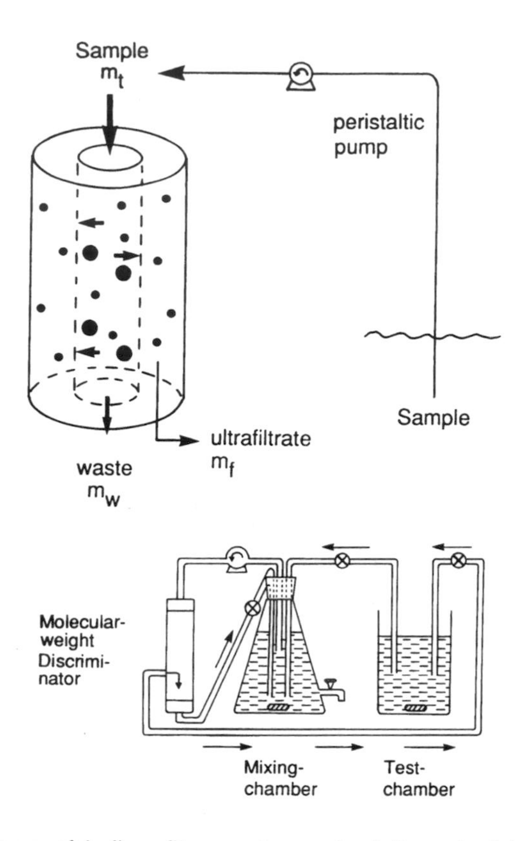

Figure 5 Schematic set-up of hollow fiber systems: (top) the principle of hollow fiber "in situ" sampling-fractionation system; (bottom) mixing experiment studying the association of LMW species with HMW compounds.

source of error, e.g., sorption to the membrane or to the walls of the dialysis cell, as previously described for conventional ultrafiltration. However, dialysis has been utilized for the fractionation of lake, river, and seawaters, as well as for soil extracts.[86]

Using hollow fibers in the dialysis mode, the solution to be desalted is circulated through the fiber interiors.[83] The dialysate circulating at a higher rate in the fiber exterior maintains the concentration gradient. The sample volume is kept constant and species are exchanged by diffusion. By withdrawing samples from the dialysate during the experiment, information on the diffusion rate of different LMW species can be obtained.

Dialysis *in situ* has proved to be a highly valuable sampling and *in situ* fractionation technique for stable water masses (e.g., lakes, rivers).[14,61,62] Dialysis bags, containing ca. 100 ml deionized water, are immersed into the waters of investigation. After dialytic equilibrium is reached (ca. 5 to 7 days, depending on convection), the bag is withdrawn for analysis of species having nominal molecular weight smaller than the membrane. Due to the large outer volume, sorption is negligible in running water. For slow moving waters (e.g., groundwater), clogging of membranes may occur due to algal growth or formation of inorganic coatings (e.g., Fe hydroxides).[16] By adding chelating resins into dialysis bags, separation according to size and charge is performed simultaneously (Section III.D).[63]

It should be underlined that size-fractionation techniques allow discrete phases to be retained, leaving LMW species relatively undisturbed. As mobility and bioavailability of trace elements are largely dependent on LMW species, differentiation between these species should be attained. Utilization of differences in diffusion rates, charge properties, or reactivity among LMW species obtained from size fractionation techniques should be further explored.

B. CHARGE FRACTIONATION TECHNIQUES

Among *in situ* techniques most applicable for fractionation of species according to charge properties are chelating, cation, and anion exchange chromatography when coupled to a pump system (on-line). Furthermore, liquid-liquid extraction or electrochemical methods (see Section V.C) can be applied at site.

As species differing in size also are fractionated according to charge, the result will to a large extent depend on experimental conditions (e.g., chemicals, time of reaction). Thus, for multicomponent and multispecies mixtures, such as natural waters, the fractions obtained using charge fractionation methods on untreated waters are operationally defined. Charge fractionation techniques have also been extensively used for preconcentration of trace metals and radionuclides, especially from marine waters (Section IV). However, if HMW forms are removed prior to charge fractionation, these techniques are most useful for speciation purposes (Section III).

1. Ion-Exchange Chromatography

Ion-exchange or chelating resins are often applied in radiochemistry[1,87,88] or for the retention of trace metals.[44,89-91] It is generally assumed that simple cations (anions) initially present in solution, or dissociated from weak complexes, react with the chromatographic reagent and are retained on the resin (electrostatic sorption or chemisorption), while opposite charged species, neutral undissociated molecules or colloidal material, are excluded and eluted. Although batch experiments (stirred cells) with resins added or filters implanted with resins can be utilized, high capacity resin columns are more efficient for retaining metal species. Pump systems with columns containing either specific resins or a mixed bed are frequently used for preconcentration or cleaning (wastewater) purposes.

Chromatographic reagents are often used in combination with chemical agents (e.g., acids) in order to differentiate species according to reactivity. Information on species retained can be obtained from a direct measurement of resin (e.g., gamma-emitters) or indirectly by the difference between total concentration and the concentration in the eluate:

$$C_{retained} = C_{tot} - C_{eluted}$$

This procedure is most frequently applied in acidified surface waters for the determination of potentially toxic forms of Al,[92] i.e., the fraction retained on Amberlite IR-120 resins is interpreted as inorganic, monomeric Al_i. The retained fraction is determined by the difference:

$$Al_i = Al_a - Al_o$$

where Al_a and Al_o are extracted fractions of the untreated sample and the elute, respectively.[93,94] The fraction of trace elements retained by resins depends on stability constants and on the equilibration time. As the interactions of naturally complexed species with resins increase with time of equilibration, results may depend upon the flow rate of solution through the column. Since it is essentially a slow fractionation procedure, a shift of chemical equilibria during separation may complicate the interpretation of results.[95] Furthermore, it has been demonstrated that neutral species of Hg,[1] and colloidal species of Al,[59,96] can be retained by ion exchange resins, i.e., physical sorption phenomena cannot be excluded. In this case, the concentration of ions interpreted from the retained fraction will be overestimated. This is observed for monomeric Al fractions in lake waters, as all fractions including the fraction retained on the resin are reduced when colloids are removed prior to ion-exchange chromatography.[96] It follows that qualitative, operationally defined conclusions, rather than reliable information on physico-chemical forms, are obtained unless HMW species are removed from waters prior to ion-exchange chromatography.

2. Extraction

Extraction techniques are well established within trace element analysis[93,97-99] and are utilized for untreated water samples (organometallic compounds) or for samples pretreated with chemicals, i.e., after addition of ligands forming strong complexes with metals (e.g., EDTA, MIBK, or crown ethers). In addition to the interaction with simple LMW ionic forms, various complexes with natural ligands may decompose, and desorption of species from the surface of colloids, pseudocolloids, or particles occurs. Thus, the fraction extracted may represent different species, and, especially if extraction is performed without the addition of chelating agents, only partial extraction may be obtained. Information on distribution constants for different species is therefore of major importance for chemical yield estimates and for interpreting results. Furthermore, analytical results will depend on experimental conditions (e.g., type and concentration of chemicals used, pH, V_{aq} to V_{org} ratio, stirring equipment and rate, contact time, temperature). As colloids or particles tend to accumulate at the phase interface, precautions should also be taken during the separation of the liquid phases.

Extractions should be performed in the field, as storage effects are avoided.[7] For practical reasons the sample volume is usually limited. For instance, a 100- to 300-ml sample is usually applied for the extraction of monomeric Al_a into MIBK using 8-hydroxy-quinoline as a complexing agent.[7,33,92,94] Being essentially a rapid separation technique (e.g., 10 to 20 seconds), distortion of equilibria during extraction can be minimized. By varying complexing agents, polarity of the organic phase, and equilibrating times, valuable information on stability constants of naturally occurring complexes and also on reactivity and kinetics involved may be obtained for modeling purposes.[1] The combination of size fractionation and extraction should be considered potentially interesting in speciation studies as reactivity can be used for differentiating LMW species.

C. SEQUENTIAL EXTRACTION

Colloids and particles act as transport agents and sinks for many trace elements in water systems. The fraction mobilized due to changing chemical conditions can be mimicked by leaching experiments. Extractions using different chemical reagents varying in displacement or dissolution strength can be performed individually on sample aliquots or sequentially on one single sample. The latter procedure has been widely used to investigate trace elements and radionuclides associated with particles in sediments and soils,[100-104] and can also provide valuable information on the potential mobility of elements associated with colloids or particles in waters.[21,66] Various sequential extraction procedures have been developed for separation of the different fractions, referred to as: water soluble, exchangeable, carbonate, hydrous metal oxides, organic matter, and mineral lattices. In order to study sorption and desorption mechanisms, many investigations include incubation with a tracer prior to extraction.[105]

Most of the extraction schemes involve leaching with strong acid as the final extraction, intended to represent the residual mineral fraction. However, mineral phases are attacked at different rates, and temperature as well as contact time is essential if complete dissolution is to be attained. If acid conservation of water samples (pH ca. 1) is used, full dissolution of particles and colloids may not be achieved.[106] If full dissolution is not attained and the methods of analysis excludes colloids and particles (e.g., electrochemical techniques), the total elemental concentration when based on extracts or acidified water samples may be underestimated.[106,107]

Like traditional extraction methods, the sequential technique is operationally defined,[101] and analytical results will depend on the experimental conditions and procedures applied (e.g., solid-to-liquid ratios). In addition, the separation of solids from liquid is critical, i.e., whether colloidal materials are included in the liquid or solid phase. However, based on replicate soil samples contaminated by ^{137}Cs from Chernobyl, the reproducibility of the technique reflecting soil waters and desorption from solid surfaces has been proved to be quite satisfactory (<10%).[104] However, the comparison with literature data is often a problem as the sequential extraction schemes vary significantly,[21] and there is a need for standardization of procedures.

D. SEPARATION OF METAL-ORGANIC SPECIES

Differentiating between trace elements associated with ligands, colloids, or particles of organic and inorganic nature is important for modeling purposes. Weak interactions may be disrupted during separations and caution should be taken when data is interpreted. UV irradiation of acidic and oxidizing (e.g., H_2O_2) media is often recommended for destroying organic material and liberating associated metals.[73,113] However, UV irradiation may also influence the size distribution pattern of clay minerals, and the elements released cannot be interpreted as associated with organic material unless the amount of clay minerals has proved to be insignificant. In freshwaters, destruction of organic coatings around colloidal iron oxides/hydroxides and a subsequent coprecipitation of metals can lead to a loss of labile species.[73]

1. GC, HPLC

Chromatographic techniques, e.g., gas chromatography (GC) or high-performance liquid chromatography (HPLC), are particularly suitable for separating organometallic compounds or metal-organic complexes,[40,43,108-110] especially when present as LMW forms. Usually, thermodynamically stable volatile species [e.g., $(CH_3)_2Se$, $(CH_3)_2AsH$, $(C_2H_5)_4Pb$] are separated according to boiling point (GC). For organometallic species, derivatization (hydride formation of As, Sb, Bi, Sn, Pb, Se, Tl, Ge) or alkylation for conversion into volatile forms is frequently applied.[110] Alternatively, chelating agents forming thermodynamically stable complexes can be introduced. However, HPLC can be utilized for nonvolatile species [e.g., $(CH_3)_3SnX$, arsenolipids], as well as for cations, anions, and metal-organic complexes when separated on ion-exchange columns.[108,110]

During recent years, extensive effort has been put into interfacing chromatographic techniques with detectors suitable for trace element determination.[43,108-112] Usually, the separated fraction obtained from LC or HPLC is injected off-line into suitable detector systems (e.g., AAS-furnace). The output from a GC can, however, be directly coupled to a flame or plasma emission source.[108,112] Combined HPLC-ETAAS is achieved by interfacing the systems through on-line volatilization of the HPLC effluent into an aerosol or through derivatization in a thermochemical hydride generation system,[110] and represents an efficient method for speciation studies, especially for the differentiation between organometallic compounds.[109-111]

E. COMBINED FRACTIONATION TECHNIQUES.

Most speciation schemes involve both size (0.45-μm filters) and charge (chelating or cation exchange resins such as Chelex-100, Amberlite IR-120) fractionation techniques, often applied sequentially. The development of combined techniques, where species are separated according to size in the colloidal range and charge simultaneously, represents an improvement within speciation studies. The use of a dialysis bag containing chelating agents[63] is a promising *in situ* technique, as species having a molecular weight lower than the pore size of the membrane are separated according to charge. Thus, the fraction retained in the resin represents the LMW cationic forms present in waters, and interferences from charged HMW forms are avoided. A similar technique has been utilized for extracts[86] and for pore waters in sediments.[65] Clogging of dialysis membranes due to algal growth in slow moving waters, or in high particle density systems such as sediments or soil, may, however, seriously influence the results.[16] Careful standardization is needed and membrane effects should be controlled.

The combination of dialysis[22,114] or ultrafiltration[96,115] with ASV[95] or chelating (Chelex-100) or cation-exchange (Amberlite IR-120) resins is most useful as the HMW forms are excluded from the sample prior to charge separation. Thus, results are easier to interpret. Electrodialysis is also a potentially interesting technique even though it is only applicable in the laboratory. In this case, charged LMW species penetrating membranes with pore sizes in the nm range (MW cut-off level of 500) migrate towards opposite charge electrodes under the influence of a low, constant, controlled potential. Using gamma-emitting nuclides as tracers, the migration of LMW charged species and retention of HMW charged forms in natural waters can be followed by gamma-camera.[116]

It should be underlined that the recently developed interfacing of separation techniques (e.g., chromatography, volatilization) with element-specific detectors of high sensitivity (e.g., ETAAS, ICPMS) represents a powerful tool for speciation purposes. Further developments are expected in the years to come (Section V).

IV. PREANALYSIS HANDLING

A. PRECONCENTATION TECHNIQUES

In principle, the concentrations of different element species in waters should be determined instrumentally with a minimum of handling. However, the concentrations of trace elements and most radionuclides are often too low to allow direct measurement. Furthermore, matrix interferences, especially in seawaters, may hinder the determination of trace elements of interest. In these cases, preconcentration techniques are needed[1,40,44,46,47,117] and should be applied in connection with on-line sampling/fractionation systems or as soon as possible after sampling and fractionation.

Preconcentration techniques act in different modes: (1) reducing large sample volumes (e.g., freeze-drying, evaporation) to increase the overall concentration of nonvolatiles; (2) selective separation of the trace elements or radionuclides of interest from the matrix; and (3) selective removal of interfering matrix components.

By reducing a large sample volume (e.g., freeze-drying,[118,119] evaporation[120]) the concentration of all constituents except volatiles (e.g., Hg) in the residue is increased. As the matrix elements as well as the elements of interest are concentrated, this method is only applicable when matrix interferences do not seriously affect the determination limits of trace elements (e.g., fresh waters). Subsampling precision is usually improved as the sample to be analyzed represents a large volume.[52]

By selective separation of trace elements and radionuclides of interest from the matrix (e.g., by using coprecipitation,[87,88,121-123] liquid-liquid extraction using organic solvents,[93,97-99] sorbents such as active carbon,[60] chelating, or ion-exchange resins[87-92,124,125]), samples enriched in the elements to be determined and depleted in interfering matrix components can be obtained.[115] Based on large volumes, the determination limits of selected trace elements can be improved.

By removing interfering elements, e.g., alkali or alkaline earth elements by using, for instance, selective ion-exchange techniques (e.g., antimony pentoxides, HAP), the determination limits of trace elements in the elute are usually improved.

For high salinity waters, scavenger techniques, based on coprecipitation of microcomponents with amorphous precipitates, e.g., hydroxides of Fe, Mn, Mg, are well-established methods for the separation of trace elements and radionuclides from seawaters.[87] The experimental design will influence the mechanism of coprecipitation and thereby the chemical yields (distribution coefficients). A rapid precipitation by vigorously stirring using concentrated reagents at low temperature will increase the yield. After dissolving the precipitate, the solution is subjected to analysis (ETAAS) or radiochemical separations prior to measurements.

More recently, ion-exchange chromatography using resins such as Chelex-100, Dowex 50, or Amberlite IR-120 are the preconcentration techniques most frequently applied for concentrating a large number of elements from large volumes of natural waters, especially seawaters.[89,90,124,125] When connected to a pump, the system represents a highly efficient high capacity on-line sampling and preconcentration device. As exchange resins tend to exclude colloids and particles while amorphous precipitates will include positively charged species in trace concentrations as well as colloidal or particulate material, variable recoveries are to be expected if HMW forms like colloids and stable complexes are present in waters.[126]

From an analytical point of view, the main problem using preconcentration techniques is to verify the chemical yields for species of interest. Most often the yields are obtained from spiked solutions. However, species of trace elements and radionuclides behaving chemically different from those added can be present in the original sample, and full exchange may not be attained. Even after a contact time of 1 year, the exchange between ionic ^{65}Zn and stable Zn in seawater was reported to be incomplete.[127] Thus, chemical yields obtained from ionic spikes may result in an underestimation of the total concentrations.

B. STORAGE EFFECTS

During storage, processes affecting the quality and stability of the sample will influence the analytical data.[87] For radionuclides, the main factor affecting the quality is loss of species due to sorption or volatilization, while for trace elements, measures must also be taken to reduce the risk of contamination (container materials, cleaning procedures, etc., Section II). The stability of a sample refers to the degree to which transformation processes influence the distribution of species originally present. Unless the species formed are actually removed from the sample (sorption, sedimentation, volatilization, and releases), the transformation processes do not affect the total concentrations in the sample, and can only be identified if fractionation techniques are applied.

During storage, samples are usually acidified, freeze-dried, or kept dark at 4°C. Sorption is reduced by acid conservation (concentration HNO_3 or HCl to reach pH ~ 1) or by freeze-drying, while algal growth is reduced when cool (4°C) and dark. Only high purity reagents should be added to the samples, and blank values can be obtained by subjecting double-distilled, deionized water to the same equipment and preanalysis treatment as the natural water sample.

Transformation processes taking place during storage exert influence on the physico-chemical forms of elements, the extent of which will depend on the actual sample and storage conditions. Loss of charged or LMW species in neutral solutions due to sorption, mobilization of HMW species due to acid conservation, dispersion due to freezing, change in pH and redox conditions due to microorganisms, particle growth in anoxic samples due to leakage of O_2, etc. will have the effect of altering the original distribution of elemental species in the water sample. Therefore, if relevant information on the distribution pattern of physico-chemical forms is to be obtained, storage prior to fractionation should be avoided. Unless the method of analysis is species-specific, analytical results obtained by different methods should be consistent. However, if particles or colloids are not fully dissolved, the measured total concentrations will depend on the method of analysis.[106]

V. METHODS OF ANALYSIS

Sensitive, element-specific methods with sufficiently high precision and accuracy are needed for determination of total trace element concentrations in untreated or fractionated water samples. Samples, especially fractionated samples, are usually preconcentrated prior to analysis. Assuming representative, noncontaminated, fractionated and preconcentrated samples, and well-defined chemical yields, the analytical results may depend on the choice of analytical method.

The analytical techniques most applicable for determination of trace elements in natural waters are (Table 1):

1. Electrothermal atomic absorption spectrometry (ETAAS)
2. Inductively coupled plasma mass spectrometry (ICPMS)
3. Electrochemical techniques such as anodic or cathodic stripping voltammetry (ASV, CSV)
4. Instrumental or radiochemical neutron activation analysis (INAA, RNAA)

Radionuclides in natural waters are measured by spectrometric techniques (gamma, beta, alpha) or liquid scintillation techniques, most often after preconcentration and radiochemical separations.

For detailed information on the general techniques and principles involved in the above techniques, an extensive number of textbooks are available. The present section will therefore focus on recent developments in relation to speciation studies.

A. ELECTROTHERMAL ATOMIC ABSORPTION SPECTROMETRY (ETAAS)

Because of its accessability, sensitivity, selectivity, and relatively rapid sample throughput, ETAAS is the analytical technique most frequently applied in aquatic trace element studies. In general, significant improvements within signal processing, furnace design (including platforms), operation procedures, and matrix effect suppression have been achieved during recent years.[128] In principle, ETAAS is a single element technique and, for many trace metals, is sensitive enough to be applied directly on acidified freshwater samples. For seawaters or when samples have been fractionated for speciation purposes, it is often necessary to preconcentrate the analyte prior to determination. The preconcentration techniques most commonly applied in connection with ETAAS are ion-exchange chromatography,[89] chelating or cation-exchange resins (e.g., Chelex-100, Amberlite IR-2) or liquid-liquid extractions.[97-99] In the latter case, a suitable complexing agent is added and pH in the solution is adjusted for optimizing extraction into the organic liquid. When combined with flow injection techniques, the on-line extraction-detection system is highly efficient. Although ETAAS is used to determine the total element concentration in untreated water, the low subsample volumes applied (20 to 100 μl) may give rise to discrimination of undissolved colloids or particles.[106] However, procedures developed for the analysis of slurry samples represent an improvement in this respect.

Various modifications of the AAS systems have been developed. The AAS-hydride generation technique has proven applicable for the determination of metalloids (e.g., Sn, As, Sb, Bi, Se, Te) forming volatile hydrides upon reduction prior to ETAAS measurement.[110] The cold vapor technique (CVAAS) primarily utilized for Hg is potentially applicable for determination of volatile organometallic hydrides. The most interesting achievements are probably the interfacing of ETAAS with chromatographic systems such as GC and HPLC for the identification of metal-organic species.[109-111,129] Results from a recent comparative study on coupled techniques for AAS showed that hydride generation with cryogenic trapping and GC-flame AAS was the most sensitive technique, while HPLC-hydride generation was the simplest.[130] Further developments with respect to interfacing separation systems with sensitive ETAAS techniques are expected in the years to come.

B. INDUCTIVELY COUPLED PLASMA-MASS SPECTROMETRY (ICPMS)

Conventional ICP emission spectrometers are not sufficiently sensitive to allow the determination of most trace elements in natural unpolluted waters. When combining a plasma source with mass spectrometry (MS), however, the sensitivity can compete with all other methods available.[131,132] By scanning through the mass range within milliseconds, the ICPMS acts as a highly sensitive multielemental technique, as about 30 elements per minute can be determined with detection limits in the range of 0.1 to 10 μg/l. Samples are introduced by means of a nebulizer, graphite furnace (electrothermal vaporization) or by means of a laser (laser ablation). It is generally assumed that the sample is fully ionized in the plasma (6000 to 8000 K) before being transferred into the mass spectrometer. However, if nebulizer systems are used for sample introduction, particle discrimination may occur.[106]

The application of ICPMS for analysis of natural waters has been rather limited even though the application of the method should be extensive. The results obtained for transuranic elements in seawaters demonstrate, however, the ability of distinguishing between low levels of ^{239}Pu and ^{240}Pu, which is impossible from radioactivity measurements.[133] Allowing the separation of different isotopes of an element, ICPMS can also be applied in tracer studies using stable as well as radioactive isotopes.

Recently, much effort has been put into the development of interface systems, particularly between separation and detection devices.[110] By interfacing HPLC with ICPMS, the extreme sensitivity of the system is well suited for speciation studies.[110,129] Even though further developments with respect to standardization, calibration, and control of interferences are still needed, major progresses are expected for trace element speciation in natural water systems.

C. ELECTROCHEMICAL TECHNIQUES

Unfortunately, ion-selective electrodes are not sufficiently sensitive for trace element speciation studies in natural waters. Furthermore, the high voltage and chemicals applied during electrophoresis may seriously influence the distribution of species. Voltammetric techniques, however, are highly useful for the determination of either total concentration (after full dissolution) or labile forms of trace metals (i.e., speciation studies), particularly in seawaters.[28,129,134-136]

Anodic stripping voltammetry (ASV, DPASV) offers high sensitivity for a limited number of trace metals capable of amalgamation with the Hg electrode (e.g., Ag, Cd, Cu, Pb, Sb, and Zn). The sensitivity of the methods can be further improved by using rotating Hg films as electrodes.

In ASV speciation analysis, the solution (pH) and electrochemical (potential) conditions are adjusted in order to obtain a labile fraction deposited on the electrode. This labile fraction is thought to be representative of the toxic metals species.[28,134] By introducing a double acidification technique, i.e., by comparing the deposition at natural pH with that at $pH < 2$, agreement with ASV-labile fraction of Cu and biotoxicity (algal assay) has been reported, while no correlation was observed for Cd, Pb, or Zn.[135] However, complexing agents present in natural waters (e.g., fulvic acid) may significantly influence the stripping peak observed. Furthermore, electrochemically labile complexes may partly dissociate during deposition, and the complexed fraction of metals may be underestimated, as demonstrated by CSV analysis of Cu in seawater.[136] In addition, sorption of species to surfaces of electrodes may occur[137-139] due to chemisorption of organic molecules[137] or covalent attachment of functional groups.[138]

Recent developments within cathodic stripping voltammetry (CSV) have increased the number of trace elements available to determination (e.g., Al, Co, Cr, Fe, Ni, Se) at ppb (mgl^{-1}) and ppt (ngl^{-1}) level in seawater.[140-142] A major advantage of CSV is that a positive deposition potential can be utilized, thus avoiding the dissolution of electrochemically labile metal complexes. Based on CSV with the addition of competing ligands, Cu was found to be fully complexed by organic material in the North Sea.[136]

When total concentrations are to be determined, total dissolution of the sample is needed. When speciation studies are performed and labile forms (e.g., LMW cations) are to be determined, HMW species or colloids should be excluded (size fractionation) prior to experiments in order to avoid interferences.

D. NEUTRON ACTIVATION ANALYSIS (NAA)

Neutron activation analysis (NAA) offers high sensitivity for the simultaneous determination of a great number of elements and the freedom of postirradiation contamination. Due to the relatively low detection limits attained by NAA for many elements, NAA was the predominant technique within trace element analysis for many years.[143] Improvements within ETAAS and, in particular, the development of ICPMS have made NAA less attractive during recent years. This is mostly due to high cost (access to a nuclear reactor), the need for highly skilled operators, and a relatively long turnover time for analysis. However, instrumental (INAA) and radiochemical (RNAA) activation analysis is still an important tool for certification of standard reference materials.

In fresh waters, INAA is capable of determining about 30 elements at concentration levels of 1 μg/l or less[144] in one sample (5 ml) and has proved suitable for trace element speciation in fresh waters.[3,145] For seawaters, freeze-drying[119] or separation of trace elements prior to irradiation (e.g., by co-precipitation[121-123] or extraction[124]) or after irradiation (e.g., controlled potential electrolysis[146]) is needed. Even 25 years ago[121] the rare earth elements in ocean water were determined by coprecipitation and RNAA at levels of 0.1 to 1 ngL^{-1}. High resolution Ge detectors are needed for analysis of the complex gamma spectra produced by INAA, and further development of computer software should reduce the few sources of errors affecting this technique.[147]

When radiochemical separations are performed after irradiation (RNAA), interferences are removed and the number of elements determined increases as the determination limits also decrease. The addition of stable isotope carriers after irradiation enable macrochemical separations of the element of interest to be utilized. After separations, the spectrum is less complex and sensitive NaI detectors with poor resolution can be used for measurements. A major advantage of RNAA is the freedom from postirradiation contamination, and technical grade reagents can be utilized. However, RNAA is a time-consuming technique and should only be utilized when INAA is not applicable.

E. MEASUREMENT OF RADIONUCLIDES

The low concentrations of most anthropogenically derived radionuclides in natural waters, especially oceans, necessitate the use of large volumes (up to more than 1000 l) and preconcentration by evaporation, coprecipitation (e.g., for transuranics, ^{90}Sr), or ion-exchange chromatography (e.g., ^{137}Cs, ^{90}Sr, ^{99}Tc) is needed.[87] Gamma-emitting radionuclides in concentrated samples can be determined directly by sodium iodide (NaI) detectors or high-purity (HPGe) and lithium-drifted (GeLi) germanium detectors. NaI, offering high sensitivity but poor resolution, allows only a few nuclides to be determined simultaneously. Many gamma-emitting radionuclides can be differentiated when high resolution Ge detectors and advanced computer software are used.[88]

Determination of beta and alpha emitters requires extensive radiochemical separation of the nuclide of interest.[88] Using low-level liquid scintillation spectrometers, beta- and alpha-emitting isotopes can be determined simultaneously (e.g., $^{239,240}Pu$ and ^{241}Pu) and the radiochemical procedure is relatively simple.[149] Still, the most sensitive techniques are based on traditional beta and alpha counters, where the sample preparation may be extensive and time consuming, and requires high competence from the operator. However, advances within ICPMS or accelerator-based MS (AMS) includes the determination of low-level, long-lived radionuclides, especially transuranics.[133]

It should be underlined that tracer techniques are useful tools within analytical chemistry, e.g., accounting for removal of species (chemical yields) and are well suited for optimization of analytical techniques.[149] By combining different fractionation techniques, the "behavior" of chemically well-defined species added to a system can be followed dynamically, and valuable information on microchemical processes and kinetics can be attained. ^{26}Al tracer has been used to study the distribution of Al species in acidic and limed river waters and to follow transformation processes influencing the species when the two waters are mixed.[8] By utilizing different gamma-emitting radioisotopes of the same element (e.g., $^{55}Fe(II)Cl_2$ and $^{59}Fe(III)Cl_3$), the behavior and processes affecting different physico-chemical forms of an element in a complex water system can be studied simultaneously. Radiotracers can also be applied in field studies, for instance, in studying the flow pattern or mixing of running water.

F. COMPARISON OF TECHNIQUES

The major advantages and disadvantages of the analytical techniques are summarized in Table 2. All methods are applicable for the direct determination of many elements in fresh waters, while in saline waters or fractionated water samples only a few methods are useful unless preconcentration techniques are used. Using electrochemical methods, a limited number of metals in a labile form only can be determined directly in marine waters. The determination limits, precision, and accuracy for different elements depend on the chosen procedure and method of analysis (Table 3).

Intercomparison tests on synthetic solutions have proved the above-mentioned methods of analysis to be of high precision and accuracy. However, in a multicomponent and multispecies system such as natural waters, variable results may be obtained. When INAA or RNAA is used, trace elements associated with particles and colloids are included in the analysis, while using ETAAS, or ICPMS particles, and colloids may be excluded.[106] Electrochemical methods applied in untreated samples give information of "labile" species only. Thus, a proper decomposing procedure is of importance for all methods, except INAA, if the fraction associated with colloids and particulates is to be included in the analysis.

G. QUALITY ASSURANCE

The quality assurance schemes which are requisite for large-scale, routine analysis are often not applied in research laboratories where relatively few samples are subject to detailed analysis of various trace elements and radionuclides. As outlined by Heydorn,[150] methods have been developed to obtain statistical control where the number of analytical results may be modest.

1. Analysis of Precision

Among the sources of errors leading to poor precision, lack of homogeneity, i.e., nonrepresen[illegible] samples or subsamples, may often be a major contributor. By comparing *a priori* precision [illegible] empirical variability obtained from replicate analysis, excess variability needs to be accounte[illegible] method relies on chi-square distribution of data (Chapter 6).

Table 2 **Comparison of methods**

Method of Analysis	Sample Type	Sample Size	Advantages	Disadvantages
ETAAS	Acidified natural water (slurry)	μl	High sensitivity, easy access, inexpensive,	Matrix interferences, single element, colloid effect
ICPMS	Acidified natural water (solid/slurry)	ml	High sensitivity, fast, multielement, and multi-isotope	Expensive investment cost, interferences from sample matrix and plasma gases, colloid effect
ASV/DPASV	Natural water	ml	Nondestructive, direct analysis of species, "multi-element"	Slow, limited number of elements, interferences from organic substances
NAA	Natural water, solid	g/ml	Nondestructive, high sensitivity, multi-element	Expensive, slow, access to reactor needed

Table 3 **Detection limits**

Element	ETAAS[a] mg/1	ICPMS[a] mg/l	ASV[b] mg/l	NAA[c] ng
As	0.05	0.02	[d]	0.02
Se	0.1	0.2	[d]	0.5
Mo		0.003	[d]	0.2
Ag		0.003		0.2
Cd	0.003	0.005	0.005	0.5
Cs		0.001		0.1
Be		0.014		
Al	0.3	0.025		5
V		0.009	[d]	0.1
Cr	0.03	0.038	0.02	2
Fe	0.5	1.0	[d]	20
Mn	0.3	0.005	0.05	0.005
Ni	0.2	0.047	0.05	0.5
Co	0.1	0.005	[d]	0.2
Cu	0.02	0.015	0.02	0.05
Zn	0.1	0.046	0.01	2
W		0.002	[d]	0.02
Au		0.012	[d]	0.05
Hg	0.1	0.005	0.1	0.1
Tl		0.002		2
Pb	0.03	0.006	0.01	2
Th		0.001		0.02
U		0.001	[d]	0.05

See Borg, Chapter 10; [b]References 28, 95, 134 to 136; Reference 144; and [d]Elements potentially analyzable by ASV.

tical methods, exhibiting different influence from interfering elements, a 1:1 e obtained. Deviation from the linear regression line demonstrates method- the analytical results.

3. Calibration control

Most methods of analysis are based on relative measurements, i.e., linear relationship between the concentration of trace elements in an unknown sample and the concentration in a high quality pure standard solution. However, matrix effects influencing signals from the real samples must be proved insignificant over the actual concentration range. The use of standard additions is applicable provided that the technique is not sensitive to variable physico-chemical forms of elemental species. Otherwise, complete exchange of the element species added with the species present in the sample must be fulfilled. This represents a major problem within speciation studies, or when preconcentration from large volumes is used, as the chemical yields obtained from spikes are only valid for equilibrated species.

4. Verification

Standard reference materials (SRM) are utilized for the verification of analytical results for biological and geological samples. For natural waters, however, the lack of suitable SRM represents a major problem. Usually, spiked, deionized water solutions containing high purity chemicals, often in unrealistic concentrations, are referred to as SRM for trace elements in water samples. However, the interferences from naturally occurring components (e.g., humic substances, clay minerals), with which elemental species in a real sample may interact, cannot be accounted for by the synthetic standard.

Analytical results will depend on the ability of the method to include all species present or to exclude specific physico-chemical forms from the analysis.[106] Variable analytical results obtained by different methods should not be readily attributed to contamination. For example, if the sample is not fully dissolved and the dissolution of particles, colloids, or metal-organic complexes is incomplete, analytical results will depend on the analytical method chosen (Figure 1). The tendency to accept the "lowest value" should be carefully considered, as data may reflect that only a few of the species present in the sample have been included in the analysis.

Intercomparison runs on natural water samples are frequently applied among institutes participating in joint international research programs, especially with respect to macrocomponents. The intercomparison on different physico-chemical forms of Al in natural waters is, however, one of the very few international quality control exercises performed for speciation in natural waters.[96] Variable analytical results for monomeric Al species were attributed to variable time of extraction, variable column flow rate, and operator practices rather than the choice of analytical method. As demonstrated in Figure 6, the monomeric Al_a and Al_o, as well as the difference $Al_i = Al_a - Al_o$ (interpreted as toxic Al species), were reduced when colloids and particles were removed from the sample by *in situ* ultrafiltration.

VI. THERMODYNAMIC EQUILIBRIUM CALCULATIONS

During recent years, emphasis has been put on the development of thermodynamic equilibrium models for speciation purposes, based on known values for metal and ligand concentrations and published metal-ligand stability constants. Recent models seem especially useful for distinguishing between LMW ionic or complexed forms.[151,152] However, several factors can contribute to invalid results:

1. The assumption of thermodynamic equilibria in samples may not be valid, as time-dependent processes may be slow and/or metastable compounds may be present. Especially for mixing zones, the calculated distribution of species is questionable.
2. The equilibria considered are usually based on thermodynamic constants derived from macrochemistry (e.g., artificial solutions at room temperature), and published data on these constants may vary considerably.[134]
3. The lack of relevant data on important microchemical processes, for instance the behavior of polymers and colloids, implies that most models describe the chemistry in the water phase only partially. The predicting power of these models in assessing mobility and bioavailability may, therefore, be questionable.
4. The quality of data will influence the calculated distribution of species. In chemical models, information on ion or ligand concentrations is needed. However, the experimental data input are often based on filtration (e.g., 0.45 μm) and may include complexes, polymers, and colloids as well. Thus, models suffer from "noisy" input data, and more effort needs to be put on the interface between model requirement and the definition of experimental data.

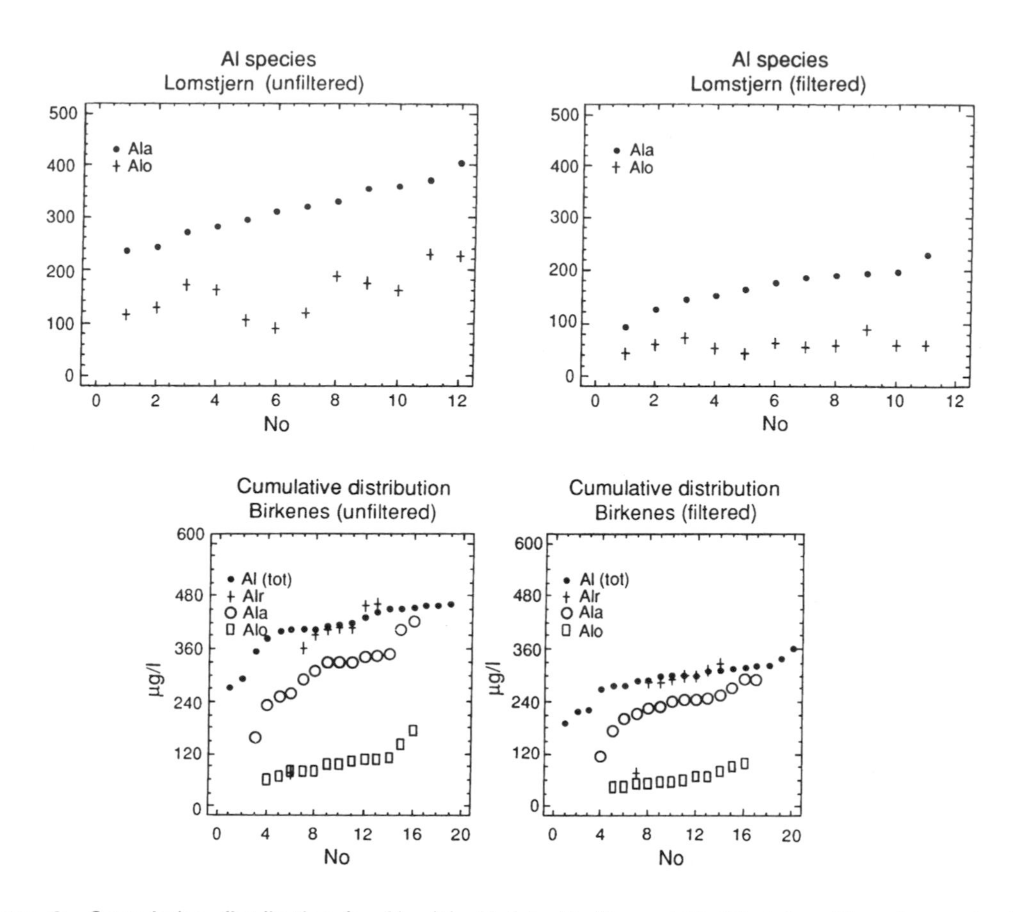

Figure 6 Cumulative distribution for Al_{tot} (●), Al_r (+), Al_a (*), and Al_o (□): (a) unfiltered sample; (b) ultrafiltered sample. Corresponding results on Al_o (●) and Al_o (+) obtained by 12 laboratories: (c) unfiltered sample; (d) ultrafiltered sample. (Redrawn from Salbu, B., Riise, G., Bjørnstad, H. E., and Lydersen, E., in *The Surface Waters Acidification Programme,* Cambridge University Press, London, 1990, 250.)

For macroelements, chemical modeling is an important tool for understanding hydrogeochemical processes. For trace elements, however, the validity of models is still questionable and information from experimental speciation studies is needed to modify models in order to increase their predicting power.

VII. CONCLUSIONS

The analytical strategy for trace elements in natural waters depends on whether total concentrations are to be determined or if information on speciation is to be obtained.

In natural waters, changing chemical and physical conditions (i.e., temperature, pH, Eh) can lead to disequilibrium systems with frequent changes in the distribution pattern of elemental species. Thus, if physico-chemical forms are to be determined, representative, uncontaminated samples must be fractionated *in situ* or shortly after sampling.

Among size fractionation techniques, *in situ* ultrafiltration (hollow fibers or cross-flow membranes) is particularly favorable, as it is essentially a rapid sampling and *in situ* fractionation technique for natural waters. Sorption can be minimized by conditioning, and large volumes of fractionated samples can be obtained for further investigations, e.g., biotest experiments.

Analytical results obtained from charge fractionation techniques are operationally defined and are to a certain extent influenced by the presence of complexes, colloids, and/or particles (e.g., desorption). Therefore, size fractionation of samples should be utilized prior to the application of charge fractionation techniques, as information on positive and negatively charged LMW species, as well as neutral species, can be attained.

As LMW species of trace metals are believed to be mobile, bioavailable, and potentially toxic to aquatic organisms, a major challenge within analytical chemistry is the development of techniques able to differentiate between LMW species without influencing the system. Furthermore, combined size and charge fractionation techniques should be utilized in:

1. Dynamic systems in order to obtain information on transformation processes and kinetics involved
2. Biotest experiments in order to assure chemically well-defined exposures when biological uptake and effects are assessed

Dynamic experiments utilizing chemically well-defined radioactive tracers should be considered highly useful for these purposes. As the basis for theoretical calculations and estimates on elemental species is still poor, experimental data is essential for improving the predicting power of models assessing mobility and biological uptake of trace elements and radionuclides in aquatic systems.

The sensitivity of ETAAS, ICPMS, ASV, and NAA is sufficiently high to allow the determination of most trace elements in water, especially when preconcentration of the element prior to analysis is performed. The combination of separation systems and ICPMS represents probably the most powerful tool in speciation studies, and major progress is expected in the years to come.

REFERENCES

1. **Beneš, P. and Majer, V.,** *Trace Chemistry of Aqueous Solutions,* Elsevier, Amsterdam, 1980.
2. **Kramer, C. J. M. and Duinker, J. C.,** *Complexation of Trace Metals in Natural Waters,* Nighoff/Junk, The Hague, 1984.
3. **Salbu, B.,** Size fractionation techniques combined with INAA for speciation studies, in *Proc. Int. Conf. Modern Trends in Activation Analysis,* Vol. 1, Heydorn, K., Ed., Risø Nat. Lab., Copenhagen, 1986, 135.
4. **Lydersen, E., Salbu, B., Poleo, A. B. S., and Muniz, I. P.,** The influences of temperature on aqueous aluminium chemistry, *Water Air Soil Pollut.,* 51, 203, 1990.
5. **Hoffman, H. R., Yost, E. C., Eisenreich, S. J., and Maier, W. J.,** Characterization of soluble and colloidal-phase metal complexes in river water by ultrafiltration. A mass balance approach, *Environ. Sci. Technol. Res.,* 15, 655, 1981.
6. **Driscoll, C. T., Jr., Baker, J. P., Bisgoni, J. J., and Schofield, C. L.,** Effects of aluminium speciation on fish in dilute acidified waters, *Nature,* 284, 161, 1980.
7. **Rosseland, B. O., Blakar, I. A., Bulger, A., Krogland, F., Kvellestad, A., Lydersen, E., Oughton, D. H., Salbu, B., Staurnes, M., and Vogt, R.,** The mixing zone between limed and acidic river waters: complex aluminium chemistry and extreme toxicity for salmonoids, *Environ. Pollut.,* 78, 3, 1992.
8. **Oughton, D. H., Salbu, B., Bjørnstad, H. E., and Day, J. P.,** Use of aluminium-26 tracer to study the deposition of aluminium species on fish gills following mixing of limed and acidic waters, *Analyst,* 117, 619, 1992.
9. **Cronan, C. S. and Schofield, C. L.,** Aluminium leaching response to acid precipitation: effects of high-elevation watersheds in the Northeast, *Science,* 204, 304, 1979.
10. **Dickson, W.,** Properties of acidified waters, in *Ecological Impacts of Acid Precipitation,* SNSF-Project, Drabløs, D. and Tollan, A., Eds., Oslo, 1980, 75.
11. **Lydersen, E., Salbu, B., and Poleo, A. B. S.,** Formation and dissolution kinetics of $Al(OH)_3$ in synthetic freshwater, *Water Resour. Res.,* 27, 351, 1991.
12. **Salbu, B.,** Radionuclides associated with colloids and particles in rainwaters, Oslo, Norway, in *Hot Particles from the Chernobyl Fallout,* Vol. 16, von Philipsborn, H. and Steinhäusler, F., Eds., Bergbau- und Industriemuseums, Theuren, 1988, 83.
13. **Salomons, W. and Förstner, U.,** *Metals in the Hydrocycle,* Springer-Verlag, Berlin, 1984, 99.
14. **Beneš, P. and Steinnes, E.,** In situ dialysis for the determination of the state of trace elements in natural waters, *Water Res.,* 8, 947, 1974.
15. **Beneš, P., Gjessing, E., and Steinnes, E.,** Interactions between humus and trace elements in fresh water, *Water Res.,* 10, 711, 1976.
16. **Salbu, B., Bjørnstad, H. E., Lindstrøm, N., Lydersen, E., Breivik, E. M., Rambæk, J. P., and Paus, P. E.,** Size fractionation techniques for the determination of elements associated with particulate or colloidal material in natural fresh waters, *Talanta,* 32, 907, 1985.

17. **Delgueldre, C., Longworth, G., Moulin, V., and Vilks, P.,** Grimsel colloid exercise: an international intercomparison exercise on the sampling and characterization of groundwater colloids, Technical Report 90-01, Nagra, Baden, 1990.
18. **Morris, A. W., Bale, A. J., and Howlands, R. J. M.,** The dynamics of estuarine manganese cycling, *Estuarine Coastal Shelf Sci.,* 14, 175, 1982.
19. **Rae, J. E. and Aston, S. R.,** The role of suspended solids in the estuarine geochemistry of mercury, *Water Res.,* 16, 649, 1982.
20. **Moran, S. B.,** The application of cross-flow filtration to the collection of colloids and their associated metals in seawater, in *Marine Particles: Analysis and Characterization,* Geophysical Monograph Series 63, Hurd, D. C. and Spencer, D. W., Eds., American Geophysical Union, Washington, D.C., 1991.
21. **Landing, W. M. and Lewis, B. L.,** Collection, processing and analysis of marine particulate and colloidal material for transition metals, in *Marine Particles: Analysis and Characterization,* Geophysical Monograph Series 63, Hurd, D. C. and Spencer, D. W., American Geophysical Union, Washington, D.C., 1991, 263.
22. **Hart, B. T. and Davis, S. H. R.,** A study of physico-chemical forms of trace elements in natural waters and waste water ultrafiltrate metal fractions, Australian Water Resources Council, Techn. Paper No. 35, 1978, 66.
23. **Salbu, B., Bjørnstad, H. E., Sværen, I., Prosser, S. L., Bulman, R. A., Harvey, B. R., and Lovett, M. B.,** Size distribution of radionuclides in nuclear fuel reprocessing liquids after mixing with seawater, *Sci. Tot. Environ.,* 130/131, 51, 1993..
24. **Salbu, B., Bjørnstad, H. E., and Brittain, J. E.,** Fractionation of cesium isotopes and ^{90}Sr in snowmelt run-off and lake waters from a contaminated Norwegian mountain catchment, *J. Radioanal. Nucl. Chem.,* 156, 7, 1992.
25. **Stumm, W. and Morgan, J. J.,** *Aquatic Chemistry,* 2nd ed., John Wiley & Sons, New York, 1981, chap. 10.
26. **James, R. O. and Parks, G. A.,** Characterization of aqueous colloids by their double layer and intrinsic surface chemical properties, in *Surface and Colloid Science,* Matijevie, E., Ed., Plemum Press, New York, 1982, 119.
27. **Salbu, B.,** The quality of analytical data for modelling purposes, BIOMOVS Tech. Rep. 3, ISSN 100-0392, National Institute of Radiation Protection, Sweden, 1988, 79.
28. **Florence, M. T.,** Electrochemical approaches to trace element speciation in waters, *Analyst,* 111, 489, 1986.
29. **Allen, H. E., Hall, R. H., and Brisbin, T. D.,** Metal speciation. Effects on aquatic toxicity, *Environ. Sci. Technol.,* 14, 441, 1980.
30. **Sunda, W. C., Klaveners, D., and Palumbo, V.,** Bioassays of cupric ion activity and copper complexation, in *Complexation of Trace Metals in Natural Waters,* Kramer, C. J. M. and Duinker, J. C., Eds., Nighoff/Junk, The Hague, 1984, 393.
31. **Shanmukhappa, H. and Neelakantan, K.,** Influence of humic acid on the toxicity of copper, cadmium and lead to the unicellular alga *Synechosystis Aquatilis, Bull. Environ. Contam. Toxicol.,* 44, 840, 1990.
32. **Baker, J. P. and Schofield, C. L.,** Aluminium toxicity to fish in acidic waters, *Water Air Soil Pollut.,* 18, 289, 1982.
33. **Lydersen, E., Poleo, A. B. S., Muniz, I. P., Salbu, B., and Bjørnstad, H. E.,** The effects of naturally occurring high and low molecular weight inorganic and organic species on the yolk-sack larvae of Atlantic salmon (Salmo salar L.) exposed to acidic aluminium-rich lake water, *Aquat. Toxicol.,* 18, 219, 1990.
34. **Florence, T. M., Lumsden, B. G., and Fardy, J. J.,** Algae as indicators of copper speciation, in *Complexation of Trace Metals in Natural Waters,* Kramer, C. J. M. and Duinker, J. C., Eds., Nighoff/Junk, The Hague, 1984, 411.
35. **Gottofrey, J. and Tjalve, H.,** Effect of lipophilic complex formation on the uptake and distribution of Hg^{2+} and CH_3-Hg^+ in brown trouts *(salmo trutta)*: studies with some compounds containing sulphur ligands, *Water Air Soil Pollut.,* 56, 521, 1991.
36. **John, J., Gjessing, E. T., Grande, M., and Salbu, B.,** Influence of aquatic humus and pH on the uptake and depuration of cadmium by the atlantic salmon *(salma salar L.), Sci. Tot. Environ.,* 62, 253, 1987.

37. **Muller, F. L. and Kester, D. R.,** Kinetic approach to trace metal complexation in seawater: application to zinc and cadmium, *Environ. Sci. Technol.,* 24, 234, 1990.
38. **Salbu, B.,** Analysis of trace elements and their physico-chemical forms in natural waters, *Mikrochim. Acta,* II, 29, 1991.
39. **Suess, M. J.,** *Design of Sampling Programmes in Examination of Waters for Pollution Control.* A Reference Handbook, Pergamon Press, Oxford, 1982.
40. **Sturgeon, R. and Berman, S. S.,** Sampling and storage of natural water for trace metals, *CRC Crit. Rev. Anal. Chem.,* 18, 209, 1987.
41. **Leppard, G. G.,** *Trace Element Speciation in Surface Waters and its Ecological Implications,* Alenium, New York, 1983.
42. **Berhard, M., Brucikman, F. E., and Sadler, P. J.,** *The Importance of Chemical "Speciation" in Environmental Processes,* Springer-Verlag, Berlin, 1986.
43. **Harrison, R. M. and Rapsomanikis, S.,** *Environmental Analysis Using Chromatography Interfaced with Atomic Spectroscopy,* Ellis Horwood, Chichester, 1989.
44. **Batley, G. E.,** *Trace Element Speciation: Analytical Methods and Problems,* CRC Press, Boca Raton, FL, 1989.
45. **Kramer, J. R. and Allen, H. E.,** *Metal Speciation, Theory, Analysis and Application,* Lewis Publishers, Chelsea, 1989.
46. **Patterson, J. W. and Passion, R.,** *Metals: Speciation, Separation and Recovery,* Lewis Publishers, Chelsea, 1990.
47. **Salomons, W. and Förstner, U.,** *Analytical Aspects of Environmental Chemistry,* John Wiley & Sons, New York, 1984.
48. **Craig, P. J.,** *Organometalic Compounds in the Environment: Principles and Reactions,* Longman, Harlow, Essex, U.K., 1986.
49. **Horowitch, A. J., Rinella, F. A., Lamothe, P., Miller, T. L., Edwards, T. K., Roche, R. L., and Rickert, D. A.,** Variations in suspended sediment and associated trace element concentrations in selected riverine cross sections, *Environ. Sci. Technol.,* 24, 1313, 1990.
50. **Hardy, J. T., Apts, C. W., Creselius, E. A., and Bloom, N. S.,** Sea-surface microlayer metal enrichment in an urban and rural bay, *Estuarine Coastal Shelf Sci.,* 20, 299, 1985.
51. **Mackay, W. A. and Pattenden, N. J.,** The transfer of radionuclides from sea to land via the air: a review, *J. Environ. Radioact.,* 12, 49, 1990.
52. **Heydorn, K. and Damsgaard, E.,** The determination of sampling constants by neutron activation analysis, *J. Radioanal. Nucl. Chem.,* 110, 539, 1987.
53. **Burrus, D., Thomas, R. L., Dominik, J., and Vernet, J. P.,** Recovery and concentration of suspended solids in the upper Rhone river by continuous flow centrifugation, *Hydrol. Proc.,* 3, 65, 1989.
54. **Schaule, B. K. and Patterson, C. C.,** Lead concentration in the Northeast Pacific: evidence for global anthropogenic perturbations, *Earth Planet Sci. Lett.,* 54, 97, 1981.
55. **Bruland, K. W., Franks, R. P., Knauer, G. A., and Martin, J. H.,** Sampling and analytical methods for the determination of Cu, Cd, Zn and Ni at the nanogram per liter level in seawater, *Anal. Chim. Acta,* 105, 233, 1979.
56. **Brugmann, L., Geyer, E., and Kay, R.,** A new Teflon sampler for trace metal studies in seawater — "Wates", *Mar. Chem.,* 21, 91, 1987.
57. **Lee, Y. H. and Iverfeldt, Å.,** Measurement of methylmercury and mercury in run-off, lake and rainwaters, *Water Air Soil Pollut.,* 56, 309, 1991.
58. **Laxen, D. P. H. and Harrison, R. M.,** Cleaning methods for polyethylene containers prior to the determination of trace elements in fresh water samples, *Anal. Chem.,* 53, 345, 1981.
59. **Lydersen, E., Bjørnstad, H. E., Salbu, B., and Pappas A. C.,** Trace element speciation in natural waters using hollow-fibre ultrafiltration, in *Speciation of Metals in Water, Sediment and Soil Systems,* Lecture Notes in Earth Sciences 11, Landner, L., Ed., Springer-Verlag, Berlin, 1987, 85.
60. **van der Sloot, H. A. and Duinker, J. C.,** Isolation of different suspended water fractions and their trace element content, *Environ. Technol. Lett.,* 2, 511, 1981.
61. **Beneš, P.,** Semicontinuous monitoring of truly dissolved forms of trace elements in streams using dialysis in situ. I. Principle and conditions, *Water Res.,* 14, 511, 1980.
62. **Borg, H.,** Metal fractionation by dialysis — problems and possibilities, in *Speciation of Metals in Water, Sediment and Soil Systems,* Lecture Notes in Earth Sciences 11, Landner, L., Ed., Springer-Verlag, Berlin, 1987, 75.

63. **Revitt, D. M. and Morrison, G. M. P.,** Metal speciation variations within separate stormwater systems, *Environ. Tech. Lett.,* 8, 793, 1987.
64. **Nisbet, A. F., Salbu, B., and Shaw, S.,** Association of radionuclides with different molecular size fractions in soil solution; implications for plant uptake, *J. Envir. Radioact.,* 18, 71, 1993.
65. **Beveridge, A., Waller, P., and Pickering, W. F.,** Evaluation of "labile" metal in sediments by use of ion-exchange resins, *Talanta,* 36, 535, 1989.
66. **Morrison, G. M. P.,** Approaches to metal speciation analysis in natural waters, in *Speciation of Metals in Water, Sediment and Soil Systems,* Lecture Notes in Earth Sciences 11, Landner, L., Ed., Springer-Verlag, Berlin, 1987, 55.
67. **Schumaker, V. M.,** Shape correction centrifugation, *Sep. Sci.,* 1, 409, 1966.
68. **Bockris, J. O. M. and Drazic, D. M.,** *Electro-Chemical Science,* Taylor and Francis, London, 1972.
69. **Shaw, D. J.,** *Introduction to Colloid and Surface Chemistry,* Butterworths, Liverpool, 1966.
70. **Lammers, W. T.,** Rep. No. K-1749, Union Carbide Corp., Oak Ridge, TN, 1968.
71. **Leppard, G. G.,** Size, morphology and composition of particulates in aquatic systems: solving speciation problems by correlative electron microscopy, *Analyst,* 117, 595, 1992.
72. **Laxen, D. P. H. and Chandler, I. M.,** Comparison of filtration techniques for size distribution in freshwaters, *Anal. Chem.,* 54, 1350, 1982.
73. **Laxen, D. P. H. and Harrison, R. M.,** A scheme for the physico-chemical speciation of trace metals in freshwater samples, *Sci. Tot. Environ.,* 19, 59, 1981.
74. **Batley, G. E. and Gardner, D.,** Sampling and storage of natural waters for trace metal analysis, *Water Res.,* 11, 745, 1977.
75. **Gardner, M. J. and Hunt, D. T. E.,** Adsorption of trace metals during filtration of potable water samples with particular reference to the determination of filtrable lead concentration, *Analyst,* 106, 471, 1981.
76. **Stumm, W. and Bilinski, H.,** Trace metals in natural waters: difficulties of interpretation arising from our ignorance on their speciation, in *Advances in Water Pollution Research,* Jenkins, S. H., Ed., Pergamon Press, Oxford, 1973.
77. **Stanley, J. S. and Stevens, R. E.,** Trace element-organic ligand speciation in oil shale wastewaters, in *Environmental Inorganic Chemistry,* Irgolic, K. J., Ed., VCH, Publishers Florida, 1985, 579.
78. **Salbu, B., Bjørnstad, H. E., Lydersen, E., and Pappas, A. C.,** Determination of radionuclides associated with colloids in natural waters, *J. Radioanal. Nucl. Chem.,* 115, 113, 1987.
79. **Mudge, S., Hamilton-Taylor, J., Kelly, M., and Bradshaw, K.,** Laboratory studies of the chemical behaviour of plutonium associated with contaminated estuarine sediments, *J. Environ. Radioact.,* 8, 217, 1988.
80. **Salbu, B. and Bjørnstad, H. E.,** Analytical techniques for determining radionuclides associated with colloids in waters, *J. Radioanal. Nucl. Chem.,* 138, 337, 1990.
81. **Orlandini, K. A., Penrose, W. R., Harvey, B. R., Lovett, M. B., and Findlay, M. W.,** Colloidal behaviour of actinides in an oligotropic lake, *Environ. Sci. Technol.,* 24, 706, 1990.
82. **Brittain, J. E., Bjørnstad, H. E., Salbu, B., and Oughton, D. H.,** Winter transport of Chernobyl radionuclides from a montane catchment to an ice-covered lake, *Analyst,* 117, 515, 1992.
83. **Salbu, B., Bjørnstad, H. E., and Pappas, A. C.,** A method for speciation of trace elements (stable and radioactive) in natural waters, in *Speciation of Fission and Activation Products in the Environment,* Bulman, R. A. and Cooper, J. R., Eds., Elsevier, London, 1985.
84. **Truitt, R. E. and Weber, J. H.,** Determination of complexing capacity of fulvic acid for copper (II) and cadmium (III) by dialysis titration, *Anal. Chem.,* 53, 337, 1981.
85. **Rona, E., Hood, D. W., Muse, L., and Buglio, B.,** Activation analysis of manganese and zinc in sea water, *Limnol. Oceanogr.,* 7, 201, 1962.
86. **Livens, F. R. and Singleton, D. L.,** Plutonium and americium in soil organic matter, *J. Environ. Radioact.,* 13, 323, 1991.
87. IAEA, *Reference Methods for Marine Radioactivity Studies,* Technical Reports No. 118, IAEA, Vienna, 1970.
88. IAEA, *Measurements of Radionuclides in Food and the Environment,* Technical Reports Series No. 295, IAEA, Vienna, 1989.
89. **Kingston, H. M., Barnes, I. L., Brady, T. J., Rains, T. C., and Champ, M. A.,** Separation of eight transition elements from alkali and alkaline earth metals in estuarine and seawater with chelating resin and their determination by graphite furnace atomic absorption spectrometry, *Anal. Chem.,* 50, 2064, 1978.

90. **McLaun, J. W., My-Kytiuk, A. P., Willie, S. N., and Berman, S. S.,** Determination of trace metals in seawater by inductively coupled plasma mass spectrometry with pre-concentration on silica-immobilized 8-hydroxy-quinoline, *Anal. Chem.,* 57, 2907, 1985.
91. **Landing, W. M., Haraldsson, C., and Paxeus, N.,** Vinyl polymer agglomerate based trace transition metal cation chelating ion-exchange resin containing the 8-hydroxyquinoline functional group, *Anal. Chem.,* 58, 3031, 1986.
92. **Driscoll, C. T.,** A procedure for the fractionation of aqueous aluminium in dilute acidic waters, *Int. J. Environ. Anal. Chem.,* 16, 267, 1984.
93. **Barnes, R. B.,** The determination of specific forms of aluminium in natural water, *Chem. Geol.,* 15, 177, 1975.
94. **Lydersen, E., Salbu, B., and Poleo, A. B. S.,** Size and charge fractionation of aqueous aluminium in dilute acidic waters: effects of changes in pH and temperature, *Analyst,* 117, 613, 1992.
95. **Figuara, P. and McDuffie, B.,** Use of chelex resin for determination of labile trace metal fractions in aqueous liquid media and comparison of the method with anodic stripping voltammetry, *Anal. Chem.,* 51, 120, 1979.
96. **Salbu, B., Riise, G., Bjørnstad, H. E., and Lydersen, E.,** Intercomparison study on the determination of total aluminium and aluminium species in natural fresh waters, in *The Surface Waters Acidification Programme,* Mason, J. B., Ed., Cambridge University Press, London, 1990, 250.
97. **Danielsson, L. G., Magnusson, B., and Westerlund, S.,** An improved metal extraction procedure for the determination of trace metals in seawater by atomic absorption spectrometry with electrothermal atomization, *Anal. Chim. Acta,* 98, 47, 1978.
98. **Armannsson, H.,** Dithizone extraction and flame atomic absorbtion spectrometry for the determination of cadmium, zinc, lead, copper, nickel, cobalt, and silver in seawater, *Anal. Chem. Acta,* 110, 21, 1979.
99. **Landing, W. M. and Bruland, K. W.,** Manganese in the North Pacific, *Earth Planet. Sci. Lett.,* 49, 45, 1980.
100. **Tessier, A., Campbell, P. C. G., and Bisson, M.,** Sequential extraction procedure for the speciation of particulate trace metals, *Anal. Chem.,* 51, 844, 1979.
101. **Chao, T. T.,** Use of partial dissolution techniques in geo-chemical exploration, *J. Geochem. Explor.,* 20, 101, 1984.
102. **An, P. and Zijian, W.,** Mercury in river sediments, in *Environmental Inorganic Chemistry,* Irgolic, K. J., Ed., VCH Publishers, Florida, 1985, 393.
103. **Livens, F. R., Baxter, M. S., and Allen, S. E.,** Association of plutonium with soil organic matter, *Soil Sci.,* 144, 24, 1987.
104. **Oughton, D. H., Salbu, B., Riise, G., Lien, H. N., Østby, G., and Nøren, A.,** Radionuclide mobility and bioavailability in Norwegian and Soviet soils, *Analyst,* 117, 481, 1992.
105. **Wilkens, B. T., Green, N., Stewart, S. P., and Major, R. O.,** Factors that affect the association of radionuclides with soil phases, in *Speciation of Fission and Activation Products in the Environment,* Bulman, R. A. and Cooper, J. R., Eds., Elsevier, London, 1985, 101.
106. **Salbu, B., Bjørnstad, H. E., Lindstrøm, N. S., Breivik., E., Rambæk, J. P., Englund, J. O., Meyer, K. F., Hovind, H., Paus, P. E., Enger, B., and Bjerkelund, E.,** Particle effects in the determination of trace elements in natural fresh waters, *Anal. Chim. Acta,* 167, 161, 1985.
107. **Oughton, D. H., Salbu, B., Brand, T., Day, J. P., and Aarkrog, A.,** Underestimations in the determination of ^{90}Sr in soils containing particles of irradiated uranium oxide fuel, *Analyst,* 118, 1101, 1993.
108. **Chan, Y. K. and Wong, P. T. S.,** Direct speciation analysis of molecular and ionic organometals, in *Trace Element Speciation in Surface Waters and its Ecological Implications,* Leppard, G. G., Ed., Plenum Press, New York, 1977, 215.
109. **Irgolic, K. J.,** Environmental inorganic analytical chemistry, in *Environmental Inorganic Chemistry,* Irgolic, K. J., Ed., VCH Publishers, Florida, 1985, 547,
110. **Chau, Y. K.,** Chromatographic techniques in metal speciation, *Analyst,* 117, 571, 1992.
111. **Brinckman, F. E., Jewett, K. L., Iverson, W. P., Irgolic, K. J., Ehrhardt, K. C., and Stackton, R. A.,** Graphite furnace atomic absorption spectrophotometers as automated element-specific detectors for high pressure liquid chromatography: the determination of arsenite, arsenate, methylarsonic acid and dimethylarsinic acid, *J. Chromatogr.,* 191, 31, 1980.
112. **Ebdorn, L., Hill, S., and Ward, R. W.,** Directly coupled chromatography-atomic spectroscopy. I. Directly coupled gas chromatography-atomic spectroscopy. A review, *Analyst,* 111, 1113, 1986.

113. **Batley, G. E. and Farrer, Y. J.,** Irradiation techniques for the release of bound heavy metals in natural waters and blood, *Anal. Chim. Acta,* 99, 283, 1978.
114. **Hart, B. T. and Davis, S. H. R.,** Trace element speciation in the freshwater and estuarine regions of the Yarra river, *Estuarine Coastal Shelf Sci.,* 12, 353, 1981.
115. **Laxen, D. P. H. and Harrison, R. M.,** The physico-chemical speciation of Cd, Pb, Cu, Fe and Mn in the final effluent of a sewage treatment works and its impact on speciation in the receiving river, *Water Res.,* 15, 1053, 1981.
116. **Salbu, B. and Rambæk, J. P.,** Fractionation of low molecular weight charged species using electrodialysis, in *Environmental Consequences of Releases from Nuclear Accidents,* Final Report NKA, Project AKTU-200, Tveten, E., Ed., Nordic Liaison Committee for Atomic Energy, Roskilde, 1990, 119.
117. **Leyden, D. E. and Wegscheider, W.,** Preconcentration for trace element determination in aqueous samples, *Anal. Chem.,* 53, 1059, 1981.
118. **Hall, A. and Godinho, M. C.,** Concentration of trace metals from natural waters by freeze-drying prior to flame atomic absorption spectrometry, *Anal. Chim. Acta,* 113, 369, 1980.
119. **Landsberger, S., Simmonds, S., Kramer, A., Drake, J. J., Vermette, S. J., Shuter, B., and Jhssen, P.,** Application of neutron activation analysis in acid precipitation studies, *J. Rad. Nucl. Chem.,* 110, 333, 1987.
120. **Görlach, U. and Boutron, C. F.,** Preconcentration of lead, cadmium, copper and zinc in water at the pg g^{-1} level by non-boiling evaporation, *Anal. Chim. Acta,* 236, 391, 1990.
121. **Høgdahl, O. T., Melsom, S., and Bowen, V. T.,** Neutron activation analysis of lanthanide elements in sea water, in *Trace Inorganics in Water,* Advances in Chemistry Series No. 73, American Chemical Society, Washington, D.C., 1968, 308.
122. **Minczewski, J., Chwastowska, J., and Dybczynski, R.,** *Separation and Preconcentration Methods in Inorganic Trace Analysis,* John Wiley & Sons, New York, 1983.
123. **Andersen, B. and Salbu, B.,** The determination of trace elements in sea water using Mg(OH)2 as scavenger, *Radiochem. Radioanal. Lett.,* 52, 19, 1982.
124. **Greenberg, R. R. and Kingston, H. M.,** Trace element analysis of natural water samples by neutron activation analysis with chelating resin, *Anal. Chem.,* 55, 1160, 1983.
125. **Van Geen, A. and Boyle, E.,** Automated preconcentration of trace metals from seawater and freshwater, *Anal. Chem.,* 62, 1705, 1990.
126. **Cox, J. A., Slonawska, K., and Gatchell, D. K.,** Metal speciation by Donnan dialysis, *Anal. Chem.,* 56, 650, 1984.
127. **Piro, A. J., Bernhard, M., Brancia, M., and Verzi, M.,** Incomplete exchange reaction between radioactive ionic zinc and stable natural zinc in seawater, in *Proc. Radioactive Contamination of the Marine Environment,* IAEA, Vienna, Austria, 1973.
128. **Sturgeon, R.,** Atomic absorption spectroscopy — present and future aspects, *Analyst,* 117, 233, 1992.
129. **van Loon, J. C. and Barefoot, R. R.,** Overview of analytical methods for elemental speciation, *Analyst,* 117, 563, 1992.
130. **Ebdon, L., Hill, S., Walton, A. P., and Ward, R. W.,** Coupled chromatography-atomic spectroscopy for arsenic speciation — a comparative study, *Analyst,* 113, 1159, 1988.
131. **Berman, S. S.,** Analysis of environmental samples for trace metals, Abstract Lecture presented at XXVII CSI Int. Conf., June 9–14, 1991, Bergen, Norway.
132. **McLaren, J. W., Lam, J. W., and Berman, S. S.,** The second generation of ICP mass spectrometry, Abstract Lecture presented at XXVII CSI Int. Conf., June 9–14, 1991, Bergen, Norway.
133. **Kim, C. K., Morita, S., Seki, R., Takaku, Y., Ikeda, N., and Assinder, D. J.,** Distribution and behaviour of ^{99}Tc, ^{237}Np, ^{239,240}Pu, and ^{241}Am in the coastal and estuarine sediments of the Irish sea, *J. Radioanal. Nucl. Chem.,* 156, 201, 1992.
134. **Florence, T. M. and Batley, G. E.,** Chemical speciation in natural water, *CRC Crit. Rev. Anal. Chem.,* 219, 1980.
135. **Florence, T. M.,** Trace element speciation by anodic stripping voltammetry, *Analyst,* 117, 551, 1992.
136. **van den Berg, C. M. G.,** Effect of the deposition potential on the voltammetric determination of complexing ligand concentrations in sea-water, *Analyst,* 117, 589, 1992.
137. **Murray, R. W.,** Chemically modified electrodes, *Acc. Chem. Res.,* 13, 135, 1980.
138. **Stickney, J. L., Soriaga, M. P., Hubbard, A. T., and Anderson, S. E.,** A survey of factors influencing the stability of organic functional groups attached to platinum electrodes, *J. Electroanal. Chem.,* 125, 73, 1981.

139. **Johnson, D. C.,** Analytical electrochemistry — theory and instrumentation of dynamic techniques, *Anal. Chem.,* 54, 9, 1982.
140. **Pihlar, B., Valenta, P., and Nürnberg, H. W.,** New high-performance analytical procedure for the voltammetric determination of nickel in routine analysis of waters, biological materials and food, *Fresenius J. Anal. Chem.,* 307, 337, 1981.
141. **van den Berg, C. M. G.,** Determination of the complexing capacity and conditional stability constants of complexes of copper (II) with natural organic ligands in seawater by cathodic stripping voltammetry of copper-catechol complex ions, *Mar. Chem.,* 15, 1, 1984.
142. **van den Berg, C. M. G.,** Determination of the zinc complexing capacity in seawater by cathodic stripping voltammetry of zinc-APDC complex ions, *Mar. Chem.,* 61, 121, 1985.
143. **Salbu, B. and Steinnes, E.,** Applications of nuclear analytical techniques in environmental research, *Analyst,* 117, 243, 1992.
144. **Salbu, B., Steinnes, E., and Pappas, A. C.,** Multielement activation analysis of freshwater using Ge(Li) gamma spectrometry, *Anal. Chem.,* 47, 1011, 1975.
145. **Salbu, B., Steinnes, E., and Bjørnstad, H. E.,** Use of different physical separation techniques for trace element speciation studies in natural waters, in *Hydrochemical Balances of Freshwater Systems,* Eriksson, E., Ed., International Association of Hydrological Sciences, University of Uppsala, Sweden (IAHS-AIHS), Publ. No. 150, 1984, 203.
146. **Jørstad, K. and Salbu, B.,** Determination of trace elements in seawater using neutron activation analysis and electro-chemical separation, *Anal. Chem.,* 52, 672, 1980.
147. **de Bruin, M. and Blaauw, M.,** Sources of error in analytical gamma-ray spectrometry, *Analyst,* 117, 431, 1992.
148. **Yu-Fu, Y., Salbu, B., and Bjørnstad, H. E.,** Recent advances in the determination of low-level plutonium in environmental and biological samples, *J. Radioanal. Nucl. Chem.,* 148, 163, 1991.
149. **Salbu, B.,** Radioactive tracer techniques in speciation studies, *Environ. Technol. Lett.,* 8, 381, 1987.
150. **Heydorn, K.,** Quality assurance and statistical control, *Microchim. Acta,* 111, 1, 1991.
151. **Schecher, W. D. and Driscoll, C. T.,** An evaluation of the equilibrium calculations within acidification models: the effect of uncertainty in measured chemical components, *Water Resour. Res.,* 23, 525, 1987.
152. **Schecher, W. D. and Driscoll, C. T.,** An evaluation of uncertainty associated with aluminium equilibrium calculations, *Water Resour. Res.,* 24, 533, 1988.

Chapter 4

Data Analysis and Statistical Methods: Environmetric Analysis of Element Data

Nils B. Vogt, Sjur Andersen, Morten Shaanning, and Rolf D. Vogt

CONTENTS

I. INTRODUCTION

Environmental and ecological systems are in their nature complex. The phenomena of chemical pollution are observed as increases in and changes of relative concentrations between chemical compounds.[1-4] To interpret and predict so that we can manage the environmental and ecological system requires the measurement and analyses of many variables simultaneously.[5-8] This is recognized by most environmental analysts when they measure a large number of variables (measured features, attributes, e.g., chemical concentrations and biological responses) on many collected samples.

Several articles, textbooks, and review articles have discussed the topic of statistical analyses in environmental work in detail focusing on particular fields. The statistical methods described range from simple univariate (mean and variance) description, through dispersion (variance and standard deviation) and linear regression and covariance/correlation analyses to multivariate analysis in all areas of environmental analyses, e.g., sediment[2,9,10] and soil,[11] water,[4,12] air,[13-15] and in the biota.[3,16]

The scope and purpose of this expository presentation is to illustrate the use of graphical, data analytical, and statistical methods for analyses of element data in environmental analyses. The presentation will describe graphical and statistical uni- and bivariate methods of data analyses which will eventually lead to an understanding of why and how multivariate methods may be used. Three data sets are used to illustrate and discuss results from statistical analysis. These are the BIOMAT data (biota), the CARE data (air), and the MEM data (marine-ecological) sets (See Appendix, Reference Tables A1, A2, and A3).

0-8493-6304-7/95/$0.00+$.50

II. SCOPE AND AIM OF ENVIRONMETRICS

The use of mathematical and statistical methods in environmental experiments, surveys, and investigations is environmetrics. The scope of environmetrics is to increase the scientific yield or result and improve the quality of environmental, ecological, and toxicological research. Environmetrics emphasizes the close connection between systematic planning and execution of experiments and investigations and analyses and interpretation of obtained data.

Environmetrics contributes to integrating fields of science by promoting systematic scientific methodology, emphasizing interdisciplinary considerations and providing quantitative methods. For an approach which may be used when intending to conduct an experiment or investigation, see Reference 17, pp. 264 to 266. The approach prescribes that a well-defined set of steps are to be followed:

1. Hypothesis: define the alternatives that are available
2. Parameterize: select which variables are controlled and which must be randomized. Determine what is to be measured
3. Design: select a lay-out of the experiment which accommodates the need for variation, control, and randomization
4. Perform: make the practical experiment
5. Data Analysis: use appropriate data analytical and statistical methods to extract significant factors and connections
6. Interpret: couple results from data analyses with knowledge of problem to make conclusion

Environmetric methods used to plan experiments and analyze data are tools; they are not the purpose of the experiment nor are they fallacy-proof guarantees for a scientific result. Environmetric methods must be used in close connection with environmental knowledge and scientific problem insight if the result of an experiment is to be valid.

The use of mathematical methods and statistical models relies on satisfying the criteria of ecological and/or environmental scope and relevance. Statistical significance is a requirement for interpretation, but not sufficient for environmental meaningfulness. Similarly, it should be recognized that although there must be consistency between mathematics and statistics in use, mathematical methods may be applied without statistical assumptions. It is appropriate to stress here that although the large majority of statistical analysis use alpha = 0.05 ($\alpha = 0.05$) as a benchmark, there is no principle that states this level to be all encompassing. A somewhat more appropriate approach in environmental analysis may be to use alpha = 0.1 as an action level warranting further inspection, alpha = 0.05 as statistically relevant, and alpha = 0.01 as safely significant.

III. SURVEYS, INVESTIGATIONS, AND CONTROLLED EXPERIMENTS

Experimental design procedures have been developed to test and ascertain explanatory variables and to measure relationships between these variables and response variables in controlled experimental situations.[18-21] Controlled experimental situations are most often found in the laboratory. Most environmental research is conducted as surveys and investigations, i.e., as observational studies.[22,23] The difference between controlled experiments and observational surveys or investigations is not the consequences of *a posteriori* statistical analysis, but the reliance on, and requirement, of additional *a priori* consideration of realism of measurement and representativity of samples during planning.[24-27]

Sampling designs for observations must balance convenience, cost, possibilities/feasibility and realism, representativity, control, and randomization. Observational studies are often weak on control, randomization, and representation. Statistical treatment of observational studies, therefore, cannot easily use the same "simple" procedures as controlled experiments. This does not imply that predictive conclusions cannot be achieved, but does emphasize additional caution and prudence in planning of such studies.

Designs and strategies for sampling in environmental analysis most often do not support the extensive conclusions proposed. When design and strategies are not optimal or even adequate, the unexpected and "non-normal" must be addressed. Analyses of survey and investigational data should emphasize exploratory — data analytical — methods of statistical analysis. This presentation will, therefore, focus on the [illegible] of such methods.

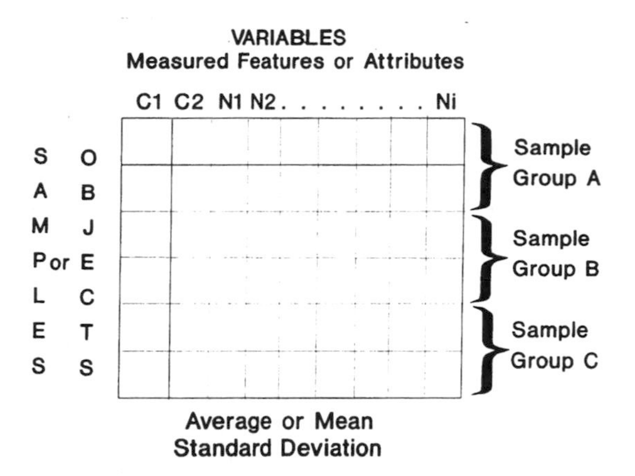

Figure 1 The data matrix is a basic element of statistical analysis of data. Samples are grouped according to categorial variables, e.g., type or station (C1 and C2). The *t*-test may be used to compare the means of numerical variables (N1 to Ni) of two sample groups. ANOVA (see text) is used to compare mean and variance of several sample groups classified according to one categorial variable, e.g., stations. MANOVA is used to compare means and variances of several sample groups classified by more than one categorial variable.

IV. THE DATA MATRIX — EXPLORATORY DATA ANALYSIS

To understand the use of statistical analysis and graphical presentation, it is necessary to have a basic understanding of the data table or matrix. Figure 1 and Table 2 illustrate the data matrix and the concepts and terms that will be discussed and used in the presentation. The data matrix has columns of measured variables or features describing the samples which are represented in rows (See Appendix, Reference Tables). The variables may be continuous numerical, binary (1/0, present/not present), or ranking variables. Each type of variable will require specific considerations during analysis (see Reference 18 and Reference 23, p. 42). The samples may be separated into groups using categorial variables, e.g., sample stations or sample type.

Univariate methods of analysis provide information on single variables (= Columns; i.e., mean, variance, standard deviation, and frequency distribution [e.g., normal {Gaussian} and log-normal distributions]). When samples (objects) are grouped according to categorial variables, analysis of variance (ANOVA) is used to compare the variance elements of variables.

Bivariate correlation and bivariate regression provide information on the relationships between two-and-two variables (= columns). Implicit in most correlation or covariance analyses is a linear hypothesis assuming that there is a straight-line relationship between two variables. Bivariate regression analyses are easily extended to encompass other relationships, although, as shown below, there may be significant problems in determining the appropriate model to apply. When multiple variables have been measured, e.g., zinc, manganese, iron, and copper, it is *not* appropriate to perform multiple univariate tests to ascertain statistical significance (see below). Instead, multivariate methods must be applied.

V. MATERIALS AND METHODS

The data used are given in Reference Tables A1 to A3 (Appendix) so that readers may attempt to copy the analyses if interested. The BIOMAT data represent fauna samples collected at seven stations in a transect along the Sandefjord fjord — samples 1, 2, and 3 are inner fjord, samples, 5, 6, and 7 are close to an industrial site, and sample 4 is outer fjord. The data set CARE (composition activity relationships) is from air samples collected in a small, predominantly wood-burning community in Norway, although there is oil heating and automobile influence. The MEM (multivariate ecotoxicological mapping) data set is from Kristiansandsfjorden. This is an open fjord with a major city and several industrial sites. The data are from sediments from which samples have been collected and analyzed chemically and the fauna diversity has been calculated. The statistical programs used in the presentation are Statgraphics,[28] SIMCA,[29] and SIRIUS.[30]

VI. UNI- AND BIVARIATE ANALYSES

Exploratory data analyses may broadly be divided into two areas: uni- or bivariate analysis and multivariate analysis. Uni- or bivariate analysis is the most common form of exploratory data analysis. The simplicity of graphical presentation and statistical analysis in one or two dimensions is obvious. The fallacy of reductionism likewise should be apparent. The presentation will discuss univariate frequency distribution, statistical inference and intervals testing, ANOVA and scatterplots, correlation, and regression.

A. FREQUENCY DISTRIBUTION AND TRANSFORMATIONS

The simplest univariate analysis is the frequency distribution histogram. This is a graphical display and statistical analysis of the distribution of a set of observations (see Reference 31, pp. 14 to 24). The frequency distribution is an important, albeit quite often forgotten, prerequisite to most statistical data analysis. The frequency distribution of variables influences statistical treatment. The most used parametric methods rely on least-squares calculations, i.e., the normal or gaussian distribution. This is true for significance testing using *t*-test and ANOVA or linear regression analysis (see Reference 32, Chapters 5.2.3 and 13.2.1) and is in fact implied in the calculation of the standard deviation. Frequency distribution analyses, both graphical and statistical, should therefore always be carried out as a preliminary analysis of environmental data when 20 or more samples are available. Outliers, special distributions, and distributional uniformity or harmony of the measured variables may be easily identified visually and by statistical parameters.

Outlier samples (data points) are important for several reasons. Single anomalous data points may quite often strongly influence statistical results and thus invalidate statistical analysis, e.g., have high leverage in regression. Outliers may also represent valuable information about the system being investigated. Although there are several useful statistical criteria for identifying outliers (see Reference 32, Chapter 16), it should be emphasized that discarding outliers based on statistical parameters alone without attempting to consider why a data point is anomalous (sampling, analysis, or other reason) is not a scientific approach.

The frequency histogram procedure first divides the data into sets of nonoverlapping intervals of equal width/interval. The number of observations within each interval is then tabulated or displayed as bar height. Statistical tests such as the Kolmogorov (often Kolmogorov-Smirnov) test (see References 32, Chapter 11.2.8 and Reference 33, p. 56) or similar are then used to investigate *departure* from an assumed distribution (see next section below).

Figure 2 shows a comparison of frequency bar histograms and distributions of two selected elements from the BIOMAT data (see Appendix, Reference Table A1). The superimposed distributions are based on parameters of the data themselves using functions describing the distribution. The histograms for Fe illustrate the effect of taking the logarithm, i.e., the distribution becomes more normal or Gaussian. Table 1 shows that although the visual inspection implies a log-normal distribution for Fe, the Kolmogorov-Smirnov statistic does not allow us to formally determine that either variable departs from either distribution.

Many variables measured in environmental chemical investigations have a log-normal distribution. This is a skewed tail distribution towards high values, i.e., the tail of the distribution on the high number side is "stretched". This distribution can be made to fit a normal distribution by taking the logarithm of the values. The reason for this may be placed on two interacting and inherent properties of environmental investigations: the reference or background values may represent a normal (Gaussian) distribution, whereas the polluted samples represent superimposed high values which when analyzed together with data from the "normal" samples stretch the high end tail of the distribution. This implies that if there are sufficient samples in each "class", i.e., similar type (group) of samples, the distribution may approach a bimodal distribution. Since statistical parameters are often calculated for separate sample groups, frequency distribution analyses must be carried out on each class separately if statistical inference analyses and hypothesis testing is to be made. The necessity of large sample numbers makes this difficult. The second property is that often the sample size, work-up, and trace analytical methods are operated at or close to the limit of quantification (LOQ) or limit of detection (LOD). The tendency to produce results looking like a log-normal or exponential distribution, but in fact equally possibly representing a cut-off (half) normal distribution is prevalent.

It is quite frequently not possible to ascertain by use of statistical tests that an observed distribution departs from a normal distribution. The extent to which it is chosen to transform a variable must then depend on both knowledge about the data and an understanding of the effect of transformation. The most used transformations in environmental analysis are the logarithmic transformation to "pull in" or make more normal skewed high value tails and the square-root transformation to contract skewed low value tails. Both must be used with respect, for the effect they may have on the result of the data analysis, e.g., logarithmic transformations, will make multiplicative relationships additive.

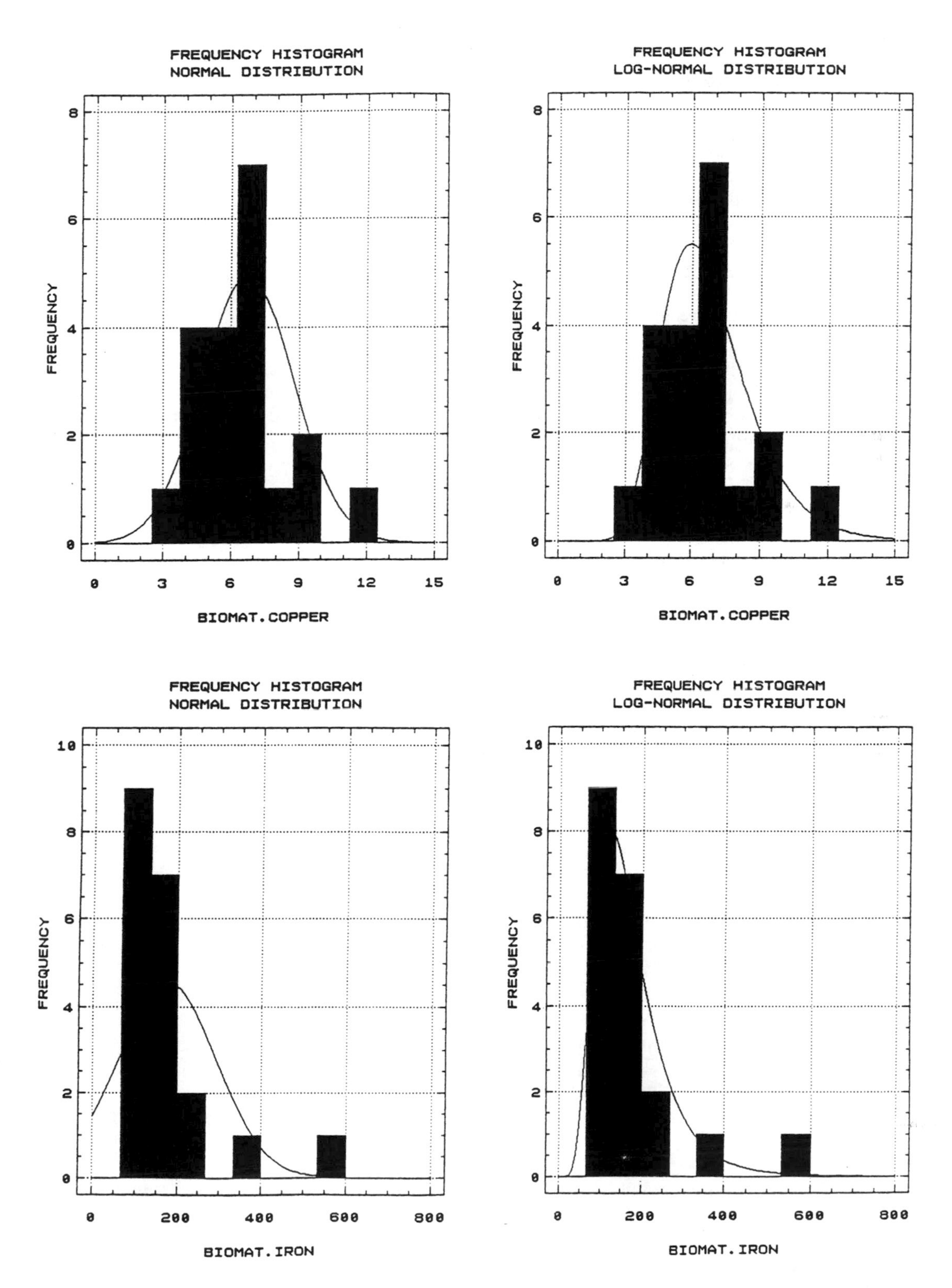

Figure 2 The frequency histograms for the elements Cu and Fe from the BIOMAT data. All 20 samples have been treated together. Superimposed are the distributions that the frequency histograms have been compared against.

Table 1 **The Kolmogorov-Smirnov test has been used to estimate departure from assumed distribution (see Figure 4)**

	Normal Distribution		Log-Normal Distribution	
Parameter	**Copper**	**Iron**	**Copper**	**Iron**
Mean	6.74	176.1	6.75	173.2
Standard deviation	2.02	116.5	2.02	88.3
K-S stat. DPLUS	0.1521[a]	0.2379[a]	0.1304[a]	0.1447[a]
K-S stat. DMINUS	0.0947	0.1975	0.1275	0.1011
K-S significance	0.7439	0.2074	0.8856	0.7966

[a] Estimated overall statistic DN.

B. STATISTICAL INFERENCE AND HYPOTHESIS TESTING

Statistical inference deals with making generalizations about population parameters from analysis of sample data. This is a situation most common to environmental analyses.[33] The two most important types of inferences are estimation of parameters and testing of hypothesis.

Estimation of the statistical parameters mean, variance, and covariance (or related parameters) is necessary for statistical analyses (see Table 2). The true value of a parameter (e.g., mean, regression coefficient, or confidence interval) is an unknown constant that can be truly measured only by considering the total population. When environmental scientists collect samples, this is not only a physical description of objects, but, in statistical terms, it also illuminates the limited extent of number of samples vs. total population. Whenever statistical parameters are calculated, it is required that the number of samples or degrees of freedom are clearly stated.

Table 2 shows the mathematical expressions used to derive basic statistical parameters. The arithmetic mean or average is one of several measures of "middleness". This constant represents an estimate of the true parameter — the population average (see Reference 19, pp. 33 to 43). To obtain a measure of the dispersion or spread of the data the variance is used. The variance and standard deviation are least-squares estimates of the dispersion of the observed data (see Reference 19, pp. 33 to 43). These parameters are used together with assumptions on distributions (although already inherent in the computation of the parameters) to test a hypothesis. The extent to which two variables covary is measured using the covariance and correlation parameters.[34] Covariance and correlation can be calculated for all pairs of variables. This is used to make the covariance and correlation matrix.

The use of statistical hypotheses, although somewhat opposite to the layperson's notion of scientific purpose (see Reference 35, p. 175), is consistent with scientific methodology and is similar to the mathematical method of proof by contradiction. The hypothesis the scientist wishes to "prove" is the alternative hypothesis (H_A). To support this hypothesis the opposite is proposed as the null hypothesis (H_0). The reason is that it is not possible to prove a hypothesis, only to contradict it. By using the inverse statement, the statistician can specify the probability of making an incorrect decision, i.e., rejecting the null hypothesis — the alpha probability — and thus get a measure of the confidence in the decision to retain the alternate hypothesis.

An example of the use of the mean and the variance for hypothesis testing is illustrated in Table 3. The data are taken from BIOMAT (see Appendix, Reference Table A3) and the purpose is to determine whether the average concentration of Cu is different in mussels *(Mytilus edulis)* (n = 7) and seaweed *(Fucus serratus)* (n = 7) in the Sandefjord. The test used is a two-sample, two-sided *t*-test (see References 33, p. 46 and 17, Chapter 3).* The *t*-test tests the difference between means. The null hypothesis (H_0) is that the difference in means is 0 vs. the alternate hypothesis (H_A) that the difference in means is different from 0. Prior to making the *t*-test it is necessary to determine whether the variances of each sample may be pooled to estimate an overall variance. If this is not the case then the analyst must use the separate estimates of variance and correct for this in the treatment (see Reference 19, p. 57). The analysis shows

* The test used is a two-sample, two-sided *t*-test (see Reference 16, p.46, and Reference 17, Chapter 3). The two-sided test is a test to determine equal to (=) or different from (<>) without considering whether this is a larger than or smaller than. A one-sided test compares equal to (=) vs. either larger than (>) or smaller than (<) (see Reference 6, p.59). The two-sample content relates to the fact that there are two samples involved. The comparison is not towards a specified constant, but between two samples, and that the comparison and degrees of freedom used must consider this.

Table 2 **Summary of statistical measures for data tables**

1. Sample Mean(s)

Mean of *i*th variable: $\bar{x}_i = 1/N \sum_{r=1}^{N} x_{ri}$

Vector of means of p variables: $\overline{x'} = \left[\bar{x}_1, \bar{x}_2, \ldots, \bar{x}_p\right]$

2. Sample Variance(s)

Variance of *i*th variable: $s_{ii} = 1/N \sum_{r=1}^{N} \left(x_{ri} - \bar{x}_i\right)^2$

3. Sample Covariance(s) and Correlation

Covariance of *i*th and *j*th variable: $s_{ij} = 1/N \sum_{r=1}^{N} \left(x_{ri} - \bar{x}_i\right)\left(x_{ri} - \bar{x}_j\right)$

Correlation of *i*th and *j*th variable: $r_{ij} = \dfrac{s_{ij}}{\sqrt{s_{ii} s_{jj}}}$

4. Sample Covariance Matrix

$$x'x = s = \begin{matrix} s_{11}s_{12}\ldots\ldots s_{1p} \\ s_{21}s_{22}\ldots\ldots s_{2p} \\ \cdot \quad s_{rr} \\ \cdot \quad\quad s_{rr} \\ s_{p1}s_{p2}\ldots\ldots s_{pp} \end{matrix}$$

The diagonal elements ($s_{11} \ldots s_{pp}$) are the variance for each variable. Substituting $\mathbf{r}_{ij}$ for $\mathbf{s}_{ij}$ gives the correlation matrix. In the correlation matrix r_{ij} where i = j is 1.00. Both matrixes are symmetric around the diagonal, i.e., $s_{1p} = s_{p1}$.

Note: N = number of objects; i = index for objects; r = counting dummy objects; P = number of variables; j = index for variables.

Table 3 **A two-sample two-sided *t*-test at a = 0.05 has been used to compare the concentrations of Cu in mussels and seaweed**

Sample Statistics	Mussel	Seaweed	Pooled
Number of observations	7	7	14
Average	6.7	7.77	7.24
Variance	3.51	5.81	4.66
Standard deviation	1.87	2.41	2.16
Median	7.5	7.4	7.45

Difference between means	-1.07
Test of equal variance	3.51/5.81 = 0.60 (95% Conf. Interval; 0.1-3.6)
Null hypothesis H_0	Difference in means = 0
Alternate hypothesis H_A	Difference in means <> 0
Computed *t*-statistic	-1.07/2.16* {1/7 + 1/7}$^{1/2}$ = -0.93
$*t*_{(2,12,0.05)}$	2.18 — Do not reject H_0

that we may pool the variances and that we cannot reject the null hypothesis; hence, it cannot be statistically determined ($\alpha = 0.05$) that there is a difference in the concentration of Cu in mussels and seaweed.

An increasingly frequent occurring situation which is not treated statistically correct is the use of multiple, or repeated, *t*-testing to test hypothesis of difference between several, or many, measured variables. In the BIOMAT data there are 11 elements (variables). Using a *t*-test to make 11 comparisons, i.e., a comparison between mussels and seaweed for all elements, reduces the confidence, i.e., power, of the test exponentially.* In this case there is a 43% chance (probability) of mistakenly rejecting the null hypothesis for one of the variables and of finding a statistically significant difference. This calculation only considers the stochastic probability. When, as is in fact very often the case in environmental analysis, the measured variables are correlated, the exercise of multiple *t*-testing is simply futile. Quite obviously if variable A has a mean in sample group X that is found to be statistically different from the mean in sample group Y and variable A is correlated with variable B, then we should expect that the mean of variable also is different in sample group B.

C. MULTIFACTOR ANALYSIS OF VARIANCE (MANOVA)

When more than two means, i.e., sample groups, are to be compared, e.g., in BIOMAT there are three different types or groups of species samples, ANalysis Of VAriance (ANOVA) may be used. ANOVA is used to separate and estimate sources of variation when samples are classified according to one controlled factor (see Reference 19, p. 65; Reference 22, Chapters 14 and 15; and Reference 33, pp. 56 to 65). When there are more than two types of samples and more than one classification variable, e.g., species type and station, multifactor ANOVA (MANOVA) may be used. Table 4 and Figure 3 illustrate the use of MANOVA to compare the mean values of Mn and Pb in the BIOMAT data. There are two factors in the model. These are three types of samples (B, S, and K) and seven stations (1 to 7). The MANOVA procedure estimates the variance (sum of squares = SS) contribution from each of the terms in the model separately. The model used here is

$$SS_{\text{Total}} = SS_{\text{Average}} + SS_{\text{Station}} + SS_{\text{Type}} + SS_{\text{Residual Error}}$$

The estimated SS is divided by the appropriate degrees of freedom (DF) to obtain the mean square (MS). The ratio between the MS of a controlled factor (e.g., station or type) and the residual (left over) MS provides an F-statistic (Fisher variance ratio). This ratio comparison of variances is used to test the significance of the size of the MS for each factor. A significant MS ratio shows that the factor accounts for a statistically significant amount of the total variation (SS_{Total}). In Table 4 the significance level (Sig.level) describes the probability of making a mistake when deciding the term is significant.

The MANOVA shows that there is a significant ($\alpha = 0.05$) effect of sample type for both Mn and Pb (probability of making mistake = 0). MANOVA does not show between which of the three types of samples there is a difference; this must be investigated subsequently by inspection of the data by a range test or comparison of groups (see Figure 3, top).

Figure 3 shows graphic presentations of the data. The trends for concentrations of Mn and Pb are different in the three types of samples; whereas seaweed has the highest Mn, the mussels have the highest Pb concentrations. The graphical presentation of the sample means and confidence intervals (top) shows that all three sample types have different concentrations of Mn, whereas for Pb only mussels is different from seaweed and krusflikk *(Chondrus crispus).* The graphical presentation also points out that although no statistically significant ($\alpha = 0.05$) effect of station could be found for either element, stations 1 to 3 have high Pb concentrations and stations 5 to 7 have high Mn concentrations. This corresponds well with what should be expected from the station locations. Samples 1 to 3 are close to the city with marine and boating activity, whereas stations 5 to 7 are from the outer part of the fjord.

Whenever a model is proposed for analysis of data it is important to check the validity of the model. The proposed model has not included any term for interaction between the factors (station and sample type). The residual plots in Figure 3 may be used to inspect the validity of this omittance. The residual

---

* We use a alpha error probability of 0.05 ($p = 0.95$) for each comparison. There is a 5% change of mistakenly rejecting the null hypothesis for each separate element when it is in fact true. The confidence of the multiple *t*-test, when all 11 elements are analyzed by t-test, is $1 - (p)^{11} = 0.431$.

Table 4 **Two-factor analysis of variance for Mn and Pb**

		Manganese			
Source of Variation	Sum of Squares	DF	Mean Squares	F-ratio	Sig.level
Between station	2371.4	6	395.2	2.61	0.080
Between type	29585.0	2	14792.5	97.62	0.000
Residual	1666.9	11	151.5		
Total (corrected)	33769.3	19			

		Lead			
Source of variation	Sum of Squares	DF	Mean Squares	F-ratio	Sig. level
Between station	17.95	6	2.99	1.53	0.257
Between type	103.33	2	51.67	26.36	0.000
Residual	21.56	11	1.95		
Total (corrected)	141.5				

Note: Sum of squares gives the squared summed difference for all samples around the respective mean station or type means. There are 7 stations (DF = 6) and 3 types of samples (DF = 2). The mean square is the sum of squares divided by DF. The residual sum of squares is the "leftover" squared summed differences not explained by the two factors station and sample type. Dividing the mean square for the factors by the residual mean square gives the F-ratio which is compared to a table of F-values to obtain significance level. The significance level provides percent probability of making a mistake when rejecting the null hypothesis.

plot for Mn (low left) shows a difference, an opposite trend, for mussels and seaweed as a function of station. For mussels, the observed values are higher than predicted for stations 1 to 4 and lower for stations 5 to 7; the opposite is the case for seaweed. This suggests that for this response there is an interaction, i.e., interdependence, between type of sample and station which the model does not account for. This interaction will be further examined in the next part.

D. SCATTERPLOTS, CORRELATION, AND REGRESSION

Using graphic presentation and inspection of data in conjunction with statistical analysis is a powerful way of determining connections between variables. Figure 4 shows a Draftsman plot where the BIOMAT samples are plotted as function of Mn, Al, and Ni concentrations together with station. The figure suggests that there are several trends in the data. When Mn and Al (middle) are used, the samples are grouped into three clusters. Inspection of the sample type shows that the three types of samples are separated, with seaweed being the sample group with high values showing a possible negative trend (correlation). When the samples are plotted as concentration of Mn vs. station, there are three linear groups with different slopes. This suggests that possibly Mn does not exhibit the same trend in the three types of samples with respect to station and illustrates the possible interaction in multifactor ANOVA. To investigate this further, correlation and regression analysis may be used.

The correlation coefficient is a measure of the degree of relationship between two variables which both have error variation. Correlation analysis is used when there is no clear reason to define one of the variables as dependent upon the other (see References 33, pp. 83 to 87; 34, Chapter 4; and 35, Chapter 10). Table 5 contains the linear correlation coefficients for all elements for both mussels and seaweed. Comparing the correlation analysis with the Draftsman plot, the negative trend between Mn and Al in seaweed is supported by a negative correlation (-0.737, $p = 0.058$ [n = 7]). There is no correlation between Mn and Al in mussels. Inspection of the correlation table also shows that whereas tin (Sn) shows negative correlations (although not significant at $\alpha = 0.05$) with all other elements in seaweed, the same element has only positive correlations (some significant at $\alpha = 0.05$) for mussels.

Correlation analysis such as illustrated here has many uses and may be helpful in pointing to connections between variables, but such tables are not very amenable to visual inspection and interpretation. We will return to the correlation analysis when discussing principal components analysis later.

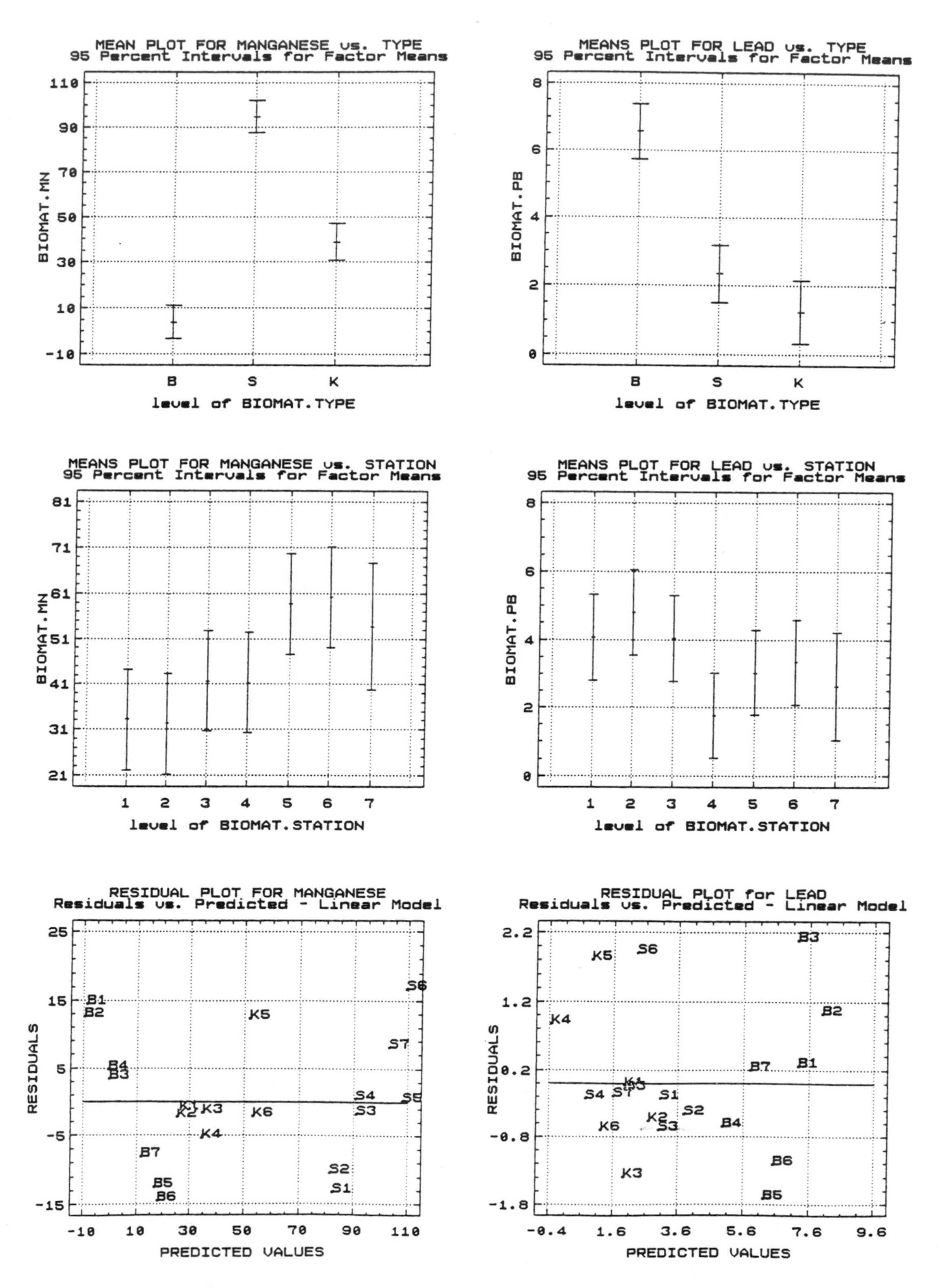

Figure 3 Graphs obtained from ANOVA of Mn (left) and Pb (right) in BIOMAT. The graphs (from top to bottom) show the average values for the three types of species together with confidence intervals, the average values for the seven stations together with confidence intervals, and the residual plot ($Y_{Obs.}$ - $Y_{Pred.}$) from the ANOVA analysis.

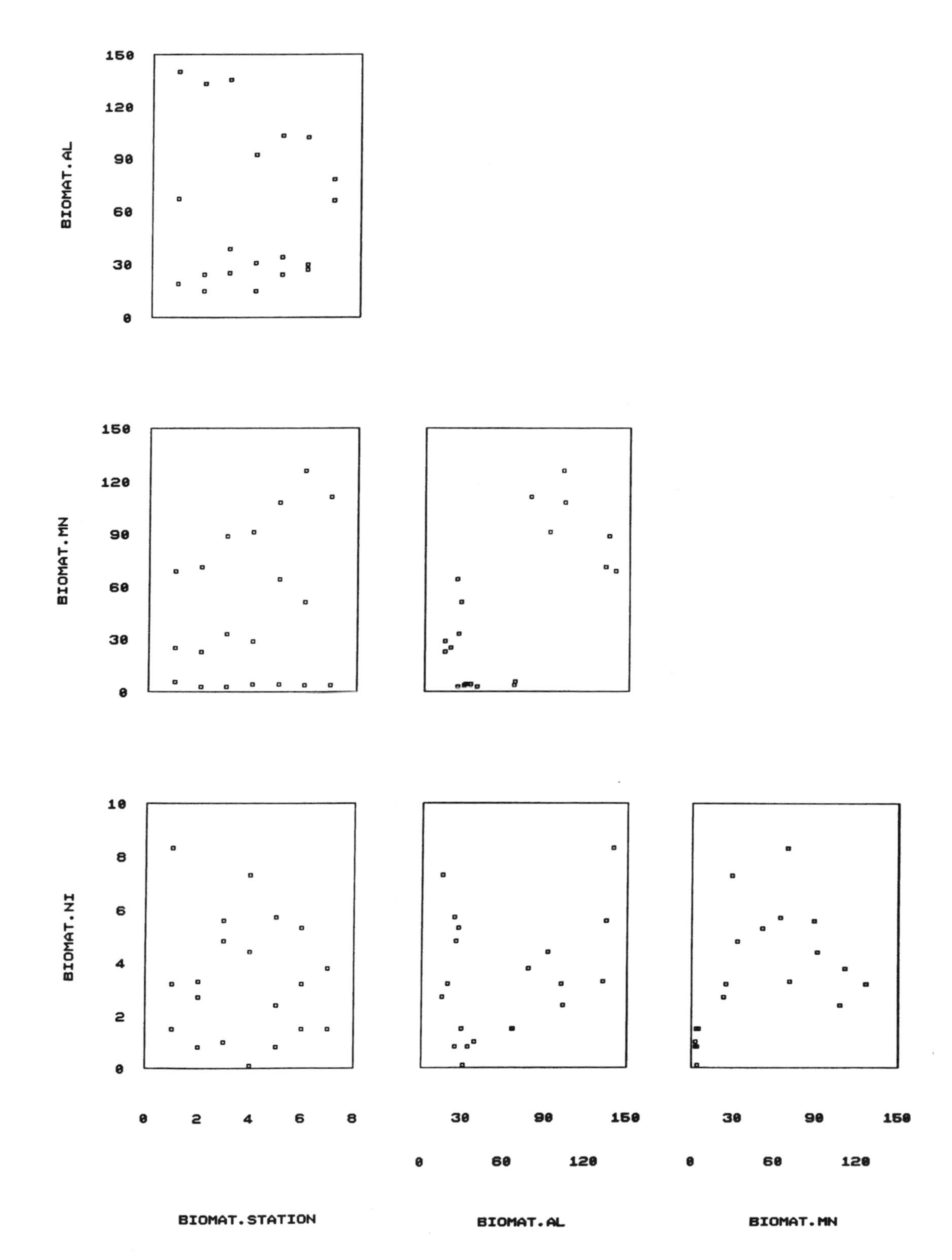

Figure 4 Draftsman plot BIOMAT samples for Al, Mn, Ni, and station.

Table 5 **Correlation coefficients between 11 elements for seaweed (lower left) and mussels (upper right)**

	Sample Correlations										
	Al	**Fe**	**Mn**	**Zn**	**Cu**	**Pb**	**Cd**	**Sn**	**Cr**	**Ni**	**Ti**
Al	1.000	.836	.583	.725	-.328	.070	.046	.846	.768	.611	.928[a]
	.000	.018	.168	.064	.471	.881	.921	.016	.043	.144	.002[b]
Fe	.805	1.000	.703	.760	.096	.066	.159	.700	.805	.586	.638
	.028	.000	.077	.047	.837	.888	.732	.079	.028	.166	.123
Mn	-.737	-.821	1.000	.745	.317	-.303	.030	.503	.844	.192	.326
	.058	.023	.000	.054	.488	.508	.948	.249	.016	.678	.474
Zn	.306	.361	.003	1.000	.052	.336	.528	.904	.975	.385	.513
	.503	.426	.994	.000	.910	.460	.222	.005	.000	.393	.238
Cu	.847	.878	-.703	.386	1.000	-.070	.244	-.256	.035	-.521	-.620
	.015	.009	.077	.391	.000	.880	.597	.578	.939	.229	.137
Pb	.485	.325	.060	.800	.368	1.000	.643	.496	.178	.234	.007
	.269	.476	.896	.030	.416	.000	.119	.257	.701	.612	.987
Cd	-.132	-.222	.580	.728	.007	.686	1.000	.444	.334	-.236	-.047
	.777	.631	.171	.063	.987	.088	.000	.317	.462	.609	.919
Sn	-.612	-.399	.480	-.417	-.557	-.171	-.027	1.000	.863	.529	.725
	.144	.374	.275	.351	.193	.712	.952	.000	.012	.221	.064
Cr	.503	.913	-.650	.337	.700	.205	-.199	-.122	1.000	.468	.547
	.249	.004	.113	.459	.079	.658	.668	.793	.000	.289	.203
Ni	.571	.780	-.614	.047	.869	-.006	-.207	-.260	.744	1.000	.607
	.180	.038	.142	.920	.010	.988	.655	.573	.054	.000	.147
Ti	.667	.771	-.695	.282	.438	.366	-.359	-.234	.687	.243	1.000
	.101	.042	.082	.539	.325	.418	.428	.613	.087	.598	.000

Note: The diagonal line with 1.000, .000 delineates the two halves. The upper right hand half describes the correlation coefficients for elements in mussel whereas the lower left half describes the correlation coefficients for seaweed. The correlation coefficient is first and below is the significance level with N = 7 for each sample set.

[a] Correlation coefficient; [b]Significance.

Many investigations are concerned with the causal relationship between an independent variable and a dependent (response) variable. Regression analysis is used for description of the functional relationship (model) between two variables, estimation of the coefficients describing the connection, and prediction of unknowns, e.g., calibration (see References 32, Chapter 13; 33, pp. 76 to 83; and 34, Chapter 3). In least-squares regression analysis it is assumed that the independent (x variable) is controlled and measured without error. Error is associated with the precision of the dependent (y variable).

Regression analysis is used to fit models to data and not vice versa. In regression analysis different types of models may be investigated. Frequently several mathematical/statistical models may fit the data equally well. Figure 5 and Table 6 give the results from regression analyses of Mn concentration vs. station for seaweed using three different models.

To compare models and choose one as best for describing the relationship between X and Y requires comparison of statistical parameters estimated for models. ANOVA of the different models shows that all three models tested equally well in describing the variation of Mn as function of station. All are statistically significant ($\alpha = 0.05$) although the linear model does have a slightly higher probability level (0.0024) for being "wrong". The upper part of each table shows the estimates for the coefficients (Intercept (a) and Slope (b)), the standard error for the estimate, the *t*-value (e.g., 58.57/7.247 = 8.08), and the probability of making a mistake if deciding the coefficient is significant. The coefficient estimates are all significant ($\alpha = 0.05$) — the models contain both intercept and slope components — and do not contribute to distinguish the models. Figure 5 shows the regression line (upper) and residuals plots (lower). The residuals plots are calculated by subtracting the predicted from the observed value for each model ($Y_{Obs.} - Y_{Pred.}$) and plotting these vs. station (predicted value or other extraneous [uncontrolled] variable may also be used). The residual plots are useful for visually determining if there are systematic trends in the data which have not been

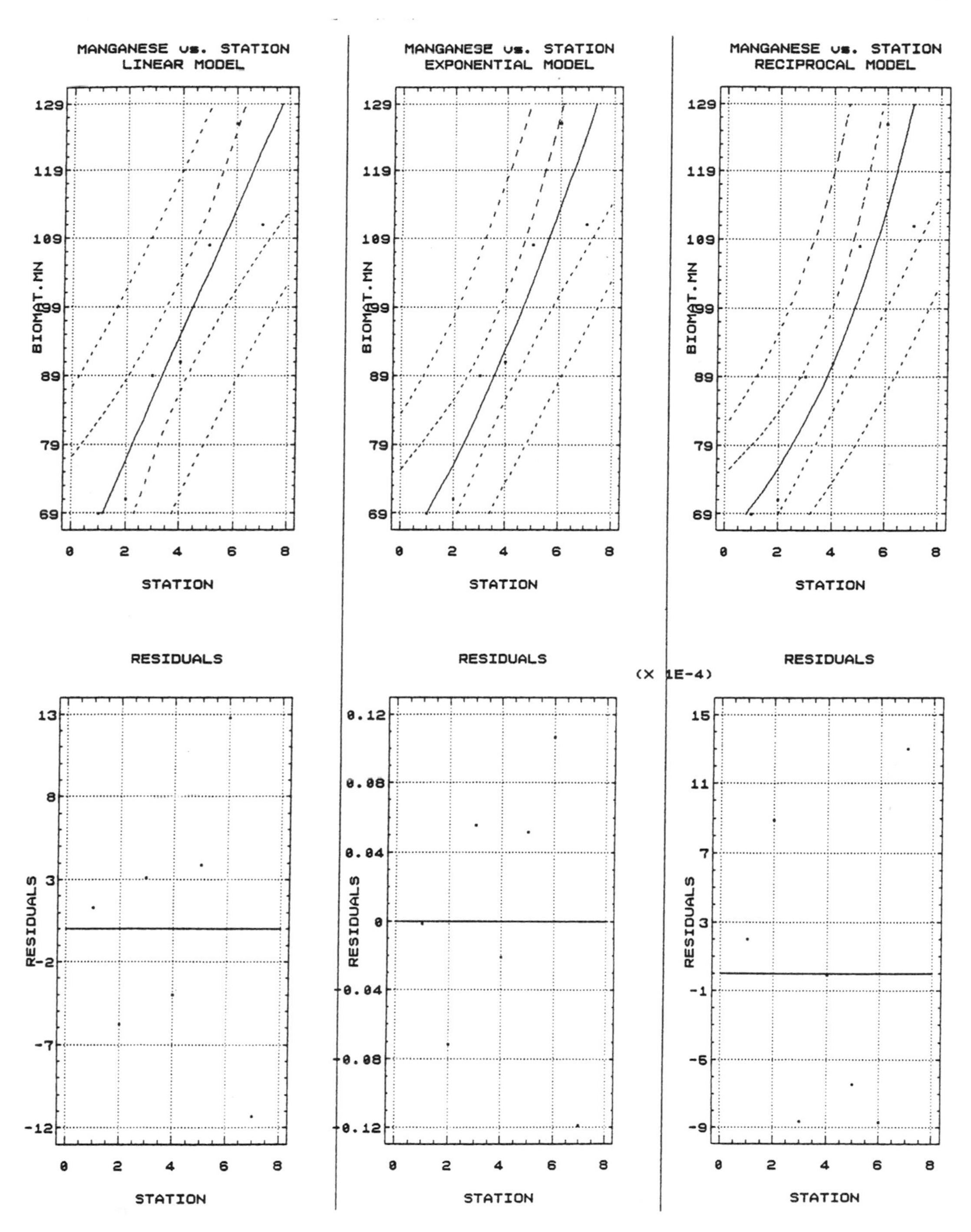

Figure 5 Regression analysis of Mn in seaweed from BIOMAT. The graphs show the station vs. concentration and regression lines together with confidence intervals (upper) and residuals plots for each of the linear model (left), the exponential model (middle), and the reciprocal model (right).

accounted for by the model (see ANOVA above). Again, with the low number of samples, there is no basis in the residual plots to select one model as being better than the others.

Disregarding the discussion of which model to choose, the Mn concentration is seen systematically to increase with increasing station number in seaweed (see also Figure 4). The situation encountered here is quite frequently observed in environmental analysis: relatively few samples and several equally

Table 6 **Tabulated data for regression analysis and ANOVA for regression analysis of Mn concentration vs. station number for seaweed**

Linear Model — **Y = a + bx**

Parameter	Coefficient Estimate	Standard Error	T Value	Probability Level
Intercept	58.57	7.247	8.08	0.0004
Slope	9.11	1.621	5.62	0.0024

Analysis of Variance

Source	Sum of Squares	DF	Mean Square	F-ratio	Probability Level
Model	2322.32	1	2322.32	31.58	0.0024
Residual	367.67	5	73.53		
Total (Corr.)	2690.00	6			

Correlation Coefficient = 0.929 R^2 = 86.33 (% squared correlation)
Stnd.Error of Estimate = 8.575

Exponential Model — **$Y = e^{(a + bx)}$**

Parameter	Coefficient Estimate	Standard Error	T Value	Probability Level
Intercept	4.1367	0.07252	57.04	0.0000
Slope	0.0988	0.01622	6.09	0.0017

Analysis of Variance

Source	Sum of Squares	DF	Mean Square	F-ratio	Probability Level
Model	0.27343	1	0.27343	37.13	0.0017
Residual	0.03681	5	0.00736		
Total (Corr.)	0.31024	6			

Correlation Coefficient = 0.938 R^2 = 88.13 (% squared correlation)
Stnd.Error of Estimate = 0.085

Reciprocal Model — **1/Y = a + bx**

Parameter	Coefficient Estimate	Standard Error	T Value	Probability Level
Intercept	0.015390	7.960E-4	19.33	0.0000
Slope	-1.097E-3	1.779E-4	-6.16	0.0016

Analysis of Variance

Source	Sum of Squares	DF	Mean Square	F-ratio	Probability Level
Model	0.000034	1	0.000034	38.00	0.001
Residual	0.000004	5	0.0000009		
Total (Corr.)	0.000038	6			

Correlation Coefficient = -0.940 R^2 = 88.37 (% squared correlation)
Stnd.Error of Estimate = 9.42E-4

Note: The upper part of each shows the coefficient estimates and tests, the lower part shows the model parameters and statistics.

well-fitting models. The selection of a model, then, depends on choosing the equation that best describes the relationship with the "simplest" expression (number of terms, interpretability, etc.). One remedy to this, if it is known that regression analysis may be used to fit a functional relationship, is to collect and analyze several parallel samples at several stations and use the lack of a fit statistic to guide the choice of model (see References 32 Chapter 13.2.7 and Reference 36, pp. 33 to 40).

The BIOMAT data set consists of 11 element variables measured on 20 samples which may be grouped according to two classification variables (species and station). Analyzing this multitude of possibilities leads to a very large number of potential models and hence an equally large number of possible problems, statistical comparisons, and decisions to be made. Remembering the discussion on multiple *t*-testing, the large number of possible treatments not only represents a practical barrier, but also includes statistical fallacies. Reducing this multitude in a systematic and efficient manner is one of the reasons for using multivariate analysis.

VII. MULTIVARIATE ANALYSES

Multivariate data-analyses (MVDA) — exploratory and statistical — covers a wide range of methods which have in common that they treat more than two variables simultaneously. The purpose of using MVDA is to identify structures, e.g., groups/clusters, or patterns, e.g., correlations, which may be of interest for classification, investigating, and interpreting relationships or prediction.[37-40] To accomplish this, the multivariate data are reduced to fewer dimensions for human perception, and statistical parameters are estimated to decide on which dimensions or groups are relevant. Many multivariate methods may be used as simple data reduction (mathematical) methods without statistical assumptions. Obviously such use of the methods does not warrant statistical interpretations and conclusions. In this presentation we will illustrate only a few of the methods used, going from the algorithmic cluster analyses and through to principal component regression (PCR).[41]

The data matrix is an important element in MVDA. Most multivariate methods are expressed more efficiently using matrix mathematics.[39] Because MVDA can be used for analysis of both samples and measured features or attributes, a nomenclature is introduced where the objects of interest (samples) are called objects and the features analyzed on these objects are called variables.

A. CLUSTER ANALYSIS

Cluster analysis is a collective name for literally hundreds of different algorithmic and mathematical methods to group data. The purpose of cluster analysis is to identify similar samples or variables and group these into clusters so that interpretation and predictive divisions can be made. Clustering algorithms are often iterative. A basic distinction is often made between hierarchical and nonhierarchical clustering methods and between agglomerative and divisive algorithms. Hierarchical clustering methods require that an object which is assigned to a group cannot be reassigned to another group, whereas nonhierarchical allows multiple assignment. Agglomerative methods start "at the bottom" with N groups (= number of samples) and build clusters by reducing the number of groups. Divisive methods start "at the top" with one big group and divide the samples into an increasing number of sample clusters.

There are (very) many metrics which may be used to cluster samples or variables. Metrics are often separated into two types: similarity measures and distance measures. This separation is often made based on types of variables, e.g., continuous numerical or binary. The best known distance measure for objects is the Euclidean; others are the Manhattan City Block (Taxi metric) and the Mahalanobis distance (see References 37, Chapter 9 and References 42 and 43). Typically, similarity measures are used for binary variables. The use of cluster analysis for exploratory data analysis to obtain guidance for interpretation of groups of samples or variables may be very beneficiary. The use of clustering methods for decisive/predictive grouping relies on understanding of the interrelationships between data transformation, scaling, metric, and procedure or algorithm used for classification.

The City Block and the Euclidean measure illustrated here compute the distance between all sample pairs using all variables:

$$\text{City Block}: CB_{ij} = \sum_{k=1}^{p} w_k \, | x_{ik} - x_{jk} |$$

$$\text{Euclidean}: E_{ij} = [\sum_{k=1}^{p} w_k (x_{ik} - x_{jk})^2]^{1/2}$$

where CB_{ij} and E_{ij} is the distance between sample *i* and sample *j* and the distance is taken over all variables k = 1 to p. The hierarchical cluster analysis using a Euclidean metric does not accommodate correlation between variables and is influenced by differences in variable range. Both of these give rise to "chaining", i.e., nondecisive connection between samples with no large separations. Scaling to unit variance removes the range problem. Scaling to unit variance is accomplished by dividing each variable by its standard deviation ($x_{ij}/\sqrt{s_{ii}}$). The effect of this is to give equal weight to each variable by representing the values relative to the standard deviation. The scale becomes the standard deviation with all numbers for all variables approximately in the range $0 \rightarrow |3|$. To remove the effect of correlation would require that a different metric or distance measure was used, e.g., Mahalanobis. The Mahalanobis distance includes the covariance matrix (see Table 2) in the calculation of distance measure (see Reference 39, pp. 15, 17, and 31). By doing this, the Mahalanobis distance eliminates correlation between variables and standardizes the variance of each. Use of the Mahalanobis distance in cluster analysis approaches the use of principal components for orthogonalization of correlated variables (see below).

Figure 6 shows the result (dendrogram) of a hierarchical cluster analysis of the BIOMAT samples. The variables have been scaled to unit variance. The dendrogram has the samples along one axes and the measure of similarity — the metric — along the other axis. The cluster analysis of BIOMAT shows two clear groups of samples separated at a similarity of approximately 0.2 with two outliers in each (samples 1 and 2, respectively). The two outliers are from station 1. This suggests that the analyst should be aware and examine these two samples closer. Within the larger cluster, consisting of seaweed and krusflikk samples, there is a separation between the two types of samples at a similarity of approximately 0.6. The cluster analysis shows that based on all measured variables it is possible to obtain a clustering according to sample type, but does not give any information on the relationship between variables or the contribution of variables in the clustering. Latent vector analysis achieves this combination.

B. LATENT VECTOR ANALYSIS

Latent vector analysis (LVA) is a collective name for mathematical and statistical methods based on decomposing the covariance or correlation matrix by using mathematical or algorithmic methods to obtain eigenvectors. Examples are principal component analysis (PCA) or factor analysis (FA), linear discriminant analysis (LDA), canonical correlation analysis (CCA), partial least-squares regression (PLS), and polynomial principal component regression (PPCR).[37-41] All these methods have as the main objective and function to develop a more parsimonious representation or summary treatment of multivariate data by using linear combinations of the original variables. Conceptually and mathematically the methods are divided in two groups: the first group analyzes one single data-matrix **[X]**, e.g., PCA and FA, and the second group analyzes the connection between two (or more) data matrices **[X]** and **[Y]**, e.g., CCA and PLS. Depending on the method and the calculation, the new latent vectors will have different properties, e.g., orthogonality (uncorrelated) or predictive parcimonicity. A short description of principal component analysis (PCA) is included here since this is used in the subsequent sections.

1. Principal Component Analysis in Environmetrics

The intention of PCA is to calculate a new set of orthogonal vectors (axes) with maximum variance as linear combinations of the original variables. Because of the variance maximizing criteria and the fact that chemical variables in the environment are most often strongly correlated, the result from PCA is a small number of independent (orthogonal) axes which describe the main, and most important, variance in the data set studied. Algebraically, the principal components (PC's) may be expressed as:

$$PC_1 = b_1 * x_1 + b_2 * x_2 + ... + b_i * x_i$$

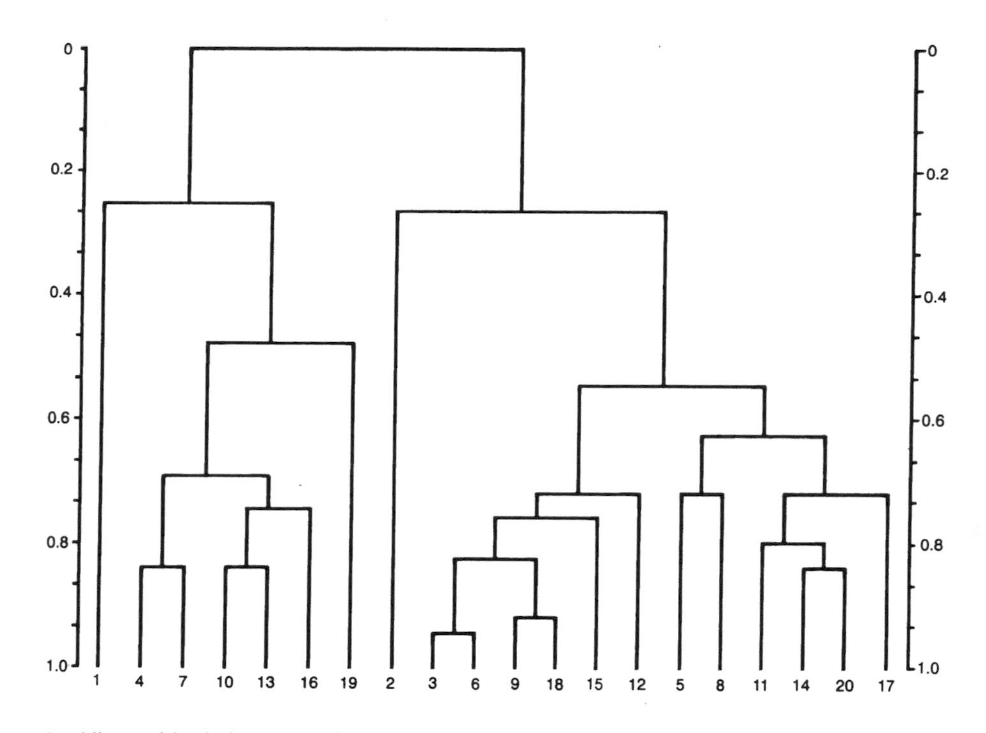

Figure 6 Hierarchical cluster analysis of BIOMAT data using variables scaled to unit variance. The numbers represent sample numbers (see Appendix, Reference Table A1).

This equation expresses that PC1 (principal component number 1) is a linear combination of the original variables (x_1 to x_i) defined by using the coefficients b_1 to b_i (eigenvectors or loadings). Each subsequent PC is calculated subject to the constraint that variance $\mathbf{PC}_i$ is maximized, and covariance $\mathbf{PC}_i$, $\mathbf{PC}_j = 0$ for $j < i$.

Because the *de facto* rotation of the original coordinate system is not unique with respect to variance, a normalization of the coefficients (b's) is included as a third constraint. Figure 7 gives a graphical presentation of the calculation of principal components. In Figure 7a, a data table with three variables (EL1 to EL3) measured on many samples is plotted in three dimensions. Sample A has the values 3.2, 1.5, and 3.1. The first PC is calculated by maximizing the variance (spread of samples) it explains. For each PC, a sample obtains a object score (vectorial projection onto PC) and each variable obtains a loading (coefficient) (Figure 7b). Subsequent PCs are calculated according to the constraints. Figure 7c shows that the data table with three variables has been transformed into two different graphical images: the loading plot and the score plot. The loading plot is used to interpret how variables are correlated with each other, and the score plot is used to interpret clusters and groups of similar samples. The combined interpretation of both the variable loading and the object score plots are used to identify factors or systematic patterns assigned to sources or processes influencing the samples.

It is customary to mean center and often to variance scale data before PCA. Mean centering eliminates a principal component which basically confounds a shift of axes coordinates with variance/covariance. Variance scaling achieves a shift from covariance which is scale dependent to correlation which is not scale dependent. Figure 8 and Table 7 show the results from SIMCA PCA of the variance scaled BIOMAT data. The two PCs account for 63.9% of the total variance in the data set. This shows that although the second PC is not significant according to crossvalidation,[29] both PCs explain a high percentage of the variance. The three sample types are separated. The first principal component (PC1) separates mussel samples from seaweed and krusflikk (compare cluster analysis, Figure 7). Interpretation of the variable loading plot together with the object score plot shows that these samples are separated because mussel samples on the average have higher concentrations of Pb, Cd, Sn, Cr, and Ti, whereas seaweed and krusflikk have higher concentrations of Mn and Ni. The second PC (PC2), in addition to separating the krusflikk and seaweed samples, exposes an opposite trend between stations and samples for krusflikk and seaweed/mussels. The general trend is that low number stations (inner fjord) for seaweed and mussels have a positive correlation between Al, Fe, Zn, and Cu. Further analysis of the data to analyze and interpret within-group correlation patterns is made by analyzing each group separately and making

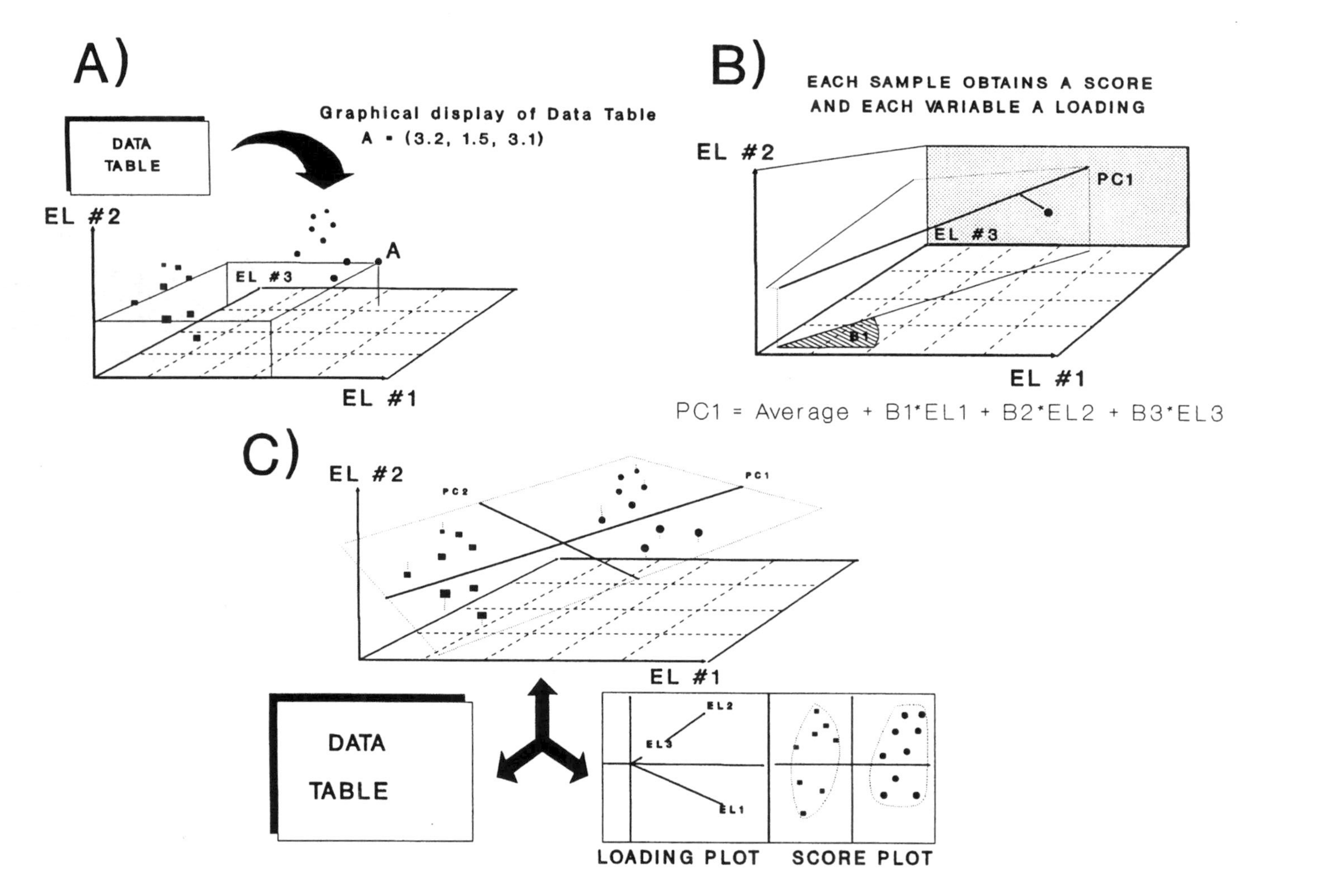

Figure 7 Latent vector analysis: (A) The data table is translated into a three-dimensional axes system (EL1 to EL3). Sample A has the coordinates 3.2, 1.5, and 3.1. (B) The first principal component is calculated so that maximum variance (spread) of the data is covered. Each sample obtains a sample score and each variable a variable loading (coefficient). (C) A two-dimensional "window" is calculated in the three-dimensional space using the first two principal components. This two-dimensional window may be used to graphically plot the variables and samples in variable loading plots and sample score plots. Variable correlation patterns and sample clustering patterns together with relationships between variables and samples may be investigated.

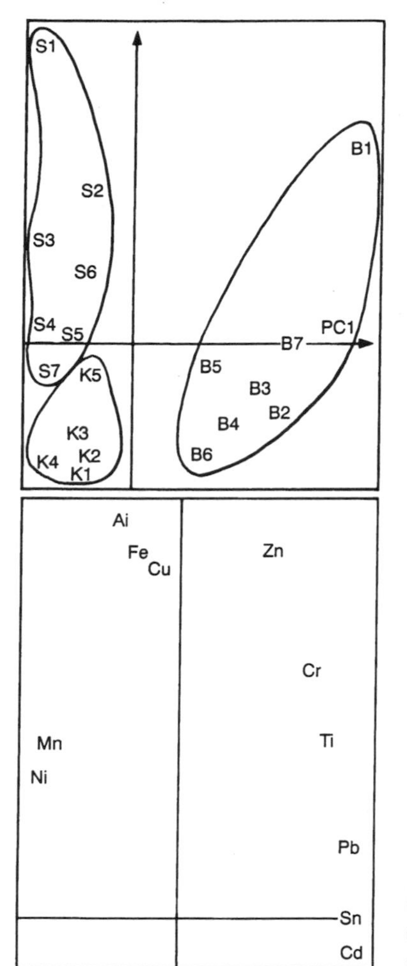

Figure 8 Principal component analysis of BIOMAT data. The object score plot (upper) and the variable loading plot (lower) of principal component 1 vs. principal component 2.

Table 7 **Results from the SIMCA PCA analysis of the BIOMAT data**

	Principal Component 1	Principal Component 2
Significance	(+)	(-)
Percent variance		
Explained	38.1%	25.8%
Sum variance	38.1%	63.9%

Note: The table shows the amount of variance described by each principal component (e.g., 38.1% for 1st PC), whether the principal component is significant according to crossvalidation (+ or -) and the accumulated variance explained (e.g., 63.9% after 2 principal components).

comparisons. To ascertain classification, the SIMCA approach[29] may be used to calculate class models for each separate group by principal components. The class models and distance measures are used to test significance of separation.[29]

Although the classification by cluster analysis or SIMCA modelling may be used to predict class membership other methods and procedures are used for multivariate regression analysis. The next part illustrates the combined use of PCA to reduce multivariate chemical data tables to a few factors or components with subsequent polynomial regression to connect these factors with biological/toxicological or ecological response variables.

2. Polynomial Principal Component Regression

In environmental analysis where the scope of the investigation is to determine the connection between multivariate chemical or, e.g., physical conditions and biological response(s), there is a need for methods which incorporate (1) the data reduction and interpretive power of latent vector methods and (2) the predictive power of regression methods. The procedures should also accommodate the need for obtaining quantitative information on higher order "nonlinear" and interaction effects. In the next two sections application of PCA in combination with polynomial regression is used to interpret and predict relationships between multivariate element data measured on different types of samples and biological response data. Polynomial regression is a linear regression method where square or higher order and interaction terms between the explanatory (independent) variables are included as regressors.

The PCA is used initially on the chemical data **[X]** and if necessary also on multiresponse **[Y]** data to obtain a reduced set of orthogonal axes[8,29,30] describing the samples. Interpretation of the latent vectors (PCs, factors) as relating to source contribution or processes influencing the samples is then made. The matrix of PC scores obtained is expanded by including square terms or higher order terms (e.g., PC_i^2) and interaction terms (e.g., $PC_i{*}PC_j$). The square terms account for "nonlinearity" or curvature and the interaction terms synergistic or antagonistic effects, i.e., interdependency. These new variables are used as independent variables in a polynomial regression. The polynomial regression model and the coefficients are then used to (1) predict the biological measurement, (2) to obtain response surfaces describing the biological response as a function of the PCs, and (3) to obtain geographical maps describing the influence the PCs have on response variables. The two examples have been described and the statistical aspects are extensively discussed in References 41 and 44 to 46.

3. Composition Activity Relationships — CARE

This example relates mutagenicity measured on air samples to the element composition of the same samples. The samples are from a small Norwegian community. The PCA of variance scaled data from columns 2 to 11 in Reference Table A2 in the Appendix showed that the main input of elements to the air samples come from wood burning, domestic oil heating, automobile emissions, and a source tentatively assigned as a copper/atmospheric transformation (smog?) source.[44-46]

The PPCR analysis showed that there was a significant polynomial regression model between PC1 and PC3, including square terms and interactions, and mutagenicity measured using TA98 with metabolic activation (+S9). The model obtained is given in the caption for Figure 9. The interested reader is referred to References 44 to 47 for detailed discussion of the data.

Figure 9 shows the combined plot used for interpretation of the mathematical/statistical analysis in CARE. The x- and y-axes are PCs. The boxes give sample number and measured mutagenicity. The response contours describe mutagenicity predicted from the PPCR model; these agree well with the measured mutagenicity. The PC1 is interpreted as a "general" pollution component, i.e., increasing or decreasing overall pollution level (size factor). This is based on the positive correlation between all measured variables and the positive correlation between these variables and the PC. The third PC (PC3) is interpreted as a contrast between samples with negative scores influenced by domestic oil heating (high loading for Ni and Cd) and samples with positive scores possibly influenced by a "smog" type transformation or chemical process (high loading for Cu and intermediate for $SO_4^=$).[44-46]

Interpretation of the contour lines shows that with increasing general pollution the mutagenicity increases. When this increase is combined with an influence from domestic oil heating (left; towards right and down), there is a synergistic effect giving rise to increased mutagenicity. When the general pollution increases and the "smog" process is dominant (left; towards right and up), the mutagenicity drops to zero. A similar situation occurs when the general pollution is low and there is mainly an influence from domestic oil heating. These antagonistic effects are interpreted as representing situations where there is a toxic effect which makes it difficult/impossible to measure mutagenicity.

4. Multivariate Ecotoxicological Mapping — MEM

This example investigates the relationships between sediment chemical composition and fauna diversity. The chemical elements used in this case comprise a representative selection of the most important chemical pollutants to the sediments in this fjord. The discussion is therefore restricted to discussing the effects of pollution in appreciation of the fact that to obtain a broader interpretation would require the use of other variables also describing the physical and marine conditions influencing the system.

Table 8 describes the results from the PCA of sediment chemical composition (see Appendix, Reference Table A3). The results again show that the sample score on PC1 reflects differences in total

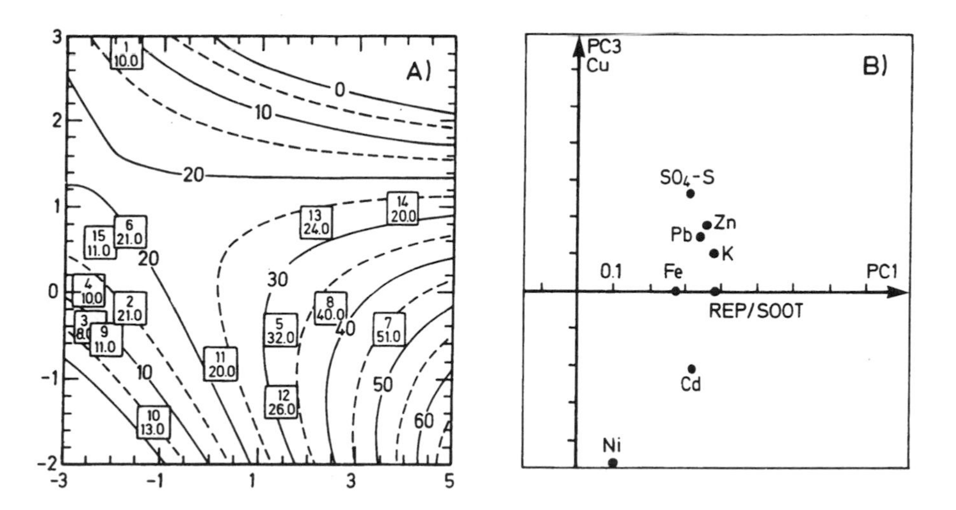

Figure 9 CARE: (A) The combined principal component sample score plot and response contour line plot. The boxes give sample numbers and measured mutagenicity. (B) The variable loading plot shows how the original measured chemistry variables correlate with the principal components. Interpretation of the axes is based on the variable loading values. The model used to obtain the contour lines is (standard error of coefficients in parenthesis):

$$\underset{}{TA98 + S9} = \underset{(0.93)}{24.45} + \underset{(0.36)}{4.67} * PC1 - \underset{(0.47)}{3.52} * PC1 * PC3 - \underset{(0.47)}{3.05} * PC3^2$$

concentration. Samples near the inner basin (samples 12 to 18) have high scores. The sample scores on PC2 reflect a difference between samples with high values and correlation between Zn, total organic carbon (TOC) and total nitrogen (TOTN), and samples with high values and positive correlation for Cd and to some extent Co and Fe. Samples from immediately outside the two factories (Falconbridge-F1 and Fiskaa-F2) have high scores on PC1 and opposite signed scores on PC2. This suggests that this component describes a difference between sediment samples influenced by the two different factories. The third PC (PC3) is interpreted as a contrast between samples with strong influence from the river/estuarine system (sample 1 and 5 have negative scores on PC3) and samples more representative of seawater.

With the empirical approach to model building used here, all models will be approximations. The final model is selected based on statistical significance of individual terms (coefficients), overall fit to describe diversity, and because it is relatively simple. Since the variables do not include important ecological measurements, e.g., salinity or depth, it is not considered appropriate to force explained variance.

Table 9 shows that the model obtained is adequate. The model is significant according to ANOVA and each term is significant. The model predicts a negative diversity for sample 14. This sample is extreme in chemical composition (see Appendix, Reference Table A3). Because of this negative value the sample has been left out in the mapping of the separate effects.

The regression model in Table 9 may be separated into the constant (average diversity) and the two regression terms. The first regression term is -3.24*PC1. The negative coefficient shows that diversity is reduced with increasing sample scores on PC1. This term therefore shows the effect on diversity from increased pollution. By calculating the sample values for each sample using the coefficient value and PC score (both with sign), the column marked "Pollution Score" in Table 9 is obtained.

Similarly for the second regression term (+2.35*PC2*PC3), each sample obtains what is called an "Interaction Score" in Table 9. A positive value for the interaction score may arise when a sample has both positive or both negative scores on PC2 and PC3 simultaneously. It is likely that the PC2*PC3 term represents in an indirect manner ecological effects which are not associated specifically or alone with pollution. Interpretation of the interaction term using sample scores and variable loadings as guides would suggest that diversity is increased when samples are either (1) both influenced by high values of TOC/TOTN and Zn and come from seawater stations or (2) when samples have negative scores on PC2 (e.g., sample 4, correlation between Cd, Fe, and Co) and are influenced by the river (freshwater) system. The negative influence from the interaction term on diversity for samples 12 and 14 is different. For sample 12 (Falconbridge), the negative value comes from a positive score on PC2 (high Zn, TOC, and TOTN) and a negative score on PC3 (high TOTN, TOC, and Cd). For sample 14 (Fiskaa), the negative value comes from a negative score on PC2 (high Cd, Fe, and Co) and a slight positive score on PC3 (high Zn

Table 8 **Results from the SIMCA PCA analysis in the MEM analysis**

	Variance Explained by Three Principal Components[a]		
	Principal Component 1	Principal Component 2	Principal Component 3
Significance	(+)	(-)	(-)
Percent variance			
Explained	57.3%	14.3%	11.2%
Sum variance	57.3%	71.6%	82.8%

Objects	PC1	PC2	PC3	Variables	PC1	PC2	PC3
	Sample Scores and Variable Loadings on Three Principal Components[b]						
1	-0.873	0.854	-1.606	Ni	0.376	-0.113	0.111
2	-0.391	1.568	-0.713	Co	0.369	-0.225	0.046
3	-3.905	-0.630	1.264	Pb	0.355	0.179	0.319
4	-3.928	-2.743	-0.093	Cd	0.255	-0.437	-0.157
5	-1.702	0.969	-0.916	Zn	0.063	0.529	0.633
6	-1.983	-0.891	-0.646	Cr	0.342	0.088	-0.039
7	-1.167	-0.033	-0.186	Fe	0.366	-0.229	0.133
8	-1.486	0.603	-0.194	Cu	0.377	-0.073	0.087
9	-1.363	-0.539	-0.667	TOTN	0.257	0.409	-0.469
10	-1.947	1.748	3.068	TOC	0.259	0.447	-0.459
11	-0.363	-0.245	-0.472				
12	2.215	1.765	-0.562				
13	2.580	1.110	0.331				
14	4.675	-2.135	0.404				
15	3.263	-0.667	0.245				
16	2.030	0.364	-0.078				
17	0.931	0.065	0.200				
18	3.414	-1.162	0.618				

[a] Percent variance explained and significance according to crossvalidation ratio; [b] Sample scores on the three principal components used for polynomial principal component regression and variable loading (eigenvectors) for the three principal components.

and possibly Pb). Figure 10 shows maps of the Kristiansandfjord. The lines are isolines for diversity based on the pollution and interaction score values from Table 9. The lines illustrate the influence on the diversity from the general pollution (concentration) gradient (PC1) obtained from the PPCR model.

VIII. SUMMARY

Scientific results in environmental analysis rely on combining knowledge about the system studied with systematic scientific strategies for planning and analyzing data. The expository presentation has illustrated the use of graphical and mathematical/statistical methods for data analysis in environmetrics. An applied approach to exploratory data analysis has been illustrated. An attempt has been made to show that although it is important to use statistical assumptions and sound scientific strategies, it is the environmental relevance which provides scientific validity in environmetrics. The examples used hopefully illustrate that application of statistical methods, graphic analysis, and multivariate methods of data analysis in environmetrics provides the scientist with possibilities that warrant closer examination.

A large number of statistical and environmental issues have not been dealt with in the presentation. Specifically, the considerations which must be placed on designing sampling schemes to assure correct results and optimum use of resources have not been discussed. This is a field of increased interest, as cost is rising and pressure for performing more effective investigations is increasing. Robust statistical methods (nonparametric methods) are also important in environmental analysis. Only seldom are enough samples collected or do variables behave as nicely as is expected in parametric tests.

Table 9 **The MEM PPCR model information. Polynomial principal component regression model to predict diversity (ES(100)) from sediment element and TOC/TOTN concentrations**

Model for Predicting Diversity[a]

Diversity (ES(100)) = 12.95 - 3.24 *PC1 + 2.35 *PC2 *PC3

(Standard Error) (1.11) (0.47) (0.71)

ANOVA for Model and Terms in Model (in Order Fitted)[b]

	Sums of Squares	DF	Mean Square	F-Ratio	P-value
Model	1025.2	2	512.6	33.8	0.000
Error	151.6	10	15.2		
Total	1176.8	12			
PC1	861.5	1	861.5	56.8	0.000
PC2 *PC3	163.7	1	163.7	10.8	0.008

Diversity (ES(100)) and the "Pollution" and "Interaction" Score[c]

Objects	Observed	Predicted	Residual	Pollution Score	Interaction Score
1	—	12.6	—	2.83	-3.22
2	—	11.6	—	1.27	-2.62
3	23.5	23.7	-0.2	12.65	-1.87
4	26.0	26.3	-0.3	12.72	0.59
5	8.8	16.4	-7.6	5.51	-2.08
6	23.3	20.7	2.6	6.42	1.35
7	20.1	16.7	3.4	3.78	0.01
8	24.7	17.5	7.2	4.81	-0.27
9	—	18.2	—	4.42	0.84
10	30.7	31.8	-1.1	6.31	12.60
11	12.2	14.4	-2.2	1.18	0.27
12	3.6	3.4	0.2	-7.18	-2.32
13	4.0	5.5	-1.5	-8.36	0.86
14	—	-4.2	—	[d]	[d]
15	5.0	2.0	3.0	-10.57	-0.38
16	5.2	6.3	-1.1	-6.58	-0.06
17	7.6	9.9	-2.3	-3.02	0.03
18	—	0.2	—	-11.06	-1.69

[a] The model together with standard error of the terms found to be significant; [b] The significance of the model and the terms according to analysis of variance (ANOVA); [c] The predicted vs* observed values of diversity together with the "pollution" and "interaction" score for each sample; [d] Left out of map projection because of negative predicted diversity.

The future will most likely require that environmental analysts move from a static descriptive stage into a dynamic predictive stage. This will require a better understanding of the effect of design strategies and data analysis in the time and space dimensions.

Understanding the effect of and how to analyze time- and space-dependent data should extend the scope of environmental analysis substantially. The MEM approach represents a step in the direction of being able to ascertain effects of influences in the geographical dimension.[48] Coupling this approach with geographic information systems (GIS) should provide a useful extension to give information on the spatial distribution of effects of multiple pollution sources in the space dimension.

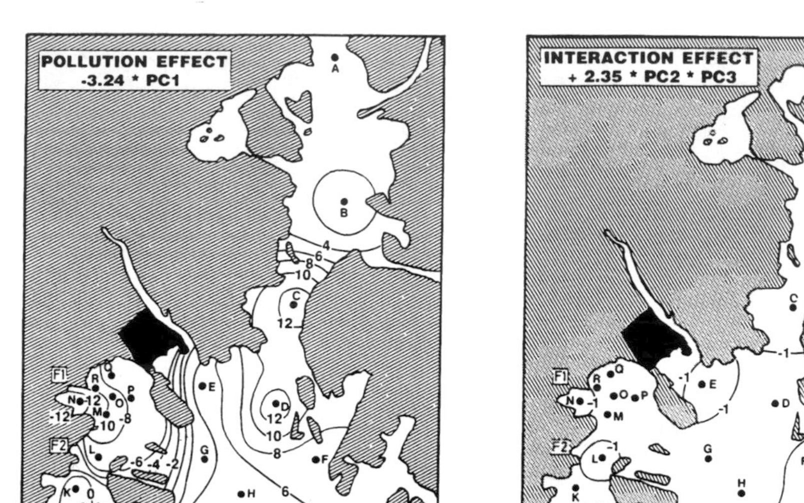

Figure 10 MEM. Map of Kristiansandsfjord: (left) pollution (-3.24*PC1) isolines and (right) interaction (+2.35*PC2*PC3) isolines. The contour isolines have been estimated by a quadratic inverse distance measure. The contour lines represent isoprincipal factor effect boundaries on diversity from each of the terms in the PPCR model (see text). Samples are marked with sample numbers. The observed values and the predicted sample values may be found in Table 9.

Being able to determine relationships between multivariate explanatory variables, e.g., chemical concentrations or physical conditions, and effect variables such as mutagenicity, diversity, or chemotoxicity represent an approach to modeling true environmental (mixture) situations. CARE provides a means to analyzing these situations using a predictive model.

Environmetrics and the use of statistics in environmental analysis, although increasing in extent, still must be considered in its infancy. The concerted effect of cooperation between environmental scientists, applied statisticians, and computer scientists will most likely provide decision makers with more meaningful and more easily available information in the future.

REFERENCES

1. **Salbu, B., Pappas, A. C., and Steinnes, E.,** Elemental composition of Norwegian Rivers, *Nord. Hydrol.,* 10(2-3), 115, 1979.
2. **Jørstad, K., Salbu, B., and Roaldset, E.,** Vertical distribution of trace elements in fresh water, saline water and sediments from lake rørholtfjorden, Norway, *Chem. Geol.,* 36, 325, 1982.
3. **Jensen, A. and Cheng, Z.,** Statistical analysis of trend monitoring data of heavy metals in flounder (Platichthys flesus), *Mar. Poll. Bull.,* 18(5), 230, 1987.
4. **Reckhow, K. H. and Chapra, S. C.,** *Engineering Approaches for Lake Management,* Vol. 1, *Data Analysis and Empirical Modelling,* Butterworths, Boston, 1983, 340.
5. National Research Council, *Complex Mixtures, Methods for In-Vivo Toxicity Testing,* National Academy Press, Washington, D.C., 1988, 227.
6. **Vouk, V. B., Butler, G. C., Upton, A. C., Parke, V., and Asher, S. C., Eds.,** Methods for assessing the effects of mixtures of chemicals, *Scope 30. SGOMSEC 3,* John Wiley & Sons, Chichester, 894, 1987.
7. **Sheenan, P. J., Miller, D. R., Butler, G. C., and Bourdeau, P.,** Effects of Pollutants at the ecosystem level. *SCOPE 22. Ed's.,* John Wiley & Sons, Chichester, 443.
8. **Thielemanns, A. and Luc Massart, D.,** The use of principal components as a display method in the interpretation of analytical chemical, biochemical, environmental and epidemiological data, *Chimia,* 39(7-8), 236, 1984.
9. **Ogugbuaja, V. O.,** Statistical analysis of heavy metal concentrations from lake sediments, *J. Environ. Sci. Health,* A20(5), 529, 1985.

10. **Nishida, H. and Suzuki, S.,** A statistical view of heavy metal pollution index of river sediment, *Bull. Environ. Contam. Toxicol.,* 32(5), 503, 1984.
11. **Davies, B. E. and Wixson, B. G.,** Use of factor analysis to differentiate pollutants from other trace metals in surface soils of the mineralized area of Madison county, Missouri, USA, *Water Air Soil Poll.,* 33, 339, 1987.
12. **VandeGinste. B. G. M., Salemink, P-J. M., and Duinker, J. C.,** Auto- and crosscorrelograms of particulate trace metals in the Rhine estuary, southern bight and Dutch Wadden Sea, *Neth. J. Sea Res.,* 10(1), 59, 1976.
13. **Hopke, P. K., Gladney, E. S., Gordon, G. E., Zolle, W. H., and Jones, A. G.,** The use of multivariate analysis to identify sources of selected elements in the Boston urban aerosol, *Atmos. Environ.,* 10, 1015, 1976.
14. **Rosocoe, B. A., Hopke, P. K., Dattner, S. L., and Jenks, J. M.,** The use of principal component factor analysis to interpret particulate compositional data sets, *J. Air Poll. Control Assoc.,* 32(6), 637, 1982.
15. **Sanchez Gomez, M. L. and Ramos Martin, M. C.** Application of cluster analysis to identify sources of airborne particulates, *Atmos. Environ.,* 21(7), 1521, 1987.
16. **Schaug, J., Rambaek, J. P., Steinnes, E., and Henry, R. C.,** Multivariate analysis of trace element data from moss samples used to monitor atmospheric deposition, *Atmos. Environ.,* 24A(10), 2625, 1990.
17. **Ostle, B.,** *Statistics in Research,* The Iowa State University Press, Ames, 1963, 585.
18. **Box, G. E. P., Hunter, W. G., and Hunter, J. S.,** *Statistics for Experimenters,* John Wiley & Sons, New York, 1978, 653.
19. **Miller, J. C. and Miller, J. N.,** *Statistics for Analytical Chemistry,* Ellis Horwood Limited, New York, 1988, 227.
20. **Box, G. E. P. and Draper, N. R.,** *Empirical Model Building and Response Surfaces,* John Wiley & Sons, New York, 1987, 669.
21. **Anderson, V. L. and MacLean, R. A.,** *Design of Experiments — A Realistic Approach,* Marcel Dekker, New York, 1974, 418.
22. **McPherson, G.,** *Statistics in Scientific Investigation,* Springer-Verlag, New York, 1990, 667.
23. **Kish, L.,** *Statistical Design for Research,* John Wiley & Sons, New York, 1987, 267.
24. **Nelson, J. D. and Ward, R. C.,** Statistical considerations and sampling techniques for ground water quality monitoring, *Ground Water,* 19(6), 617, 1981.
25. **Kratochvil, B., Goewie, C. E., and Taylor, J. K.,** Sampling theory for environmental analysis, *Trends Anal. Chem.,* 5(10), 253, 1986.
26. **Kratochvil, B. and Taylor, J. C.,** Sampling for chemical analysis, *Anal. Chem.,* 53(8), 924A, 1981.
27. **Cooper, D. W.,** Sequential sampling statistics for evaluating low concentrations, *J. Environ. Sci.,* Sept./Oct., 33, 1988.
28. *Statgraphics,* Version 5, STSC, Inc., Rockville, MD., 1991.
29. **Wold, S., Albano, C., Dunn, W. J., Edlund, U., Esbensen, K., Geladi, P., Hellberg, S., Johannson, E., Lindbergh, W., and Sjøstrøm, M.,** Multivariate data-analyses in chemistry, in *Proc. NATO Adv. Study Institute of Chemometrics,* Kowalski, B. Ed., Reidel Publishing, Dordrecht, 1983, 17. (Chemical/Multivariate).
30. **Kvalheim, O. M. and Karstang, T.,** Finner ikke referert i noen base. Men det gjør vel ikke noe siden funnet årstallet allerede, *SIRIUS,* Dept. of Chemistry, University of Bergen, Allegaten, Bergen, Norway, 1992.
31. **Bhattacharyya, G. K. and Johnson, R. A.,** *Statistical Concepts and Methods,* John Wiley & Sons, New York, 1977, 637.
32. **Wadsworth, H. M.,** *Statistical Methods for Engineers and Scientists,* McGraw-Hill, New York, 1990.
33. **Liteanu, C. and Rica, I.,** *Statistical Theory and Methodology of Trace Analysis,* Ellis Horwood Limited, Chichester, England, 1980, 446.
34. **Edwards, A. L.,** *An Introduction to Linear Regression and Correlation,* W. H. Freeman, San Francisco, 1976, 213.
35. **Mendenhall, W.,** *Introduction to Probability and Statistics,* Duxbury Press, North Scituate, MA, 1979, 594.
36. **Draper, N. R. and Smith, H.,** *Applied Regression Analysis,* John Wiley & Sons, New York, 1981, 709.
37. **Everitt, B. S.,** *Statistical Methods for Medical Investigations,* Oxford University Press, New York, 1989, 195.
38. **Flury, B. and Riedwyl, H.,** *Multivariate Statistics: A Practical Approach,* Chapman and Hall, London, 1988, 296.

39. **Mardia, K. V., Kent, J. T., and Bibby, J. M.,** *Multivariate Analysis,* Academic Press, London, 1980, 521.
40. **Morrison, D. F.,** *Multivariate Statistical Methods,* McGraw-Hill, New York, 1967, 338.
41. **Vogt, N. B.,** Polynomial principal component regression: an approach to analysis and interpretation of complex mixture relationships in multivariate environmental data, *Chemometrics Intelligent Lab. Sys.,* 7, 117, 1989.
42. **Anderberg, M.,** *Cluster Analysis for Applications,* Academic Press, New York, 1973, 359.
43. **Gordon, A. D.,** *Classification,* Chapman and Hall, London, 1981, 193.
44. **Vogt, N. B., Bye, E., Thrane, K. E., Jacobsen, T., and Benestad, C.,** Composition activity relationships — CARE. I. Exploratory multivariate analysis of elements, polycyclic aromatic hydrocarbons and mutagenicity in air samples, *Chemometrics Intelligent Lab. Sys.,* 6, 31, 1989.
45. **Vogt, N. B., Bye, E., Thrane, K. E., Jacobsen, T., and Benestad, C.,** Composition activity relationships — CARE. II. Indirect and direct mutagens and multivariate dose-response regression, *Chemometrics Intelligent Lab. Sys.,* 6, 127, 1989.
46. **Vogt, N. B. and Kolset, K.,** Composition activity relationships — CARE. III. Polynomial principal component regression and response surface analysis of mutagenicity in air samples, *Chemometrics Intelligent Lab. Sys.,* 6, 221, 1989.
47. **Vogt, N. B.,** Multivariate ecotoxicological mapping of the relationships between sediment chemical composition and fauna diversity, *Sci. Total Environ.,* 90, 149, 1990.
48. **Vogt, N. B.,** Multivariate CARE and MEM in environmetrics, in *Multivariate Statistical Techniques for Environmental Analysis,* Seip, K. L. and Vigerust, B., Eds., Water Pollution Report 22 of ERP/EEC, 152, 1990.

APPENDIX

Reference Table A1 **Data from BIOMAT — a survey of elements in mussels *(Mytilus edulis),* krusflikk *(Chondrus crispus),* and seaweed *(Fucus serratus)* from the Sandefjord fjord**

	Station & Type	Sample	TS	Al	Fe	Mn	Zn	Cu	Pb	Cd	Sn	Cr	Ni	Ti
1	1 B	B1	17.5	66	218	5	185	7.5	7.5	1.14	3.8	12.0	1.5	7.8
2	1 S	S1	8.9	140	590	69	108	12.1	2.8	0.045	0.1	2.4	8.3	3.6
3	1 K	K1	19.7	19	161	25	44	4.6	1.9	0.17	0.23	0.5	3.2	1.1
4	2 B	B2	19.1	24	94	2	93	6.3	9.0	1.32	2.58	1.5	0.8	3.3
5	2 S	S2	10.4	133	360	71	103	8.2	3.3	0.035	0.21	1.2	3.3	4.1
6	2 K	K2	18.5	15	186	23	51	5.0	2.1	0.1	0.24	0.5	2.7	6.5
7	3 B	B3	18.5	38	120	3	77	7.5	9.3	0.99	2.73	1.0	1.0	4.3
8	3 S	S3	12.0	135	230	89	70	9.2	2.3	0.025	0.27	0.5	5.6	2.1
9	3 K	K3	21.9	25	195	33	46	4.9	0.5	0.11	0.68	0.5	4.8	1.6
10	4 B	B4	19.2	30	77	4	73	7.5	4.3	0.94	2.21	1.3	0.1	3.8
11	4 S	S4	11.5	92	111	91	81	7.1	0.5	0.035	0.13	0.5	4.4	1.2
12	4 K	K4	17.1	15	122	29	45	6.2	0.5	0.15	0.3	0.5	7.3	0.5
13	5 B	B5	20.7	34	163	4	70	9.3	4.5	0.92	1.95	1.3	0.8	3.5
14	5 S	S5	12.4	103	115	108	87	5.4	1.9	0.025	0.18	0.5	2.4	2.7
15	5 K	K5	16.9	24	138	64	89	7.4	2.7	0.2	0.36	0.5	5.7	1.4
16	6 B	B6	20.5	29	97	3	56	5.3	5.3	0.63	2.09	1.0	1.5	4.0
17	6 S	S6	10.8	102	107	126	122	7.4	4.2	0.23	0.32	0.5	3.2	1.5
18	6 K	K6	20.7	27	159	51	45	5.3	0.5	0.18	0.73	0.5	5.3	1.7
19	7 B	B7	17.6	66	165	3	96	3.5	6.0	0.89	3.12	3.5	1.5	11.0
20	7 S	S7	15.9	78	114	111	70	5.0	1.4	0.025	0.92	1.0	3.8	2.1

Note: Seven stations and three species at each station, except station 7, are tabulated; data reported as less than (<) have been analyzed as equal to the detection limit. "B" is blueskjell, "S" is seaweed, and "K" is krusflikk; TS is dry matter (% of weight); all numbers are given as mg/g*TS.

Reference Table A2 **Data from CARE — measurement of element concentrations and mutagenicity in air samples from Elverum**

Obj.	Rep	Soot µg/m³	Sulph	Pb	Cd	K	Ni	Zn	Cu	Fe	PC1	PC3	TA98 +S9
			NG/M3										
1	30	18	290	130	0.19	120	1.5	35	8	10	-1.81	2.83	10
2	39	22	220	145	0.62	70	2	30	2	5	-1.39	-0.20	21
3	23	16	240	30	0.79	40	1	55	1	10	-2.52	-0.37	8
4	29	12	470	65	0.39	65	1	30	0.5	5	-2.68	-0.09	10
5	58	40	430	110	1.63	140	1	120	1	5	1.70	-0.38	32
6	45	22	380	100	0.41	90	0.3	45	1	5	-1.52	0.51	21
7	76	41	810	200	1.11	225	2	105	0.5	20	3.60	-0.33	51
8	49	49	960	165	0.72	225	2	75	0.5	20	2.50	-0.14	40
9	31	26	440	50	0.33	65	1.5	30	0.5	5	-2.24	-0.50	11
10	41	23	210	85	0.72	85	3	30	1	15	-1.11	-1.45	13
11	53	29	430	105	0.78	110	2	50	0.5	30	0.42	-0.94	20
12	60	30	700	145	1.04	130	3	60	0.5	15	1.20	-1.29	26
13	52	37	1230	155	1.13	145	0.5	130	1	10	2.11	0.90	24
14	77	40	1880	155	1.22	165	1.5	105	3	35	4.02	1.16	20
15	27	19	820	40	0.36	70	0.5	35	0.5	10	-2.26	0.31	11

Note: The scores for the samples used in the polynomial principal component regression (PPCR) are tabulated.

Reference Table A3 **Data from MEM — measurement of elements in sediments and diversity of sediment fauna from the Kristiansand fjord**

Obj.	ES100	Spec.	Individ.	Ni	Co	Pb	Cd	Zn	Cr	Cu	Fe	TOC	TOTN
							mg/g					%	
1	-99[a]	2	8	65	23	100	-99	151	42	48	2.2	0.42	4.2
2	-99	1	3	77	11	202	0.25	289	64	69	2.8	0.42	4.1
3	23.5	40	297	32	5	88	0.03	168	20	28	1.4	0.1	1.1
4	26	53	532	45	11	36	-99	50	18	25	1.4	0.08	0.9
5	8.8	44	3499	72	7	74	0.19	204	38	64	1.8	0.29	4.4
6	23.3	57	1141	83	19	71	-99	97	34	42	2.2	0.21	2.5
7	20.1	40	434	290	13	138	0.07	106	71	186	2.3	0.21	2.9
8	24.7	46	414	100	10	124	0.1	175	54	66	2.7	0.24	-99
9	-99	-99	-99	114	22	86	-99	113	47	62	2.6	0.23	3
10	30.7	55	466	90	10	262	0.07	1700	30	48	2.1	0.17	1.9
11	12.2	40	1624	445	19	100	0.36	178	44	367	2.6	-99	-99
12	3.6	15	3303	1620	64	415	0.16	267	114	1515	4.4	0.44	6.2
13	4	15	1767	1520	68	668	0.13	228	446	1539	6	0.31	4.8
14	-99	3	18	5440	360	706	0.76	138	318	3940	30.6	-99	-99
15	5	7	274	3680	168	488	0.5	208	118	3857	11.2	0.35	3.6
16	5.2	23	4314	1815	78	405	0.19	168	108	1928	6.7	0.35	4.1
17	7.6	28	6942	1160	52	274	0.26	206	50	1153	4.1	0.27	3.9
18	-99	5	21	8200	354	366	0.5	232	74	5778	14.0	0.32	3.3

Note: All elements except iron (Fe) are given in mg/g; Iron, total organic carbon (TOC), and total nitrogen (TOTN) are given in %; the number of species (Spec.) and the total number of individuals (Individ.) are tabulated together with the diversity. The sample numbers follow those on the map and are the same as the sample numbers in the SIMCA analysis and the polynomial principal component regression.

[a] -99 is missing value identificator.

Chapter 5

Precipitation

Howard B. Ross and Stephen J. Vermette

CONTENTS

I. PROCESSES GOVERNING TRACE METALS IN PRECIPITATION

Trace substances, both gases and particles, are removed from the atmosphere by precipitation (wet deposition), cloud or fog impaction (cloud deposition), or by impaction onto a surface in the absence of precipitation and clouds (dry deposition). For many trace metals (e.g., As, Cd, Cr, Cu, Hg, Pb, Se, V, and Zn) the atmosphere is the major medium for transport to the biosphere, and wet deposition is the predominate removal mechanism. For example, wet deposition is an important source of nutrient limiting dust, which supplies essential trace metals to the Pacific Ocean,[1] and a wide variety of potentially toxic metals (e.g., Pb, Cd, and As) to forest ecosystems in rural regions of North America and Europe.[2,3] Thus, to know how trace elements are cycled in the environment, accurate information on their atmospheric wet deposition is essential. Unfortunately, this has proven difficult because atmospheric water samples are easily contaminated. Ross[4] and Barrie et al.[5] have pointed out that many reported values in the literature are in error because proper precautions to avoid sample contamination were not followed.

A. EMISSIONS TO THE ATMOSPHERE

With the advent of industrialization, man has greatly altered the biogeochemical cycling of a variety of trace metals. For many metals man-made emissions to the air are now significantly larger than natural emissions (Table 1). A measure of the anthropogenic influence on metal global cycling is the *interference factor* (IF) which is defined as the ratio of natural and anthropogenic emissions.[6] For Cr, Cu, Ni, Hg, V, and As, natural and anthropogenic emissions are about the same (IF $\approx$ 1). For Zn, Cd, and especially Pb (IF $\approx$ 18), man's activity now dominates these metals' global atmospheric cycling.

Major anthropogenic emissions are from fossil fuel combustion and the processing of sulfide ores to produce various metals and alloys. For Pb, automobile emissions dominate all other sources, while for

0-8493-6304-7/95/$0.00+$.50

Table 1 **Estimate of global emissions of trace metals to the atmosphere from anthropogenic and natural sources**

Anthropogenic Sources (Units: 10^6 kg yr^{-1})	**As**	**Cd**	**Cr**	**Cu**	**Hg**	**Mn**	**Ni**	**Pb**	**Se**	**V**	**Zn**
Coal combustion	2	0.53	11	5.2	2.1	11	14	8.2	3	7.9	11
Oil combustion	0.06	0.14	1.4	1.9		1.4	27	2.4	0.51	76	1.4
Nonferrous metal production	13	5.5		23	0.13	18.2	8.7	49	1.7	0.06	72
Steel and iron manufacturing	1.4	0.16	16	1.5			3.7	7.6	0.05	0.74	20
Refuse incineration	0.31	0.75	0.84	1.6	1.2		0.36	2.4	0.21	1.2	5.9
Phosphate fertilizers		0.17		0.41		8.2	0.41	0.16			4.1
Cement production	0.53	0.27					0.49	7.1			9.8
Wood combustion	0.18	0.12	1.3	0.9	0.18		1.2	2.1	0.37		3.6
Automobiles								253			
Others	2								0.19		3.2
Total	19.5	7.6	30.5	34.5	3.6	38.8	55.9	332.0	6.0	85.9	131.0
Natural Sources (Units: 10^6 kg yr^{-1})	**As**	**Cd**	**Cr**	**Cu**	**Hg**	**Mn**	**Ni**	**Pb**	**Se**	**V**	**Zn**
Windblown dust	2.6	0.2	27	8	0.05	221	11	3.9	0.18	33	19
Sea salt	1.7	0.06	0.7	3.6	0.02	0.86	1.3	1.4	0.55	3.1	0.44
Volcano	3.8	0.82	15	9.4	1	42	14	3.3	0.95	5.6	9.6
Wild forest fires	0.19	0.11	0.09	3.8	0.02	23	2.3	1.9	0.26	1.8	7.6
Continental biosphere — particulates	0.26	0.15	1	2.6	0.02	27	0.51	1.3	0.12	0.92	2.6
Continental biosphere — volatile	1.3	0.04	0.05	0.32	0.61	1.3	0.1	0.2	2.6	0.13	2.5
Marine biosphere — volatile	2.3	0.05	0.06	0.39	0.77	1.5	0.12	0.24	4.7	0.16	3
Total	12.15	1.43	43.9	28.11	2.49	316.66	29.33	12.24	9.36	44.71	44.74
Interference factor (anthropogenic/natural emissions)	1.60	5.34		1.23	1.45	0.12	1.90	27.12	0.64	1.92	2.93

Data is adapted, except for Se, from Nriagu, J.O. and Pacyna, J. M., *Nature,* 333, 6169, 1988. For Se, data is from Mosher, B. W. and Duce, R. A., *J. Geophys. Res.,* 92, 13289, 1987.

V oil combustion is the major source. The uncertainties in anthropogenic emission estimates are generally smaller than those of natural emissions, since detailed records are kept in the production of fossil fuels and the mining and smelting of ores. The major uncertainties are thus due to the variability of metal content in the various raw materials and the emission factors for the various industrial processes.

The errors associated with the natural emissions are due to the large uncertainties in the global production of soil dust and the large variations of metal concentration in soils. Volcano emissions are also highly uncertain due to the variability in metal concentrations in the volcanic ash and the frequency of eruptions. Other possible natural sources of trace metals are biological mobilization in the terrestrial and marine biospheres.[7] These fluxes have not been studied in great detail with the exception of Se,[8,9] where it has been shown that gaseous emissions of Se from the ocean, presumably from biogenic mobilization, are an extremely important source of Se to the atmosphere. If the preliminary estimates in Table 1 are correct, it would appear that biogenic sources of trace metals in some instances dominate metal cycling in the natural atmosphere. However, it is important to note that the subsequent release of trace metals by biological activity may also be a detoxification processes, which would not be as large if global metal pollution had not occurred.

B. PARTICLE SIZE CONSIDERATIONS

Once in the atmosphere, the mass of most trace metals will be associated with atmospheric aerosol particles. The major exception is Hg, where >95% of the atmospheric Hg is found in the gas phase.[10] In addition, trace metals in the ambient atmosphere will make only a small contribution to the total aerosol burden, generally less than 5% of the particle mass (dry wt.). Hence, for understanding trace metals cycling in the atmosphere, knowledge of the properties of atmospheric aerosols is important.

Atmospheric particles range from <0.01μm to over 10 μm in diameter and their mass may vary between 10^{-18} to 10^{-9} g. The smallest particles termed *Aikten nuclei* (<0.2μm in diameter) have combustion processes and gas-to-particle conversion as their major sources. Their main sinks are coagulation and scavenging in clouds. Atmospheric lifetimes are generally less than 1 h. The next class of particles termed the *accumulation mode* (0.2 to 2 μm in diameter) arises principally from combustion, cloud droplet evaporation, and the coagulation of Aikten nuclei. The major sinks for these particles are precipitation scavenging and to some extent dry deposition. The *coarse particles* (>2.0μm) are produced by mechanical processes, such as soil erosion, and by inefficient combustion processes (e.g., fly ash). The major sink for coarse particles is dry deposition since their gravitational settling velocity is so large; generally, lifetimes of coarse particles are on the order of minutes to hours.

A review of particle size distributions for trace metals in urban, rural, and remote atmospheres indicates that elements from natural and anthropogenic combustion sources are associated with submicron aerosols.[11] This is probably due to the fact that during the combustion processes, many trace metals will be volatilized. Since the surface area of an aerosol ensemble is skewed toward the submicron range, the elements will condense onto particles in this size interval.[12] Elements which are released into the atmosphere by mechanical processes (e.g., soil erosion and sea salt formation) are associated with larger particles. Elements which have a wide variety of sources generally do not have a distinct mass median diameter.

Metals which are associated with accumulation mode particles have the ability to be transported long distances from their sources, since the atmospheric lifetime of these particles are on the order of days. Metals associated with larger particles generally have short lifetimes and are usually associated with local sources. In addition, the particle mass is proportional to the third power of its radius. Therefore, one must be careful in the collection of precipitation to avoid local sources, because the incorporation of dry-deposited large particles will dominate and mask the regional metal composition of the precipitation.

The residence time of atmospheric particles is much shorter than the characteristic time for interhemispheric mixing; hence, anthropogenic metal pollution of the atmosphere, like acidification, is a regional problem. This is clearly exemplified in Figure 1, which shows the concentration of Se and Pb in atmospheric submicron particulate matter. The data is taken from the compilation of Wiersma and Davidson,[13] and has been divided into four regions: rural sites from industrialized countries in North America and Europe, the north Atlantic, the remote Pacific, and Antarctica. For Pb, large variations in the atmospheric burden are observed; concentrations in Antarctica are about 1000 times lower than in rural regions. Selenium variations are much less pronounced, the difference between maximum and minimum concentrations being of the order of 15. These data are consistent with observations and processes which govern the cycling of these metals in the atmosphere:[14,15]

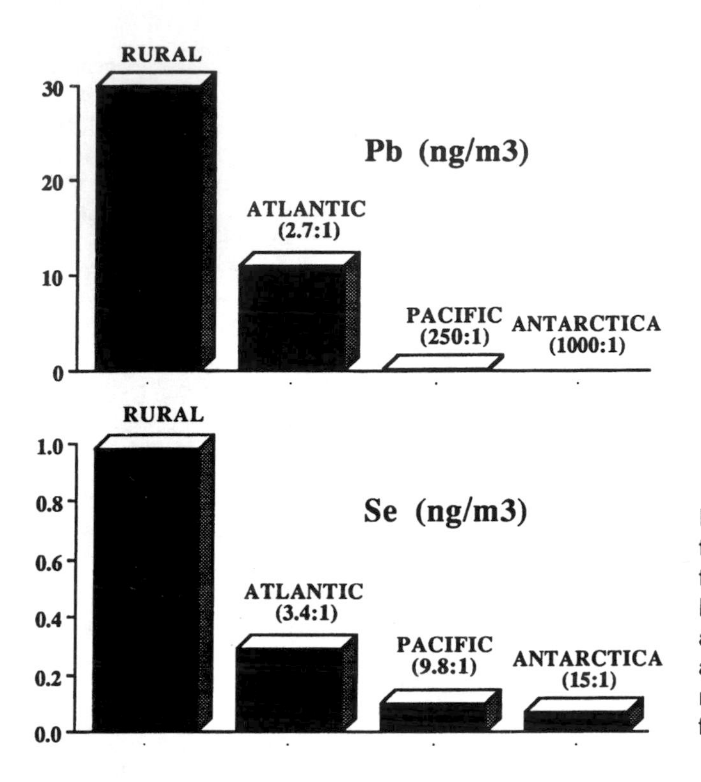

Figure 1 Average particle concentrations of Pb and Se at rural locations in North America, Europe, the North Atlantic, the Pacific, and Antarctica. The numbers in parenthesis are the ratio of the rural data with respect to the regions. (Data are taken from Reference 13.)

1. Anthropogenic emissions (mostly in the northern hemisphere) are the dominant source of airborne Pb
2. Nearly all airborne Pb is found on submicron particles
3. There are large, relatively uniform, natural sources of Se from biogenic sources in the oceans and on land

C. INCORPORATION OF TRACE METALS INTO ATMOSPHERIC PRECIPITATION

In the formation of precipitation, particles and gases are scavenged into the falling precipitation both in and out of cloud. Precipitation formed by large-scale synoptic systems (i.e., the passage of a warm front) is associated with the lifting of warm moist air over colder surface air. The falling precipitation will then scavenge particles and gases from the lower air masses (termed *below cloud scavenging*). In convective storm cells, warm surface air rises and penetrates into the free troposphere. Due to the continued lifting of warm air, convective storms are "fed" particles, gases, and vapors.

During a precipitation event, the concentration of soluble ions will vary considerably.[16,17] The concentration of pollutants in precipitation will depend on a large number of factors, including scavenging efficiencies (both in and out of cloud), the origin of the air mass, the type of precipitation, and precipitation intensity. Concentrations of ions generally first decrease through the course of the storm. Towards the end, when rain intensity is lower, concentrations may begin to increase due to evaporation of the rain drops. Hence, for estimating atmospheric wet deposition, it is extremely important that the initial and final segments of the storm are collected.[16]

In midlatitude locations in Europe and North America, one generally observes seasonal variations of solutes in precipitation.[18,19] This is exemplified in Figure 2, which presents the concentration of Cd in precipitation at a site in southern Sweden.[19] The atmospheric Cd cycle in Europe is dominated by anthropogenic sources[20] which are fairly constant (±20%) during the year. Hence, the differences in concentrations during winter and summer are primarily due to meteorological factors. These include differences in scavenging efficiencies of snow and rain and the height of the boundary layer. During winter the level of the atmosphere where air is uniformly mixed is approximately one half of the height where it occurs during the summer.

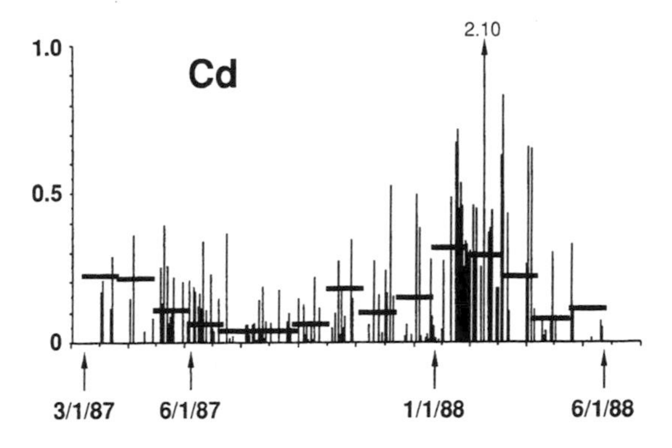

Figure 2 Concentration of Cd in precipitation collected on a daily basis at a site in southern Sweden. The solid horizontal lines represent monthly volume weight mean concentrations. (Adapted from Reference 36).

II. SAMPLING AND ANALYTICAL CONSIDERATIONS

A. CONTAMINATION — GENERAL CONSIDERATIONS

Sample contamination of precipitation samples can occur during the collection, the subsequent shipping and handling, and the chemical analysis. Problems with contamination are further exasperated because many trace metals will adsorb on to the surface of the storage bottle. Studies indicate that the adsorption is dependent on the element, its concentration, sample pH, and the container material.[21-23] Increasing adsorption of metals with increasing pH has been observed and is thought to be due to the adsorption of hydroxo or carbonato metal complexes which predominate at a higher pH. It has also been observed that when samples are acidified to a pH $\approx$ 1, there is no metal adsorption onto the container walls if they are constructed of polyethylene, Teflon, or Pyrex glass.

When samples are acidified, any metals which are on the container surface will be released into solution, and if the bottles are not scrupulously cleaned beforehand, sample contamination will occur. Therefore, several protocols have been established for cleaning bottles, collectors, and labware. Haraldsson and Magnuson[24] and Ross[25] have developed a cleaning procedure for the routine monitoring of trace metals in precipitation in Sweden where samplers and collection bottles are allowed to soak in a series of acids for 1 month (Table 2). This procedure is similar to the procedures developed for the sampling of polar snow and ice.[26] Tramontano et al.[27] have developed a somewhat less rigorous procedure in which sampleware cleaning takes about 1 week. The authors have tested their procedure and found that it is acceptable for the determination of trace metals in precipitation along the U.S. eastern seaboard. Common to all of the procedures is that the equipment is first rinsed in acetone to remove any organic contaminants and then exposed to acids at different strengths. Generally, acids of gradually increasing quality are introduced during the cleaning procedure while the acid strength is reduced. Both HCl and HNO_3 are used since it is thought that some metal compounds have different solubilities in different acids. During the final stages, the equipment is allowed to soak in ultra-pure 0.1 *N* HNO_3 for an extended period, rinsed in ultra-clean deionized water, dried in a particle-free environment, and then stored in plastic bags. Using these procedures it has been shown that polyethylene equipment, which costs substantial less then Teflon, can be used in the routine monitoring of trace metals in precipitation. However, there is a trade-off between the cost of Teflon bottles and the cost and time to clean polyethylene bottles. For example, Vermette et al.[28] were able to follow a less rigorous cleaning protocol for Teflon bottles — a series of 24-h HNO_3 and deionized water soaks/rinses over a period of 3 days.

The importance of cleaning sampling equipment and labware can not be over-emphasized. Figure 3 presents the concentrations of trace metals (Cd, Cu, Fe, Mn, Pb, and Zn) in rain water at four locations in Sweden.[4] Samples were collected and stored in acid-washed equipment and in equipment which was only washed in deionized water. Subsequent sample handling and analysis were identical. These results show that the acid washing of the bottles can have a profound effect on concentrations. The largest effects were observed for Cd, Cu, and Zn. For Cu, differences in concentrations could be up to a factor of 100. The propensity of Cu contamination is probably due to the fact that Cu is used as a catalyst when forming low density polyethylene. While the degree of contamination is highly variable, further analysis of the data indicates that there is an inverse relationship between sample volume (i.e., precipitation amount) and contamination. This would verify the hypothesis that contamination mainly arises from the inside surface of the container.

Table 2 **Methodology for the cleaning of labware and precipitation collectors for the routine monitoring of precipitation in Sweden**

1. Rinse in acetone
2. Rinse in deionized water
3. Soak 1 week in concentrated HCl (reagent grade; diluted 1:1 with deionized water)
4. Rinse in deionized water
5. Soak 1 week in concentrated HNO_3 (reagent grade; diluted 1:1 with deionized water)
6. Rinse in deionized water
7. Soak 1 week in 0.1 *N* HNO_3 (analytical grade) (bath 3)
8. Rinse in deionized water and transfer to the clean room
9. Rinse in high purity deionized water (maximum linear resistivity of 18 mW cm)
10. Soak 1 week in 0.1 *N* HNO_3 (suprapur; Merck)
11. Rinse in high purity deionized water
12. Dry under particle-free air and store in double plastic bags

Note: Equipment which is reused is washed starting with step 6.

Adapted from Ross, H. B., Report CM-67, University of Stockholm, Department of Met., 35 pp., 1984.

B. COLLECTION METHODOLOGIES

1. Siting Criteria

Wet deposition (D_w) is the product of precipitation amount (P) and concentration (C). The determination of P in itself is not trivial and requires special care. Errors in the determination of P are due to deformation of the windfield above the collector's rim, wetting of the gauge, evaporation, splashing of rain drops, and the blowing of snow. An evaluation of several standard meteorological precipitation gauges used by meteorological services around the world showed that P was 3 to 30% systematically underestimated.[29] The largest errors were due to the blowing of raindrops and snowflakes. In general, the greater the wind speed the greater the uncertainty in precipitation estimates. It is therefore recommended that P should only be determined with calibrated rain and snow gauges and not by the amount of water in the precipitation sampler used for determining concentrations.

The location of the sampling site is critical for obtaining reliable data on trace metal deposition, and certain basic siting criteria have been established for avoiding local contamination.[30,31] For regional precipitation chemistry networks, the collector should be located in an area that typifies a region but minimizes local point or area sources. Regardless of the size of the network, consistent siting criteria should be followed. Important points to remember are

1. All objects higher than the collector should be at a distance two times the objects' height. Buildings should be at least 30 m away and not within a 30° cone of the mean wind direction.
2. The area surrounding the collector (radius of 30 m) should be flat and covered with vegetation; the vegetation should be lower than the collectors.
3. Local access roads should be sparsely trafficked and be at least 300 m away.

2. Collection Methods

Falling precipitation for chemical analysis can be collected in bulk or in wet-only collectors. A bulk collector is a container that is open to the environment continuously; in its most simple form it is a bucket or a funnel attached to a bottle. Wet-only collectors have lids which open during the precipitation event and close during dry periods; the movement of the lid is controlled by a precipitation sensor. The advantage of the wet-only collector is that dry deposition is excluded from the sample. The major disadvantage is that the sampling site must be equipped with electricity, the collector is inherently more complicated, and there is greater risk for sample contamination. There are a number of spring operated quasi wet-only collectors (lids which open during the precipitation event but do not close after the event) described in the literature; these usually do not require electricity.[32,33]

Ross[25,34] has developed sampling techniques for the routine monitoring of precipitation in Sweden, using funnel and bottle (bulk) collectors constructed from low-density polyethylene. Collection periods are for 1 month and at each location there are 3 collectors. The advantage of collecting samples in triplicate is that outliers and contaminated samples are easily observed. Special care is taken to avoid contamination during the placement and removal of the collectors. For example, station managers wear plastic gloves when handling the collectors, and all equipment is shipped in doubly sealed plastic bags.

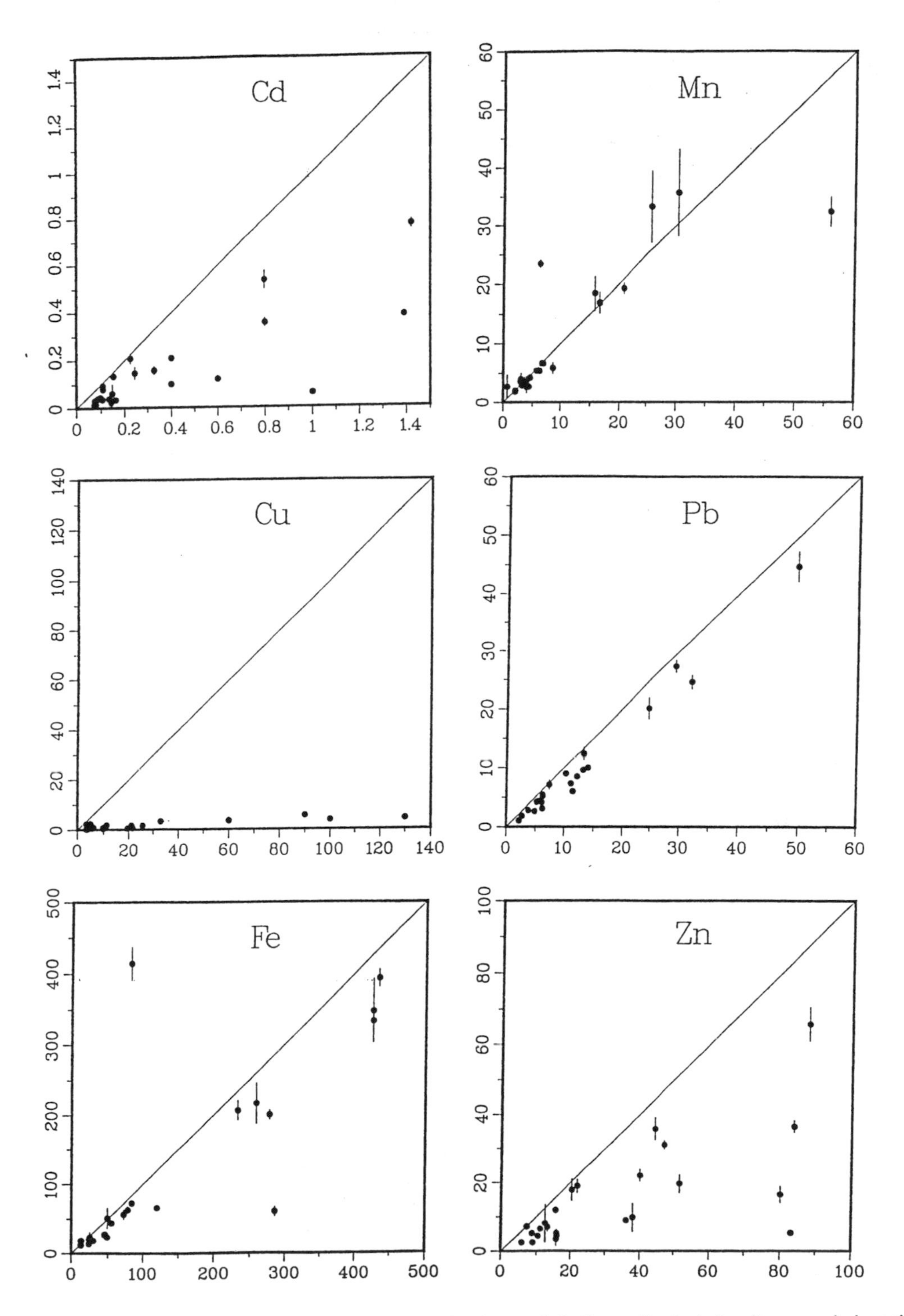

Figure 3 Metal concentrations (μg l⁻¹) in atmospheric precipitation collected simultaneously in acid-washed collectors (x-axis) and in collectors only rinsed in distilled water (y-axis). Collection handling, and analysis were identical. The error bars represent the standard deviation of metal concentrations from samples collected and stored in acid-washed containers. (Adapted from Reference 4).

The sampling equipment is washed according to the procedures in Table 2 and the sampling bottles are used for storage of the samples in the laboratory. The collectors relative to other precipitation samplers are quite small (the diameter of the funnel is 75 mm) and a typical monthly sample of 50 mm of precipitation is around 220 ml. The advantage of the small size is that it facilitates acid cleaning of the sampling equipment. In addition, the small sample volume is not a problem since chemical analysis is performed by graphite furnace atomic absorption spectroscopy. Uncertainties due to the collection procedures have been analyzed and are on the order of ± 20%, and are dependent on the metal and time of year. The largest differences in sample concentrations are found during the spring and summer.

Chan et al.[35] and Tramontano et al.[27] report procedures for the routine monitoring of precipitation for trace metal analysis using wet-only samplers. In the first study, a commercially available wet-only collector is used with plastic bag inserts. Tramontano et al. modified their collector so that the lid and supporting arms were made of plastic or coated with Teflon and plastic bag inserts rigorously washed in acids. Haraldsson and Magnuson[24] report that the routine monitoring of precipitation in wet-only collectors requires special care since the collector can contaminate the sampler. For example, certain colored plastics can have large amounts of Cd and Zn. They recommend that wet-only collectors be constructed of colorless low density polyethylene and that lids be washed in strong acids.

Vermette et al.[28] also report procedures for the routine monitoring of precipitation at continental sites, using a modified commercially available wet-only sampler. In addition to changes with the lid and support arms, as described by Tramontano et al.,[27] the sampler was further modified to use a Teflon funnel, tubing, and collection bottle. The funnel, tubing, and bottle were routinely replaced with acid-cleaned ones prior to each sampling period. Furthermore, the base of the sampler was enclosed to minimize contamination and maintain a constant temperature while the sample was in the field. Exposure of the funnel to the heated enclosure allowed for the efficient melting of collected snow.

Ross[36] reports the collection of precipitation using a specially constructed wet-only collector where the lid, the funnel, and the bottles are removed periodically for acid washing. Results from 15 months of sampling in Sweden indicated that sample contamination for the more pristine rain events could be a problem. However, since precipitation concentrations could vary over a factor of 3 during the month, the effect on monthly deposition estimates was marginal. Monthly bulk and daily wet-only precipitation samples were also compared for trace metals at two sites. It was observed that for metals which primarily arise from anthropogenic sources (e.g., Cd, Pb, Cr, and V), bulk collectors provide a fairly accurate estimate of wet deposition. Differences between bulk and wet-only collectors are <30% (Table 3). For metals which have an appreciable source from crustal material (e.g., Fe, Zn, Ni, and Cr), their wet deposition can be determined within an accuracy of approximately 50% with bulk collectors. One more than likely could expect higher uncertainties in bulk measurements for these metals in areas where local dust levels are higher. For Mn, wet deposition fluxes could only be estimated with the wet-only collectors because dry deposition of biogenic material from nearby trees contaminated the bulk samples.

For the determination of trace metals in precipitation from remote regions in the Pacific[1,37-40] and in polar snow and ice,[26,41-43] sample procedures have been developed which limit the risk of sample contamination. These include constructing all sampling equipment out of Teflon and/or low-density polyethylene plastic, washing it in strong acids, and performing the chemical analyses in ultra-free metal environments. Samples are collected manually, on an event basis with highly trained personnel, and in most instances wearing clean-room protective clothing. With such techniques, metal concentrations at the picogram l^{-1} have been successfully determined.

In Europe and North America, the sampling of ground snow or "snowpack" has been shown to be an effective technique for determining trace metal deposition during winter months providing the snow has not melted.[44-46] The advantage of snowpack sampling is that a large number of locations can be measured relatively easily, the influence of local point sources can be determined, storm-to-storm variations are smoothed out, and input of trace metals to ecosystems during the spring melt can be estimated. When sampling snowpack, special care must be taken to avoid underbrush and ground vegetation, and in pristine regions clean-room protective clothing is usually worn. Samples are either taken by digging a pit or by using an acid-washed hollow cylinder.

C. LABORATORY FACILITIES — GENERAL CONSIDERATIONS

The concentration of trace metals in precipitation is usually in the range of $\mu g\ l^{-1}$ or below. The low level of metals dictates that samples, labware, reagents, and standards be treated with the utmost concern for contamination. Laboratory facilities should be free from external sources of contamination. A Class 100 clean room or the use of a laminar-flow clean work station is highly recommended; if unavailable, though, maintenance of a clean work area and careful handling practices are necessary, including capping between

Table 3 **A comparison of wet deposition estimated as measured in bulk collectors and wet-only collectors at a site 100 km south of Stockholm, Sweden**

	Wet Only (mg m² time⁻¹)	Bulk (mg m² time⁻¹)	Relative Difference (%)
# of samples	162	15	
Precipitation (mm)	619	657	6
Cd	77	88	14
Cu	1,400	1,500	7
Fe	17,000	22,000	29*
Mn	1,500	2,900	93*
Pb	2,300	2,300	<0
Zn	6,300	8,200	30*
Cr	99	130	24
Ni	270	280	4
V	720	860	19

Note: The bulk collector samples were collected for 1 month and are the average from three collectors. The wet-only collector samples are daily samples. The asterisk indicates differences which are significantly different than 0 to a 95% confidence level.

reagent introductions and analysis.[28] Sample handling and addition of reagents should be kept to a minimum, as metal concentrations tend to increase by 1 to 3% with each "handling."[47] Labware must be kept impeccably clean and considerable effort may be necessary to find the "right" labware. The water system should be constructed of nonmetal components, and ASTM type 1 water is recommended. Similarly, chemicals added to the samples should be of reagent grade or better.

D. TECHNIQUES FOR METAL DETERMINATIONS

Many high-precision analytical techniques are available for determining metals at ppb and subppb levels. As discussed in Chapter 3, the more commonly used instruments are atomic adsorption spectroscopy (AAS), electrothermal atomic adsorption (ETAAS), inductively coupled plasma-atomic emission spectrometry (ICP-AES), inductively coupled plasma-mass spectroscopy (ICP-MS), instrumental neutron activation analysis (INAA), proton-induced X-ray emission (PIXE), and anodic stripping voltammetry (ASV).

There are also numerous variations to these instruments. Each has its own advantages and disadvantages, and many investigations use the instruments in combination to maximize advantages. In general, INAA and PIXE are better suited for the analysis of solid samples (e.g., insoluble fractions of wet deposition), whereas atomic adsorption, both flame and GF, and ICP (both AES and MS) are best adapted for solutions. Matrix effects (interferences from other constituents in solution with the target metals) are negligible for INAA. The matrix effects, while present for AA and ICP, are less troublesome due to the relatively simple chemical matrix of precipitation.

In choosing instrumentation, consideration must be given to instrument availability, performance (detection limits, method precision, and bias), sample costs, analytical throughput, prechemistry required, operator experience, and treatment of samples. For example, it may be necessary to weigh the time and effort needed to preconcentrate ICP-AES samples with the time saved by the ICP-AES multielement capabilities. Sample treatment is another concern: samples preserved with HCl cannot be analyzed by INAA. Consideration must also be given to the individual facility and its dedication to low-level analysis. For example, running solutions from a zinc plating firm in the instrument used for trace metal analysis will contaminate the cleaner sample. A distinction must also be made between the capabilities of a research or methods development laboratory in comparison to routine analysis. For example, a "methods development" laboratory may report determination limits of 0.2 ppb, while reported determination limits in the "production" laboratory (same instrument and protocol) are restricted to 1.0 ppb. Much also depends on the experience of the instrument operator. If the operator has no experience with trace levels or precipitation samples (precipitation samples do differ from other natural waters), a learning curve will be needed.

It is absolutely critical to analyze laboratory blanks, field blanks, and quality control standards which are certified at levels comparable to the precipitation samples, and which are run along with the samples.

This offers a continuous check on trace-metal contamination and potential problems with the instrumentation. A number of known standards and previously run duplicates should be run blind to the instrument operator. Unacceptable are published results that omit QA/QC data. Similarly unacceptable is the repeated use of "generic" QA/QC data passed on from investigation to investigation.

1. Atomic Absorption Spectroscopy

The use of AAS and ETAAS are routine in the analysis of precipitation.[19,48-57] ETAAS offers increased sensitivities, often one and two orders of magnitude improved over AAS, and has been recognized as the standard to which other techniques are compared. Although sample preconcentration is required for AAS, it remains the most readily available instrument for the analysis of precipitation and is often used with other instruments (e.g., Pb determinations to complement INAA). A limitation of ETAAS, in comparison to multielement instruments, is its reliance on single (sometimes dual) element determinations. In addition, ETAAS presents extreme difficulties in the routine determination of highly refractory elements (e.g., As and V). Scudlark et al.[58] describe the need for specialized techniques in the analysis of precipitation, including compensating for the suppression of the Pb signal by salts and increasing the analytical sensitivity for Al and Fe by the addition of cirtic acid.

2. Inductively Coupled Plasma Techniques

The principal advantage of ICP techniques over ETAAS is their multielement capabilities. Major drawbacks of ICP-AES are the higher detection limits (about an order of magnitude) and a large number of spectral interferences. Keller et al.[59] have demonstrated the successful use of ICP-AES in precipitation studies with preconcentration techniques, and Scudlark et al.[58] have shown advantages over ETAAS in the detection of particle-bound materials.

The addition of a mass spectrometer (ICP-MS) improves detection limits to levels equal to or better than GFAA, with fewer spectral interferences. ICP-MS is ideally suited to the analysis of precipitation due to its minimal matrix interferences. It combines the low detection limits of ETAAS (no need for preconcentration) with the sample throughput of ICP. ICP-MS also offers the added potential of providing isotope ratio data. Interferences still exist with precipitation, and each determination of a particular element in a particular matrix must be evaluated. Internal standards such as In or Mo are needed to track calibration drift, but recent work showed that the addition of Mo interfered with the determination of Cd.[28] Interferences are also caused by the analyte reacting with oxygen and by the carrier gas (e.g., Ar interferes with Ca and Fe determinations). In a recent methodology paper,[60] ICP-MS sensitivity and accuracy is shown as acutely affected by the proper regulation of operating conditions for plasma, vacuum, and voltages, thus implying the need for higher maintenance demands.

3. Instrumental Neutron Activation Analysis

A significant advantage of INAA is the use of specific gamma-ray energies that are usually subject to very few interferences. The method provides multielement analysis, is nondestructive, and particularly well suited for halogens (Br, Cl, and I) and rare-earth elements. While no preirradiation chemistry is needed for many elements (precipitation is often poured directly into vials for analysis), traditional evaporation and freeze-drying protocols have been used to lower detection limits, especially for the longer-lived nuclides. For detection in the presence of particles, INAA is often preferred, as no particle discrimination occurs. There are some disadvantages: a long turn-around time, constraints on the size of the sample load, and limitations on or an inability to detect Pb, S, and other elements. Related techniques, such as epithermal neutron activation analysis (ENAA), fast neutron, activation analysis (FNAA), and prompt gamma activation analysis (PGAA), each bring unique improvements to activation analysis, although they have seldom been applied to precipitation samples.

At present more than 300 research nuclear reactors are in operation in 54 countries, and most of these provide facilities for neutron activation analysis. While extensively used in aerosol research, INAA has not been as fully exploited for precipitation studies.[49,61-65] For an excellent review of INAA sample preparation and a listing of earlier precipitation studies, refer to Landsberger et al.[66]

4. Proton-Induced X-Ray Emission

PIXE provides multielement determinations and offers significant advantages for low Z elements. Other advantages are high speed and small sample size requirements. In contrast to other techniques, precipitation samples cannot be analyzed in solution with PIXE. Sample preparation usually takes the form of preconcentrating (evaporation or freeze drying), and samples are deposited onto a filter or adsorbed onto activated carbon. Filter contamination is a concern, as are matrix effects if the sample is too thick (although this should not be a problem with precipitation).

The use of PIXE has been limited in precipitation investigations. Published works include a study on urban snow,[67] a review of PIXE methods and precipitation-related applications prior to 1984,[66] and a rainwater analysis comparing PIXE and ETAAS.[68]

5. Anodic Stripping Voltammetry

Low instrument costs, multielement capability, and low detection limits are advantages offered by ASV. However, analytical throughput is limited by the relatively long exposure times required to achieve low detection limits, and determinations are restricted to those metals that can dissolve in metallic Hg (e.g., Zn, Cd, Pb, and Cu). Significant problems also arise with interferences from the formation of intermetallic and organic compounds (even in a relatively pure matrix such as rainwater). Corrections are made to account for these interferences, and methods have been developed to overcome interferences (e.g., UV irradiation or active carbon impregnated filter papers to overcome organic interferences). While successfully used in the analysis of natural waters, ASV has been scarcely used for precipitation studies. Lingerak et al.[69] provide a detailed methodology description for the analysis of precipitation samples, and Slanina et al.[70] report determinations achieved using ASV.

III. CONCENTRATION LEVELS

An excellent and often cited review of trace metals in precipitation is given by Galloway et al.[71] However, the measurement of trace elements in precipitation prior to the early 1980s may have been compromised by gross contamination or by losses to container walls and should be interpreted with caution.[5]

A survey of recent precipitation chemistry literature[72] highlights the regional differences as shown by the varying concentration/deposition over the world's oceans, ranked in order as the North Atlantic > North Indian > North Pacific > South Atlantic > South Indian > South Pacific. This ranking highlights the elevated concentrations found in the northern hemisphere in relation to the southern hemisphere. In the same survey, rural and urban concentrations were shown to range from 10 to 25 times and 35 to 150 times higher than for remote sites, respectively. An excellent review of urban precipitation chemistry is given by Gatz.[73] Two clear trends appear in precipitation chemistry: (1) the decline of toxic metals from developed countries (i.e., the decline in Pb levels in many parts of the world coincident with its phase out in gasoline;[74] and (2) increasing levels in developing countries, attributed to a sharp increase in metal emissions.[72]

Table 4 lists trace-metal concentrations in precipitation and deposition as reported after 1982. The list is by no means exhaustive and is provided as an illustration of levels reported in the literature. The data have been divided into the following regions: rural/remote, marine, and polar. No distinction has been made between rural and remote sites as their definition varies from country to country. Marine concentrations include values from cruises and from island and continental coastal sites.

IV. METAL SPECIATION

Speciation, as applied to trace metals, is generally defined in the environmental literature as "the identification of a specific form of an element as it actually occurs in the sampling medium." Total-element determinations dominate the trace-element literature, in step with improvements in instrument sensitivity and multielement capabilities. Very little information on metal speciation in precipitation is available in the literature. This is unfortunate, as the characterization of specific forms of the elements, especially toxic and/or mobile forms, is important to understanding the effects of metals on ecosystems and transport mechanisms.

A. SOLUBILITY

Precipitation includes both dissolved and particulate forms of chemical constituents, but the soluble/insoluble partitioning is often disregarded in precipitation chemistry studies. Samples are most often acidified upon collection to avoid wall losses from the samples. Acidification artificially releases metals from insoluble particles in the samples. "Acid extractable", referring to the amount of dissolved plus particulate metals released to solution following 24-h equilibration with dilute acid, is often used as an operational definition. The acid-extractable metal would then represent an estimate of the maximum biologically available metal concentration.

Table 4 **Survey of selected trace element concentrations in precipitation (volume-weight, $\mu g\ l^{-1}$) and in wet deposition flux (values in parenthesis, $mg\ m^{-2}\ yr^{-1}$)**

Location	As	Cd	Cr	Cu	Pb	Hg	Ni	V	Zn
Rural/Remote									
Canada									
Eastern[63]	0.23	<0.2	<1.0	1.3	15.4	—	—	0.41	4.7
	(0.25)	—	—	—	(16.92)	—	—	(0.45)	(5.16)
Ontario (north)[35]	—	0.12	—	1.5	4.8	—	—	—	—
Ontario (south)[35]	—	0.12	—	1.6	7.0	—	—	—	4.9-10
Great Lakes[50]	—	0.2-0.8	—	—	2-11	—	—	—	—
	—	(0.1-0.6)	—	—	(2-10)	—	—	—	—
United States									
Eastern[54]	—	0.12	—	—	1.1	—	—	—	3.1
	—	(0.24)	—	—	(2.16)	—	—	—	(6.0)
Eastern[51]	0.096	0.31	0.14	0.95	4.5	—	0.75	1.1	3.7
Eastern[65]	0.141	—	—	—	—	—	—	1.27	8.3
Central[28]	0.15	0.23	—	2.0	2.0	—	—	—	3.0
Central[55]	—	—	0.6	2.9	1.9	—	1.0	<0.1	9.3
Northern[86]	—	—	—	—	—	0.0187	—	—	—
						(0.0126)			
Northern[87]	—	—	—	—	—	0.0105	—	—	—
	—	—	—	—	—	(0.0045)	—	—	—
Sweden									
North[19]	—	0.041	0.064	—	1.84	—	0.13	0.27	4.1
	—	(0.024)	(0.048)	—	(1.32)	—	(0.10)	(0.24)	(2.88)
South[19]	—	0.125	0.160	—	3.75	—	0.44	1.16	10.2
	—	(0.072)	(0.084)	—	(1.92)	—	(0.24)	(0.60)	(5.4)
Holland[56]	0.5	0.5	1.4	6.7	13	—	2.4	6.7	21
	(0.29)	(0.24)	(0.66)	(4.54)	(7.4)	—	(0.98)	(4.27)	(13.7)
Central Italy[88]	—	—	—	—	—	0.01	—	—	—
Germany[89]	—	0.10-0.30	—	0.74-3.40	4.27-14.1	—	—	—	6.93-33.4
Northern India[64]	0.27	—	5.40	—	—	—	—		14.22

Table 4 (Continued) **Survey of selected trace element concentrations in precipitation (volume-weight, μg l^{-1}) and in wet deposition flux (values in parenthesis, mg m^{-2} yr^{-1})**

Location	As	Cd	Cr	Cu	Pb	Hg	Ni	V	Zn
Marine									
Atlantic									
Ireland[57]	—	0.04	—	0.86	0.51	—	—	—	8.05
Scotland[52]	—	0.68	—	2.3	4.0	—	—	—	13
	—	(0.39)	—	(1.3)	(2.3)	—	—	—	(7.6)
Bermuda[48]	—	0.062	—	0.32	0.722	—	0.167	0.096	1.53
East U.S.[90]	—	0.10	—	0.76	1.9	—	1.12	—	5.16
	—	(0.108)	—	(0.836)	(2.09)	—	(1.23)	—	(5.68)
Cruise[91]	—	0.07-0.95	—	0.13-0.67	0.13-3.65	—	—	—	1.23-4.75
Cruise[92]	—	0.031	—	0.5	0.471	—	—	—	2.1
Pacific									
Northeast[93]	—	—	—	—	—	0.009	—	—	—
Samoa[49]	—	—	<0.010	0.042	0.040	—	—	<0.050	0.96
	—	—	—	(0.04)	(0.03)	—	—	—	(0.3)
Enewetak[49]	—	—	<0.010	0.020	0.038	—	—	<0.050	0.088
Enewetak[1]	—	0.002	—	0.013	<0.04	—	—	0.018	0.05
NW U.S.[94]	—	0.012	—	0.14	0.15	—	—	—	0.99
Polar									
E. Arctic Ocean[41]	—	0.005	—	0.097	0.185	—	0.197	—	—
W. Can. Arctic[53]	—	0.012	—	0.26	0.63	—	0.18	0.72	0.89
Antarctic[95]	—	—	—	—	—	<0.00096	—	—	—

Part of the reluctance to separate dissolved and particulate phases before analysis involves the potential contamination from a filter and the excess handling. Soluble/insoluble separation is most often done by passing the collected precipitation through a filter, either as part of the collection device or in the laboratory. The separation is operationally defined by what passes through either a 0.40-mm pore size polycarbonate or a 0.45-mm cellulose acetate filter. Another approach is to run an unfiltered sample through both INAA and through an instrument that requires particles to be dissolved before measurement (e.g., AAS, GFAA, ICP-AES, or ICP-MS). The former detects the total metal content, while the latter group is somewhat less efficient in detecting particle-bound metals. Agreement between the two techniques suggests a predominately soluble metal, and a higher INAA determination suggests a nonsoluble state. While somewhat crude, this approach has been used by researchers as a first approximation method.[58,63] This distinction supports a concern that the latter group of instruments should be used on separate samples of the dissolved and nondissolved fractions (properly dissolved for analysis) of the precipitation samples.[75]

Soluble/insoluble partitioning has been found to be highly variable between metals although consistent between studies. For Zn, Cd, Cu, and Pb, Gatz[75] found solubility in precipitation similar to that in other natural waters, decreasing with increases in pH and the concentrations of total insoluble mass. Others find a limited relationship with pH, and caution should be taken in the extrapolation of results from one medium to another. The variability in partitioning has been attributed to the source of the parent aerosol. For example, Al (crustal sources) is generally more insoluble, while Pb (combustion sources) is highly soluble.[58] Dependence on parent aerosol and scavenging preferences may also explain soluble/insoluble partitioning with rain amount (sample volume). Heavy rain events exhibit higher metal solubility than lighter rains, which are more efficient below-cloud scavengers of mostly alkaline crustal materials.

For precipitation collected in Rhode Island,[65] the halides, the alkali, the alkaline earth elements, Se and Zn, were essentially completely dissolved. Arsenic, Sb, Mn, and V were mostly dissolved, while the crustal elements Fe, Sc, and Al exhibited less solubility. For Illinois, Gatz[75] reported Zn and Cd the most soluble, followed in order by Cu and Pb (all >80% soluble).

B. LEAD

Aside from inorganic Pb, a variety of alkyllead compounds are present in the atmosphere.[76] These compounds have been attributed to volatilization of tetraalkyllead (R_4Pb) used as a gasoline antiknock additive. Tetraalkyllead compounds decompose in the atmosphere, forming more stable trialkyllead (R_3Pb) and dialkyllead (R_2Pb) species. These compounds are scavenged by raindrops and have been measured in precipitation, although ratios of total alkyllead to inorganic lead indicate that organic lead is only a minor component.[77-79] These same studies showed that R_4Pb is not found in rainwater due to its rapid decomposition in collected rainwater. Lead speciation offers a valuable tool to distinguish gasoline emissions from soil or industrial emissions and to characterize its toxic effects. For example, R_3Pb has been blamed for forest damage in Germany,[80] although this conclusion is questioned by some.[81]

C. MERCURY

Mercury speciation is important because its most toxic form (methyl-Hg) accumulates in aquatic organisms. Accurate speciation is difficult to achieve and reliable data are scarce from a wide variety of sources. Aside from soluble/insoluble partitioning, Hg in precipitation is occasionally reported as elemental Hg, Hg(II), and methyl-Hg. Hg(II) is operationally defined in some studies as Hg reduced by $NaBH_4$. Hg(II) is thought to be an environmentally active phase of Hg consisting of unstable inorganic, organic, and particulate associations. Hg(II) can further be fractionated into inorganic-Hg if reduced by Sn^{2+}, and organo-Hg, by difference. Organo-Hg is important. It is thought to be the result of anthropogenic activities and one form of Hg that is transported in the atmosphere over long distances.[82] However, the primary long-range transport species is elemental Hg vapor. Precipitation is dominated by the Hg(II) form, and decreasing levels of Hg(II) during a storm event suggest an association with particles.[83] Furthermore, the absence of any difference between filtered and unfiltered samples suggests that Hg may be associated with colloidal particles. In Wisconsin, levels of total-Hg were substantially higher than could be attributed to measured inorganic-Hg and methyl-Hg. The difference was attributed to particulates strongly bound to organic substances.[84] Unlike Hg(II), methyl-Hg may occur throughout a precipitation event at relatively constant levels, suggesting a different phase than other Hg forms.[83]

REFERENCES

1. **Arimoto, R., Duce, R. A., Ray, B. J., and Uni, C. K.,** Atmospheric trace elements at Enewetak Atoll. II. Transport to the ocean by wet and dry deposition, *J. Geophys. Res.,* 90, 2391, 1985.
2. **Berqvist, B.,** Metal fluxes in spruce and beech forest ecosystems of south Sweden, Ph.D. Thesis, dept. of Ecology, University of Lund, Sweden, 1986.
3. **Lindberg, S. E. and Turner, R. R.,** Factors influencing atmospheric deposition, stream export, and landscape accumulation of trace metals in forested watersheds, *Water Air Soil Pollut.,* 39, 123, 1988.
4. **Ross, H. B.,** The importance of reducing sample contamination in routine monitoring of trace metals in atmospheric precipitation, *Atmos. Environ.,* 20, 401, 1986.
5. **Barrie, L. A., Lindberg, S. E., Chan, W. H., Ross, H. B., Arimoto, R., and Church, T. M.,** On the concentration of trace metals in precipitation, *Atmos. Environ.,* 21, 1133, 1987.
6. **Lantzy, R. J. and Mackenzie, F. T.,** Atmospheric trace metals: global cycles and assessment for man's impact, *Geochim. Cosmochim. Acta,* 43, 511, 1979.
7. **Nriagu, J. O.,** Natural versus anthropogenic emissions of trace metals to the atmosphere, in *Control and Fate of Atmospheric Trace Metals,* Pacyna, J. M. and Ottar, B., Eds., Kluwer Academic Publ., Dordecht, 1988, 3.
8. **Mosher, B. W. and Duce, R. A.,** Vapor phase selenium in the marine atmosphere, *J. Geophys. Res.,* 88, 6761, 1983.
9. **Mosher, B. W. and Duce, R. A.,** Atmospheric selenium: geographical distribution and ocean to atmosphere flux in the Pacific, *J. Geophys. Res.,* 92, 13277, 1987.
10. **Lindqvist, O. and Rodhe, H.,** Atmospheric mercury — a review, *Tellus,* 37B, 3, 1985.
11. **Milford, J. B. and Davidson, C. I.,** The sizes of particulate trace elements in the atmosphere: a review, *J. Air Pollut. Control. Assoc.,* 35, 1249, 1985.
12. **Rahn, K. A.,** Sources of trace elements in aerosols — an approach to clean air, Technical Rep., Col. of Engineering, Dept. of Met. and Ocean., U. of Michigan, 089030-9-T,309 pp., 1976.
13. **Wiersma, G. B. and Davidson, C. I.,** Trace metals in the atmosphere of rural and remote areas, in *Toxic Trace Metals in the Environment,* Nriagu, J. O. and Davidson, C., Eds., John Wiley & Sons, New York, 201, 1986.
14. **Nriagu, J. O.,** Global inventory of natural and anthropogenic emissions of trace metals to the atmosphere, *Nature,* 279, 409, 1979.
15. **Mosher, B. W. and Duce, R. A.,** A global atmospheric selenium budget, *J. Geophys. Res.,* 92, 13289, 1987.
16. **de Pena, R. G., Carlson, T. N., Takas, J. F., and Holian, J. O.,** Analysis of precipitation collected on a sequential basis, *Atmos. Environ.,* 12, 2665, 1984.
17. **Ames, D. L., Roberts, L. E., and Webb, A. H.,** An automatic rain gauge for continuous, real time determination of rainwater chemistry, *Atmos. Environ.,* 21, 1947, 1987.
18. **Lindberg, S. E. and Turner, R. R.,** Factors influencing atmospheric deposition, stream export and landscape accumulation of trace metals in four forested watersheds, *Water Air Soil Pollut.,* 39, 123, 1988.
19. **Ross, H. B.,** Trace metal deposition in Sweden: insight gained from daily wet only collection, *Atmos. Environ.,* 24A, 1929, 1990.
20. **Pacyna, J. M., Semb, A., and Hanssen, J. E.,** Emissions and long-range transport of trace elements in Europe, *Tellus,* 36B, 163, 1984.
21. **Chan, W., Tomassini, F., and Loescher, B.,** An evaluation of sorption properties of precipitation constituents on polyethylene surfaces, *Atmos. Environ.,* 17, 1779, 1983.
22. **Moody, J. and Lindstrom, R.** Selection and cleaning of plastic containers for storage of trace element samples, *Anal. Chem.,* 49, 2264, 1977.
23. **Subramanian, K. S., Chakrabarti, C. L., Sueiras, J. E., and Maines, I. S.,** Preservation of some trace metals in samples of natural waters, *Anal. Chem.,* 50, 444, 1978.
24. **Haraldsson, C. and Magnuson, B.,** Heavy metals in rainwater; collection, storage and analysis of samples, in *Proc. Int. Conf. Heavy Metals in the Environment,* Heidelberg, CEP Associates, 1983.
25. **Ross, H. B.,** Methodology for the collection and analysis of trace metals in atmospheric precipitation, University of Stockholm, Department of Met., Report CM-67, 35 pp., 1984.
26. **Boutron, C.,** Reduction of contamination problems in sampling of Antarctic snows for trace element analysis, *Anal. Chim. Acta.,* 106, 1, 1979.

27. **Tramontano, J. M., Scudlark, J. R., and Church, T. M.,** A method for the collection, handling, and analysis of trace metals in precipitation, *Environ. Sci. Technol.,* 21, 749, 1987.
28. **Vermette, S. J., Peden, M. E., Willoughby, T. C., Lindberg, S. E., and Weiss, A. D.,** A pilot network for metals in wet deposition, in Proc. 85th Annual Meeting of the Air & Waste Management Association, Kansas City, June 21 to 26, 1992.
29. **Sevruk, B.,** Methods of correction for systematic error in point precipitation for operational use, World Meteorological Organization, Operational Hydrology 21, Geneva, Switzerland, 91 pp., 1982.
30. **Granat, L.,** Siting Criteria, in *Proc. Symp. on Monitoring and Assessment of Airborne Pollutants with Special Emphasis on Long-range Transport and Deposition of Acidic Materials,* Pierce, R. C., Whelpdale, D. M., and Scheffer, M. G., Eds., NRCC No. 20642, 1982, 281.
31. NADP, Instruction manual: NADP/NTN site selection and installation, Natural Resource Ecology Laboratory, Colorado State University, Fort Collins, CO, 23 pp., 1984.
32. **Vermette, S. J. and Drake, J. J.,** Simplified wet-only sequential fraction rain collector, *Atmos. Environ.,* 21, 715, 1987.
33. **Vermette, S. J. and Drake, J. J.,** Modifications to the McMaster wet-only rain Collector, *Atmos. Environ.,* 22, 195, 1988.
34. **Ross, H. B.,** Trace metals in precipitation in Sweden. *Water Air Soil Pollut.,* 36, 349, 1988.
35. **Chan, W. H., Tang, A. J. S., Chung, D. H. S., and Lusis, M.,** Concentration and deposition of trace metals in Ontario-1982, *Water Air Soil Pollut.,* 29, 373, 1986.
36. **Ross, H. B.,** Trace metal deposition in Sweden: insight gained from daily wet only collection, *Atmos. Environ.,* 24A, 1929, 1990.
37. **Settle, D. M., Patterson, C. C., Turekian, K. K., and Cochran, J. K.,** Lead precipitation fluxes at tropical oceanic sites determined from ^{210}Pb measurements, *J. Geophys. Res.,* 87, 1239, 1982.
38. **Arimoto, R., Duce, R. A., Ray, B. J., Hewitt, A. D., and Williams, J.,** Trace elements in the atmosphere of American Samoa: concentrations and deposition to the tropical south Pacific, *J. Geophys. Res.,* 92, 8465, 1987.
39. **Arimoto, R., Ray, B. J., Duce, R. A., Hewitt, A. D., Boldi, R., and Hudson, A.,** Concentrations, sources, and fluxes of trace elements in the remote marine atmosphere of New Zealand, *J. Geophys. Res.,* 95, 8465, 1990.
40. **Vong, R. J., Hansson, H. C., Ross, H. B., Covert, D. S., and Charlson, R. J.,** Northeastern Pacific submicrometer aerosol and rainwater composition: a multivariate analysis, *J. Geophys. Res.,* 93, 1625, 1988.
41. **Mart, L.,** Seasonal variations of Cd, Pb, Cu and Ni levels in snow from the eastern Arctic Ocean, *Tellus,* 35B, 131, 1983.
42. **Boutron, C. F. and Patterson, C. C.,** Relative levels of natural and anthropogenic lead in recent Antarctic snow, *J. Geophys. Res.,* 92, 8454, 1987.
43. **Peel, D. A. and Wolf, E. W.,** Concentrations of cadmium, lead and zinc in snow near Dye 3 in south Greenland, *Ann. Glaciol.,* 10, 193, 1988.
44. **Barrie, L. A. and Vet, R. J.,** The concentration and deposition of acidity, major ions and trace metals in the snowpack of the eastern Canadian shield during the winter of 1980-1981, *Atmos. Environ.,* 18, 1459, 1984.
45. **Batifol, F. M. and Boutron, C. F.,** Atmospheric heavy metals in high altitude surface snows from Mont Blanc, French Alps, *Atmos. Environ.,* 18, 2507, 1984.
46. **Ross, H. B. and Granat, L.,** Deposition of atmospheric trace metals in northern Sweden as measured in the snowpack, *Tellus,* 38B, 27, 1986.
47. **Slanina, J., Baard, J. H., Broersen, B. C., Mols, J. J., and Voors, P. I.,** The stability of precipitation samples under field conditions, *Int. J. Environ. Anal. Chem.,* 28, 247, 1987.
48. **Church, T. M., Tramontano, J. M., Scudlark, J. R., Jickells, T. D., Tokos, J. J., Jr., Knap, A. H., and Galloway, J. N.,** The wet deposition of trace metals to the western Atlantic Ocean at the mid-latitude coast and on Bermuda, *Atmos. Environ.,* 18, 2657, 1984.
49. **Arimoto, R., Duce, R. A., Ray, B. J., Hewitt, A. D., and Williams, J.,** Trace elements in the atmosphere of America Somoa: concentrations and deposition to the tropical South Pacific, *J. Geophys. Res.,* 92, 8465, 1987.
50. **Gatz, D. F., Bowersox, V. C., and Su, J.,** Lead and cadmium loadings to the Great Lakes from precipitation, *J. Great Lakes Res.,* 15, 246, 1989.
51. **Muhlbaier-Dasch, J. and Wolff, G. T.,** Trace inorganic species in precipitation and their potential use in source apportionment studies, *Water Air Soil Pollut.,* 43, 401, 1989.

52. **Balls, P. W.,** Trace metal and major ion composition of precipitation at a North Sea coastal site, *Atmos. Environ.,* 23, 2751, 1989.
53. **Gorzelska, K.,** Locally generated atmospheric trace metal pollution in Canadian Arctic as reflected by chemistry of snowpack samples from the Mackenzie Delta region, *Atmos. Environ.,* 23, 2729, 1989.
54. **Petty, W. H. and Lindberg, S. E.,** An intensive 1-month investigation of trace metal deposition and throughfall at a mountain spruce forest, *Water Air Soil Pollut.,* 53, 213, 1990.
55. **Cadle, S. H., Vandekopple, R., Mulawa, P. A., and Muhlbaier-Dash, J.,** Ambient concentrations, scavenging ratios, and source regions of acid related compounds and trace metals during winter in northern Michigan, *Atmos. Environ.,* 24A, 2981, 1990.
56. **Daalen, J. V.,** Air quality and deposition of trace elements in the province of South-Holland, *Atmos. Environ.,* 25A, 691, 1991.
57. **Lim, B., Jickells, T. D., and Davies, T. D.,** Sequential sampling of particles, major ions and total trace metals in wet deposition, *Atmos. Environ.,* 25A, 745, 1991.
58. **Scudlark, J. R., Church, T. M., Conko, K. M., Moore, S. M.,** A method for the automated collection, proper handling and accurate analysis of trace metals in precipitation, in *Proc. The Deposition and Fate of Trace Metals in our Environment,* Verry, E. S. and Vermette, S. J., Eds., U.S. Department of Agriculture, Forest Service, North Central Forest Experiment Station, St. Paul, MN, 1992.
59. **Keller, B. J., Peden, M. E., and Skowron, L. M.,** Development of standard methods for the collection and analysis of precipitation: trace metals, Illinois State Water Survey Contract Report 438, Champaign, IL, 1988.
60. **Boomer, D. W. and Powell, M. J.,** The analysis of acid precipitation samples by inductively coupled plasma mass spectrometry, *Can. J. Spectroscopy,* 31, 104, 1986.
61. **Vermette, S. J. and Bingham, V. G.,** Trace elements in Frobisher Bay rainwater, *Arctic,* 39, 177, 1986.
62. **Landsberger, S., Simmons, S., Kramer, A., Drake, J. J., Vermette, S. J., Shuter, B., and Ihssen, P.,** Application of neutron activation analysis in acid precipitation studies, *J. Nucl. Radio Chem.,* 110, 333, 1987.
63. **Barrie, L. A.,** Aspects of atmospheric pollutant origin and deposition revealed by multi-element observations at a rural location in eastern Canada, *J. Geophys. Res.,* 93, 3773, 1988.
64. **Mahadevan, T. N., Negi, B. S., and Meenakshy, V.,** Measurement of elemental composition of aerosol matter and precipitation from a remote background site in India, *Atmos. Environ.,* 23, 869, 1989.
65. **Heaton, R. W., Rahn, K. A., and Lowenthal, D. H.,** Determination of trace elements, including regional tracers, in Rhode Island precipitation, *Atmos. Environ.,* 24A, 147, 1990.
66. **Landsberger, S., Jervis, R. E., and Monaro, S.,** Trace analysis of wet atmospheric deposition by nuclear methods, in *Trace Analysis,* Lawrence, J. F., Ed., Academic Press, New York, 1985, chap. 4.
67. **Landsberger, S., Jervis, R. E., Kajrys, G., and Monaro, S.,** Characterization of trace element pollutants in urban snow using proton induced x-ray emission and instrumental neutron activation analysis, *Int. J. Environ. Anal. Chem.,* 16, 91, 1983.
68. **Hansson, H. C., Ekholm, A. K. C., and Ross, H. B.,** Rainwater analysis: a comparison between proton induced x-ray emission and graphite furnace atomic absorption spectroscopy, *Environ. Sci. Technol.,* 22, 527, 1988.
69. **Lingerak, W. A., Wensveen-Louter, A. M. V., and Slanina, J.,** The determination of zinc, cadmium, lead, and copper in precipitation by computerized differential pulse voltammetry, *Int. J. Environ. Anal. Chem.,* 19, 85, 1985.
70. **Slanina, J., Baard, J. H., and Zijp, W. L.,** Tracing the sources of the chemical composition of precipitation by cluster analysis, *Water Air Soil Pollut.,* 20, 41, 1983.
71. **Galloway, J. N., Thornton, J. D., Norton, S. A., Volchok, H. L., and McLean, R. A. N.,** Trace metals in atmospheric deposition: a review and assessment, *Atmos. Environ.,* 16, 1677, 1982.
72. **Nriagu, J. O.,** Worldwide contamination of the atmosphere with toxic metals, in *Proc. The Deposition and Fate of Trace Metals in our Environment,* Verry, E. S. and Vermette, S. J., Eds., U.S. Department of Agriculture, Forest Service, North Central Forest Experiment Station, St. Paul, MN, 1992.
73. **Gatz, D. F.,** Urban precipitation chemistry: a review and synthesis, *Atmos. Environ.,* 25B, 1, 1991.
74. **Eisenreich, S. J., Metzer, N. A., and Urban, N. R.,** Response of atmospheric lead to decreased use of gasoline, *Environ. Sci. Technol.,* 20, 171, 1986.
75. **Gatz, D. F., Warner, B. K., and Chu, L. C.,** Solubility of metal ions in rainwater, in *Deposition Both Wet and Dry,* Hicks, B. B., Ed., Butterworths, Boston, 1984, chap. 8.

76. **Harrison, R. M. and Perry, R.,** The analysis of tetraalkyllead compounds and their significance as urban air pollutant, *Atmos. Environ.,* 11, 847, 1977.
77. **Van Cleuvenbergen, R. J. A., Chakraborti, D., and Adams, F. C.,** Occurrence of tri- and di-alkyllead species in environmental water, *Environ. Sci. Technol.,* 20, 589, 1986.
78. **Radojevic, M. and Harrison, R. M.,** Concentration, speciation, and decomposition of organolead compounds in rainwater, *Atmos. Environ.,* 21, 2403, 1987.
79. **Allen, A. G., Radojevic, M., and Harrison, R. M.,** Atmospheric speciation and wet deposition of alkyllead compounds, *Environ. Sci. Technol.,* 22, 517, 1988.
80. **Faulstich, H. and Stournaras, C.,** Potentially toxic concentrations of trialkyllead in Black Forest rainwater samples, *Nature,* 317, 714, 1985.
81. **Unsworth, M. H. and Harrison, R. M.,** Tree death — is lead killing German forests, *Nature,* 317, 674, 1985.
82. **Brosset, C.,** The behavior of mercury in the physical environment, *Water Air Soil Pollut.,* 34, 145, 1987.
83. **Bloom, N. S. and Watras, C. J.,** Observations of methylmercury in precipitation, *Sci. Tot. Environ.,* 87/88, 199, 1989.
84. **Fitzgerald, W. F., Mason, R. P., and Vandal, G. M.,** Atmospheric cycling and air-water exchange of mercury over mid-continent lakes, *Water Air Soil Pollut.,* 56, 745, 1991.
85. **Nriagu, J. O. and Pacyna, J. M.,** Quantitative assessment of worldwide contamination of air, water and soil by trace metals, *Nature,* 333, 6169, 1988.
86. **Sorensen, J. A., Glass, G. E., Schmidt, K. W., Huber, J. K., and Rapp, G. R., Jr.,** Airborne mercury deposition and watershed characteristics in relation to mercury concentrations in water, sediments, plankton, and fish of eighty northern Minnesota lakes, *Environ. Sci. Technol.,* 24, 1716, 1990.
87. **Fitzgerald, W. F., Mason, R. P., and Vandal, G. M.,** Atmospheric cycling and air-water exchange of mercury over mid-continent lakes, *Water Air Soil Pollut.,* 56, 745, 1991.
88. **Ferrara, R., Maserti, B., Petrosino, A., and Bargagli, R.,** Mercury levels in rain and air and the subsequent washout mechanism in a central Italian region, *Atmos. Environ.,* 20, 125, 1986.
89. **Nguyen, V. D., Merks, A. G. A., and Valenta, P.,** Atmospheric deposition of acid, heavy metals, dissolved organic carbon, and nutrients in the Dutch Delta area in 1980-1986, *Sci. Tot. Environ.,* 99, 77, 1990.
90. **Church, T. M. and Scudlark, J. R.,** Trace elements in precipitation at the middle Atlantic coast: a successful record since 1982, in *Proc. The Deposition and Fate of Trace Metals in our Environment,* Verry, E. S. and Vermette, S. J., Eds., U.S. Department of Agriculture, Forest Service, North Central Forest Experiment Station, St. Paul, MN, 1992.
91. **Church, T. M., Tramontano, J. M., Whelpdale, D. M., Andreae, M. O., Galloway, J. N., Keene, W. C., Knap, A. H., and Tokos, J., Jr.,** Atmospheric and precipitation chemistry over the north Atlantic Ocean: shipboard results, April-May 1984, *J. Geophys. Res.,* 96, 18, 795, 1991.
92. **Church, T. M., Veron, A., Patterson, C. C., Settle, D., Erel, Y., Maring, H. R., and Flegal, A. R.,** Trace elements in the north Atlantic troposphere:shipboard results of precipitation and aerosols, *Global Biogeochem. Cycles,* 4, 431, 1990.
93. **Fitzgerald, W. F.,** Atmospheric and oceanic cycling of mercury, in *Chemical Oceanography Vol. 10, SEAREX: The Sea/Air Exchange Program,* Duce, R. A., Riley, J. P., and Chester, R., Eds., Academic Press, London, 1989, chap 57.
94. **Vong, R. J., Hansson, H. C., Ross, H. B., Covert, D. S., and Charlson, R. J.,** Simultaneous observations of rainwater and aerosol chemistry at a remote mid-latitude site, Symp. Acid Rain: Sources and Atmospheric Processes, presented at the 191st meeting of the American Chemical Society, New York, 1986.
95. **Sheppard, D. S., Patterson, J. E., and McAdam, M. K.,** Mercury content of Antarctic surface ice and snow: further results, *Atmos. Environ.,* 25A, 1657, 1991.

Chapter 6

Interstitial Waters

Jihua Hong, Wolfgang Calmano, and Ulrich Förstner

CONTENTS

I. INTRODUCTION

Although interstitial waters occupy only a small part of the total volume of waters in the earth, recent studies[1-4] have led to an increased awareness of the important role played by interstitial waters in ecosystems. During the last two decades, a significant progress has been achieved on both sampling and analytical techniques, particularly for the study of trace elements in the aquatic environment.

Interstitial water exists in solid materials of the earth surface, such as rock, soil, and sediment. Compared with other solid materials, sediment has been given more concern in most previous research.

0-8493-6304-7/95/$0.00+$.50

In this chapter the discussion will focus on distribution, transport, and cycle of trace elements in sediment interstitial waters, as well as, to a lesser extent, in soil interstitial waters.

Interstitial water studies presently focus on the following subjects: early diagenesis, impact assessment of soil, sediment and other solid materials, and sediment criteria development. For the chemical understanding of early diagenetic processes, the knowledge about trace elements in sediment interstitial waters is very important. In particular, the amount of trace elements deposited in the sediment fraction can more accurately be predicted by analyzing the trace element content and speciation in the interstitial water as a function of depth rather than by measuring the overlying water. Sediment interstitial water represents an important phase in the transport of trace elements from the sediment to the overlying water and *vice versa*.

The studies of interstitial water, which is an important medium for trace element reactions, have provided a link between water transport processes and sedimentary accumulation by showing evidence for the release of trace elements associated with the degradation of biogenic detritus in bottom sediments.[5] The most obvious variation in sediment profiles is the change in redox potential. This results in a redistribution of trace elements between solid and solution phases. From a chemical viewpoint, the sediment-water exchange processes and complexation processes of trace elements in interstitial waters seem to be the most important.

In practical application, interstitial water studies on trace elements have become an important tool for sediment criteria development. The composition of interstitial water is the most sensitive indicator of the types and extent of reactions that take place between pollutants on waste particulate and the aqueous phase which contacts them. Particularly for fine-grained material, the large surface area related to the small volume of its entrapped interstitial water ensures that minor reactions with the solid phases will be indicated by major changes in the composition of the aqueous phase.

In order to assess the trace elements in interstitial waters it is necessary to respect some important factors:

- The effective sampling techniques for avoiding possible chemical changes and problems due to adsorption and contamination
- Distribution characteristics and speciation of trace elements in interstitial waters
- Possible influence of solid "speciation" of trace elements on their concentrations in interstitial waters
- Relationship between the distribution of trace elements and physico-chemical parameters such as pH, Eh, etc.

II. SAMPLING AND ANALYSIS

A. GENERAL CONSIDERATIONS

Because of the complex nature of sediment/interstitial water systems and the particular behavior of trace elements, sampling and preserving of interstitial water should be done under conditions as close as possible to the natural state before analysis. The following points should be considered during the sampling.[1,2,6]

Original redox state. Before interstitial water is extracted from wet sediment, keeping the original redox state of the sediment/interstitial water is considered the most important point due to three reasons: (1) some trace element species are sensitive to the change in redox potential.[7] (2) Interstitial waters show different intensities of anaerobic processes in bottom sediments. Oxygen can diffuse through gas-liquid equilibrium processes into the interstitial water or can directly react with the substances in the interstitial water, if the samples are exposed to the atmosphere. (3) The change in redox state can indirectly lead to a change in concentrations of trace elements. Ferric hydroxide formed during oxidation, for example, can coprecipitate some trace metal ions such as Cu^{2+}, Pb^{2+}, and Zn^{2+}.

Original pressure, temperature, and components. Prior to extraction of interstitial water from sediment solids, the pressure in the system and the components of the interstitial water should be kept stable or changed as little as possible. Evaporation and release of carbon dioxide, hydrogen sulfide, and other volatile substances will lead to changes in pH and solubility of relevant components. Trace element speciation is greatly affected by these changes. Temperature should be kept stable, if the analysis for gases (e.g., O_2, CO_2, and H_2S) is required.

Contamination. Any trace element contamination introduced by sampling, handling, storage, and analysis should be avoided. In general, tools for sampling and handling of the interstitial water should be made with nonmetallic materials.

B. SEDIMENT SAMPLING METHODS

Except for *in situ* dialysis, the sediment sample has to be collected in order to get an interstitial water sample for trace element analysis. A number of reports have introduced various sampling techniques and applications.[8-38] In many nearshore environments, sediments may be sampled directly by shoveling or scooping at low tide, or simply by an open plastic tube. The latter operation can be performed from a small boat or by scuba divers,[1,8] thus allowing considerable flexibility in use. Nevertheless, many, perhaps most, sampling programs cannot be conveniently carried out with handheld equipment, but rely on samplers lowered to the bottom on cables or allowed to free-fall with flotation devices for automatic return to the surface. Without expensive camera or television monitors, the investigator cannot see a remote sampler at work and this leads to questions as to the representativeness of the recovered sample.

The increased interest in the reconstruction of historical developments of pollution influences has led to widespread sampling of sediment cores. Such profiles may cover the last 200 years of industrial development, and in accordance with an average sedimentation rate for lacustrine and marine coastal environments of approximately 1 to 5 mm per year, which is applicable to moderately humid climates, the procedure entails sampling of a core of 20 cm to 1 m in length. A number of devices have been designed to "core" an undisturbed small section of bottom sediments.[8-20] Gravity cores and box cores are often used for sampling the sediments. The sampling devices depend on sediment types, operation conditions, and quantity of sample required. Metal devices can be modified with plastic liners to avoid sample contamination. Large, undisturbed samples for studies of trace elements in interstitial waters are best collected with a box corer.[5,12,19-30] These devices usually recover a core having a surface area of 30 × 40 cm or 20 × 30 cm and a depth of 30 to 40 cm from bottom sediment surface.

The interstitial water should be separated from the sediments as soon as possible. If it is impossible to do this, the samples should be stored at 4°C under nitrogen atmosphere.

C. SEPARATION OF INTERSTITIAL WATER

Several methods have been developed for separating interstitial water. These methods can be classified as follows: (1) indirect methods including squeezing and centrifugation; (2) *in situ* methods including dialysis and suction, mainly dialysis.

1. Squeezing

Squeezing is a traditional technique for separation of interstitial water from sediments and soils.[5,14,16,27,28,35-46] Cores can be extruded using gas pressure, then sectioned and stored in plastic liners which can be cut open. In trace element studies this operation is performed in an inert atmosphere, e.g., in a glove box, to avoid oxidation of any anoxic material and possible changes in speciation distribution.[47-49] For the study of trace elements, especially on shallow, unconsolidated sediments, Reeburgh Squeezers using inert gas (e.g., nitrogen and helium gases; pressure up to 1 MPa) are commonly applied.[50] Some modifications for the original Reeburgh Squeezer have been reported.[51-57]

Another technique, the high-pressure squeezing,[38] is effective and convenient in removing interstitial water from consolidated low porosity sediments or soils. Pioneering studies by Ramann et al.[58] required 40 kg soil in 11 squeezings with a "hydrostatic press" at 300 kg/cm^2 to obtain 1 l of interstitial water. Also using a hydraulic press, Lipman[59] extracted interstitial water at 3000 kg cm^{-2}. These studies have been criticized by Northrup,[60] who pointed out that such high pressure may change physico-chemical equilibria and chemical composition of the interstitial waters. Application of self-sealing free gaskets to steel squeezers by Kriukov[61] greatly improves the technique of hydraulic pressure filtration; these simple and effective squeezers can be used both for soil and sediment samples. Further modifications[38,62] were successfully employed on investigations of interstitial waters from ocean sediments. Thick-walled steel squeezers, permitting pressures up to 10,000 kg cm^{-2}, have obtained interstitial water from even dense sedimentary rocks.[63] However, the most widely used devices[63] produce pressures from 200 to 700 kg cm^{-2}.

2. Centrifugation

a. Direct Centrifugation

Centrifugation, commonly used for extraction of soil water, is now increasingly applied on sediment.[5,13,17,18,25,26,30,64-72] It has the advantage that only little sample manipulation is required. The sediment can be placed directly in precleaned polyethylene centrifuge tubes, and water will be separated above the solid material for fine-grained compressible sediments. However, for coarse-grained sediments, water is collected at the bottom of the tube, and a basal collection cup is required. In order to avoid oxidation, sediments are extruded from the core liners and filled into centrifuge bottles in a glove box under nitrogen gas.

b. *Solvent-Displacement Centrifugation*

A number of displacements, mostly halogenated hydrocarbons, can be used for the extraction of interstitial water. Desirable properties are high density, low water solubility, low volatility, low toxicity, and high chemical inertness. Scholl[73] introduced the use of an immiscible liquid extraction procedure for removing interstitial water from coarse-grained sediments. The procedure utilizes a liquid that is less dense and more viscous than the interstitial water. It is placed above the sediment in a filter press and forced through the sediment by gas pressure displacing the water. High molecular-weight epoxy plasticizers were used as immiscible liquids for this purpose. Carbon tetrachloride has been used but has the disadvantages that its vapor is quite toxic and it attacks most of the plastics (e.g., polypropylene) employed in centrifuge tubes.

Batley and Giles[6,74] studied the use of a water-immiscible fluorocarbon solvent to displace interstitial water during centrifugation with no loss of trace metals. The samples were centrifuged at *in situ* temperatures. The solvent Fluorinert FC-78 is an inert dense liquid (density = 1.7 g/cm^3), having a low dielectric constant (1.8). Because FC-78 has a low boiling point (50°C), it is readily separated from the interstitial water by vacuum distillation. Kinniburgh and Miles[75] used trifluoroethane (F113) as a displacement, which is compatible with polypropylene.

3. *In Situ* Dialysis

The use of *in situ* dialysis methods has become increasingly popular for the study of trace elements in sediment interstitial water since its introduction by Hesslein[76] and Mayer.[77] The principle operation of the dialysis sampler is the equilibrium of a contained quantity of water with the surrounding water through a dialysis membrane. The contained water is then removed from the system and analyzed.[76] A major advantage of this method is that it avoids changes in temperature and pressure in the sediment during extraction of interstitial water. Otherwise, the escape of gases might result in trace element precipitation and pH variation. This technique has been applied in investigations on the geochemical distribution and cycling of trace elements in estuarine, bay, and other sediment profiles.[68,77-79] The samplers have been modified and used by different authors.[80-87] After the assembled sampler is inserted into sediment, the equilibrated time required is estimated experimentally. Carignan et al.[85] suggested that 20 days for cold (4 to 6°C) and 15 days for warm (20 to 25°C) sediments appear to provide time periods long enough for equilibration. Hesslein[76] and Schwedhelm et al.[79] found about 1 week was needed for the equilibration with interstitial water. Mayer[77] and Bottomley and Bayly[81] recommended 100 h and 10 days, respectively.

D. STORAGE AND ANALYSIS

Storage and analysis of interstitial water is similar to surface water samples described in Chapter 3. Generally, the interstitial water samples acidified to a pH of 1 to 2 are suitable for storage and analysis for most trace elements. The solution should be frozen unless the samples are analyzed immediately after the separation. Species analysis for trace elements of the samples kept under nitrogen atmosphere, however, should be performed as soon as possible after sampling without acidification.[70]

III. MOBILIZATION OF TRACE ELEMENTS IN SOLID/INTERSTITIAL WATER SYSTEMS

A. INTRODUCTION

Interstitial water is formed by the entrainment of water during sedimentation.[6] This results in its eventual trapping and isolation from the overlying water, where it may be considered to be in equilibrium with the sediment of which it is a part. This environment usually differs considerably from that of the surface water. Oxygen is exhausted near the sediment surface by microorganisms in the sediments. Microorganisms use the oxidized forms present in the sediment as electron acceptors, thereby decreasing the redox potential of the sediment. In many sediments this zone is readily identified by the presence of black reduced iron sulfide. The pH of the water is also generally lower than that of the surface water. In this anaerobic, generally organic-rich environment, both deposition of reduced trace element species and element mobilization as organic complexes are possible reactions. At shallow depths (above 1 m), the porosity of the sediment will permit diffusive transport of trace elements in the interstitial water, with possible chemical transformations and eventual release into the overlying water. This plays an important role for the ecological significance. Generally, trace elements are sensitive to environmental conditions such as pH, Eh, temperature, and the change in other chemical components (e.g., volatilization of dissolved gases such as H_2S and CO_2). Of particular interests in the understanding of these processes are studies of the distribution of trace elements and the chemical speciation of dissolved elements in the interstitial waters.

Solubility and mobility of particulate-bound trace elements can be changed by five major factors in solid/interstitial water systems:[4,89-93] lowering pH, increasing occurrence of natural and synthetic complexing agents, increasing salt concentrations, changing redox conditions, and decomposing organic matter containing trace elements. Here, particular attention will be given to pH, redox changes, and organic matter as well as their effects on trace element mobility in solid/interstitial water systems.

B. ACIDIFICATION AND ALKALIZATION

1. pH Variation

In the presence of molecular oxygen, compounds containing Fe, N, and S in lower oxidation states in interstitial waters are ultimately oxidized to Fe^{3+}, NO_3^-, and SO_4^{2-}, thereby releasing equivalent amounts of hydrogen ions into the environment. Most of these oxidation reactions are microbially mediated.[94] Proton transfer processes which are important in the oxidation of solid/interstitial water system are summarized in Table 1. The three elements S, Fe, and N are most important in the redox processes of the solid/interstitial water system. It is not only due to their chemical activity but also to their abundance in natural waters and sediments. For example, in tidal marsh sediments, pyrite (FeS_2) contents are often in the order of 1 to 5 mass percent[95] or even higher.[96] If the contents of chemical components in the sediments are known, the acid-producing potential (APP)[97] can be calculated. The ratio of the mole amount of the reducing species to the mole amount of the produced hydrogen ion, which is called the acid-producing coefficient *f*, is listed in Table 1.

Periodical redox processes can cause a variation in AAP or pH of a soil/interstitial water system. Several types are shown in Figure 1.[98] Permanent acidification in alternating aerobic and anaerobic systems has been investigated.[99,100] It was found that the acidification of the systems due to a "split" of sulfate undergoes two stages (Figure 1). The first stage is characterized by the reduction of sulfate, in particular in tidal flats or seabottom sediments. It leads to an increase in $APP_{(s)}$, while ANC(acid neutralization capacity)$_{(aq)}$ (HCO_3^-) formed during sulfate reduction (Equations 1,2) is removed by tidal turbulence or by diffusion into the overlying waters. As a result mobile $ANC_{(aq)}$ (HCO_3^-) and immobile potential acidity (FeS_2) are separated or "split". The increase in $APP_{(s)}$ leads to a permanent decrease in the $ANC_{(aq)}$ after the next aeration and oxidation and results in extreme acidification of the solid/interstitial water system.

$$CO_2 + H_2O = HCO_3^- + H^+ \tag{1}$$

$$SO_4^{2-} + 2\,H^+ = H_2S + 2\,O_2 \tag{2}$$

If during the reduced stage $M(HCO_3)_2$ were retained in the system, e.g., as precipitated carbonate, the acidity formed in sulfide oxidation would be neutralized exactly by the carbonate formed, without permanent change in ANC and APP. A similar process involving formation of FeS in young nonacid marine clay sediments may also lead to a rapid acidification in the solid/interstitial water system.[101]

Ferrolysis ("dissolution by iron") in periodical redox processes proceeds also in two stages (Figure 1). When the system becomes reduced, part of the Fe^{2+} formed is mobilized and displaces other cations such as Ca^{2+} and Mg^{2+}. The displaced cations together with the anions that appear simultaneously with dissolved Fe^{2+} (mainly bicarbonate and other organic anions[105]) can be removed by percolation or by diffusion into the surface water. During aeration of the system, exchangeable Fe^{2+} is oxidized to essentially insoluble Fe^{3+} oxide and H^+ takes the place of adsorbed Fe^{2+} to the same extent that formerly adsorbed base cations have been leached:[106,107]

$$Fe^{2+}_{(ads)} + 1/4\,O_2 + 3/2\,H_2O = 2\,H^+_{(ads)} + FeOOH \tag{3}$$

In ferrolysis, exchangeable ferrous iron takes the place of ferrous sulfide as the immobile potentially acid substance formed during reduction, while exchangeable H^+ is the acidic product formed after oxidation of exchangeable ferrous iron. Periodic redox processes, ferric iron reduction, and relevant displacement

Table 1 **Acid producing coefficient *f* and main oxidation reactions in interstitial waters**

Element Oxidized	Reaction Equations	*f*
	Inorganic	
S	$H_2S + 2\ O_2 = SO_4^{2-} + 2\ H^+$	2
S	$S^o + 3/2\ O_2 + H_2O = SO_4^{2-} + 2\ H^+$	2
S, Fe	$FeS + 9/4\ O_2 + 3/2\ H_2O = FeOOH + SO_4^{2-}$	2
S, Fe	$FeS_2 + 15/4\ O_2 + 5/2\ H_2O = FeOOH + 2\ SO_4^{2-} + 4\ H^+$	4
Fe	$Fe^{2+} + 1/4\ O_2 + 5/2\ H_2O = Fe(OH)_3 + 2\ H^+$	2
N	$NH_4^+ + 2\ O_2 = NO_3^- + H_2O + 2\ H^+$	2
N	$NO_X + (5 - 2X)/4\ O_2 + 1/2\ H_2O = NO_3^- + H^+$	1
	Organic	
N	$R\text{-}NH_2 + 2\ O_2 = R\text{-}OH + NO_3^- + H^+$	1
S	$R\text{-}SH + H_2O + 2\ O_2 = R\text{-}OH + SO_4^{2-} + 2\ H^+$	2

reactions, volatilization of H_2S, denitrification, oxidation of organic matter, and their effects on pH in solid/interstitial water have been discussed elsewhere.[97]

For anoxic marine sediments, Ben-Yaakov[108] has proposed a model to calculate pH values of the interstitial waters based on four processes: (1) the presence of high concentrations of weak acids and bases which are byproducts of organic decomposition; (2) transfer of charge from the nonprotonlytic species to the protonlytic ions ($SO_4^{2-} \rightarrow HS^-$); (3) precipitation of metal sulfides; and (4) precipitation of calcium carbonate. The model predicts that the pH of the interstitial water in anoxic marine sediments should remain in the range of 6.9 to 8.3.

Typical changes in pH values in interstitial water profiles have been observed in salt marsh sediment.[109,110] In Great Sippewissertt Marsh (U.S.), sediments of the control site and the experiment site treated with 150 g m^{-2} of a sewage had similar pH distribution profiles.[109] Both plots exhibited a minimum pH value at 4 to 8 cm (Figure 2). The profile of sediments treated with sewage sludge differed from the control site in absolute pH values, but not their distribution patterns. A similar subsurface pH minimum was observed in another sediment interstitial water profile[110] and can be explained by sulfide oxidation. The evidence of pH decrease originating from sulfide oxidation in interstitial water was also obtained by Jørgensen and Revsbech in the vicinity of a *Beggiatoa* mat located at the interface of a Danish coastal sediment.[93]

2. pH Influence on Trace Element Mobility

The mobility of trace elements in solid/interstitial water systems is strongly affected by pH. While there is a predominance of simple mineral-solution equilibria for the major elements in the aquatic environment, the behavior of many trace elements is more complex. Trace element equilibrium is determined by such mechanisms as coprecipitation, surface effects, and interaction with inorganic and organic phases.[111,112]

It can be expected that changes from reducing to oxidizing conditions, which involve transformation of sulfide and a shift to more acidic conditions, will increase the mobility of typical "chalcophilic" elements such as Hg, Zn, Pb, Cu, and Cd. On the other hand, the mobility is characteristically lowered for Mn and Fe under oxidizing conditions. Geochemical mobility, in response to environmental acidification, will significantly increase, if Mn, Zn, and, to a lesser degree, Cd, Co, and Ni are also present in the active fraction of the sediments. The elements forming anionic species, such as S, As, Se Cr, and Mo, are appreciably solubilized at neutral to alkaline pH conditions.[113]

The study of Carignan and Nriagu on Clearwater Lake and McFarlane Lake located in the Sudbury smelter area of Ontario provided an example of trace element mobility in a low pH environment.[89] In the core-top 5 cm interstitial waters, Cu concentration reached about 0.8 μM in Clearwater Lake (pH 4.5), while it was not more than 0.26 μM in McFarlance Lake (pH 8). Even stronger pH effect could be found for Al in Clearwater Lake, where interstitial water Al concentrations was found as high as 8 μM compared to <0.5 μM in McFarlane Lake. Similar results for Zn were found in the other eight lakes with different pH background.

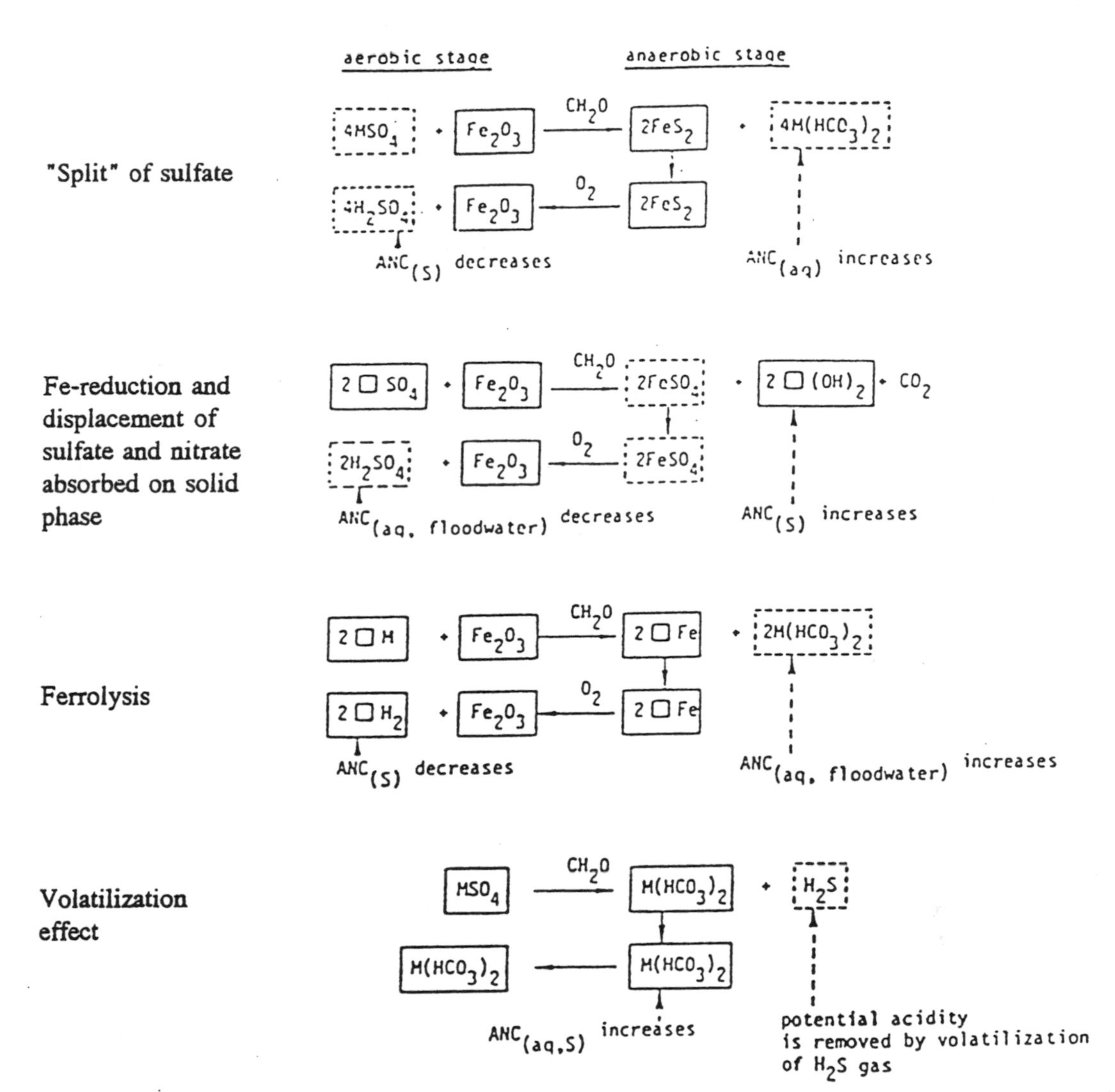

Figure 1 Diagram illustrating the chemical changes in periodical aerobic (left) and anaerobic (right) samples. Components in broken-line rectangles are mobile. Small squares denote exchangeable sites. M stands for a divalent cation (for example, Ca^{2+} or Mg^{2+}) and CH_2O refers to oxidizable organic matter. (From Breemen, N. van, *Neth. J. Agr. Sci.*, 35, 271, 1987. With permission.)

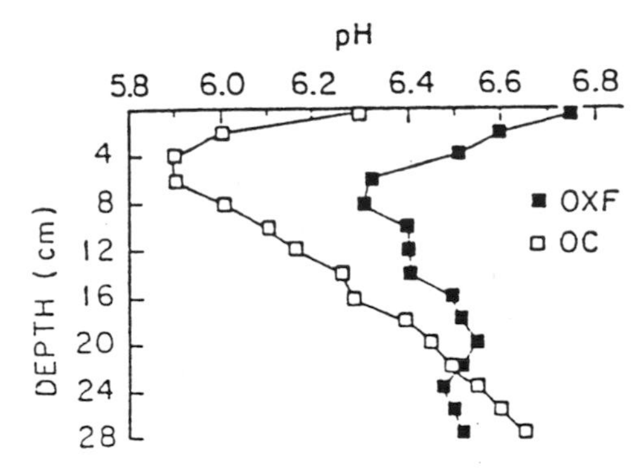

Figure 2 The pH values in interstitial waters in Salt Marsh sediments. (From Giblin, A. E., Luther, G. W., III, and Valiela, I., *Estuarine Coastal Shelf Sci.*, 23, 477, 1986. With permission.)

C. REDOX ENVIRONMENT IN INTERSTITIAL WATERS

The most intensive trace element reactions are usually found in sediments where variations of redox conditions take place.[3] Examples of redox processes involving the elements C, N, O, H, Fe, and Mn are arranged in Figure 3 in the sequence of reactions observed in an aqueous system at various Eh values.[114] Since the reactions considered (with the possible exception of the reduction of MnO_2 and FeOOH) are biologically mediated, the chemical reaction sequence is paralleled by an ecological succession of microorganisms — aerobic heterotrophs, denitrifiers, fermentors, sulfate reducers, and methane bacteria. The driving force of reduction processes is the decomposition of organic matter by nonphotosynthetic organisms, thereby obtaining a source of energy for their metabolic needs. The reduction processes involving oxidation of organic matter follow the sequence "aerobic transpiration", "denitrification", "nitrate reduction", "sulfate reduction", and "methane fermentation".

1. Redox Indicators

Ever since the classic work of Krumbein and Garrels,[115] the geochemical environment of trace element mobility in interstitial waters has been characterized in terms of the hydrogen ion activity (pH) and the oxidation-reduction potential (Eh) which sometimes is expressed as the electron activity (pe). The measurement of redox potential is often used as a redox indicator.[116-118] Many of the species which are involved in important oxidation-reduction reactions in sediment/interstitial water system (e.g., SO_4^{2-}, NO_3^-, N_2, NH_4^+, CH_4) are not electroactive, or, in other words, they do not readily take up or give off electrons at the surface of the platinum or gold electrodes used to measure Eh.[117-119] As a result, the measured electrode potential sometimes does not agree with the value of Eh calculated from thermodynamic data and independent measurements of oxidized and reduced species.[7]

Berner developed a classification of interstitial water environment based on the presence or absence of dissolved oxygen and dissolved sulfide in the sediment/interstitial water system.[7] The sediment/interstitial water systems were first of all divided into *oxic* and *anoxic* depending on the dissolved oxygen concentration. Anoxic environments, in turn, were subdivided into *sulfidic* and *nonsulfidic* depending on the presence of measurable dissolved sulfide. Anoxic-nonsulfidic conditions comprise both postoxic milieus resulting from oxygen removal without sulfate reduction (weakly reducing), and methanic environments resulting from complete sulfate reduction with consequent methane formation (strongly reducing). Dissolved oxygen, sulfide in interstitial waters, as well as iron and manganese mineral characteristics in solids, are listed in Table 2. Similar to Berner's work, Froelich et al.[120] defined three classes of marine sediments based on the investigated interstitial water redox profiles: *oxic, suboxic,* and *anoxic.* The suboxic milieu type from Froelich et al. corresponds to the postoxic environment of Berner.[7] The anoxic environment includes the sulfidic and methanic sediment/interstitial water systems.

In practical investigations, Mn is often chosen as a redox indicator in interstitial waters[5,21,79,121] because of its sensitivity for changes in redox conditions and its abundance which makes it easy to be measured. Based on the distribution of Mn in interstitial waters taken from different areas, less reducing, intermediate reducing, and strongly reducing interstitial waters are defined.[13] In other studies for interstitial water profiles, the Mn-reducing zone, as a redox indicator, is discussed with respect to the cycling of trace elements during early diagenesis.[5] Oxygen concentration,[5,21] nitrate,[5,21,79,121] sulfur species, and ammonia[79] are also used as indicators or redox parameters.

2. Vertical Distribution of Redox Zones

The interstitial space of sediments in the aquatic environment and of wetland soils is permanently or intermittently saturated with water. Under these conditions, oxygen diffuses slowly through the water and is exhausted by microorganisms within the first few millimeters of the sediment or soil surface. Below the aerobic sediment surface, facultative anaerobic and pure anaerobic microorganisms use the oxidized forms present in the sediment as electron acceptors, thereby decreasing the redox potential of the system. In this interstitial water system, the redox evolution processes are characterized by various redox zones vertically distributed in the interstitial waters.

Vertical reaction zones for trace elements in interstitial water profiles of suboxic pelagic sediments have been observed by Klinkhammer (Figure 4).[121] The depth interval where interstitial water manganese is undetectable ($<0.02\ \mu M$) is referred to as the *oxidized zone.* Any zone below the oxidized zone where the Mn-depth profile is linear or shows positive curvature and where Fe is undetectable is the *manganese reduction zone.* Depth where Fe is detectable ($>0.2\ \mu M$) is referred to as the *iron reduction zone.* While Cu was released in the oxidized zone, Ni, Fe, and Mn were trapped efficiently. The Ni concentration

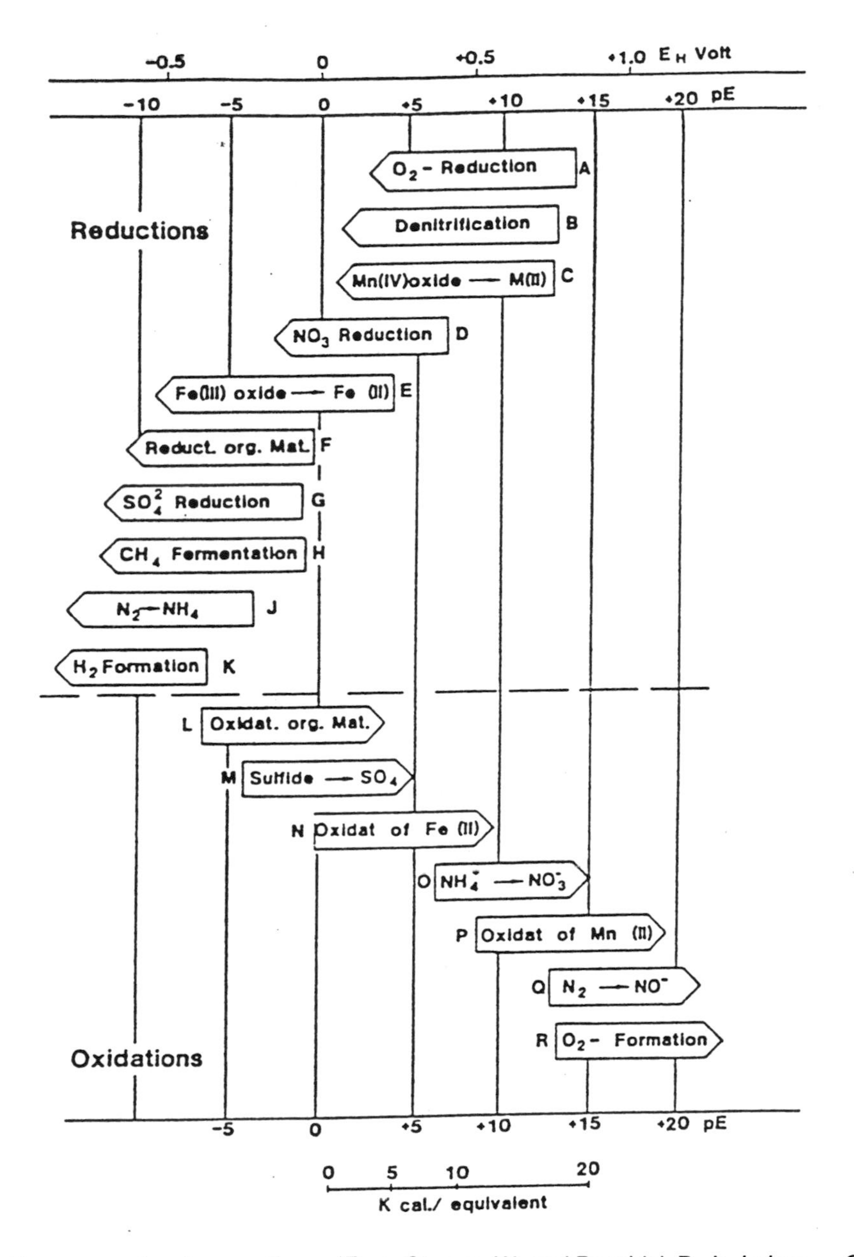

Figure 3 Sequence of redox reactions. (From Stumm, W. and Bacchini, P., in *Lakes — Chemistry, Geology, Physics,* Lerman, A., Ed., Springer-Verlag, New York, 1978, 91. With permission.)

reached a maximum in the Mn reduction zone. In fact, the Ni/Mn ratio in the interstitial waters of the Mn reduction zone were nearly the same in all samples. A similar vertical distribution of redox zones was found by Froelich et al.[120]

Presley and Trefry[1] summarized the typical manganese profiles reported in the literature, which fall into four categories (Figure 5). Case I is observed in a stratified water column with anoxic bottom waters, as is the case for the Black Sea and some lakes. Here, Fe and Mn may be reduced and solubilized in the water column. Case II represents oxic waters overlying an anoxic sediment. Maximum metal remobilization occurs at the sediment-water interface. In Case III, mildly reducing conditions allow the development of a well-defined oxic zone, wherein remobilized trace elements may be trapped and strongly concentrated over natural levels. Well known is the increase in manganese and iron in the top layer of these sediments.[1,122] Robbins and Callender[123] showed that for Lake Michigan sediments, the negative interstitial manganese gradient below the AB horizon may be controlled by the precipitation of rhodochrosite ($MnCO_3$). Case IV is an example of a slowly accumulating sediment where a thick oxidized zone occurs

Table 2 **Classification of sediment/interstitial water systems**

Type	Indicator	Characteristic Phase
Oxic	$[O_2] \geq 1\ \mu M$	Hematite (Fe_2O_3), goethite (FeOOH), MnO_2-type minerals
Anoxic	$[O_2] < 1\ \mu M$	
Sulfidic	$[H_2S] \geq 1\ \mu M$	Pyrite (FeS_2), marcasite rhodochrosite ($MnCO_3$), alabandite (MnS)
Nonsulfidic	$[H_2S] < 1\ \mu M$	
Postoxic		Glouconite and other Fe^{2+}-Fe^{3+} silicates (also siderite, vivianite, rhodochrosite); no sulfide minerals
Methanic		Siderite ($FeCO_3$), vivianite ($Fe_3(PO_4))_2$ rhodochrosite, earite formed sulfide minerals

Modified from Berner, R. A., *J. Sediment. Petrol.*, 51, 359, 1981.

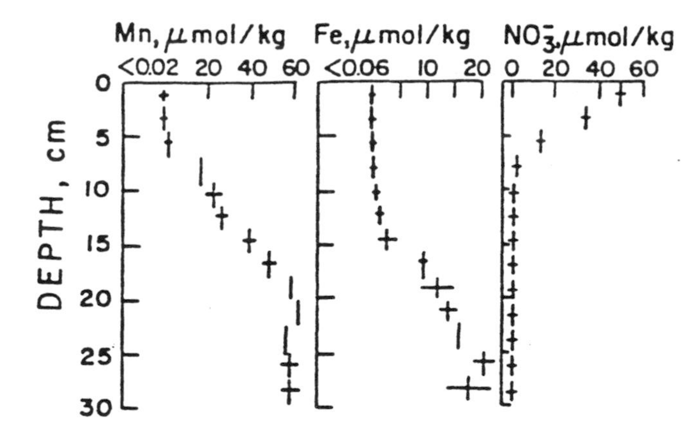

Figure 4 Distributions of interstitial water Mn, Fe and NO_3^- in the redox zone (see text). (From Klinkhammer, G. P., *Earth Planet. Sci. Lett.*, 49, 81, 1980. With permission.)

with possible discrete anoxic sediments. Mn reduction always occurs before iron reduction in the interstitial water profiles.[21] The stronger reducing zone is characterized by the presence of sulfide. A schematic representation in the presence of sulfide has been shown in Figure 6.

Shaw et al.[5] studied trace element mobilization through interstitial water in different redox zones of some sediments. The processes leading to the release and/or removal of the trace elements from the interstitial fluids are represented in summary fashion in Figure 7a and b. Where as in the Patton Escarpment sediments there is a more distinct separation of redox-zone (Figure 7a), in the outer and inner offshore basins these zones are more compact (Figure 7b), often leading to overlaps.

3. Oxic Milieu

Under oxidizing conditions, the controlling solid earth materials may change gradually from metal sulfides to carbonate hydroxides, oxyhydroxides, or silicates, thus changing the solubility of trace elements. Discrepancies have been found between experimental data and calculated equilibrium concentrations: Cd, Cu, Ni, and Pb are far below, while Fe and Mn are far above the equilibrium values. Possible explanations are the scavenging effects of the solids and the formation of humic complexes.[124]

Generally, trace element concentrations in oxic interstitial waters are controlled by adsorption to a wide extent. Mn(IV) and Fe(III) oxides as well as organic matter are often considered as adsorbents for trace elements.[125-127] They tend to form a "coating" on mineral surfaces[128] which have a high affinity to some trace elements. Investigations on the adsorption of trace elements on oxides[126,129-131] indicate that Cd, Co, Cu, Ni, Pb, and Zn are adsorbed by Mn(IV) oxides more strongly than by Fe(III) oxides. The order of adsorption stability and affinity of goethite (α-FeOOH) and amorphous Fe(III) oxyhydrate for trace elements is as follows:[129,131-134]

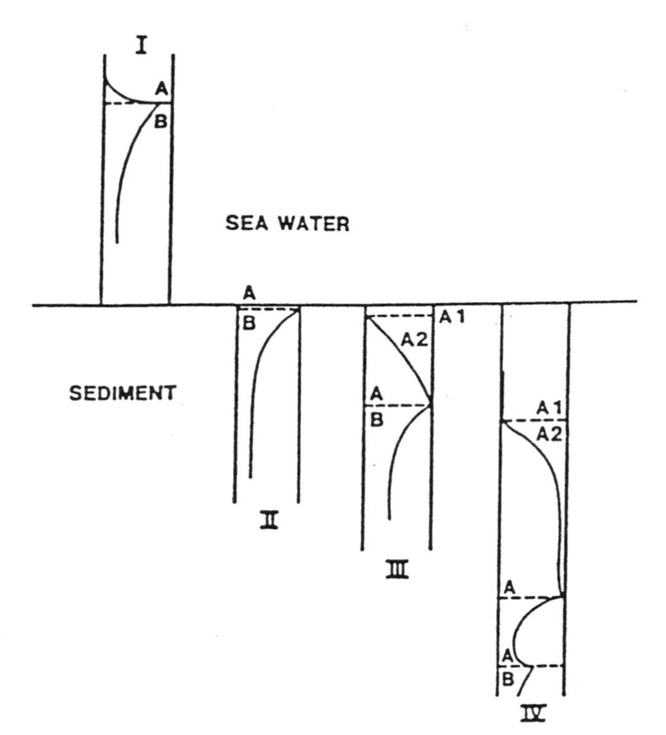

Figure 5 Typical interstitial Mn profiles in sediments. (From Presley, B. J. and Trefry, J. H., in *Chemical and Biogeochemistry of Estuaries,* Olausson, E. and Cato, I., Eds., John Wiley & Sons, New York, 1980, 187. With permission.)

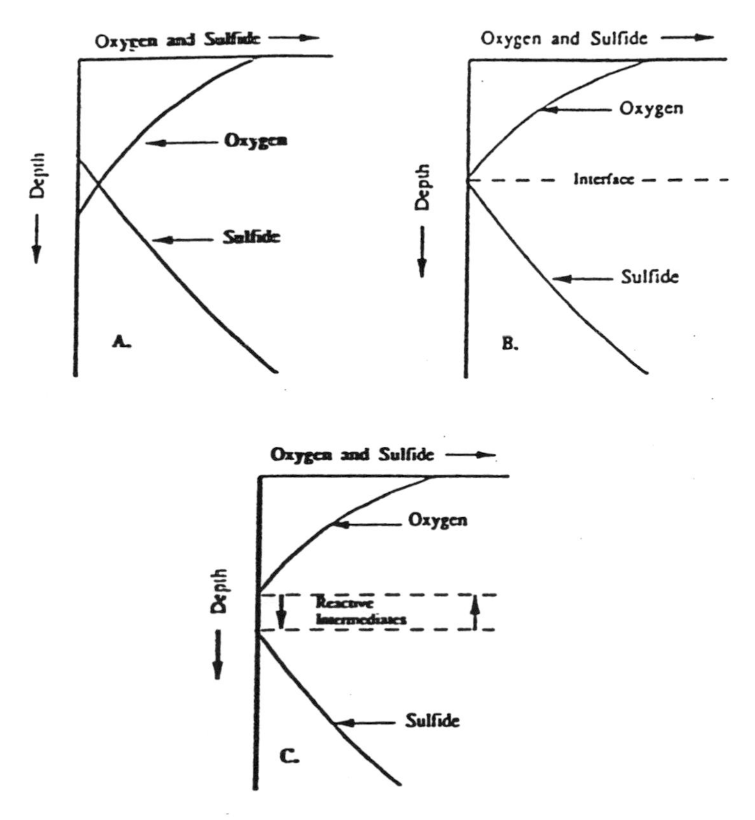

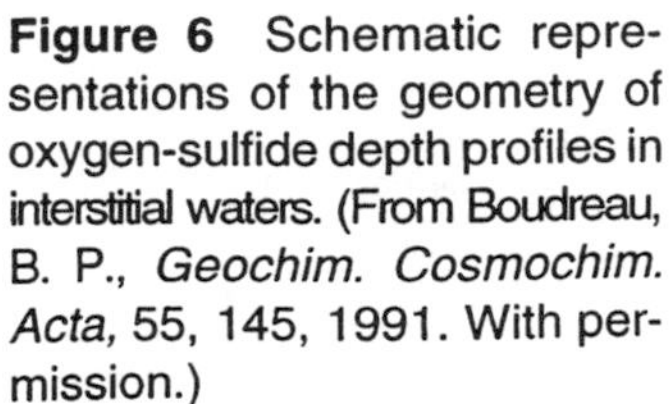

Figure 6 Schematic representations of the geometry of oxygen-sulfide depth profiles in interstitial waters. (From Boudreau, B. P., *Geochim. Cosmochim. Acta,* 55, 145, 1991. With permission.)

$$Cd \approx Ni < Co < Zn << Cu \approx Pb$$

Similarly, for birnessite (δ-MnO_2) and Mn(IV) oxyhydrate, the order is[129,131,135]

$$Ni < Zn, Co < Cd < Cu < Pb$$

The affinity of trace elements with marine oxic sediments with high Mn(IV) content is as follows:[133]

$$Cd < Ni < Zn < Cu < Co < Pb$$

Co(II) has a high affinity to Mn(IV) minerals. It is oxidized by Mn(IV) and transformed into the insoluble trivalent form.[136] Investigations carried out by Balistrieri and Murray[133] on metal adsorption to marine sediments with similar cation exchange capacity but strongly different contents of Mn(IV) oxides suggest that binding of the trace elements Co, Ni, Ba, and Cd increases with Mn(IV) content. In contrast, adsorption of Pb and Zn is influenced only slightly by Mn(IV) content. Arsenic exists as oxyanion (arsenate) in an oxic environment and is adsorbed preferentially on Fe(III) oxides.[127]

The sediment/interstitial water systems in oxic environments contain a large variety of dissolved decomposition products of solid organic matter. Investigations using voltametric methods have shown that Ni,[137] Zn,[138] Cd,[22] Cu[127], and Co[137] in oxic interstitial waters exist largely as inert complexes or bound to colloids. It has been demonstrated that the highest Cd concentration in interstitial water of marine sediments occurs at the interface between oxic and suboxic zones.[22,139] In contrast, Ni,[5,18,21,121] Co,[15] and As[140] concentrations in oxic interstitial water normally do not differ much from those in overlying water or in the suboxic layer of some sediments.

Early sediment changes and trace element mobilization from interstitial water in a man-made estuarine marsh have been investigated by Darby et al.[141] This study exemplifies both mechanisms of release of metals via interstitial water extraction and subsequent changes by the effect of oxidation. Compared to the river water concentration, the channel sediment interstitial water is enriched by a factor of 200 for Fe and Mn, 30 to 50 for Ni and Pb, approximately 10 for Cd and Hg, and 2 to 3 for Cu and Zn.

An example of oxidizing remobilization of cadmium and other trace elements through interstitial water has been studied in a tidal freshwater flat in the upper Elbe estuary near Hamburg.[139,142] Elevated cadmium contents in the rhizomes of the emerged macrophytes indicated a high proportion of bioavailable Cd species in the root zone. The results of the sequential extractions of the core sediment samples separated at 22-cm levels (Figure 8) indicated that in the anoxic zone 60 to 80% of the Cd was associated with the sulfidic/organic fraction. In the upper oxic and transition zone, the association of Cd in the carbonatic and exchangeable fractions increased simultaneously up to 40% of total Cd. Thus, high proportions of mobile cadmium forms correlated with the reduction in total cadmium contents. Higher amounts of labile cadmium forms were accompanied with a marked depletion in the total content of metal compared to that in the anoxic sediment zone. Comparison of the fraction patterns and total contents of other diagenetically less mobile examples indicated that a significant proportion of cadmium was leached from the surface sediment through interstitial water transfer by a process of *"oxidative pumping"* by tidal water drainage in this high-energetic environment. This could result in migration of the mobilized trace elements into either the deeper anoxic zone, where it could precipitate again to contribute to the enhanced oxidizable sulfidic/organic fraction, or to the surface water, from where it could be exported to the outer estuary.

4. Suboxic Milieu

In suboxic environments, dissolved nitrates in interstitial waters, and Mn(IV) and Fe(III) oxides in the solid phase, act as electron acceptors for microbial decomposition of organic matter. Since oxides are important binding sites for many trace elements, their dissociation leads to the release of other trace elements. In contrast to oxic environments, not only oxidation of organic matter but also reduction of terminal electron acceptors contributes to mobilization of trace elements. Higher concentrations of Co and Mn in suboxic sediment layers compared to oxic conditions have been found in many investigations.[15,43,143] Similar effects have been observed for Ni.[5,18,21,43,121] Although in different sediments, the Mn reduction zone in interstitial waters occurs at different depths,[13] there is always a strong correlation between the Mn and Ni concentrations in the interstitial waters. Arsenic concentration in suboxic

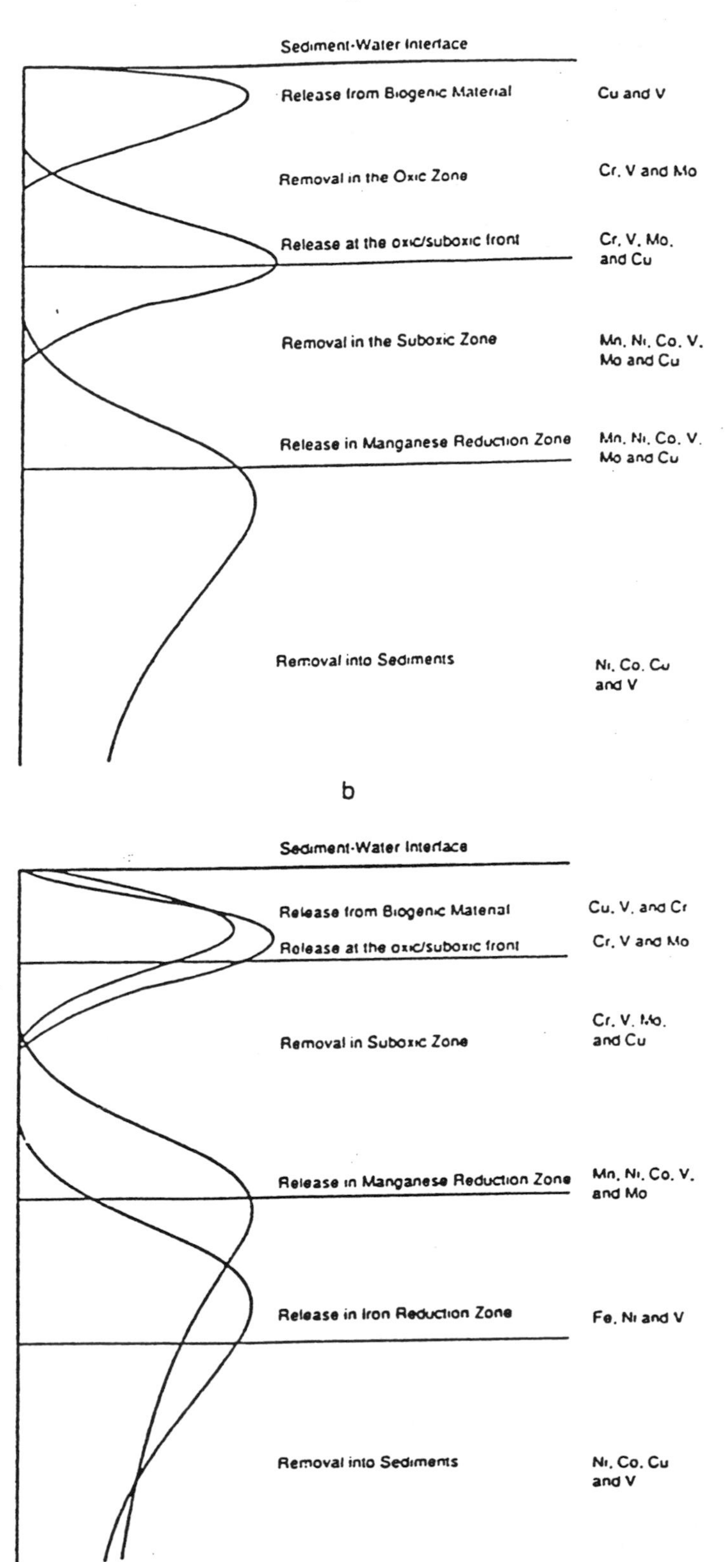

Figure 7 Schematic representations of interstitial water trace element profiles in the Southern California Borderland for two types (a) and (b) showing characteristic zones of release and removal. (From Shaw, T. J., Gieskes, J. M., and Jahnke, R. A., *Geochim. Cosmochim. Acta,* 54, 1233, 1990. With permission.)

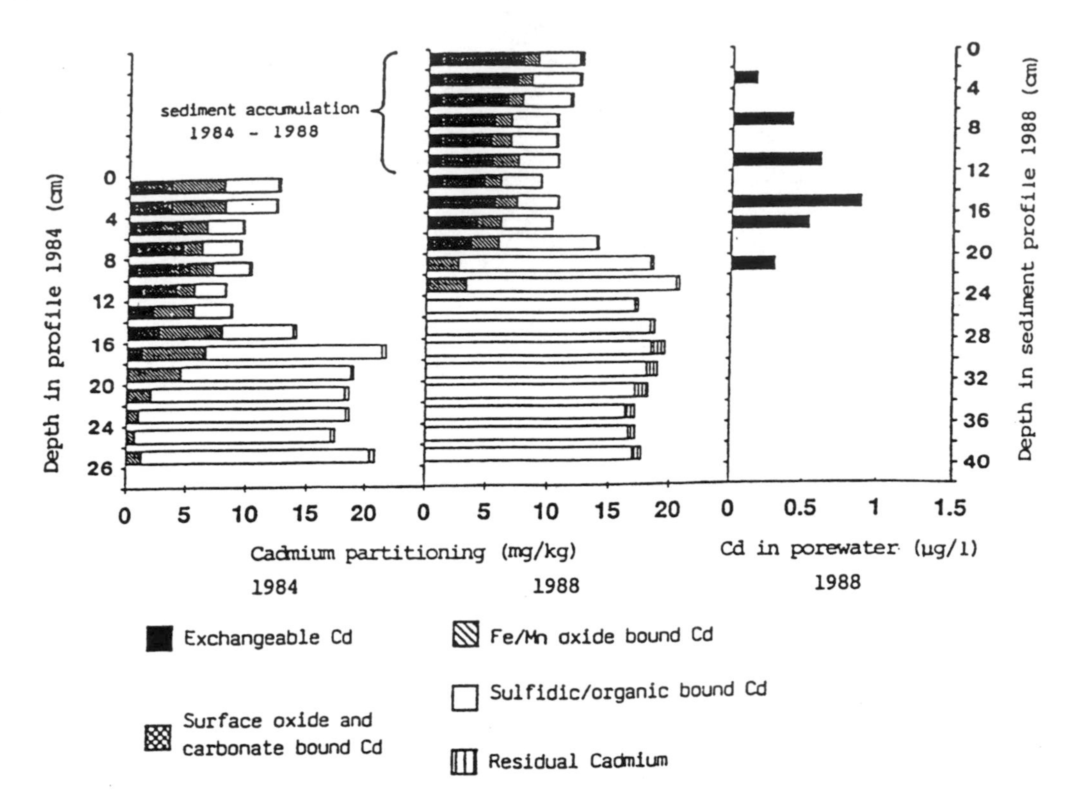

Figure 8 Interstitial water concentrations and partitioning of cadmium in a tidal flat sediment profile in the Heuckenlock area sampled in 1984 and 1988. (From Förstner, U., et al., in *Metal Speciation in the Environment,* Broekaert, J. A., et al., Eds., Springer-Verlag, Berlin, 1990, 1. With permission.)

interstitial waters is significantly higher than that in oxic interstitial waters.[144,145] It is therefore suggested that As release is resulting from mobilization of iron compounds.

In contrast to Co, Ni, and As, some other metals such as Cd and Zn were not found to be mobilized in suboxic environments.[18,43] Investigations by Shaw et al. on the influence of Mn(IV) oxides on Co, Ni, and Cu mobility indicate that there is a constant comobilization of Co with Mn(IV) in interstitial water and release of Ni accompanying reduction of Mn(IV) oxides. However, the Cu concentration in suboxic interstitial waters is always lower than that in the oxic zone. Similarly, Klinkhammer[121] found a decrease of Cu concentration at the oxic-suboxic interface of some sediments.

With the extraction method, reducible Fe and Mn can be found as their main form. A correlation of Co and Ni contents with Mn content in the reducible fraction of suboxic and oxic sediments has been reported.[43,147-150] However, Cu is less significantly correlated with Mn,[149,150] while Pb[150] and Cd[148] are not affected by Mn mobilization. On the other hand, mobilization of As[146,147] and Cd[151] occurs at the time of Fe mobilization.

5. Anoxic Sulfidic Milieu

In anoxic environments, the metal distribution is likely to be controlled primarily by the formation of (highly insoluble) sulfide in marine sediment interstitial waters.[152-163] At lower free S^{2-} concentrations such as found in freshwater system, it should rather be controlled by the stability of carbonate and phosphate phases.[160] The behavior of interstitial water trace elements in anoxic sulfidic environment is very different from that in other environments. In most cases, sulfide ions will "titrate" trace elements,[122] resulting in their removal from interstitial water. Moreover, pyrite (FeS_2) formed during the "titration" in anoxic sulfidic environment can adsorb or coprecipitate trace elements.[152] Adsorption experiments show that the pyrite adsorption of divalent cations of trace elements decreases in the order $Co^{2+} < Cd^{2+} < Mn^{2+} < Ni^{2+} < Zn^{2+}$.[152] On the other hand, trace elements can be mobilized by the formation of organic and inorganic complexes.

The concentrations of trace Cu and Fe in interstitial water of sediments from a salt marsh along the Delaware estuary have been found to be controlled mainly by sulfide formation.[42] Microprobe studies of

Zn in anoxic sediments performed by Elderfield et al.[153] have suggested that it is more likely for metals to be scavenged by Fe sulfides than to form discrete sulfide minerals. The investigations by Luther et al.[154] on sediments of Newark Bay, New Jersey, using scanning electron microscopy with energy dispersive X-ray analysis evidenced zinc sulfide minerals; Ni and Mn, however, were incorporated into the structure of pyrite and iron monosulfide crystals, respectively.

Hallberg[155-157] has expressed the opinion that both precipitation processes and the influence of organic chelators are major counteracting factors controlling trace element distribution in anoxic interstitial solutions. Presley et al.[39] described the mobilization of Zn, Ni, and Cu from reducing sediments to the reconstitution of organic complexes and the stabilization of trace elements initially leached from silicates and oxides. In addition, bacterial action may lead to a further decrease in pH values, thereby increasing the solubility of various trace elements. Elderfield and Hepworth[158] found that the concentrations of most trace elements in the interstitial waters were significantly higher than could be predicted from their sulfide solubility. This effect is most striking for copper, lead, and zinc, which diverge by factors 10^{21}, 10^{12}, and 4×10^{8}, respectively, from the computed figure.

D. EFFECTS OF ORGANIC SUBSTANCES ON MOBILITY OF TRACE ELEMENTS

There is a long discussion about the effect of organic matter on the binding and mobilization of trace elements in natural waters. Reaction models of the degradation of organic matter in marine sediments have been developed,[164,165] and a basic assumption is that organic carbon is oxidized continuously by a sequence of energy-yielding reactions with decreasing energy production per mole of organic matter oxidized. Cu is often considered as a trace element typically associated with organic carbon.[18,121] The calculated ratios of the Cu/C flux support the argument for copper remobilization during organic carbon oxidation.

Sediment humic substances have the ability to chelate trace elements.[45,162,166-173] Nissenbaum and Swaine[162] found that the elements concentrated in the interstitial solutions are those which are also enriched in sedimentary humus. The transport of dissolved metals in the interstitial waters, therefore, is strongly influenced by the vertical gradient of dissolved humic substances.[45] In general, the mechanism is thought to involve the leaching of trace elements from various mineralogical phases and their chelation by low molecular weight "dissolved humic compounds". These are considered to be precursors of high molecular weight organics, which, in turn, are diagenetically altered to sedimentary humus, thereby providing a pathway for the accumulation of humic substances and their associated trace elements in organic-rich marine sediments. It has been proposed by Elderfield[30] that most of the organically associated trace elements are colloidal entities, equivalent to the high molecular weight organics. Significant enrichment of trace elements in interstitial waters has been explained by effects of *complexing* by organic substances.[16,24,39] In this respect, Jonasson[170] has established a probable order of binding strength for a number of metal ions onto humic or fulvic acids: $Hg^{2+} > Cu^{2+} > Pb^{2+} > Zn^{2+} > Ni^{2+} > Co^{2+}$. The transport of dissolved trace elements in the interstitial waters is, therefore, strongly influenced by the vertical gradient of dissolved humic substances.[45]

Experimental data of Lu and Chen[124] indicated that under reducing conditions Fe and Ni concentrations were controlled by complexing organic substances. The high values measured for interstitial copper in comparison with the calculated values may be explained by the presence of humic-copper complexes. Similar results have been obtained by Piemontesi and Baccini[44] for Cu, Tills and Alloway[169] for Pb, and Huynh-Ngoc et al.[137] for Ni and Co. The data of Elderfield[30] indicated that approximately 80% of the total Fe and Cu, approximately 40% of the total Ni, and 18% of the total Mn were organically bound in the interstitial water of surface sediments. In interstitial waters from Loch Duich, Scotland,[45] 74 to 84% of the dissolved Fe was retained by an ultrafiltration membrane (mol wt > 1000) in the high molecular weight (HMW) phase, compared with 3 to 17% of the dissolved Mn. From GFC (gel filtration chromatography) separation of interstitial water samples using Sephadex G50, the complexing capacity of the organic matter was found to decrease with depth in the sediment profiles of Narrows Inlet (Canada).[46] Cu and to a lesser extent Pb were associated with organic matter of 500 to 10,000 mol wt in the interstitial water obtained from the upper 4.6 cm of the sediment profiles. However, Cu and Pb in deeper interstitial water samples were assumed to be in either ionic or inorganic forms because the observed complexing capacity of the organic matter was low. Dissolved organic copper and chromium complexes were measured in interstitial and overlying waters of Narragansett Bay (U.S.) by Douglas et al.[173] Of total Cu, 22 to 67% and 23 to 55% of total Cr were organically bound. The correlation between total and organically bound concentrations is given in Figure 9. Van den Berg and Dharmvanij[138] studied the organic complexation capacity for zinc in interstitial water samples collected from the upper Gulf of

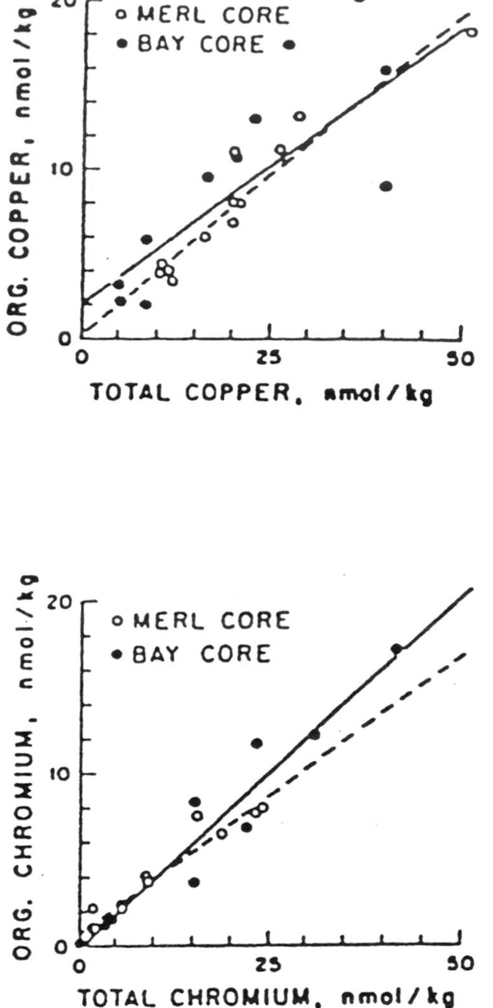

Figure 9 Organic copper versus total copper and organic chromium versus total chromium in interstitial waters for two cores. (From Douglas, G. S., Mills, G. L., and Quinn, J. G., *Mar. Chem.*, 19, 161, 1986. With permission.)

Thailand and the estuary of River Mersey in England. The authors concluded that 93 to 98% of zinc was complexed by organic material in the interstitial water as a result of higher ligand concentrations. The strong correlation of the dissolved concentrations of Mo, V, and Cr with "yellow substances" reflects the importance of dissolved organic matter for the mobility of these trace elements in sediment geochemistry.[20]

Hart and Davies[171] indicated that the major portion of cadmium in the interstitial water from a sediment sample of an urban creek at Melbourne, Australia was associated with colloidal labile species and more strongly bound complexes; to the major proportion, zinc and lead were associated with the colloidal phase, 20% in labile, and 10% in more tightly bound (organic) forms. Water samples from the Woronora River in New South Wales (Australia) indicated an increasing solubility of lead and copper species in the anoxic interstitial water (mangrove sediment) compared with oxic surface water;[6] this may be the result of organic complexation, with which the increase in "labile bound" trace element species is also associated.

E. MIGRATION OF TRACE ELEMENTS BETWEEN OVERLYING WATERS AND SEDIMENT INTERSTITIAL WATERS

Transport of trace elements between sediment and overlying water or between top and deeper layers of the sediment column can be calculated quantitatively on the basic assumption that the diffusive fluxes of trace elements are driven by the interstitial water gradients of these elements. The trace element flux is estimated from Fick's first law of diffusion:

$$F_j = -D_j \, dC_j/dx \tag{4}$$

The assumptions inherent in this simplified equation have been discussed in detail by many authors.[121,174-176] F_j is the flux of a trace element j, which is calculated from the product of the effective diffusion coefficient (D_j) and the concentration gradient *(dC_j/dx)*.

Trace element fluxes into the Mn oxidation zone are referred to as *oxidation zone fluxes*. The trace element fluxes across the sediment/water interface are referred to as *benthic fluxes*.[21] These migration processes can change the distribution of trace elements in the sediment interstitial waters and in the solid phase. For most pelagic sediments the bottom water is oxic; usually, the top of these sediment column is also oxic. Figure 10 illustrates some examples of the upward migration of Mn. This process shows Mn^{2+} transfer from the deeper section in the sediment column to the top by Mn^{2+} interstitial water gradient. Transferred Mn^{2+} will be precipitated in the surface sediment as Mn(IV) oxide. While these processes considerably consume Mn^{2+} and solid bound Mn in the deeper sediment section, freshly formed Mn(IV) oxides accumulate in the surface sediment. As a long-term result of these interactions, the distribution of trace elements between the sediment and the interstitial water is changed strongly (Figure 10a,b).

Another pattern of trace elements described as benthic flux is illustrated in Figure 10c and d. In this way, Cu and Cd, which are released by decomposition of biologic materials, can diffuse upwards into overlying water or downwards into deeper sediment sections. As can be seen in Figure 10c and d, Cd shows high concentrations in the interstitial water of the surface sediment while the solid bound Cd in this sediment section is depleted.

Some investigations have demonstrated that the benthic fluxes of some trace elements are large enough to influence their distribution not only in a local area but also in whole deep ocean waters and bulk sediments. The relative significance of this bottom source can be estimated by comparing it to the flux required to maintain the deep ocean concentrations of the elements.[121] For example, Cu fluxes between 2.5 and 1.05 nmol cm^{-2} yr^{-1} would be required to maintain the distribution of Cu in deep waters in the Pacific.[121,177] Several investigators estimated the migration fluxes of Cu to overlying waters in the Pacific[21,121] to be from 1.8 to 2.82 nmol cm^{-2} yr^{-1}. This agreement suggests that the Cu in deep waters is partially supplied by migration from interstitial waters into overlying waters.

Some migration differences of trace elements have been found in types of sediments with respect to redissolution of the trace elements deposited with particulate matter into overlying waters by upward diffusion of interstitial water. For example, while 2.5% of deposited Mn in the pelagic clay sediments was redissolved into overlying water, 39% of deposited Mn was found in the carbonate ooze. In the siliceous ooze an average percentage was 9% (Figure 11). A greater difference is observed between different elements. For the pelagic clay and siliceous ooze of the northeastern equatorial Pacific Ocean, the average regenerated fraction of deposited manganese was calculated to be less than 5%, whereas more than 88% of deposited Cu was returned to overlying waters. The authors concluded that Mn was associated with terrigenous particles, whereas the input of Cu was predominantly regulated by biogenic phases such as organic matter, skeletal calcium carbonate, or silica. During diagenesis, there was a minimal transport of Mn from sediment to overlying waters through interstitial water. Dissolution of biogenic materials releases Cu, which was partly incorporated in manganese micronodules (Figure 11) actively growing in surface sediments and partly lost into the overlying water column by diffusive transport. Similarly, after calculating the fluxes of copper in the interface of the Resurrection Fjord in Alaska, Heggie[40] found that more than 90% of deposited Cu was returned to the overlying waters. Estimated from Cu concentration changes in deep and bottom waters during the period of bottom isolation the benthic flux of Cu was 32 nmol cm^{-2} yr^{-1}. For Cd, similar migration was observed in Laurentian Trough sediments (Canada) by Gobeil et al.[22] Dissolved Cd was released into the oxidized zone of the sediment, probably due to aerobic degradation of fresh organic matter. About 80% of the total Cd flux to the sediments was returned into the water column by upward diffusion. About a quarter of the remaining 20% Cd buried with the sediment may be explained by downward diffusion and precipitation, perhaps as a sulfide, below the oxidized zone. The downward fluxes in 3 cores were between 10 and 29 ng cm^{-2} yr^{-1}.

Migration of Mn under reducing sediments has been well documented.[27,123,179-182] For example, a study on the migration of dissolved Mn from Mississippi Delta sediments into overlying water column has been performed by Trefry and Presley.[12] It was found that concentrations of Mn were lower in the Mississippi Delta sediments than in Mississippi River particulate. Probably this results from a diffusive transfer of reduced Mn^{2+} species from the sediment interstitial water to the overlying seawater. A hypothetical reaction for the reduction of Mn(IV) can be written as:

$$CH_2O + 2\,MnO_2 + 3\,H^+ = 2\,Mn^{2+} + HCO_3^- + 2\,H_2O \qquad (5)$$

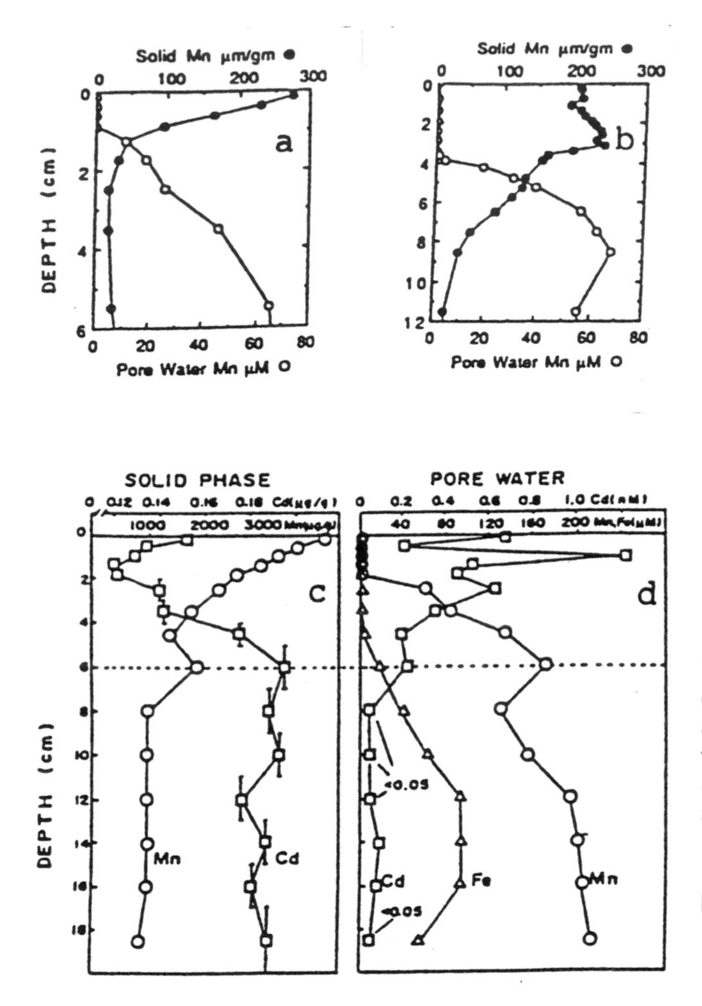

Figure 10 The vertical distributions of solid sediment phase Mn and Ca and interstitial water Mn and Cd profiles. (a) and (b) California Borderland; (c) and (d) Laurential Trough. (From Shaw, T., Gieskes, J. M., and Jahnke, R. A., *Geochim. Cosmochim. Acta,* 54, 1233, 1990; and Gobeil, C., Silverberg, N., Sundby, B., and Cossa, O., *Geochim. Cosmochim. Acta,* 51, 589, 1987. With permission.)

The amount of available organic matter and the sedimentation rate control the extent to which this reaction occurs. The conventional picture is a downward flux of O_2 balanced by an upward flux of Mn^{2+} with little or no flux of the reduced species into the overlying waters. A Mn^{2+} flux to the overlying waters, however, does provide a mechanism for depleting Mn in nearshore sediment and allowing Mn to be transported to the deep sea. The same authors[12] calculated the Mn flux values and changes from nearshore, through inner delta and outer delta to slope sediments in the Mississippi Delta (Table 3). For nearshore sediments, the calculated Mn flux across the sediment-water interface was 840 $\mu g\ cm^{-2}\ yr^{-1}$. The net flux of Mn^{2+} to the overlying waters was significantly reduced in the outer delta. Calculated Mn^{2+} fluxes from the slope sediment was less than 1 $\mu g\ cm^{-2}\ yr^{-1}$ and provided a representative picture of the very low Mn release rate from these slowly accumulating sediments. The order of the Mn^{2+} fluxes was as follows: nearshore > inner delta > outer delta > slope sediments.

Seasonal variation of the migration flux of trace elements through interstitial waters has been observed.[178] Mn^{2+} flux into overlying waters from the interstitial waters of Long Island Sound is highest in summer and lowest in winter. Mn^{2+} fluxes in this area can be predicted with a factor of 2 to 6 by use of Fick's first law of diffusion and assuming a linear concentration gradient from the top 0 to 1 cm of sediment to overlying waters.[178]

F. RELATIONSHIP OF TRACE ELEMENT CONCENTRATIONS BETWEEN SOLID PHASE AND INTERSTITIAL WATERS

It is important to know whether the concentration of a trace element in the interstitial waters is associated with its total content in the sediment solid. The correlation depends on adsorption/desorption processes as well as on precipitation/dissolution processes occurring in the sediment/interstitial water system. If the latter is the case, the trace element concentrations in the interstitial waters will be independent of the

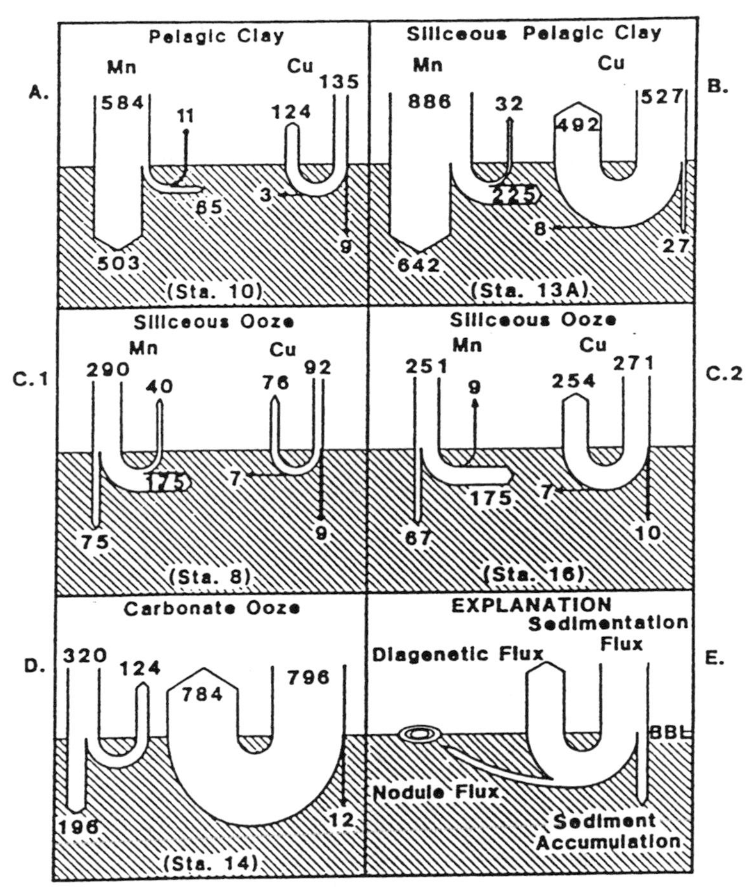

Figure 11 Fluxes of manganese and copper in northeast equatorial Pacific Ocean. All fluxes are expressed in $\mu g\ cm^{-2}\ ky^{-1}$. (From Callender, E. and Bowser, C. J., *Am. J. Sci.*, 280, 1063, 1980. With permission.)

concentrations in the solid phase. The difference between these two processes is schematically shown in Figure 12.[183]

In the studies of marine and freshwater sediments, some examples relating to the mechanisms have been found in arsenic distribution of interstitial waters.[145] The interstitial As concentrations were found to coincide with solid-phase As concentrations in marine sediments at the depth of the interstitial water peak, indicating an As solid-solution partitioning controlled by adsorption/desorption mechanisms. On the other hand, lacustrine interstitial water As peaks were not found to show the same relationship, which possibly indicates control by dissolution/precipitation mechanisms.

Comparisons between Cd, Cu, and Zn levels in anoxic interstitial waters and respective sediment concentrations show that no direct relation is found between Cd, Cu, and Zn concentrations in the sediment and in the interstitial water. In addition, if one calculated distribution coefficients from these data, the values would be much higher than the ones normally found for oxic sediment/water systems. Similar effects were observed by Hoshika et al.[26] for the concentrations of Cu, Pb, and Cu in the interstitial water of sediments at Hiro Bay (Japan). The general result from these studies is that trace element distributions in interstitial waters cannot simply be explained on the basis of the total contents of these elements in the solid phase. This experience has considerable consequences for the application of interstitial water data as sediment quality criteria.

IV. DISTRIBUTION AND SPECIATION OF TRACE ELEMENTS IN INTERSTITIAL WATERS

A. SEDIMENTS

1. Pelagic and Hemipelagic Sediment/Interstitial Water Systems

Normally the surface of pelagic sediments is oxic, which strongly differs from nearshore sediments with reducing surface. Studies of trace element biogeochemical cycles include observations of their depletions

Table 3 **Manganese flux values for Mississippi Delta sediments with supporting data**[12]

Place	Station	S^a ($g\ cm^{-2}\ yr^{-1}$)	D_j ($cm^2\ yr^{-1}$)	dc/dx ($\mu g\ cm^4$)	F_{Mn2+} ($\mu g\ cm^{-2}\ yr^{-1}$)
Nearshore	8	>2	108	7.8	840
Inner delta	7,10,12,31	1.3->2	110-130	4.6-6.4	510-830
Outer delta	14,15,16	0.08-0.5	80-104	0.3-3.2	30-330
Slope	24	0.06	80	0.05	<1

[a] Sedimentation rate.

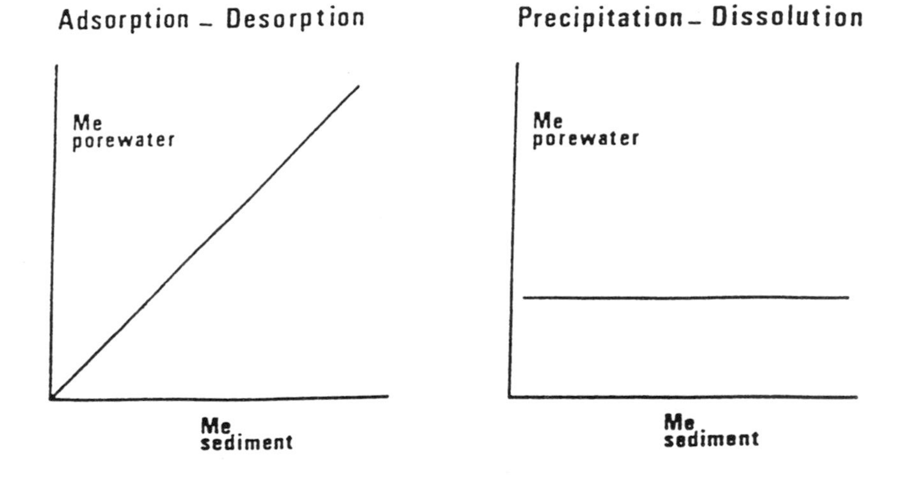

Figure 12 Interstitial water concentrations for adsorption/desorption vs. precipitation/dissolution controlled processes. (From Salomons, W., *Environ. Technol. Lett.*, 6, 315, 1985. With permission.)

and enrichment in deep marine pelagic and hemipelagic sediment/interstitial water systems.[14,18,21,121,165,166,184-196]

The data show that the interstitial water profiles of Mn, Fe, Ni, and Cu are controlled by processes involving the major oxidants (O_2, NO_3^-), Mn(IV), and Fe(III) oxides.[121] The vertical distribution of Cu shows a regular pattern: concentrations are highest in the upper 10 cm of the sediment and then decrease to a minimum concentration at deeper sediment sections (Figure 13). Cu release in the boundary layer interface of the pelagic sediment produces a maximum of dissolved concentration which is 10 to 40 times higher than that in ambient seawater.[18] Cu, V, Cr, Cd, Ni, and Mn enrichment was found[25] in surface interstitial waters in a depth < 1 cm. Similar results have been obtained by Sawlan and Murray.[21] Trace element concentrations in the bottom water (C_{BW}) and in the core-top interstitial water (C_{IW}) are listed in Table 4.[25] Some trace elements, e.g., Cd and Ni,[184-188] Cr,[189,190] and Zn[187,191] are correlated with oceanic nutrient profiles. The general reaction scheme for oxidation processes of organic material has been observed in a variety of deep-sea sediments over a wide range of organic carbon oxidation.[165,166,192-196]

2. Bays and Sounds

Compared with pelagic sediments, bay and sound sediment environments appear more reducing.[13,17,31,40,45,78,178,182,197,200,201] Frequent exchange/transfer of trace elements occurs at the bay sediment-water interface in the organic-rich environment. For the interstitial water Cu, Pb, Cd, and Cr, significant diffusive fluxes toward the overlying waters and downward into the deeper sediment have been observed in Villefranche Bay (France) by Gaillard et al.[78] The authors calculated the net diffusive fluxes (F1) and compared them to the biogenic sedimentary fluxes (F2). The results (Table 5) showed that 15 to 30% of the biogenic sedimentary trace elements would return to the overlying waters or downward into deeper sediments through interstitial waters. In another investigation, most Cu remobilized in surface sediments was returned to bottom waters and little (about 3%) was removed by subsequent diagenetic reaction in the buried sediments in Resurrection Fjord, Alaska (U.S.).[40] Similar work has been reported by Fernex

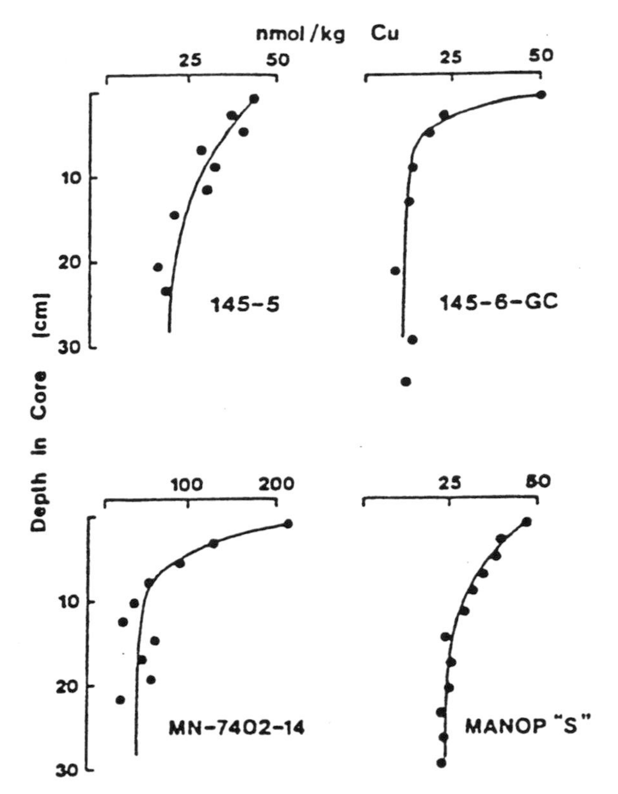

Figure 13 Interstitial copper profiles of red clay (145-5 and -6), carbonate ooze (MN-7402-14), and siliceous ooze (MANOP "S") sediments. (From Sawlan, J. J. and Murray, J. W., *Earth Planet. Sci. Lett.*, 64, 213, 1983. With permission.)

Table 4 **Bottom water (C_{BW}) and measured core-top interstitial water (C_{IW}) trace element concentrations (n*M*)**

Trace element	C_{BW}	C_{IW}	C_{IW}/C_{BW}
Cu	8.0 ± 1.0	160-165	20
Ni	12.4 ± 1.6	38-67	3-5
V	30.8 ± 2.8	287-478	9-15
Cd	1.02 ± 0.09	2.3-3.2	2-3
Cr	5.8	15.4-19.9	2.5-3.5
Mn	6.3 ± 2.1	54-112	9-18

From Heggie, D., Kahn, D., and Fischer, K., *Earth Planet. Sci. Lett.*, 80, 106, 1986. With permission.

et al.[29] and Rapin et al.[19] The studies of diagenetic processes near the sediment-water interface of Long Island Sound showed that the Mn^{2+} flux from the interstitial waters was highest in the summer and lowest in the winter.[178] The seasonal range encompassing values in this area was 0.01 to 4 mmol/m^2·d and the yearly average flux range was 0.23 to 2.6 mmol/m^2·d in the central sound. The author concluded that the flux of Mn^{2+} from the sediments through interstitial waters was sufficiently high to influence both small- and large-scale distribution of Mn in the ocean basins. The lithologic Fe background prevents easy recognition of Fe deposition patterns resulting from diagenetic remobilization, but it is likely that an absolute quantity of Fe is similar to or greater than that of Mn.

Different speciations of Fe and Mn have been determined in anoxic interstitial waters from the sediments of Loch Duich, Scotland.[45] The total concentration of trace element in interstitial water showed a large surface enrichment for Mn (0.727 to 13.5 mg l^{-1}) which decreased to a concentration of 40 to 60 mg l^{-1} and 14 to 40 mg l^{-1} at depths of approximately 50 cm in 2 cores, respectively. After ultrafiltration of the untreated interstitial water, the fraction of manganese present in the high molecular weight (HMW) fraction (7 to 12%) was considerably lower than that of iron (56 to 57%). After oxidation and ultrafiltration, the percentages of trace elements present in the HMW fraction decreased for both elements to 3 to

Table 5 **Benthic fluxes of Cr, Cu, and Cd through interstitial waters in Villefranche Bay and comparison to their biogenic fluxes to the sediment**

	Cr	Cu	Cd
$\Delta C/\Delta x$ to the overlying water (M cm^{-4})	5.0×10^{-13}	1.0×10^{-11}	2.6×10^{-13}
$\Delta C/\Delta x$ downcore (M cm^{-4})	14.0×10^{-13}	1.3×10^{-11}	1.4×10^{-13}
F_1 (μM m^{-2} yr^{-1})	6	72	1.3
F_2 (μM m^{-2} yr^{-1})	24	235	8.9
F_1/F_2 (%)	25	30	15

Modified from Gaillard, J. F., Jeandel, C., Michard, G., Nicolas, E., and Renard, D., *Mar. Chem.*, 18, 233, 1986.

5% (Mn) and 27 to 44% (Fe). The significant amount of iron which was retained by the ultrafiltrater after oxidation might be due to the existence of strong organic-Fe associations in the interstitial waters. When a sample of the interstitial water was acidified to pH 2 and left to stand, a brown precipitate was formed which was "humic" in nature.[168] This phase contained between 15 and 40% of the dissolved iron, but only 1% of the DOC. Similar evidence for organically associated iron in nearshore interstitial waters of Great Bay (U.S.) was observed by Lyons et al.[31] Interstitial water Fe, Mn, Cu, Cd, and Ag concentrations and speciation in the tidal flat sediments from Branford Harbor, Long Island Sound have been investigated and calculated by Lyons and Fitzgerald (Table 6).[201]

3. Estuaries and Rivers

In estuaries there is a similar reducing environment to bays and similar influence on trace element distribution in the interstitial waters.[68,79,116,158-160,201,202] Enrichment of trace elements in estuarine sediment interstitial water was found. Trace element concentrations in the interstitial waters of Conway Estuary indicate a characteristic increase of Mn and Co by factors of 1000 and 10,000, respectively, and a 50- to 100-fold increase of Fe, Pb, Cu, and Ni.[116,158,159]

The interstitial water results from the Rhine estuary showed that between 25 to 27% of Cd was present as nonlabile complexes.[160] In River Mersey estuary in England, Zn concentrations from 0.068 to 0.642 μM were observed; 31 to 98% of total Zn concentrations were found as organic complexes.[138] A comparison of trace element species in surface and interstitial waters of the Woronora River, New South Wales (Australia) was made by Batley and Giles.[6] Table 7 indicates an increased solubility of trace elements in the anoxic interstitial water compared to the surface water. For Pb and Cu this increase is related to both the labile and stable bound (organic) fraction, while the difference for Cd in the organically complexed form is less significant. Trace elements in sediment interstitial water from Mystic River estuary in Long Island Sound were studied by Lyons and Fitzgerald.[201] Thermodynamic calculations indicate that metal-polysulfide and bisulfide complexes may control the distribution of Fe, while Mn and Cd are controlled by other inorganic species.

4. Lakes

Enrichment of trace elements in interstitial water of lake sediments has been found.[3,203,204] A comparison of trace element concentrations in the interstitial water of Lake Ijsselmeer sediment with the overlying water indicates a marked increase of Mn and Fe by factors 100 and 1,000, respectively, and a 10- to 20-fold increase of Cr and As (Table 8). The studies of Frevert and Sollmann,[203] however, show that the interstitial water concentrations of Pb, Cd, Zn, and Cu in Lake Kinneret (Israel) differ only slightly from the overlying waters.

The concentrations of Cu, Ni, and Al in the interstitial waters of Clearwater Lake (Canada) vary inversely with pH.[89] Interstitial water Al increases gradually to 5 to 10 μM at 35 cm. All profiles show a pronounced decrease in levels of dissolved Ni and Cu in the zone of active sulfate reduction. The

Table 6 **Major trace element species in the interstitial waters of Branford Harbor**

Trace Element	Branford Harbor
Fe	$FeCl^+$, Fe^{2+}
Mn	$MnCl^+$, Mn^{2+}, $MnCl_2^o$
Cu	$Cu(S_4)_2^{3-}$, CuDOM?
Cd	$Cd(OH)^+$, $CdHS^+$?
Ag	$AgHS^o$?

Modified from Lyons, W. B. and Fitzgerald, W. F., in *Trace Elements in Sea Water,* Wong, C. S., Bruland, K. W., Burton, J. D., and Goldberg, E. D., Eds., Plenum Press, New York, 1983, 621.

Table 7 **Trace element species in interstitial waters from the river sediments and in surface waters**

	Cadmium		Lead		Copper		
Number	Labile	Bound	Labile	Bound	Labile	Bound	Reference
	(μg l-1)	(μg l-1)	(μg l-1)	(μg l-1)	(μg l-1)	(μg l-1)	
1[a]	0.11	0.09	0.28	0.32	0.74	1.15	6
2[b]	0.29	0.11	0.50	0.60	1.35	1.80	6
3[c]	<0.01-0.12	0.09-0.44	0.62-2.05	0.35-1.86	3.61-13.4	2.9-11.1	202
4[d]	0.04	0.20	1.44	1.10	6.93	5.60	202

[a] Surface water (oxic) from the Woronora River, New South Wales, Australia; [b] Interstitial water of mangrove sediment from the Woronora River; [c] Concentration range of interstitial water trace elements in the sediment core from the river Elbe, Hamburg, Germany; [d] Average concentrations of 10 section samples in Number 3.

Table 8 **Trace element concentrations in the interstitial waters (top layer) and the overlying waters in the Ijsselmeer**

Trace Element	Interstitial Water (μg l-1)	Overlying Surface Water (μg l-1)	Enrichment Factor
Cr	6.3	0.6	11
Cd	1.2	<0.2	>6
As	21	0.9	23
Fe	21,000	20	1050
Mn	9,160	50	183
P	5,380	145	37

Modified from Salomons, W. and Förstner, U., *Metals in the Hydrocycle,* Springer-Verlag, Berlin, 1984.

oxidation of upward diffusing Fe^{2+} to insoluble Fe^{3+} within the oxidized surface layer or in the water column may at least partly explain the observed Fe enrichment (up to 10%) in the surface sediments of both lakes.[89]

A study of As speciation and distribution in the sediment interstitial water of Lake Washington (U.S.) showed that the concentrations of As^{3+}, As^{5+}, and total As were between 23 to 560 n*M,* 31 to 550 n*M,* and 82 to 800 n*M,* respectively.[145] As^{3+}/As^{5+} ratios in the interstitial water were usually between 1 and 4. This was strongly different from the sediments of the Washington coastal environments with As^{3+}/As^{5+} ratios < 1 in the interstitial waters. The authors pointed out that the interstitial water As exhibited a similar behavior to Fe^{2+} and Mn^{2+}, suggesting adsorption or coprecipitation with Fe and Mn oxides. The final expression of As release to the interstitial water would be more closely associated with an increase of dissolved Fe^{2+}.

B. SOILS

In contrast to the porous sediments it is much more difficult to get enough volume of interstitial water sample for determining trace elements from soil. Therefore, the information on trace element distribution in soil interstitial water seems to be limited.

Interstitial trace element concentrations in different soil types are very different. Kinniburgh and Miles[75] investigated Mn, Fe, Cu, Zn, Sr, Ba, and Al concentrations in eight types of soil. The concentration differences between different soil types were very large and reached >170-, >68-, and 23-fold for Mn, Al, and Fe, respectively. However, the range was only 5-, 7-, and 12-fold for Cu, Sr, and Ba, respectively (Table 9).

The seasonal variation of interstitial water Cd in river foreland soils in the Rhine-Meuse estuary (The Netherlands) has been reported by Palsma and Loch.[151] When in winter the groundwater level reached the soil face, Cd concentrations in the interstitial water were relatively low (<0.2 μM). In spring the groundwater fell and reached the lowest level (50 cm depth). As the groundwater level rose in the fall again, the redox potential decreased and the Cd concentration in the interstitial water increased up to 1.0 µm. This high Cd concentration persisted for about 2 months. Later it decreased again to a value of <0.2 μM. The investigation showed that only at low Fe concentrations high Cd concentrations could be found. At increased Fe concentrations, Cd concentrations were always low. Cd concentrations in interstitial water appeared high at redox potentials above +200 mV (nonequilibration), but as soon as the redox potential dropped below +200 mV (reductive dissolution of Fe-(hydr)oxides), Cd concentrations were low.

Tills and Alloway[169] investigated the concentration and the speciation of Pb in soil solution from strongly polluted soils. A fractionation scheme employing ion-exchange chromatography was used to determine the speciation of Pb in soil solution from five samples. In acidic soils, Pb was found to be mainly in cationic form with some organic complexes present. In a calcareous soil, neutral complexes were dominant with some cationic Pb species also present (Table 10).

C. OTHERS

The distribution of total Se, Se(IV), Se(VI), and organic Se as well as their seasonal variation in interstitial waters of the Great Marsh in Delaware (U.S.) was observed.[41] Profiles of dissolved Se exhibited distinct seasonal variation. Changes in the concentration of Se appear related to the redox characteristics of the marsh environment. Maximum concentrations were observed in June 1985 and 1986 when the marsh was most oxic. Only low or undetectable concentrations were found when the marsh was reducing (April and December 1985 and March 1986). Dissolved Se(IV + VI) comprised ca. 50% of the total interstitial water Se at the surface of the marsh, and this fraction decreased in concentration with depth. In the same site, Boulegue et al.[42] studied Fe and Cu species in the interstitial waters. The measurement and calculation emphasized inorganic species. Iron species were mainly $FeCl_2^0$, $FeCl^+$, Fe^{2+}, and $FeSO_4^0$. In the presence of polysulfide ions Cu speciation should primarily be $Cu(S_4)_2^{3-}$ and $Cu(S_4S_5)^{3-}$. In the samples which contained hydrogen sulfide and thiosulfate, the dominant species should be $Cu(HS)_3^-$ and $CuS(HS_4)_2^{3-}$. The concentrations of dissolved Fe and Cu in the interstitial water are controlled by the formation of metal sulfides (FeS, Fe_3S_4, CuS, and/or coprecipitates). The destabilization of polysulfides and polysulfide complexes by bacterial sulfate reduction has shown that Cu is most probably strongly complexed by organic matter. A calculation of the apparent stability constant of such a complex indicates that it may be a (Cu)(I)-organo-thiol association.

In a study to investigate the effects of nutrient and trace element pollution in salt marshes by measuring the concentration of trace elements, nutrients, and soluble sulfides in the interstitial waters, a sewage sludge fertilizer was applied to experimental plots in Great Sippewisset Marsh (U.S.).[109] Concentrations of dissolved Fe, Mn, Zn, Cu, and Cd were always higher in interstitial waters from fertilized plots than those from control plots (Figure 14). The difference in the average dissolved trace element concentration between the plots was greatest near the surface. In all plots, trace elements concentrations decreased with depth. The decrease in trace element concentration with depth represented the removal of trace elements by precipitation and adsorption reactions.

V. SUMMARY AND OUTLOOK

One of the most important characteristics of interstitial waters is a remarkable vertical variation of redox potential and relevant indicators in the interstitial profiles. Redox condition can be indicated by the presence or absence of different "marker components". Trace element distribution and mobility are strongly influenced by these "marker components", both in interstitial waters and in solid phases. These

Table 9 **Trace element concentrations in interstitial water extracted from ten topsoils (μg l⁻¹)**

Soil Series	Soil Type	pH	Cu	Zn	Sr	Ba	Mn	Fe	Al
Harwell	Brown earth	5.3	19	26	120	25	341	300	350
Icknield	Rendzina	7.8	6	6	80	8	2	150	100
Grove	Gleyed Calcareous	8.1	15	8	200	12	2	60	<10
Rowsham	Surface water gley	8.0	15	8	150	12	<2	50	<10
Fyfield	Brown earth	7.9	8	7	60	9	8	140	70
Thames	Groundwater gley	[a]	11	18	290	96	<2	50	<30
Marcham	Brown calcareous	8.3	5	55	110	11	3	460	160
Denchworth	Surface water gley	7.0	18	12	200	17	73	880	570
Southampton	Podzol	4.3	25	137	40	67	55	1140	680
Berkhamsyed	Gleyed brown earth	7.1	26	23	80	22	24	510	300
Median			15	15	12	15	16	230	300

[a] Insufficient sample for measurement.

Modified from Kinniburgh, D. G. and Miles, D. L., *Environ. Sci. Technol.,* 17, 362, 1983.

Table 10 **Speciation of lead in the soil solution phase**

		Pb	% Pb in the Solution Phase			
Site	**pH**	**(mg l⁻¹)**	**Cationic**	**Anionic**	**Neutral**	**Less Polar Organics**
1[a]	4.24	9.888	96.50	0.075	0.865	2.256
2[b]	3.96	2.040	77.44	4.92	3.83	8.89
3[c]	4.11	12.368	98.77	0.090	0.12	0.81
4[d]	7.47	0.192	27.12	1.939	67.67	3.25
5[e]	4.71	0.090	82.4	7.9	0.2	9.7

[a] Snertingdal (a), Norway; [b] Snertingdal (b), Norway; [c] Dyfed, Wales; [d] Velet, England; [e] Leiester, England.

From Tills, A. R. and Alloway, B. J., *Environ. Technol. Lett.*, 4, 529, 1983. With permission.

processes result in transformations of trace element in interstitial profiles and then lead to concentration gradient formation. The gradients of trace element distribution in vertical direction enable trace element diffusion and migration.

Recently, interstitial water studies have become a more and more important tool for understanding early diagenesis of trace elements, assessing trace element mobility in polluted soils and mine waste, and developing sediment quality criteria.[4,109,205,206]

The study of early diagenesis of trace elements in interstitial waters is not only for the understanding of pure geochemistry but also for practical application. For example, in the "MANOP" program the studies on interstitial water trace elements have revealed the truth about transformation and migration of Mn and other trace elements between sediment and aqueous phases in the sediment column and then, finally, sequential formation of manganese micronodules, which is considered as an industrial mine resource. This discovery provides theoretical basis for exploration and mining in the deep sea.

Although water quality criteria have been developed by many countries, sediment quality criteria have not been established yet. New objectives regarding the improvement of water quality as well as problems with the resuspension and deposition of dredged materials require a standardized assessment of sediment quality. Biological criteria integrate sediment characteristics and pollutant loads, while generally not indicating the cause of effects. With respect to chemical-numerical criteria, immediate indications on biological effects are lacking; major advantages lie in their easy application and amendment to modeling approaches. On the other hand, numerical approaches are based on (1) accumulation, (2) interstitial water assessment, (3) solid/liquid equilibrium partition, and (4) elution properties of trace elements. The second term "interstitial water assessment" in the scheme would also include characteristics of the solid substrate, in particular buffer capacity against pH depression.[205] Interstitial water studies have become a necessary mean for assessing sediment quality.

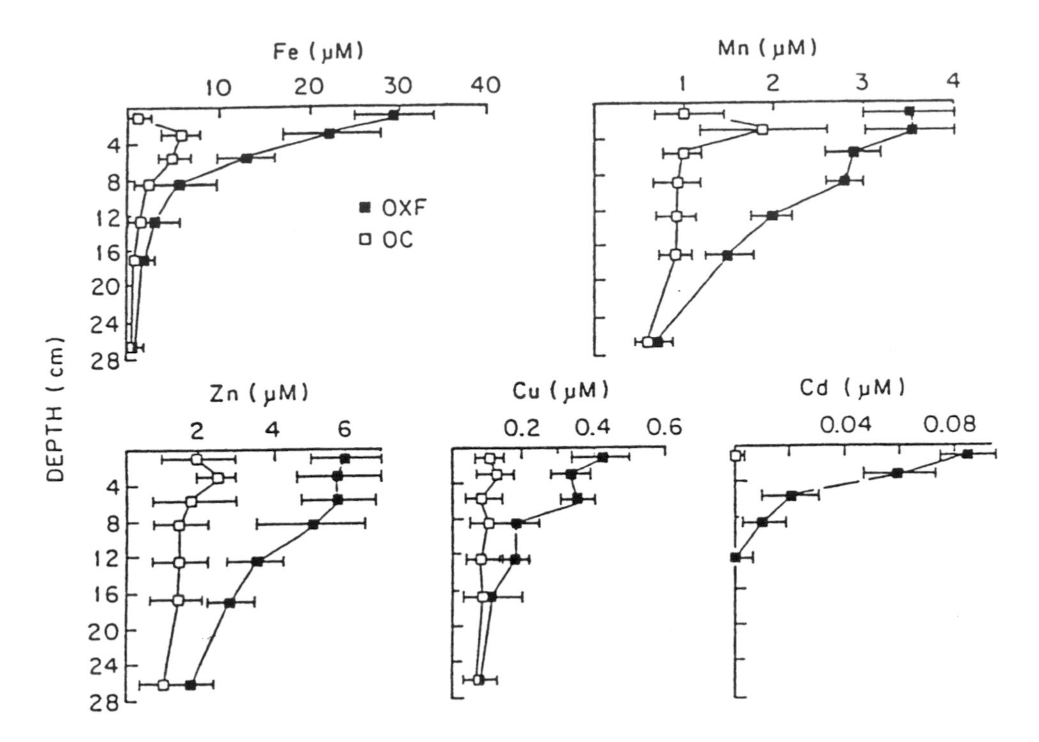

Figure 14 The average concentration of iron, manganese, zinc, copper, and cadmium in interstitial waters from the experiment site treated with sewage sludge (OXF) and the control site (OC). (From Giblin, A. E., Luther III, G. W., and Valiela, I., *Estuarine Coastal Shelf Sci.,* 23, 477, 1986. With permission.)

The bioavailability of trace elements in sediments is related to the chemical activity of the trace elements in the sediment interstitial water system. Di Toro et al. have found relations between AVS (acid volatile sulfide), AVM ("acid volatile" metal) and mortality of aquatic organisms.[207,208] Normally, aquatic organisms react with AVM through interstitial waters acting as links between solid AVM and organisms. This discovery stresses the importance of the interstitial water approach to sediment criteria development.

In general, the practical applicability of sediment quality criteria has been developed to a somewhat limited extent. Assessments on the basis of equilibrium calculations and interstitial water composition in the sediment/interstitial water system therefore still seem to require further studies and discussion.

REFERENCES

1. **Presley, B. J. and Trefry, J. H.,** Sediment-water interactions and the geochemistry of interstitial waters, in *Chemistry and Biogeochemistry of Estuaries,* Olausson, E. and Cato, I., Eds., John Wiley and Sons, New York, 1980, 187.
2. **Manheim, F. T.,** Interstitial waters of marine sediments, in *Chemical Oceanography,* 2nd ed., Riley, J. P. and Chester, R., Eds., Academic Press, London, 1976, 115.
3. **Salomons, W. and Förstner, U.,** *Metals in the Hydrocycle,* Springer-Verlag, Berlin, 1984, 349.
4. **Förstner, U. and Kersten, M.,** Assessment of metal mobility in dredged material and mine waste by pore water chemistry and solid speciation, in *Chemistry and Biology of Solid Waste — Dredged Material and Mine Tailings,* Salomons, W. and Förstner, U., Eds., Springer-Verlag, Berlin, 1989, 214.
5. **Shaw, T. J., Gieskes, J. M., and Jahnke, R. A.,** Early diagenesis in differing depositional environments: the response of transition metals in pore water, *Geochim. Cosmochim. Acta,* 54, 1233, 1990.
6. **Batley, G. E. and Giles, M. S.,** A solvent displacement technique for the separation of sediment interstitial waters, in *Contaminants and Sediments,* Vol. 2, Backer, R. A., Ed., Ann Arbor Science, Ann Arbor, MI, 1980, 101.
7. **Berner, R. A.,** A new geochemical classification of sedimentary environments, *J. Sediment. Petrol.* 51, 359, 1981.

8. **Aller, R. C. and Cochran, J. K.,** $^{234}Th/^{238}U$ disequilibrium in nearshore sediment: particle reworking and diagenetic time scales, *Earth Planet. Sci. Lett.,* 29, 1976, 37.
9. **Bouma, A. H.,** *Methods for the Study of Sedimentary Structures,* Wiley-Interscience, New York, 1969, 458.
10. **Duxbury, A. C.,** *The Earth and Its Ocean,* Addison-Wesley, Reading, MA, 1971, 318.
11. **McQuillin, R. and Ardus, D. A.,** *Exploring the Geology of Shelf Seas,* Graham and Trotman, London, 1977, 234.
12. **Trefry, J. H. and Presley, B. J.,** Manganese fluxes from Mississippi Delta sediments, *Geochim. Cosmochim. Acta,* 46, 1715, 1982.
13. **Shimmield, T. M., MacKenzie, A. B., and Price, N. B.,** ICP-MS analysis of trace element concentrations in interstitial waters of Scotland sea loch sediments, in *Heavy Metals in the Environment,* Vol. 1, Farmer, J. G., Ed., CEP Consultants Ltd., Edinburgh, 1991, 457.
14. **Callender, E. and Bowser, C. J.,** Manganese and copper geochemistry of interstitial fluids from manganese nodule-rich pelagic sediments of the northeastern equatorial Pacific Ocean, *Am. J. Sci.,* 280, 1063, 1980.
15. **Heggie, D. and Lewis, T.,** Cobalt in pore waters of marine sediment, *Nature,* 311, 453, 1984.
16. **Duchart, P., Calvert, S. E., and Price, N. B.,** Distribution of trace metals in the pore waters of shallow water marine sediments, *Limnol. Oceanogr.,* 18, 605, 1973.
17. **Pedersen, T. F.,** Early diagenesis of copper and molybdenum in mine tailings and natural sediments in Rupert and Holberg inlets, British Columbia, *Can. J. Earth Sci.,* 22, 1474, 1985.
18. **Klinkhammer, G., Heggie, D. T., and Graham, D. W.,** Metal diagenesis in oxic marine sediments, *Earth Planet. Sci. Lett.,* 61, 211, 1982.
19. **Rapin, F., Nembrini, G. P., Förstner, U., and Garcia, J. I.,** Heavy metals in marine sediment phases determined by sequential chemical extraction and their interaction with interstitial water, *Environ. Technol. Lett.,* 4, 387, 1983.
20. **Brumsack, H. J. and Gieskes, J. M.,** Interstitial water trace-metal chemistry of laminated sediments from the Gulf of California, Mexico, *Mar. Chem.,* 14, 89, 1983.
21. **Sawlan, J. J. and Murray, J. W.,** Trace metal remobilization in the interstitial waters of red clay and hemipelagic marine sediments, *Earth Planet. Sci. Lett.,* 64, 213, 1983.
22. **Gobeil, C., Silverberg, N., Sundby, B., and Cossa, D.,** Cadmium diagenesis in Laurentian Trough sediments, *Geochim. Cosmochim. Acta,* 51, 589, 1987.
23. **McCorkle, D. C. and Klinkhammer, G. P.,** Porewater cadmium geochemistry and the porewater cadmium: $\delta^{13}C$ relationship, *Geochim. Cosmochim. Acta,* 55, 161, 1991.
24. **Brooks, R. R., Presley, B. J., and Kaplan, I. R.,** Trace elements in the interstitial waters of marine sediments, *Geochim. Cosmochim. Acta,* 32, 397, 1968.
25. **Heggie, D., Kahn, D., and Fischer, K.,** Trace metals in metalliferous sediments, MANOP Site M: interfacial pore water profiles, *Earth Planet. Sci. Lett.,* 80, 106, 1986.
26. **Hoshika, A., Takimura, O., and Shiozawa, T.,** Determination of cadmium, lead and copper in interstitial water by anodic stripping voltammetry, *J. Oceanogr. Soc. Jpn,* 33, 161, 1977.
27. **Presley, B. J., Brooks, R. R., and Kaplan, I. R.,** Manganese and related elements in the interstitial water of marine sediments, *Science,* 158, 906, 1967.
28. **Heggie, D., Klinkhammer, G., and Cullen, D.,** Manganese and copper fluxes from continental margin sediments, *Geochim. Cosmochim. Acta,* 51, 1059, 1987.
29. **Fernex, F. E., Span, D., Flatau, G. N., and Renard, D.,** Behavior of some metals in surficial sediments of the Northwest Mediterranean continental shelf, in *Sediment and Water Interactions,* Sly, P. G., Ed., Springer-Verlag, Berlin, 1986, 353.
30. **Elderfield, H.,** Metal-organic associations in interstitial waters of Narragansett Bay sediment, *Am. J. Sci.,* 281, 1184, 1981.
31. **Lyons, W. B., Gaudette, H. E., and Armstrong, P. B.,** Evidence for organically associated iron in nearshore pore fluids, *Nature,* 282, 202, 1979.
32. **Shaw, T. J.,** An apparatus for fine-scale sampling of pore waters and sediments in high porosity sediments, *J. Sediment. Petrol.,* 59, 633, 1989.
33. **Addy, S. K. and Ewing, M.,** A new box core design for the investigation of manganese-nodule distribution in a sediment column, *Mar. Geol.,* 17, M17, 1974.
34. **Watson, P. G., Frickers, P. E., and Goodchild, C. M.,** Spatial and seasonal variations in the chemistry of sediment interstitial waters in the Tamar estuary, *Estuarine Coastal Shelf Sci.,* 21, 105, 1985.

35. **Richards, L. F.,** A pressure-membrane extraction apparatus for soil solutions, *Soil Sci.,* 51, 377, 1941.
36. **Siever, R.,** A squeezer for extracting interstitial water from modern sediments, *J. Sediment. Petrol.,* 32, 329, 1962.
37. **Hartmann, M.,** An apparatus for the recovery of interstitial water from recent sediments, *Deep Sea Res.,* 12, 225, 1965.
38. **Manheim, F. T.,** A hydraulic squeezer for obtaining interstitial water from consolidated and unconsolidated sediments, Geol. Survey Prof. Paper 550-C, 1966, 256.
39. **Presley, B. J., Kolodny, Y., Nissenbaum, A., and Kaplan, I. R.,** Early diagenesis in a reducing fjord, Saanich Inlet, British Columbia. II. Trace element distribution in interstitial water and sediment, *Geochim. Cosmochim. Acta,* 36, 1073, 1972.
40. **Heggie, D.,** Copper in the Resurrection Fjord, Alaska, *Estuarine Coastal Shelf Sci.,* 17, 613, 1983.
41. **Velinsky, D. J. and Cutter, G. A.,** Geochemistry of selenium in a coastal salt marsh, *Geochim. Cosmochim. Acta,* 55, 179, 1991.
42. **Boulegue, J., Lord C. J., III, and Church, T. M.,** Sulfur speciation and associated trace metals (Fe, Cu) in the pore waters of Great Marsh, Delaware, *Geochim. Cosmochim. Acta,* 46, 453, 1982.
43. **Addy, S. K., Presley, B. J., and Ewing, M.,** Distribution of manganese, iron and other trace elements in a core from the northwest Atlantic, *J. Sediment. Petrol.,* 45, 813, 1976.
44. **Piemontesi, D. and Baccini, P.,** Chemical characteristics of dissolved organic matter in interstitial waters of lacustrine sediments and its influence on copper and zinc transport, *Environ. Technol. Lett.,* 7, 577, 1986.
45. **Krom, M. D.,** On the association of iron and manganese with organic matter in anoxic marine pore waters, *Geochim. Cosmochim. Acta,* 42, 607, 1978.
46. **Sugai, S. F. and Healy, M. L.,** Voltammetric studies of the organic association of copper and lead in two canadian inlets, *Mar. Chem.,* 6, 291, 1978.
47. **Bray, J. T., Bricker, O. P., and Troup, B. N.,** Phosphate in interstitial waters of anoxic sediments: oxidation effects during sampling procedure, *Science,* 180, 1362, 1973.
48. **Troup, B. N., Bricker, O. P., and Bray, J. T.,** Oxidation effect on the analysis of iron in the interstitial water of recent anoxic sediments, *Nature,* 249, 237, 1974.
49. **Lyons, W. B., Gaudette, H. E., and Smith, G. M.,** Pore water sampling in anoxic carbonate sediments: oxidation artifacts, *Nature,* 277, 48, 1979.
50. **Reeburg, W. S.,** An improved interstitial water sampler, *Limnol. Oceanogr.,* 12, 163, 1967.
51. **Presley, B. J., Brooks, R. R., and Kappel, H. M.,** A simple squeezer for removal of interstitial water from ocean sediments, *J. Mar. Res.,* 25, 355, 1967.
52. **Kalil, E. H. and Goldhaber, M.,** A sediment squeezer for removal of pore waters without air contact, *J. Sediment. Petrol.,* 43, 553, 1973.
53. **Robbins, J. A. and Gustinis, J.,** A squeezer for efficient extraction of pore water from small volumes of anoxic sediment, *Limnol. Oceanogr.,* 16, 905, 1977.
54. **Barnes, R. D.,** An in situ interstitial water sampler for use in unconsolidated sediments, *Deep Sea Res.,* 20, 1125, 1973.
55. **Sayles, F. L., Manheim, F. T., and Waterman, L. S.,** Interstitial water studies on small core samples, Leg. 15, in *Initial Reports Deep Sea Drilling Project,* Heezen, B. C., Ed., U. S. Government Printing Office, Washington, D.C., 1973, 783.
56. **Sayles, F. L., Wilson, T. R. S., Hume, D. N., and Mangelsdorf, P. C., Jr.,** In situ sampler for marine sedimentary pore waters: evidence for potassium depletion and calcium enrichment, *Science,* 181, 154, 1973.
57. **Sayles, F. L., Mangeldorf, P. C., Jr., Wilson, T. R. S., and Hume, D. N.,** A sampler for the in situ collection of marine sedimentary pore waters, *Deep Sea Res.,* 23, 259, 1976.
58. **Ramann, E., März, S., and Bauer, H., Über** Boden-Press-Säfte, *Int. Mitt. Bodenkd.,* 6, 27, 1916.
59. **Lipman, C. B.,** A new method of extracting the soil solution, *Univ. California, Pub. in Agr. Sci.,* 3, 131, 1918.
60. **Northrup, Z.,** The true soil solution, *Science,* 47, 638, 1918.
61. **Kriukov, P. A.,** Recent methods for physicochemical analysis of soils: methods for separating soil solutions, in *Rukovodstvo dlya polevykh i laborotornykh issledovanii pochv,* Moscow Izdat. Akad. Nauk, SSSR, 1947, 3.
62. **Manheim, F. T. and Sayles, F. L.,** Composition and origin of interstitial waters of marine sediments, based on deep sea drill cores, in *The Sea,* Vol. 5, Goldberg, E. D., Ed., Wiley-Interscience, New York, 1974, 527.

63. **Kriukov, P. A.,** *Interstitial Waters of Soils, Rocks and Sediments,* Izdat. Nauka, Novosibirsk, 1971, 219.
64. **Edmunds, W. M. and Bath, A. H.,** Centrifuge extraction and chemical analysis of interstitial waters, *Environ. Sci. Technol.,* 10, 467, 1976.
65. **Holdren, G. C., Jr., Armstrong, D. E., and Harris, R. F.,** Interstitial inorganic phosphorus concentrations in lakes Mendota and Wingra, *Water Res.,* 11, 1041, 1977.
66. **Adams, D. D., Darby, D. A., and Young, R. J.,** Selected analytical techniques for characterizing the metal chemistry and geology of fine-grained sediments and interstitial water, in *Contaminants and Sediments,* Vol. 2, Baker, R. A., Ed., Ann Arbor Science, Ann Arbor, MI, 1980, 3.
67. **Adams, D. D. and Darby, D. A.,** A dilution-mixing model for dredged sediments in freshwater systems, in *Contaminants and Sediments,* Vol. 2, Baker R. A., Ed., Ann Arbor, MI, 1980, 374.
68. **Emerson, S., Jahnke, R., and Heggie, D.,** Sediment-water exchange in shallow water estuarine sediments, *J. Mar. Res.,* 42, 709, 1984.
69. **Lyons, W. B., Wilson, K. M., Armstrong, P. B., Smith, G. M., and Gaudette, H. E.,** Trace metal pore water geochemistry of nearshore Bermuda carbonate sediments, *Ocean. Acta,* 3, 363, 1980.
70. **Batley, G. E.,** Physicochemical separation methods for trace element speciation in aquatic samples, in *Trace Element Speciation: Analytical Methods and Problems,* Batley, G. E., Ed., CRC Press, Boca Raton, FL, 1989, 1.
71. **Bischoff, J. L., Gree, R. E., and Luistro, A. D.,** Interstitial waters of marine sediments: temperature of squeezing effects, *Science,* 167, 1245, 1970.
72. **Fanning, K. A. and Pilson, M. E. Q.,** Interstitial silica and pH in marine sediments: some effects of sampling procedures, *Science,* 173, 1228, 1971.
73. **Scholl, D. W.,** Techniques for removing interstitial water from coarse-grained sediment for chemical analysis, *Sedimentology,* 2, 156, 1963.
74. **Batley, G. E. and Giles, M. S.,** Solvent displacement of sediment interstitial waters before trace metal analysis, *Water Res.,* 13, 879, 1979.
75. **Kinniburgh, D. G. and Miles, D. L.,** Extraction and chemical analysis of interstitial water from soils and rocks, *Environ. Sci. Technol.,* 17, 362, 1983.
76. **Hesslein, R. H.,** An in situ sampler for close interval porewater studies, *Limnol. Oceanogr.,* 21, 912, 1976.
77. **Mayer, L. M.,** Chemical water sampling in lakes and sediments with dialysis bags, *Limnol. Oceanogr.,* 21, 909, 1976.
78. **Gaillard, J. F., Jeandel, C., Michard, G., Nicolas, E., and Renard, D.,** Interstitial water chemistry of Villefranche Bay sediments: trace metal diagenesis, *Mar. Chem.,* 18, 233, 1986.
79. **Schwedhelm, E., Vollmer, M., and Kersten, M.,** Bestimmung von Konzentrationsgradienten gelöster Schwermetalle an der Sediment/Wasser-Grenzfläche mit Hilfe der Dialysetechnik, *Fresenius Z. Anal. Chem.,* 332, 756, 1988.
80. **Gunkel, G., Heller, S., and Sztraka, A.,** Ein modifizierter Dialysekammerstab (n. Ripl) zur Ermittlung der Mikroschichtung gelöster Schwermetalle in der Gewässersediment-Kontaktwasser-Zone, *Vertz.-Ges. Ökol.,* 13, 211, 1985.
81. **Bottomley, E. Z. and Bayly, I. L.,** A sediment porewater sampler used in root zone studies of the submerged macrophyte, *Myriophyllum spicatum, Limnol. Oceanogr.,* 29, 671, 1984.
82. **Carignan, R.,** Interstitial water sampling by dialysis: methodological notes, *Limnol. Oceanogr.,* 29, 667, 1984.
83. **Howes, B. L., Dacey, J. W. H., and Wakeham, S. G.,** Effects of sampling technique on measurements of porewater constituents in salt marsh sediments, *Limnol. Oceanogr.,* 30, 221, 1985.
84. **Reeburgh, W. S. and Erickson, R. E.,** A "dipstick" sampler for rapid, continuous chemical profiles in sediments, *Limnol. Oceanogr.,* 27, 556, 1982.
85. **Carignan, R., Rapin, F., and Tessier, A.,** Sediment porewater sampling for metal analysis: a comparison of techniques, *Geochim. Cosmochim. Acta,* 49, 2491, 1985.
86. **Martens, C. S. and Klump, J. V.,** Biogeochemical cycling in an organic-rich coastal marine basin. I. Methane sediment-water exchange processes, *Geochim. Cosmochim. Acta,* 44, 471, 1980.
87. **Hopner, T.,** Design and use of a diffusion sampler for interstitial water from fine grained sediments, *Environ. Technol. Lett.,* 2, 187, 1981.
88. **Fagerström, T. and Jernelöv, A.,** Aspects of the qualitative ecology of mercury, *Water Res.,* 6, 1193, 1972.

89. **Carignan, R. and Nriagu, J. O.,** Trace metal deposition and mobility in the sediments of two lakes near Sudbury, Ontario, *Geochim. Cosmochim. Acta,* 49, 1753, 1985.
90. **Boudreau, P. B.,** Modelling the sulfide-oxygen reaction and associated pH gradient in porewaters, *Geochim. Cosmochim. Acta,* 55, 145, 1991.
91. **Boudreau, P. B.,** A steady-state diagenetic model for dissolved carbonate species and pH in the porewater of oxic and suboxic sediments, *Geochim. Cosmochim. Acta,* 51, 1985, 1987.
92. **Boudreau, P. B. and Canfield, D. E.,** A provisional diagenetic model for pH in anoxic porewaters: application to the FOAM Site, *J. Mar. Res.,* 46, 429, 1988.
93. **Jørgensen, B. B. and Revsbech, N. P.,** Colorless sulfur bacteria, *Beggiatoa* spp. and *Thiovulum* spp., in O_2 and H_2S micro-gradients, *Appl. Environ. Microbiol.,* 45, 1261, 1983.
94. **Morel, F. M. M.,** *Principles of Aquatic Chemistry,* John Wiley & Sons, New York, 1983, 446.
95. **Breemen, N. van,** Soil forming processes in acid sulfate soil, in *Acid Sulphate Soils,* Dost, H., Ed., IL RI Publ. 18, Wageningen, The Netherlands, 1973, 66.
96. **Postma, D.,** Pyrite and siderite oxidation in swamp sediments, *J. Soil Sci.,* 34, 163, 1983.
97. **Hong, J., Förstner, U., and Calmano, W.,** Effects of redox processes on the acid-producing potential and metal mobility in sediments, in *Bioavailability: Physical, Chemical and Biological Interactions,* Hamelink, J. L., Landrum, P. F., Bergman, H. L., and Benson, W. H., Eds., CRC Press, Boca Raton, FL., 1994, 129.
98. **Breemen, N. van,** Effects of redox processes on soil acidity, *Neth. J. Agr. Sci.,* 35, 271, 1987.
99. **Breemen, N. van, Mulder, J., and Driscoll, C. J.,** Acidification and alkalinization of soils, *Plant Soil,* 75, 283, 1983.
100. **Akittrick, J. A., Fanning, D. S., and Hossner, L. R., Eds.,** *Acid Sulfate Weathering,* SSSA Special Publication 10, Soil Society of America, Madison, WI, 1982, 234.
101. **Breemen, N. van,** Acidification and deacidification of coastal plain soils as a result of periodic flooding, *Soil Sci. Soc. Am. Proc.,* 39, 1153, 1975.
102. **Brinkman, R.,** Ferrolysis, a hydromorphic soil forming process, *Geoderma,* 3, 199, 1979.
103. **Brinkman, R.,** Ferrolysis, a soil-forming process in hydromorphic conditions, Agricultural Research Reports 887, PURDOC, Wageningen, 1979.
104. **Eaqub, M. and Blume, H. P.,** Genesis of a so-called ferrolysed soil of Bangladesh, *Z. Pflanzenernähr. Bodenkd.,* 145, 470, 1982.
105. **Breemen, N. van,** Effects of seasonal redox processes involving iron on the chemistry of periodically reduced soils, in *Iron in Soils and Clay Minerals,* Stucki, J. W., Goodman, B. A., and Schwertman, U., Eds., Reidel Publishing, Dordrecht, The Netherlands, 1988, 197.
106. **Breemen, N. van,** Long-term chemical, mineralological, and morphological effects of iron-redox processes in periodically flooded soils, in *Iron in Soils and Clay Minerals,* Stucki, J. W., Goodman, B. A., and Schwertmann, U., Eds., Reidel Publishing, Dordrecht, The Netherlands, 1988, 811.
107. **Eeckman, J. P. and Laudelout, H.,** Chemical stability of hydrogen-montmorillonite suspensions, *Kolloid Z.,* 178, 99, 1961.
108. **Ben-Yaakov, S.,** pH buffering of pore water of recent anoxic marine sediments, *Limnol. Oceanogr.,* 18, 86, 1973.
109. **Giblin, A. E., Luther, G. W., III, and Valiela, I.,** Trace metal solubility in salt sediments contaminated with sewage sludge, *Estuarine Coastal Shelf Sci.,* 23, 477, 1986.
110. **Lord, C. J., III,** The chemistry and cycling of iron, manganese, and sulfur in salt marsh sediments, Ph.D. thesis, University of Delaware, 1980.
111. **Plant, J. A. and Raiswell, R.,** Principles of environmental geochemistry, in *Applied Environmental Geochemistry,* Thornton, I., Ed., Academic Press, London, 1983, chap. 1.
112. **Förstner, U.,** Chemical forms and reactivities of metals in sediments, in *Chemical Methods for Assessing Bioavailable Metals in Sludges and Soils,* Leschber, R., Davis, R. D., and L'Hermite, P., Eds., Elsevier, London, 1985, 1.
113. **Dreesen, D. R., Gladney, E. S., Owens, J. W., Perkins, B. L., Wienke, C. L., and Wangen, L. E.,** Comparison of levels of trace elements extracted from fly ash and levels found in effluent waters from a coal-fired power plant, *Environ. Sci. Technol.,* 11, 1017, 1977.
114. **Stumm, W. and Baccini, P.,** Man-made perturbation of lakes, in *Lakes — Chemistry, Geology, Physics,* Lerman, A., Ed., Springer-Verlag, New York, 1978, 91.
115. **Krumbein, W. C. and Garrels, R. M.,** Origin and classification of chemical sediments in terms of pH and oxidation-reduction potentials, *J. Geol.,* 60, 60, 1952.

116. **Förstner, U. and Wittmann, G. T. W.,** *Metal Pollution in the Aquatic Environment,* 2nd ed., Springer-Verlag, Berlin, 1983, 486.
117. **Stumm, W.,** Redox potential as an environmental parameter: conceptual significance and operational limitation, presented at 3rd Int. Conf. Water Pollution Res., Munich, Germany, No. 13, 1966, 1.
118. **Stumm, W. and Morgan, J. J.,** *Aquatic Chemistry,* 2nd. ed., John Wiley & Sons, New York, 1981, 1.
119. **Berner, R. A.,** *Principles of Chemical Sedimentology,* New York, McGraw-Hill, 1971, 240.
120. **Froelich, P. N., Klinkhammer, G. P., Bender, M. L., Luedtke, N. A., Heath, G. R., Cullen, D., Dauphin, P., Hammond, D., Hartman, B., and Maynard, V.,** Early oxidation of organic matter in pelagic sediments of the eastern equatorial Atlantic: suboxic diagenesis, *Geochim. Cosmochim. Acta,* 43, 1075, 1979.
121. **Klinkhammer, G.,** Early diagenesis in sediments from the eastern equatorial Pacific. II. Porewater metal results, *Earth Planet. Sci. Lett.,* 49, 81, 1980.
122. **Berner, R. A.,** *Early Diagenesis — A Theoretical Approach,* Princeton University Press, Princeton, NJ, 1980, 241.
123. **Robbins, J. A. and Callender, E.,** Diagenesis of manganese in Lake Michigan sediments, *Am. J. Sci.,* 275, 512, 1975.
124. **Lu, C. S. J. and Chen, K. Y.,** Migration of trace metals in interfaces of seawater and polluted surficial sediments, *Environ. Sci. Technol.,* 11, 174, 1977.
125. **Lion, L. W., Altmann, R. S., and Leckie, J. O.,** Trace-metal adsorption characteristics of estuarine particulate matter: evaluation of contribution of Fe/Mn oxide and organic surface coatings, *Environ. Sci. Technol.,* 16, 660, 1982.
126. **Davies-Colley, R. J., Nelson, P. O., and Williamson, H. J.,** Copper and cadmium uptake by estuarine sedimentary phase, *Environ. Sci. Technol.,* 18, 491, 1984.
127. **Wallmann, K.,** Die Frühdiagenese und ihr Einfluß auf die Mobilität der Spurenelemente As, Cd, Co, Cu, Ni, Pb und Zn in Sediment- und Schwebstoff-Suspensionen, Ph.D. thesis, Technical University of Hamburg-Harburg, Hamburg, 1990.
128. **Jenne, E. A.,** Trace element sorption by sediments and soil-sites and processes, *Molybdenum in the Environment,* Vol. 2, Chappell, W. R. and Peterson, K. K., Eds., Marcel Dekker, New York, 1977, 425.
129. **Mc Kenzie, R. M.,** The adsorption of lead and other heavy metals on oxides of manganese and iron, *Aust. J. Soil Res.,* 18, 61, 1980.
130. **Oakley, S. M., Nelson, P. O., and Williamson, K. J.,** Model of trace-metal partitioning in marine sediments, *Environ. Sci. Technol.,* 15, 474, 1981.
131. **Gerth, J.,** Untersuchungen der Adsorption von Nickel, Zink und Cadmium durch Bodenfraktionen unterschiedlichen Stoffbestands und verschiedene Bodenkomponenten, Ph.D. thesis, Christian-Albrecht University, Kiel, 1985.
132. **Tessier, A., Rapin, F., and Carignan, R.,** Trace metals in oxic lake sediment: possible adsorption onto iron oxyhydroxides, *Geochim. Cosmochim. Acta,* 49, 183, 1985.
133. **Balistrieri, L. S. and Murray, J. W.,** The surface chemistry of sediments from the Panama Basin: the influence of Mn oxides on metals adsorption, *Geochim. Cosmochim. Acta,* 50, 2235, 1986.
134. **Leckie, J. O.,** Adsorption and transformation of trace element species at sediment/water interface, in *The Importance of Chemical "Speciation" in Environmental Processes,* Dahlem Konferenzen 1986, Bernhard, M., Brinckman, F. E., and Sadler, P. J., Eds., Springer-Verlag, Berlin, 1986, 237.
135. **Murray, J. W.,** The interaction of metal ions at the manganese dioxide-solution interface, *Geochim. Cosmochim. Acta,* 39, 505, 1975.
136. **Murray, J. W. and Dillard, J. G.,** The oxidation of cobalt(II) adsorbed on manganese dioxide, *Geochim. Cosmochim. Acta,* 43, 781, 1979.
137. **Huynh-Ngoc, L., Whitehead, N. C., and Boussemart, M.,** Dissolved nickel and cobalt in the aquatic environment around Monaco, *Mar. Chem.,* 26, 119, 1989.
138. **Van den Berg, C. M. G. and Dharmvanij, S.,** Organic complexation of zinc in estuarine interstitial and surface water samples, *Limnol. Oceanogr.,* 29, 1025, 1984.
139. **Kersten, M.,** Mechanismen und Bilanz der Schwermetallfreisetzung aus einem Süßwasserwatt der Elbe, Ph.D. thesis, Technical University of Hamburg-Harburg, Hamburg, 1989.
140. **Andrear, M. O.,** Arsenic speciation in seawater and interstitial waters: the influence of biological-chemical interactions on the chemistry of a trace element, *Limnol. Oceanogr.,* 24, 440, 1979.
141. **Darby, D. A., Adams, D. D., and Nivens, W. T.,** Early sediment changes and element mobilization in a man-made estuarine marsh, in *Sediment and Water Interactions,* Sly, P. G., Ed., Springer-Verlag, Berlin, 1986, 343.

142. **Kerner, M., Kausch, H., and Kersten, M.,** Der Einfluß der Gezeiten auf die Verteilung von Nährstoffen und Schwermetallen in Wattsedimenten des Elbe-Aestuars, *Arch. Hydrobiol.,* 75 (Suppl.), 118, 1986.
143. **Renard, D., Michard, G., and Hoffert, M.,** Compartement geochimique du Cuivre, du Nickel et du Cobalt a l'interface eau-sediment — Application a l'enrichissement en ces elements dans les formation ferro-manganesiferes, *Mineralum Deposita,* 11, 380, 1976.
144. **Edenborn, H. M., Belzile, N., Mucci, A., Lebel, J., and Silverberg, N.,** Observations on the diagenetic behavior of arsenic in a deep coastal sediment, *Biogeochemistry,* 2, 359, 1986.
145. **Peterson, M. L. and Carpenter, R.,** Arsenic distribution in pore water and sediment of Puget Sound, Lake Washington, the Washington Coast and Saanich Inlet, B. C., *Geochim. Cosmochim. Acta,* 50, 353, 1986.
146. **Clement, W. H. and Faust, S. D.,** The release of arsenic from contaminated sediments and mud, *J. Environ. Sci. Health,* 16, 87, 1981.
147. **Brannon, J. M. and Partrick, W. H., Jr.,** Fixation, transformation, and mobilization of arsenic in sediments, *Environ. Sci. Technol.,* 21, 450, 1987.
148. **Gendron, A., Silverberg, N., Sundby, B., and Lebel, J.,** Early diagenesis of cadmium and cobalt in sediments of the Laurentian Trough, *Geochim. Cosmochim. Acta,* 50, 741, 1986.
149. **Graybeal, A. L. and Heath, G. R.,** Remobilization of transition metals in surficial pelagic sediments from the eastern Pacific, *Geochim. Cosmochim. Acta,* 48, 965, 1984.
150. **Tsungai, S., Yonemaru, I., and Kusakabe, M.,** Postdepositional migration of Cu, Zn, Ni, Co, Pb and Ba in deep sea sediments, *Geochem. J.,* 13, 239, 1979.
151. **Palsma, A. J. and Loch, J. P. G.,** Seasonal Variation of Cd in porewater of river foreland soils in the Rhine-Meuse estuary, in *Heavy Metals in the Environment,* Vol. 2, Farmer, J. G., Ed., CEP Consultants Ltd., Edinburgh, 1991, 20.
152. **Kornicker, W. K. and Morse, J. W.,** Interactions of divalent cations with the surface of pyrite, *Geochim. Cosmochim. Acta,* 55, 2159, 1991.
153. **Elderfield, H., Hepworth, A., Edwards, P. N., and Holliday, L. M.,** Zinc in the Conway River and estuary, *Estuarine Coastal Mar. Sci.,* 9, 403, 1975.
154. **Luther, G. W., Meyerson, A. L., Krajewski, J. J., and Hires, R.,** Metal sulfides in estuarine sediments, *J. Sediment. Petrol.,* 50, 1117, 1980.
155. **Hallberg, R. O.,** The microbiological C-N-S cycles in sediments and their effect on the ecology of the sediment-water interface, *Oikos,* 15, 51, 1973.
156. **Hallberg, R. O.,** Paleoredox conditions in the eastern Gotland Basin during the recent centuries, *Merentutkimuslaitoksen Julk. Havsforskninginst. Skr.,* 238, 3, 1974.
157. **Hallberg, R. O.,** Metal distribution along a profile of an intertidal area, *Estuarine Coastal Mar. Sci.,* 2, 153, 1974.
158. **Elderfield, H. and Hepworth, A.,** Diagenesis, metals and pollution in estuaries, *Mar. Pollut. Bull.,* 6, 85, 1975.
159. **Elderfield, H., Thornton, I., and Webb, J. S.,** Heavy metals and oyster culture in Wales, *Mar. Pollut. Bull.,* 2, 44, 1971.
160. **Paalmann, M. A. A., Van de Meent-Olieman, G. C., and Van der Weijden, C. H.,** Porewater chemistry of estuarine Rhine-Meuse sediment: (re)mobilization of Cu, Zn, and Ni, in *Heavy Metals in the Environment,* Vol. 1, Farmer, J. G., Ed., CEP Consultants Ltd., Edinburgh, 1991, 449.
161. **Salomons, W. and Geritse, R. G.,** Some observations on the occurrence of phosphorus in recent sediments from western Europe, *Sci. Total Environ.,* 17, 37, 1981.
162. **Nissenbaum, A. and Swaine, D. J.,** Organic matter metal interactions in recent sediments: the role of humic substances, *Geochim. Cosmochim. Acta,* 40, 809, 1976.
163. **Nriagu, J. O. and Coker, R. D.,** Trace metals in humic and fulvic acids from Lake Ontario sediments, *Environ. Sci. Technol.,* 14, 433, 1980.
164. **Galoway, F. and Bender, M.,** Diagenetic models of interstitial nitrate profiles in deep sea suboxic sediments, *Limnol. Oceanogr.,* 27, 624, 1982.
165. **Jahnke, R., Heggie, D., Emerson, S., and Grundmanis, V.,** Pore waters of the central Pacific Ocean: nutrient results, *Earth Planet. Sci. Lett.,* 61, 233, 1982.
166. **Nissenbaum, A., Baedecker, M. J., and Kaplan, I. R.,** Studies on dissolved organic matter from interstitial water of a reducing marine fjord, in *Advances in Organic Geochemistry,* van Gaertner, H. R. and Wehner, H., Eds., Pergamon Press, Oxford, 1972, 427.

167. **Bender, M. and Heggie, D.,** The fate of organic carbon on the sea-floor: a status report, *Geochim. Cosmochim. Acta,* 48, 977, 1984.
168. **Krom, M. D. and Sholkovitz, E. R.,** Nature and reactions of dissolved organic in the interstitial water of marine sediments, *Geochim. Cosmochim. Acta,* 41, 1565, 1977.
169. **Tills, A. R. and Alloway, B. J.,** The speciation of lead in soil solution from very polluted soils, *Environ. Technol. Lett.,* 4, 529, 1983.
170. **Jonasson, I. R.,** Geochemistry of sediment/water interactions of metals, including observations on availability, in *The Fluvial Transport of Sediment-Associated Nutrients and Contaminants,* Shear, H. and Watson, A. E. P., Eds., IJC/PLUARG, Windsor, Ontario, 1977, 255.
171. **Hart, B. T. and Davies, S. H. R.,** A new dialysis-ion exchange technique for determining the forms of trace metals in water, *Aust. J. Mar. Freshwater Res.,* 28, 105, 1977.
172. **Förstner, U., Nähle, C., and Schöttler, U.,** Sorption of metals in sand filters in the presence of humic acids, *DVWK Bull.,* 13, 95, 1979.
173. **Douglas, G. S., Mills, G. L., and Quinn, J. G.,** Organic copper and chromium complexes in the interstitial waters of Narragansett Bay sediments, *Mar. Chem.,* 19, 161, 1986.
174. **Anikkouchine, W. A.,** Dissolved chemical substances in compacting marine sediments, *J. Geophys. Res.,* 71, 505, 1967.
175. **Manheim, F. T.,** The diffusion of ions in unconsolidated sediments, *Earth Planet. Sci. Lett.,* 9, 307, 1970.
176. **Tzur, Y.,** Interstitial diffusion and advection of solute in accumulating sediments, *J. Geophys. Res.,* 76, 4208, 1971.
177. **Boyle, E. A., Sclater, F. R., and Edmound, J. M.,** The distribution of dissolved copper in the Pacific, *Earth Planet. Sci. Lett.,* 37, 38, 1977.
178. **Aller, R. T.,** Diagenetic processes near the sediment-water interface of Long Island Sound. II. Fe and Mn, in *Advances in Geophysics,* Vol. 22, Academic Press, London, 1980, 351.
179. **Lynn, D. C. and Bonatti, E.,** Mobility of manganese in diagenesis of deep sea sediment, *Mar. Geol.,* 3, 457, 1965.
180. **Li, Y., Bischoff, J., and Mathieu, G.,** The migration of manganese in the Arctic Basin sediment, *Earth Planet. Sci. Lett.,* 7, 265, 1969.
181. **Calvert, S. E. and Price, N. B.,** Diffusion and reaction profiles of dissolved manganese in the pore waters of marine sediments, *Earth Planet. Sci. Lett.,* 16, 245, 1972.
182. **Holdren, G. R., Jr., Bricker, O. P., III, and Matstoff, G.,** A model for the control of dissolved manganese in the interstitial waters of Chesapeake Bay, in *Marine Chemistry in the Coastal Environment,* Church, T. M., Ed., A. C. S. Symposium Series No. 18, American Chemical Society, Washington, D.C., 1975, 364.
183. **Salomons, W.,** Sediments and water quality, *Environ. Technol. Lett.,* 6, 315, 1985.
184. **Boyle, E. A., Sclater, F., and Edmond, J. M.,** On the marine geochemistry of Cd, *Science,* 263, 42, 1976.
185. **Bruland, K. W., Knauer, G. A., and Martin, J. H.,** Cadmium in NE Pacific Waters, *Limnol. Oceanogr.,* 23, 42, 1978.
186. **Slater, F. R., Boyle, E., and Edmond, J. M.,** On the marine geochemistry of nickel, *Earth Planet. Sci. Lett.,* 31, 119, 1976.
187. **Bruland, K. W.,** Oceanographic distribution of cadmium, zinc, nickel and copper in the North Pacific, *Earth Planet. Sci. Lett.,* 47, 176, 1980.
188. **Boyle, E. A., Huested, S. S., and Jones, S. P.,** On the distribution of copper, nickel and cadmium in the surface waters of the North Atlantic and North Pacific Ocean, *J. Geophys. Res.,* 86, 8048, 1981.
189. **Campbell, J. A. and Yeats, P. A.,** Dissolved chromium in the northwest Atlantic Ocean, *Earth Planet. Sci. Lett.,* 53, 427, 1981.
190. **Cranston, R. E.,** Chromium in Cascadia Basin, northeast Pacific Ocean, *Mar. Chem.,* 13, 109, 1983.
191. **Bruland, K. W., Knauer, G. A., and Martin, J. H.,** Zinc in northeast Pacific, *Nature,* 271, 741, 1978.
192. **Emerson, S., Jahnke, R., Bender, M., Froelich, P., Klinkhammer, G., Bowser, C., and Setlock, G.,** Early diagenesis in sediments from the eastern equatorial Pacific. I. Pore water nutrients and carbonate results, *Earth Planet. Sci. Lett.,* 49, 57, 1980.
193. **Bruland, K. W., Bertine, K., Koide, M., and Goldberg, E. D.,** History of metal pollution in southern California coastal zone, *Environ. Sci. Technol.,* 8, 425, 1974.
194. **Finney, B. P. and Huh, C.,** History of metal pollution in the southern California bight: an update, *Environ. Sci. Technol.,* 23, 294, 1989.

195. **Bertine, K. K.,** The deposition of molybdenum in anoxic waters, *Mar. Chem.,* 1, 43, 1972.
196. **Bertine, K. K. and Turekian, K. K.,** Molybdenum in marine deposits, *Geochim. Cosmochim. Acta,* 37, 1415, 1973.
197. **Murray, J. W., Grundmanis, V., and Smethie, W. M.,** Interstitial water chemistry in the sediments of Saanich Inlet, *Geochim. Cosmochim. Acta,* 42, 1011, 1978.
198. **Heggie, D. and Burrell, D.,** Depth distribution of copper in interstitial waters and the water column of an Alaskan Fjord, *Trans. Am. Geophys. Un.,* 56, 1006, 1975.
199. **Armstrong, P. B., Lyons, W. B., and Gaudette, H. E.,** Application of formaldoxine colorimetric method for the determination of manganese in estuarine pore water, *Estuaries,* 2, 198, 1979.
200. **Contreas, R., Fogg, T. R., Chasteen, N. D., Gaudette, H. E., and Lyons, W. B.,** Molybdenum in porewaters of anoxic marine sediments by electron paramagnetic resonance spectroscopy, *Mar. Chem.,* 6, 365, 1978.
201. **Lyons, W. B. and Fitzgerald, W. F.,** Trace metals speciation in nearshore anoxic and suboxic pore waters, in *Trace Elements in Sea Water,* Wong, C. S., Bruland, K. W., Burton, J. D., and Goldberg, E. D., Eds., Plenum Press, New York, 1983, 621.
202. **Dethleff, D.,** Frühdiagenetische Mineralisationsprozesse und Mobilität der Spurenmetalle Cd, Pb, und Cu im Porenwasser eines Elbwatt-Sediments, Diplom-Arbeit, Christian Albrechts Universität, Kiel, 1990.
203. **Frevert, T. and Sollmann, C.,** Heavy metals in Lake Kinneret (Israel). III. Concentrations of iron, manganese, nickel, cobalt, molybdenum, zinc, cadmium, lead and copper in interstitial water and sediment dry weights, *Arch. Hydrobiol.,* 109, 181, 1987.
204. **Wang, Z. J., Ghobary, H. El., Giovanoli, F., and Favarger, P.-Y.,** Interpretation of metal profiles in a sediment core from Lake Geneva: metal mobility or pollution, *Schweiz. Z. Hydrol.,* 48, 1, 1986.
205. **Förstner, U., Ahlf, W., Calmano, W., and Kersten, M.,** Sediment criteria development, in *Sediments and Environmental Geochemistry,* Heling, D., et al., Eds., Springer-Verlag, Berlin, 311, 1990.
206. **Förstner, U., Ahlf, W., Calmano, W., Kersten, M., and Schoer, J.,** Assessment of metal mobility in sludges and solid wastes, in *Metal Speciations in the Environment,* Broekaert, J. A., Guecer, S., and Adams, F., Eds., Springer-Verlag, Berlin, 1990, 1.
207. **Di Toro, D. M., Mahony, J. D., Hansen, D., Scott, J. K., Carlson, A. R., and Ankley, G. T.,** Acid volatile sulfide predicts the acute toxicity of cadmium and nickle in sediments, *Environ. Sci. Technol.,* 26, 96, 1992.
208. **Di Toro, D. M., Mahony, J. D., Hansen, D., Scott, J. K., Hicks, M. B., Mayr, S. M., and Redmond, M. S.,** Toxicity of cadmium in sediments: the role of acid volatile sulfide, *Environ. Toxicol. Chem.,* 9, 1487, 1990.

Chapter 7

Groundwater

Bert Allard

CONTENTS

I. INTRODUCTION

It is important to distinguish between subsurface water and groundwater. The former comprises all the water that exists below the surface of the earth, i.e., water in the soil zone, in the capillary fringe, in the bedrock, etc. *Groundwater* is the part of the subsurface water which is in the zone of saturation, i.e., under the water table where all pores are filled with water. The groundwater will receive contributions from infiltrating surface water and soil water, which have a flow direction from the surface downwards, but also from deeper flowpaths, representing groundwater from other underground locatio
direction upwards or horizontal. It is important to realize that the groundwater at a
represents a mixture of waters of different origin (surface waters as well as groundwater)
also with different composition in terms of dissolved constituents, etc. The dissolved con
the various reactions between water and solid geologic phases (weathering, dissoluti

0-8493-6304-7/95/$0.00+$.50

Table 1 **Element concentration levels (μg l⁻¹) in groundwater systems**

>1000	Ca, Cl, K, Mg, Na, S, Si
100–1000	Al, F, Fe
10–100	B, Ba, Br, N, P, Sr, Zn
1–10	Cu, I, Li, Mn, Rb, Ti
0.1–1	Ag, As, Be, Ce, Co, Cr, La, Mo, Nd, Ni, Pb, Sb, Se, U, V, Zr
0.01–0.1	Bi, Cd, Cs, Ga, Hf, Hg, Sc, Sm, Th, W, Yb
<0.01	Au, Eu, Lu, Sn, Ta, Tb

redox reactions, complexation, etc.) that are continuously taking place along the water pathway. Thus, the water at any individual sampling location is generally not at equilibrium with the various geologic phases at the sampling place, with few exceptions. This is, of course, due to the dynamic mixing of water from various sources, but also due to the fact that true chemical equilibria are rarely achieved between the common silicate components in the bedrock and the water phase.

The composition of groundwaters of various origin and age, as well as the chemical processes that determine the concentration levels of the major constituents, is described in several comprehensive textbooks[1–9] and will not be discussed in this chapter. The emphasis will be put on sources of some selected trace elements, contributing processes, concentration levels and speciation in deep pristine groundwater systems from primarily crystalline bedrock, and with special reference to complexation and redox processes (cf. Chapter 2). Some of the processes that lead to a removal of trace elements, notably adsorption and coprecipitation phenomena (Chapter 2), will be briefly reviewed. With respect to sampling and preanalysis handling of water samples with low concentrations of metals, etc., as well as analytical procedures (Chapter 3), only additional comments pertinent to groundwater systems are given here.

The concentrations of elements in groundwater reflect the general composition of the solid geologic phases that have been interacting with the water. The most common elements in groundwater systems are compiled in decreasing order of abundance in Table 1. Detailed chemical information, besides total concentrations, is scarce for most of the elements at concentrations below 0.1 μg l^{-1}. The information on the chemical properties of these elements in undisturbed groundwater environments, as well as in fact on most of the other elements present at higher concentration levels, is primarily based on common chemical knowledge and rarely on direct observations in field systems. Very few published studies of groundwater systems include careful analyses of the trace components, and, consequently, there are only a few accurate trace element concentration data representative of uncontaminated groundwater systems. This fact limits the number of elements that can be described, except for general discussions of chemical properties as well as of predicted behavior in groundwater systems. Only a few elements with *background concentrations* largely in the range of 0.1 to 100 μg l^{-1} under natural conditions are selected for discussion in this chapter (Cr, Ni, Cu, Zn, As, Cd, Hg, and Pb). Cases of local groundwater contamination, pollution effects, etc. are discussed elsewhere.[10–16]

II. SOURCES AND CONTRIBUTING PROCESSES

A. MACROCOMPONENTS IN GROUNDWATER

Subsurface waters interacting with the ground can be considered as a multicomponent multiphase system, however, with a fairly limited number of *major* components. Dominating minerals in crystalline rocks are feldspars, quartz, micas (muscovite and biotite), amphiboles, pyroxenes, and olivins.[17] Besides, various accessory minerals would be present, as well as organic material originating in the biosphere. Processes within the aqueous phase, exchanges with the atmosphere and biosphere, and interactions with the exposed solid mineral phases determine the composition of the groundwater in terms of dissolved constituents. Some of these processes are illustrated in Figure 1.

The most important hydrochemical parameter is generally pH, which is in many cases determined by the carbonate system (HCO_3^-–CO_2 as well as Ca^{2+}–CO_3^{2-}). The Ca–CO_3 system is actually one of the few systems which frequently reach local equilibrium. Calcite is a common fissure mineral in crystalline rocks even in the absence of any primary carbonate phases. A near to neutral pH (6.2 to 8.6) would be obtained, ong as solid carbonate phases are present, the carbonate being an efficient pH buffer. Other pH-ing processes are the weathering of silicates (pH 5 to 6.2), ion exchange reactions (pH 4.2 to 5),

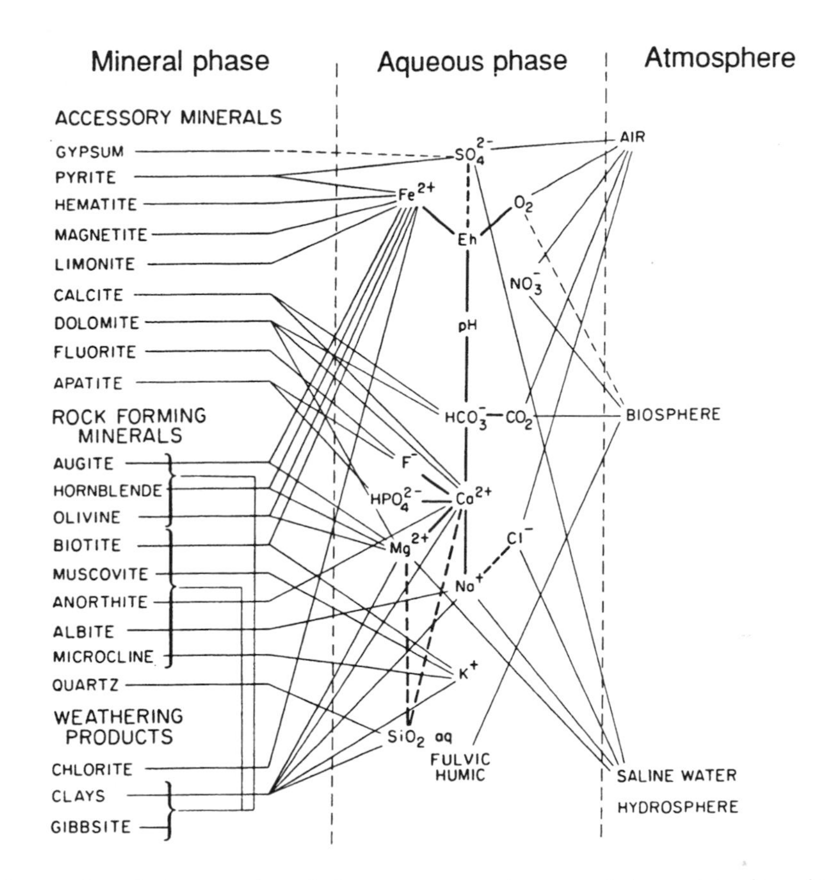

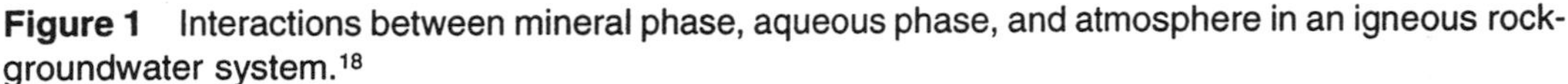
Figure 1 Interactions between mineral phase, aqueous phase, and atmosphere in an igneous rock-groundwater system.[18]

and the dissolution of metal hydroxides (pH 3 to 4.2 for aluminum; pH < 3 for iron, etc.), as well as the dissociation of natural organic acids at moderate or high concentration levels (pH 3.5 to 5.5; in surface waters). These natural buffers would counteract changes in pH, for instance due to mixing of waters of various pH with different origin.

The redox potential is governed by the presence of oxygen, as well as by the Fe(III)/Fe(II) and SO_4^{2-}/S^{2-} systems, the degradation of organics, etc. Most of these processes are involving exchanges of protons, which means that they are governed by pH — or influencing pH of the system. The calcium concentration is dependent of the carbonate equilibria which set an upper concentration limit. However, calcium is, together with magnesium, sodium, and potassium, also a product in slow diffusion-controlled weathering reactions involving feldspars, etc. and subsequent fast ion exchange equilibria with clay minerals and organic matter with high exchange capacities and large surface/volume ratios.

Chloride, when present at substantial concentration levels in fresh waters, often has a marine origin. Other anions (sulfate, fluoride, phosphate) may originate from the weathering of relevant mineral phases, which would set an upper solubility-related concentration limit. However, these systems rarely reach equilibrium, and the concentration levels of the corresponding anions are generally below saturation. All the various processes are discussed in detail in common textbooks.

The degradation of biogenic matter in soil and surface waters yields a soluble organic fraction of high molecular weight, humic and fulvic acids, as well as simple carboxylic acids, etc.[19–22] Concentrations of dissolved organics are substantial even in deep and old groundwater.[23,24]

Thus, the chemical evolution of natural waters is governed by a few fundamental key processes. The general mixing of waters of different origin has the effect that most terrestrial waters are dominated by a few major species in solution. The concentrations of these species usually fall within a fairly narrow range (typically about 1.5 to 2 orders of magnitude), despite the complexity of the solid heterogenous ground and bedrock system (cf. Figure 1). This is illustrated in Figure 2, which shows the CO_3^{2-} concentration vs. pH, as calculated from measured HCO_3^- concentrations in groundwater from deep crystalline granitic bedrock (72 samples from 8 different locations). The inherent inertia in the composition

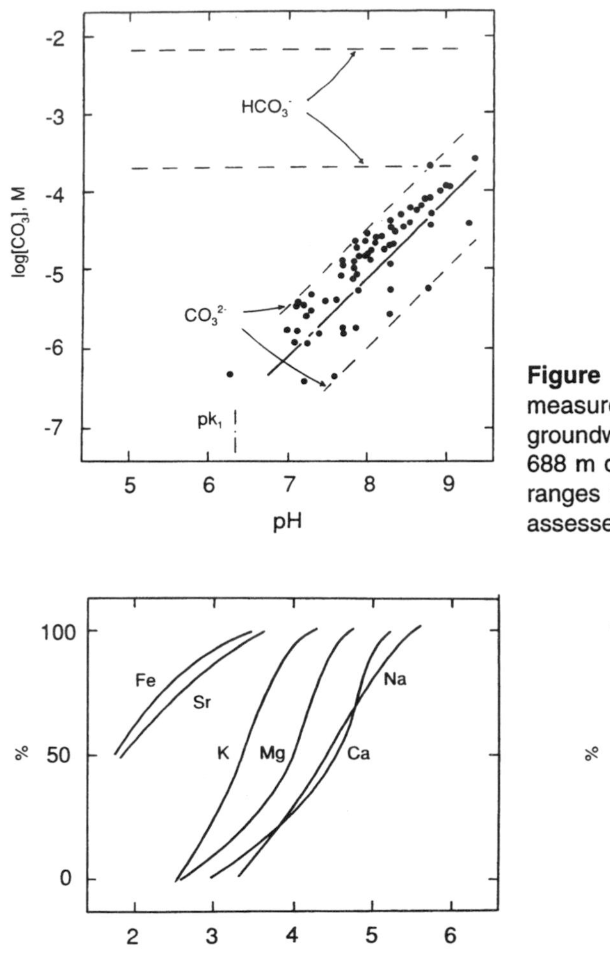

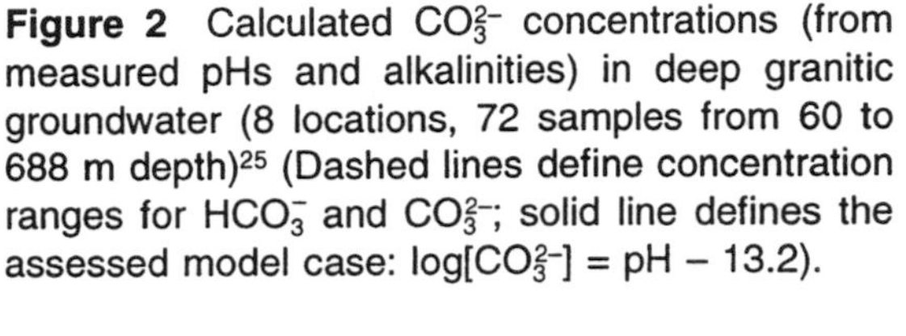

Figure 2 Calculated CO_3^{2-} concentrations (from measured pHs and alkalinities) in deep granitic groundwater (8 locations, 72 samples from 60 to 688 m depth)[25] (Dashed lines define concentration ranges for HCO_3^- and CO_3^{2-}; solid line defines the assessed model case: $\log[CO_3^{2-}] = pH - 13.2$).

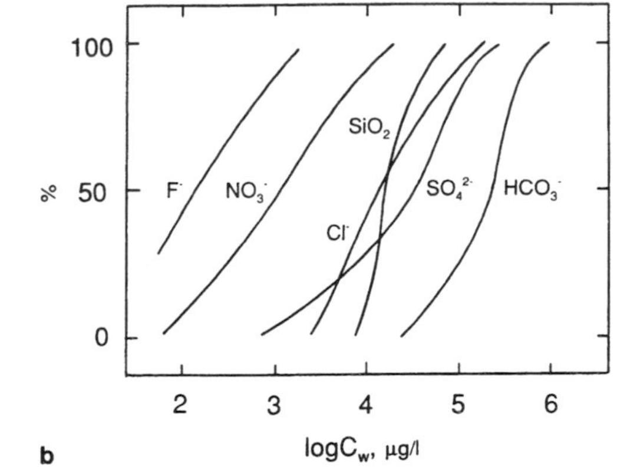

Figure 3 Cumulative curves showing the frequency distribution of the concentrations (C_w, µg l^{-1}) of the dominating cationic (a) and anionic (b) constituents in terrestrial waters (modified after Stumm and Morgan[2]).

of groundwater and other terrestrial freshwater systems is further illustrated in Figure 3, which shows the concentration ranges (frequency distribution) of the dominating cations and anions with the exception of organic acids in terrestrial waters.

The most common potential determining redox systems in groundwater environments are illustrated in Figure 4. Four pe/pH domains of importance for groundwater systems can be distinguished:

Domain 1: Groundwater with free oxygen; can indicate short residence time or the absence of metabolizable organics

Domain 2: Groundwater with Fe(II) and Mn(II) and with pe determined by equilibria with solid phases such as MnO_2, $Fe(OH)_3$, or Fe_2O_3

Domain 3: Groundwater with H_2S–HS^- and with pe determined by the SO_4^{2-} reduction

Domain 4: Groundwater/sediment with pe determined by decomposition of organics

Thus, the groundwater as a reaction medium can be defined in terms of a few dominating major constituents that are to some extent interrelated, either chemically (notably pH and carbonate) or through mixing processes, despite the heterogeneity and complexity of the geosphere that provides the various components to the aquatic systems.

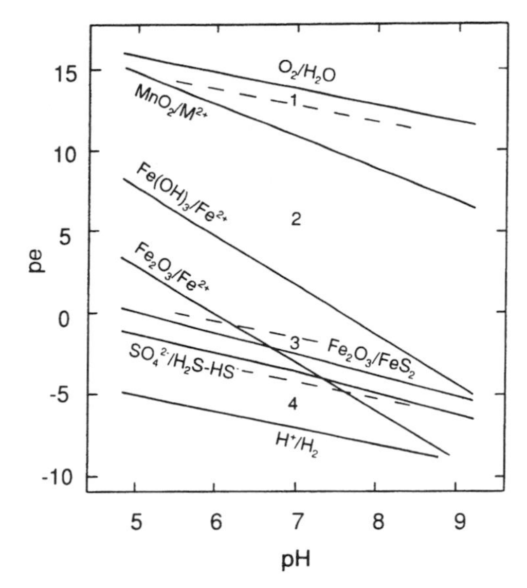

Figure 4 Potential/pH diagram with some common potential determining redox systems (modified after Drever[7]). (Domains 1 to 4: See text for explanation).

Table 2 **Concentrations of trace elements in various rocks and soils**

	Concentration[a] (mg kg^{-1})							
Material	**Cr**	**Ni**	**Cu**	**Zn**	**As**	**Cd**	**Hg[b]**	**Pb**
Basaltic	40–600	45–410	30–160	48–240		0.006–0.6		2–18
rocks[26]	220	140	90	110		0.2		6
Granitic	2–100	2–20	4–30	5–140	0.2–13.8[27]	0.003–0.18	5–250[28]	6–30
rocks[26]	20	8	15	40	1.5	0.15		18
Shales,	30–590	20–250	18–120	18–180	0.3–500[27]	<0.3–11[29]	5–3250[28]	16–50
clays[26]	120	68	50	90	14.5	1.4		20
Black	26–1000	10–500	20–200	34–1500		<0.3–8.4		7–150
shales[26]	100	50	70	10		1.0		30
Lime-	<1–120[30]	7–20[31]	0.6–13[32]	<1–180[17]	0.1–20.1[27]	0.05–0.1[29]	40–220[28]	5–9[29]
stones	10		6	20	2.6		40	
Sand-	35–90[29]	2–9[31]	6–46[32]	5–170[17]	0.6–120[27]	0–0.05[29]	<10–300[28]	<1–31[26]
stones			30	30	4.1	<0.03	55	7
Soils[31]	5–1500	2–750	2–250	1–900	0.1–55[27]	0.01–2.0	20–150[28]	2–200[33]
	70	50	30	90	7.2	0.35	70	20
Soils[34]	1–2000	<5–700	<1–700	<5–2900				<10–700
	54	19	25	60				19

[a] Upper value, ranges; lower values, averages; [b] In µg/kg.

B. TRACE ELEMENTS IN GROUNDWATER

The abundance of trace elements in various rock types as well as soils, sediments, and surficial materials varies within a broad range, as illustrated in Table 2. Thus, concentrations may differ by up to three orders of magnitude between adjacent mineral grains in the bedrock (e.g., alkali feldspars and amphiboles/pyroxenes) as illustrated in Table 3.[17] Average values can be entirely misleading when used for interpretation of local gradients and concentration differences, as well as of effects on groundwater composition. Still, there is an evident correlation between the average concentration of elements in accessible geologic phases and natural freshwater systems, as illustrated in Figure 5. The differences in distribution between solid mineral phases and solution phase between anionic species (Cl, Br, S, etc.), cationic elements at low

Table 3 **Distribution of Cu, Zn, and Pb in rock-forming minerals[17]**

Mineral	Abundance in Earth's Crust (%)	Concentration[a] (mg kg^{-1})		
		Cu	Zn	Pb
Alkali feldspars	22	1–20	10–24	2–700
		4	15	53
Plagioclase	62	8–700	1–50	1–70
		62	17	20
Muscovite	2	5–152	24–200	6–70
		36	59	20
Biotite	4	1–480	34–4000	7–95
		86	527	21
Amphibole	5	1–300	34–690	1–70
		78	196	11
Pyroxene	4	4–1000	16–200	0.3–20
		120	97	6
Olivine	1	6–960	50–82	0.2–7.2
		115	63	2
Quartz	18		4–11	0.1–3
		2	7	1

[a] Upper value, ranges; lower value, averages.

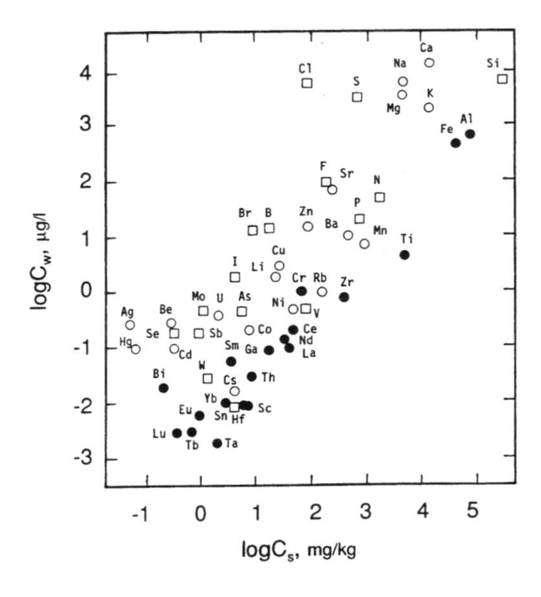

Figure 5 Concentration levels (median values) of elements in fresh waters (C_w, μg l^{-1}) vs. concentration in soils and bedrock (C_s, mg kg^{-1}): ● cations, M(III), M(IV); ○ cations, M(I), M(II); □ anions.

oxidation states (I and II; Na, K, Mg, Ca, etc.), and of high oxidation state (III, IV) are evident and reflect different solubilities and weathering characteristics of the corresponding solid phases as well as adsorption processes and chemical states in solution, notably hydrolysis behavior (see Section III).

There are several reviews of heavy metal pollution in the aquatic environment, including groundwater systems.[10,11,13] There are, however, only a few studies of what can be considered as original pristine groundwaters.[35] The background levels representative of groundwaters and other freshwater systems suggested by Bowen[31] are frequently quoted and considered as "natural" levels. Concentrations of trace elements in soils and biota, as well as in waters, are compiled and summarized by Adriano.[36] Some concentration ranges that are claimed to represent groundwater levels are summarized in Table 4. It is not possible to distinguish any "natural" range or background level representative of uncontaminated groundwaters solely from these data. The reported ranges and individual values vary considerably. This may indicate differences in natural background concentration levels related to the bedrock and soil composition, but also various artifacts. Analytical techniques and handling procedures have been improved, and many of the early measurements have probably yielded too high values[62] (see also Chapter 3).

Table 4 **Concentrations of trace elements in groundwaters of various origin**

Concentration (µg l⁻¹)								
Cr	**Ni**	**Cu**	**Zn**	**As**	**Cd**	**Hg**	**Pb**	**Ref.**
				0.1–3.4				37
				0.01–800				38
1–5	0.5–2	2–5	4–10		0.1		0.1–1	39
0.5–20	8–22	8–470	5–740		0.2–1.0		5–124	40
<1	<4	<10	<10		<7	0.03	<10	41
2.5	4.4	1.1	17.8		0.3		1.5	42
	0.3	1.8						43
		1.5	5.7					44
0.1–6	0.02–27	0.2–30	0.2–100	0.3–230	0.01–3	0.0001–2.8	0.06–120	31
						0.01–0.10		45
0.6	1–15	18–27	190–570		2.9–5.1		5.1–6.3	46
				0.0–6.1				47
						0.01–0.46		48
		0.4–2.9	0.9–80		0.1–1.2		0.3–3.0	49
<5		<5–17	<5–52		<5		<10	50
						0.0005–0.0013		51
				3–69				52
0.1–0.5	0.1–1	0.1–2	0.2–5	0.1–0.5	0.005–0.05		0.05–0.5	53
				1.9–75				54
						<0.001		55
<0.2–7.6		<0.3–18	<5–390		<0.05–2		<0.1–1.4	56
<0.2–1.1		<0.3–10	5–21		<0.05–0.11		<0.1	57
	0.1–1							58
				1.1–180				59
0.04–0.7		0.3–4	2–40		0.006–0.1		0.02–0.3	35
						0.0006–0.052		60
				0.6–1.5				61

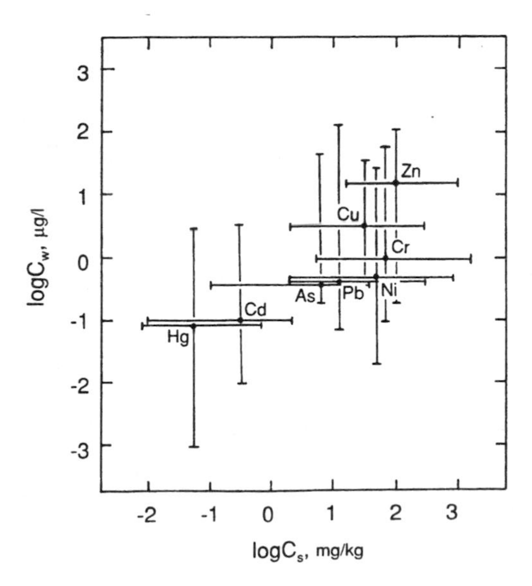

Figure 6 Concentration ranges for Cr, Ni, Cu, Zn, As, Cd, Hg, and Pb (C_w, µg l⁻¹) in groundwaters vs. concentration ranges in granitic rocks (C_s, mg kg⁻¹) (modified after Ledin et al.[35]).

However, the general correlation between concentration ranges in groundwaters and in the corresponding solid geologic phases (Figure 6) confirms the overall correlation (Figure 5).

Background concentration levels in mineralized areas could be far above the levels indicated in Table 4, due to natural weathering processes of metal-rich phases, low natural pH, etc. Metal concentrations in unpolluted groundwaters from mineralized areas of up to 850 μg l^{-1} Cu, 2.6 mg l^{-1} Zn, 27 μg l^{-1} Cd, and 390 μg l^{-1} Pb have been reported.[63] Similar high concentrations were recorded in the shallow groundwater of an area with metal-contaminated soil during recharge conditions, while the levels under discharge conditions were within "normal" ranges.[64] A variation of "natural" groundwater concentration levels within two orders of magnitude must be expected, also in areas without local metal-rich mineralizations. The assessment of background levels in groundwater is further discussed in Section V.

III. CHEMICAL REACTIONS AND REMOVAL PROCESSES

A. CHEMICAL SPECIATION IN SOLUTION

The chemical speciation of trace elements in a groundwater system is determined by

- The hydrogen ion availability (defined by pH), determining the concentration of hydroxide ions
- The presence and concentrations of inorganic ligands (CO_3^{2-}, SO_4^{2-}, S^{2-}, Cl^-, F^-, and HPO_4^{2-}, besides OH^-)
- The presence and concentrations of organic complexing agents (primarily humic and fulvic acids)
- The free electron availability (defined by pe)
- The concentration of the element as well as the ionic strength and cation distribution

Chromium exists predominantly in the trivalent state under groundwater conditions but can be oxidized to the hexavalent state (anionic CrO_4^{2-}) under oxidizing conditions. Arsenic is tri- or pentavalent in solution, predominantly as the anionic AsO_3^{3-} or AsO_4^{3-}. Mercury is divalent or exists in elementary form, while all of Ni, Cu, Zn, Cd, and Pb are solely divalent in aquatic systems.

The distribution of chemical species of Cr, Ni, Cu, Zn, As, Cd, Hg, and Pb is discussed below for each element on the basis of the calculated speciation in a reference groundwater (Table 5), representative of crystalline granitic bedrock.[23] Formation constants for the inorganic species (complexes with natural ligands of importance in a groundwater environment, OH^-, CO_3^{2-}, SO_4^{2-}, Cl^-, F^-, and HPO_4^{2-}) have been assessed from standard compilations, Table 6.[65–67] Conditional stability constants have been assessed for complexes with fulvic acids, based on reported data from studies with different techniques and various humic materials,[69–83] but also considering the expected differences between the elements as well as the heterogeneous character of the humic substances.[81–84] Significantly higher formation constants would be expected for the metal complexation at low total metal concentrations, if the organic acid contains some element specific binding sites. This may be the case for mercury (see below). Further general chemical information on the elements is given in several handbooks, etc.[31,36,85–94]

1. Chromium

Trivalent chromium, which would dominate in anoxic waters, forms strong complexes with oxygen ligands, and hydrolysis is generally the dominating process in solution at pH above 5 to 6. Strong complexes are consequently also formed with natural organic humic and fulvic acids as well as with carbonate. The cationic hydrolysis product $Cr(OH)_2^+$ is the expected dominating chemical species in a groundwater system, Figure 7a, with significant fractions of $CrOH^{2+}$ at pH < 6.5 and $Cr(OH)_3$ at pH > 7.5. Complex formation with fluoride may be significant at low pH and high fluoride concentration; however, it is not likely in most groundwater systems. Sulfate and chloride in nonsaline groundwaters would not significantly affect the speciation in solution.

The concentrations of organics in groundwater would not be expected to significantly affect the speciation, assuming the conditional stability constant given in Table 6 (extrapolated from data for europium). This constant may, however, be highly underestimated; no reliable measurements of Cr(III) humate complexation are reported.

Under oxic conditions, in the pe/pH domain 1 and upper part of 2, Figure 4, the hexavalent oxidation state will dominate at equilibrium, Figure 8a. The anionic CrO_4^{2-} dominates at pH > 6.5, and $HCrO_4^-$ could exist at a lower pH. The anionic $Cr_2O_7^{2-}$ can exist at pH > 6 and at total chromium concentrations of 10^{-2} *M* or above,[66] which would not be encountered in undisturbed groundwaters.

2. Nickel

Nickel, which is entirely divalent in aquatic systems, forms fairly weak complexes with oxygen ligands, but strong complexes with sulfur. The noncomplexed Ni^{2+} ion is the dominating nickel species expected in a groundwater system in the whole relevant pH interval, Figure 7b, with significant contributions of $NiSO_4$ at pH < 6 to 7 (at moderate or high sulfate concentration) and $NiCO_3$ at pH > 8, however. Chloride complexes are of little importance.

Table 5 **Composition of reference groundwater**

Species and Concentration Range[a]	Selected Concentration in Reference Water
pH = 5–9	log[OH] = pH–14
$(CO_3)_{tot}$ = 90–275 mgl^{-1}	$\log[CO_3]$ = pH–13.2[b]
SO_4^{2-} = 0.5–15 mg l^{-1}	$\log[SO_4]$ = –3.81 (15 mg l^{-1})
Cl^- = 4–15 mg l^{-1}	log[Cl] = –3.37 (15 mg l^{-1})
F^- = 0.5–5 mg l^{-1}	log[F] = –3.58 (5 mg l^{-1})
$(PO_4)_{tot}$ = 0.01–0.2 mg l^{-1}	$\log[HPO_4]$ = pH – 23.6[c]
HA = 0–1 mg l^{-1}[d]	$\log[HA]_{tot}$ = –5.30[e] (1 mg l^{-1})

[a] Typical for a nonsaline groundwater from a granitic bedrock;[23] [b] See Figure 2; [c] Assuming equilibrium with $CaHPO_4$(s), where Ca is limited by $CaCO_3$(s); [d] Soluble humic/fulvic acid; [e] Total acid capacity, assuming 5 × 10^{-3} eq/g.

Table 6 **Formation constants for metal complexes used in the model calculations**

	$\log\beta_1$[a]						
Species	**Cr(III)**	**Ni**	**Cu**	**Zn**	**Cd**	**Hg**	**Pb**
MOH^{z-1}	10.0	4.1	6.5	5.0	3.9	10.4	6.3
$M(OH)_2^{z-2}$	18.3[b]	9.0	11.8	11.1	7.7	21.7	11.0
$M(OH)_z(s)$	–29.8	–15.2	–19.3	–16.2	–14.4	–25.4	–14.9
MCO_3^{z-2}	(8.2)	(4.4)	6.7	(5.1)	4.0	12.1	7.0
$MCO_3(s)$		–6.9	–9.6	–10.0	–13.7	–16.1	–13.1
MSO_4^{z-2}	2.8	2.3	2.4	2.1	2.5	2.5	2.7
$MSO_4(s)$							–7.7
MS(s)		–26.0	–46.7	–22.6	–27.3	–52.7	–27.5
MCl^{z-1}	0.1	0.6	0.5	0.4	1.7	6.0[c]	1.5
MF^{z-1}	5.2	1.1	1.5	1.2	1.0	1.6	2.0
$MHPO_4^{z-2}$		3.1	4.2	3.4	(3.2)		3.2
MA[d]	(5.1)	(3.5)	4.0	3.7	3.1	(10)	4.1

[a] Reaction $M^{z+} + iL^{n-} = ML_i^{z-in}$; extrapolated values within parenthesis; [b] $\log\beta$ = 24.0 for $Cr(OH)_3$; [c] $\log\beta$ = 12.5, 14.5, 15.1, and 16.6 for $HgCl_2$, $HgCl_3^-$, $HgCl_4^{2-}$ and HgOHCl, respectively; [d] Complex with natural fulvic acid; conditional constant at pH 5, $\log\beta^*$ = 0.2 pH + const., is assumed (with $\log k_a$ = –0.6 pH – 1.8).[68]

The fraction bound to humic and fulvic acids will be minor, but significant, in surface waters and also in many groundwaters.

3. Copper

Copper, which is entirely divalent in environmental aquatic systems, forms strong complexes with oxygen ligands as well as with sulfur. Complexes with natural organic ligands are strong with oxygen, but also with nitrogen as ligand atoms. Hydrolysis would be significant at pH > 6, but the uncharged carbonate $CuCO_3$ rather than $CuOH^+$ or $Cu(OH)_2$ is the expected predominant species in solution at pH > 6.5 (Figure 7c). The noncomplexed Cu^{2+} ion is dominating at a lower pH, and the contributions from chloride or sulfate would be minor, except at very high sulfate concentrations.

A significant or high fraction of organic complexes with humic materials must be expected, not only in surface waters with high concentrations of dissolved organic carbon. The conditional stability constant used in the speciation calculation (Table 5) is probably underestimated. Direct measurements of the speciation of trace metals (cationic and anionic/organic species) in shallow groundwaters have shown that the fraction bound to organic acids could be entirely dominating.[96]

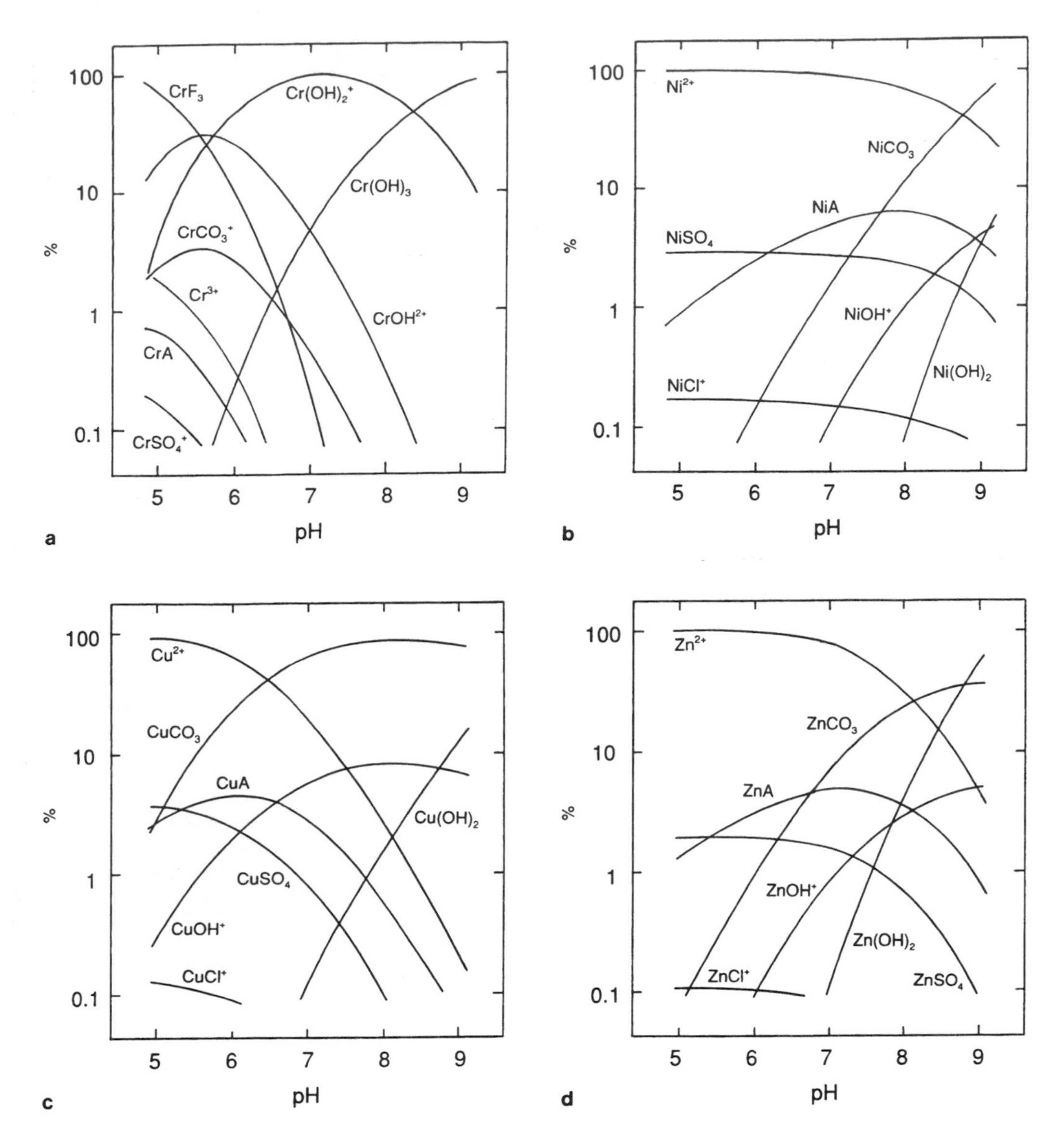

Figure 7 (a) to (g) Distribution of metal species (Cr(III), Ni, Cu, Zn, Cd, Hg, and Pb) in solution vs. pH in a reference groundwater (Table 5).

4. Zinc

Zinc, which is divalent in aqueous solutions, forms moderately strong complexes with oxygen containing ligands. The noncomplexed Zn^{2+} ion dominates at pH < 8 and uncharged $ZnCO_3$ at a higher pH (Figure 7d). Hydrolysis becomes significant at a pH > 7.5, although the hydroxy complexes $ZnOH^+$ and $Zn(OH)_2$ would not dominate in the presence of carbonate at groundwater concentration levels. The sulfate complex $ZnSO_4$ can be significant at low pH and high sulfate concentrations, while chloride complexes are of little importance.

The formation of a minor but significant organically complexed fraction must be considered in surface waters and probably also in most groundwaters.

5. Arsenic

Arsenic exists as trivalent arsenous acid H_3AsO_3 or as corresponding anions and salts (arsenites) under reducing conditions and in the pentavalent state as arsenic acid H_3AsO_4 and corresponding anions and salts (arsenates) under oxic conditions (Figure 8b). Dominating species in groundwater are the anionic arsenates $H_2AsO_4^-$ and $HAsO_4^{2-}$ under conditions corresponding to the pe/pH domains 1 and 2 (Figure 4).

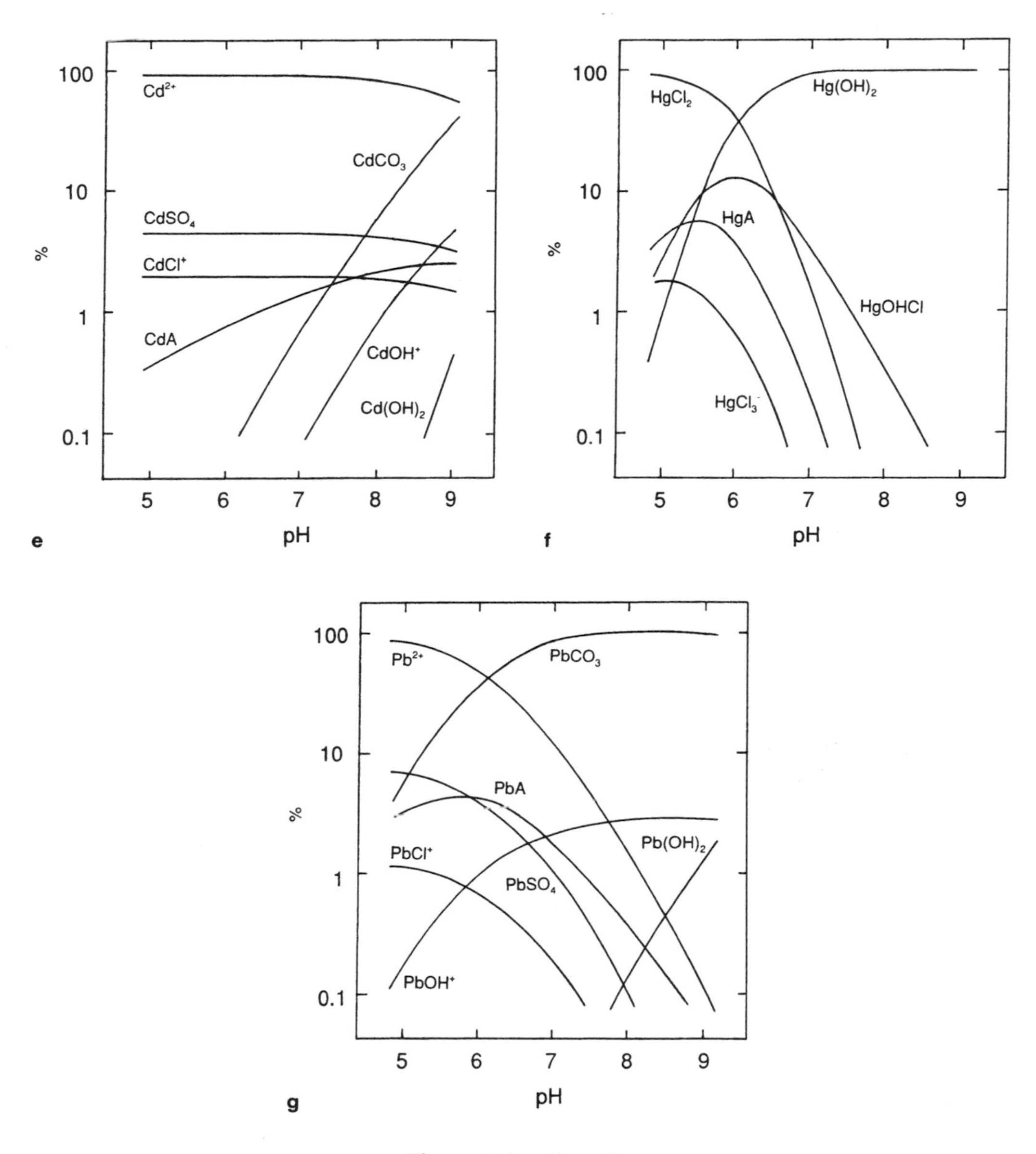

Figure 7 (continued).

The undissociated arsenious acid is dominating in domain 3 and 4, where pe is determined by the sulfur system and by degradation of organics, respectively. Since arsenic exists as an anion, it does not form complexes with other simple anions (chloride, sulfate, phosphate) but rather has a ligand character itself.

Organoarsenicals can be formed with both tri- and pentavalent arsenic during microbial processes in sediments and soil systems,[97] notably monomethylarsenic acid $H_2AsO_3CH_3$ (MMAA) and dimethylarsinic acid $HAsO_2(CH_3)_2$ (DMAA), as well as trimethylarsineoxide $AsO(CH_3)_3$ (TMA), phenylarsonic acid $H_2AsO_3C_6H_5$ (PAA), and others.[98] The release of MMAA and DMAA, produced by anaerobic microorganisms in sedimentary pore waters, into subsurface water columns has been proposed by Wood.[99] No such processes, leading to the existence of organoarsenicals in anoxic groundwaters, have in fact been confirmed.[100] A high arsenic concentration originating from mineralogic processes is not expected to lead to the formation of organic forms under anoxic groundwater conditions.[61]

6. Cadmium

Cadmium, which is divalent in aqueous solutions, forms fairly weak complexes with oxygen containing ligands but moderately strong complexes with chloride, as well as strong complexes with sulfide. The

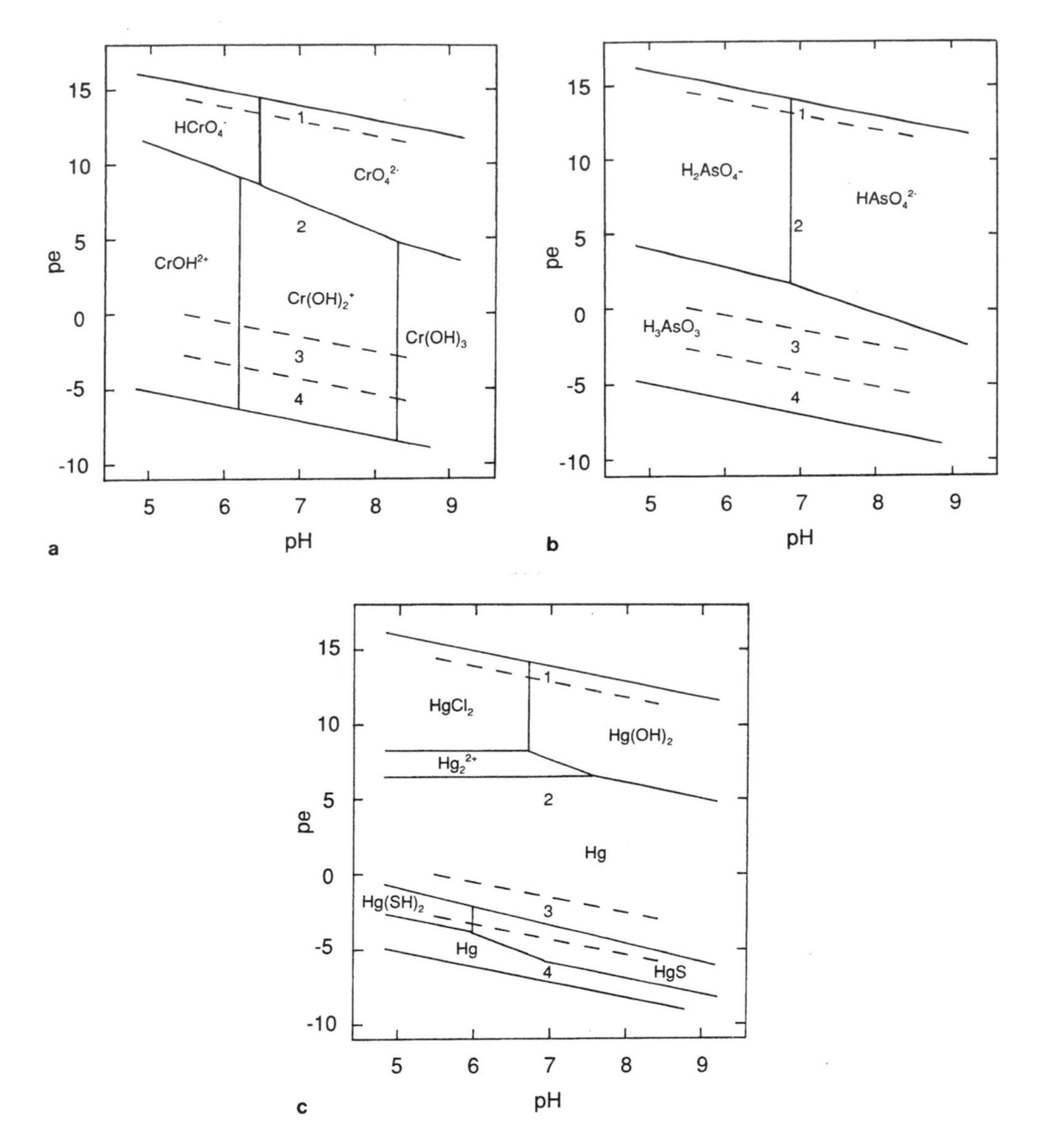

Figure 8 (a) to (c) Distribution of redox sensitive elements (Cr, As, and Hg) in solution vs. pe and pH (modified after Pourbaix[95]).

noncomplexed Cd^{2+} ion dominates in the environmental pH range, however, with significant contributions of the uncharged carbonate $CdCO_3$ at pH > 8.5 in a groundwater environment (Figure 7e). Hydrolysis is significant only at pH > 8. Both the neutral $CdSO_4$ and the cationic $CdCl^+$ could be significant species in groundwater at a pH up to 7.5 to 8.

Formation of complexes with natural organics can be significant in surface waters but not in groundwater, unless sulfate and chloride concentration levels are low and the dissolved organic carbon high.

7. Mercury

The divalent state of mercury would dominate in most natural waters, however, with elemental Hg^o as the most stable form in a broad pe/pH range. Divalent mercury forms very strong complexes with oxygen and sulfur ligands, as well as with chloride. A significant hydrolysis starts already at pH > 1 and dominates at pH > 2, in the absence of other complexing agents. The chloride complexes $HgCl^+$ and $HgCl_2$ would dominate over hydroxide species at pH < 5 and chloride concentrations >1 to 3 mg/l, and $HgCl_2$ would dominate in solution up to pH 6 in most groundwater systems under oxic or mildly reducing conditions (pe/pH domains 1 and 2) (Figure 7f). A significant fraction of the anionic $HgCl_3^-$ can be expected at high chloride levels and pH < 5 to 6, as well as of HgOHCl at pH 5 to 7. The uncharged $Hg(OH)_2$ is the main species at pH > 6 in nonsulfidic groundwaters as well as in surface waters, unless the chloride concentration is very high.

Mercury forms very strong complexes with humic and fulvic acids, and these complexes can entirely dominate in humic rich surface waters at pH > 5 to 6.[101] A significant humic fraction would also be expected in groundwaters. Published conditional stability constants for mercury-humate complexes range from 10^5 to 10^{21}, which can reflect differences in techniques and conditions, etc., as well as artifacts, but also different mercury affinities related to the origin of the organic matter and the presence of specific binding sites. The conditional stability constant used for the calculated speciation in Figure 7f (Table 5, in agreement with the constant suggested by Lövgren and Sjöberg[80]) is extrapolated from reported formation constants for cadmium and zinc complexes with humic materials as well as with simple well-defined carboxylic acids.[102]

The distribution of mercury species in solution in the various pe/pH domains is illustrated in Figure 8c. Elemental mercury, Hg^o, dominates in a broad range of domain 2. The strong complexes with humic substances would generally be expected to stabilize mercury in the divalent state, but the effect may in fact be the opposite. A reduction of the mercury to elemental state can take place through the reversible quinone-hydroquinone redox couple in the humic substance, once the complex has been formed.[103]

Dominating species in solution in domain 3 and 4, governed by the sulfur system and organic degradation, respectively, would be $Hg(SH)_2$ at pH < 6, HgS_2H^- at pH 6 to 9, and HgS_2^{2-} at pH > 9, assuming logβ = 42.7, 51.0, 20.0, and 37.7 for the complexes HgS, HgS_2^{2-}, $Hg(SH)^+$, and $Hg(SH)_2$, respectively, and logk = 31.5 for the reaction $Hg^{2+} + 2SH^- = HgS_2H^- + H^+$ (derived from data by Dyrssen and Wedborg;[104] see also Stumm and Morgan[2]).

Formation of organomercury compounds, particularly monomethyl mercury CH_3Hg^+, but also the uncharged dimethyl mercury $(CH_3)_2Hg$, can be achieved through microbial processes in soils and lake sediments. The precursor Hg^{2+} can be methylated by both aerobic and anaerobic bacteria and subsequently demethylated by other bacteria. Craig[105] lists 12 pure bacterial strains capable of methylating $HgCl_2$ and 17 capable of demethylating CH_3Hg^+. It is also claimed that mercuric ions can be abiotically methylated, possibly by both enzymatic and nonenzymatic reactions. The formation of organomercury species is affected by parameters such as pH, redox conditions, mercury concentration, microbial population, temperature, etc. The mercury cycle and the transformations between inorganic and organomercury species are discussed in several comprehensive texts.[36,87,105]

Monomethyl mercury behaves as a cation capable of forming strong complexes with ligands containing O, S, Cl, etc.[2] at a high rate of complexation, and the CH_3Hg^+ ion is kinetically inert towards breaking of the C–Hg bond. A large fraction of the total mercury in sediments can be organic. However, organomercury concentration levels in natural waters are generally rather low, at 1 to 10% of the total mercury,[106] and probably even lower in groundwater. The presence of methylmercury in anoxic groundwaters cannot, however, be excluded, since an abiotic methylation by humic and fulvic acids appears to be feasible.[107]

8. Lead

Lead is divalent in environmental aquatic systems. Strong complexes are formed with oxygen and sulfur ligands, and, consequently, strong complexes are also expected for natural organic ligands. Hydrolysis would be significant at pH > 6, but $PbCO_3$, rather than hydrolysis products, is the expected dominating species in solution at pH > 6 (Figure 7g), with the noncomplexed Pb^{2+} ion dominating at lower pH. Significant fractions of $PbSO_4$ and $PbCl^+$ can be present in groundwaters at pH < 6 to 7.

A significant fraction of organic complexes with humic materials is expected in surface waters, but likely also in groundwater, just as for copper.

B. PRECIPITATION AND SOLID SPECIES

Trace elements are released as the result of mineral degradation and alteration, but the corresponding trace element concentrations in solution are rarely at equilibrium with discrete solid crystalline mineral phases. Dissolved trace elements in aquatic systems may precipitate under oxic conditions, for instance, when the solubility product for sparingly soluble hydroxides or carbonates is exceeded due to a sudden pH increase. This may be the case when metal-enriched effluents (including subsurface waters) at low pH from an ore body, etc. encounter an alkaline groundwater or surface water.[64,108] Frequently, however, a precipitation of Fe(III) as $Fe(OH)_3(s)$ can be expected when anoxic groundwater with high total iron concentrations (for instance, in the concentration range 0.1 to 1 mg/l as Fe(II)) encounters and mixes with oxic waters.

The fresh iron hydroxide precipitate will act as a coprecipitation agent and scavenger for hydrolyzable trace metals at levels far below their solubility limits.[109,110]

A precipitation and true solubility control by the formation of solid sulfides may take place when the redox potential is changing from domain 1 or 2 to 3 or 4 (sulfidic domains); however, it probably is coupled with the precipitation of iron sulfides and corresponding control of the sulfide concentration in solution.

1. Chromium

Chromium is predominantly trivalent in geologic phases and occurs mainly as oxides of the chromite type ($(Fe,Cr)_2O_3$). Natural concentrations in groundwaters arise largely from mineral weathering processes.

The precipitation of $Cr(OH)_3(s)$, or rather coprecipitation with $Fe(OH)_3(s)$, limits the solubility of Cr(III) in natural anoxic aquatic systems. The formation of chromates $M(II)CrO_4(s)$ with divalent metals (Fe, Ba, Pb) is likely to determine the solubility under oxic conditions.[110]

2. Nickel

Nickel is entirely divalent and occurs mainly as oxides (laterite) or sulfides (pentlandite) in association with other metal sulfides in geologic systems. Oxidative weathering of the sulfide would release nickel as well as other sulfide-bound metals into natural waters.

The hydroxide $Ni(OH)_2(s)$ rather than the carbonate may be the solubility-limiting solid form in aquatic systems under oxic/anoxic conditions, while NiS(s) would limit the solubility under reducing conditions (domain 3 and 4).

3. Copper

Copper is mono- or divalent in geological phases. It is present as oxides or sulfides (covellite, chalcosite), particularly in association with iron as in chalkopyrite ($CuFeS_2$) and bornite (Cu_5FeS_4), but also as carbonates (azurite, malachite) and sulfosalts.

Solubility-limiting species in aquatic systems would be the divalent hydroxide or carbonate hydroxycarbonates ($Cu_2(OH)_2CO_3(s)$ and others) under oxic/anoxic conditions and sulfidic phases under reducing conditions (domain 3 and 4).

4. Zinc

Zinc is entirely divalent and is present mainly as a sulfide (sphalerite), oxide (zincite), and carbonate (smithsonite) in the geosphere. Oxidative weathering of the sulfides would release zinc and associated metals into aquatic systems.

Solubility-limiting species in aqueous systems would be the carbonate $ZnCO_3(s)$ under oxic/anoxic conditions and the sulfide ZnS(s) under reducing conditions (domain 3 and 4).

5. Arsenic

Arsenic exhibits the oxidation states III, 0, III, and V in the environment. Arsenic is usually associated with sulfides, and principal arsenic-bearing minerals are mixed sulfides of the type M(II)AsS (arsenopyrite, niccolite, cobaltite, with M(II) = Fe, Ni, and Co, respectively, and others).

The formation of $As_2S_3(s)$ could be solubility limiting in a sulfidic environment, while the formation of solid arsenates of barium ($Ba_3(AsO_4)_2(s)$), and possibly of other metals (Fe(III), Al, Ca, Pb, etc.), would limit the solubility of As(V) under oxic conditions.[111]

6. Cadmium

Cadmium is entirely divalent in the environment. It is present mainly as a sulfide, associated with zinc (in sphalerite), with amounts ranging from 0.1 to 5% of the zinc,[112] but also together with other heavy metal sulfides (Pb, Cu).

The carbonate $CdCO_3(s)$ would be solubility-limiting species in aqueous systems under oxic/anoxic conditions and the sulfide CdS(s) under reducing conditions (domain 3 and 4).

7. Mercury

Mercury exhibits the oxidation states 0, I, and II in the environment. It occurs mainly as the sulfide (cinnabar) or associated with chalcophilic elements (Cu, Cd, Pb, etc.), but also in elemental form as Hg^0.

Solubility-limiting species under oxic conditions are probably the hydroxides $Hg(OH)_2(s)$. However, the solubility of Hg^0 is around 1 mg/l and above 10 mg/l for $Hg(OH)_2(aq)$. The sulfide HgS(s) limits the solubility to the ng/l level in sulfidic environments, although Hg^o otherwise would be the expected stable species in domain 3 and 4, in the absence of sulfide (Figure 8c).

8. Lead

Lead is mainly divalent in geologic systems and is present mainly as a sulfide (galena), but also as carbonate (cerussite) and sulfate (anglesite) as well as tetravalent oxide (plattnerite). Several silicate minerals (feldspars) are notable accumulators of lead, which replaces other cations in the lattice.

Solubility-limiting species in aquatic environments would be carbonates $PbCO_3(s)$ or mixed hydroxycarbonates ($(Pb_3(OH)_2(CO_3)_2(s)$ and others) and possibly $PbSO_4(s)$ at high sulfate concentrations and low pH.

C. ADSORPTION ON SOLID SURFACES AND COPRECIPITATION

Trace elements in aquatic systems in contact with solid phases are interacting with exposed surfaces in adsorption processes. Ideal sorption mechanisms and removal processes can be distinguished, as discussed in several comprehensive books, etc.[2,113–116] These mechanisms can arbitrarily be divided into

- *physical adsorption,* due to nonspecific forces of attraction (van der Waals forces) involving the entire electron shells of the dissolved trace element and the adsorbent
- *electrostatic adsorption* (ion exchange), due to the coulombic interaction between charged solute species and the adsorbent
- *specific adsorption* (chemisorption), due to the action of chemical forces of attraction leading to bonds between specific surface sites and dissolved metal species
- *coprecipitation;* precipitation with a sparingly soluble carrier substance that exceeds its solubility limit
- *chemical substitution;* replacement of ions within a solid lattice
- *mineralization;* alteration of mineral composition or structure, phase transformation, etc.

Physical adsorption is rapid, reversible, nonselective, and largely independent of concentrations and ionic strength of the solution. Electrostatic interactions are generally nonselective and dependent on factors such as pH, ionic strength, and concentrations of competing ions. Chemisorption processes can be selective, slow, and partly irreversible and highly dependent on concentrations and solution composition.

The sorption of a particular trace element is generally dependent on its chemical state. Thus, all hydrochemical parameters of importance for speciation in solution would also affect the sorption process (chemical state of the element as well as surface properties of the adsorbent). The mechanisms behind the overall adsorption of a dissolved trace element can rarely be unambiguously characterized in detail. Surface processes could be determined by the chemical state of the solutes and adsorbents (thermodynamic control), while other processes are related to concentration gradients and mass transfer (kinetic control). Usually, a rapid uptake lasting in the time-scale minutes-hours (physical adsorption and electrostatic interactions on primary surfaces) is followed by a slower process lasting days-months (diffusion-controlled transport, slow phase transformations, etc.). The various theoretical models that have been suggested to explain and predict adsorption behavior are discussed in several textbooks (reviewed by Stumm[116]) and are outside the scope of this chapter.

There are numerous articles reporting measurements of trace element adsorption on various soil constituents and geologic phases (see compilations in comprehensive books on metal behavior in the environment[11,36,92,113,114,116,117]). However, there appears to be no field measurements that establish the importance of sorption processes for the metal distribution and apparent concentration in a groundwater environment, except for studies in soil systems and in the unsaturated zone, etc. Still, the same mechanisms that are in operation in soil systems and in controlled laboratory systems would be expected to affect the metal distribution also in a deep groundwater environment. Important hydrochemical parameters that have direct influence on adsorption behavior and thereby the apparent metal concentrations in solution are

- pH of the system
- The presence of complexing agents in solution
- The occurrence of particulate adsorbent phases (colloids)

The adsorption of hydrolyzable cations will generally increase from near 0 to near 100% on oxide/hydroxide and silicate surfaces as pH increases over a critical range 1 to 2 units wide, as examplified in

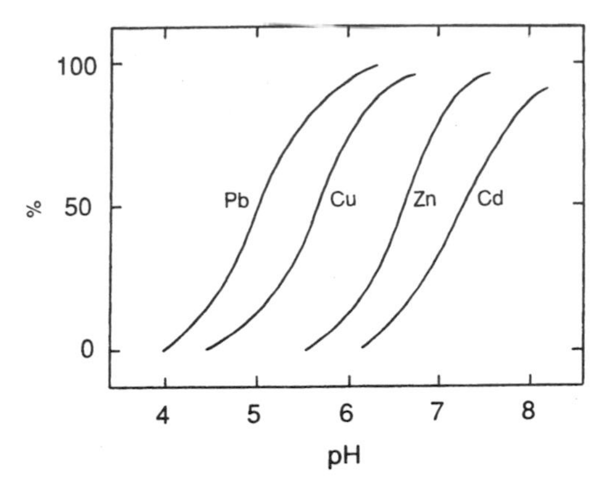

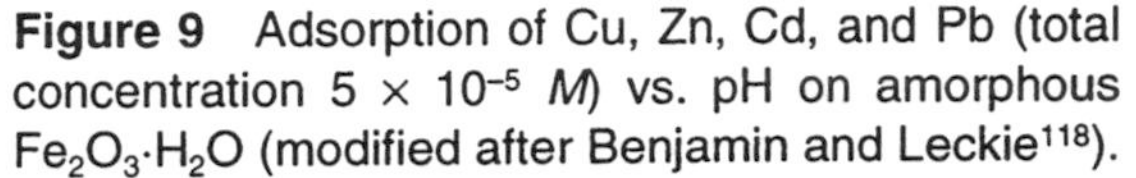

Figure 9 Adsorption of Cu, Zn, Cd, and Pb (total concentration 5×10^{-5} *M*) vs. pH on amorphous $Fe_2O_3 \cdot H_2O$ (modified after Benjamin and Leckie[118]).

Figure 9. The pH of adsorption generally increases with decreasing hydrolysis constants, that is, in the order Cr < Pb, Cu < Zn < Ni < Cd. Anionic species (for instance, arsenates) would exhibit a reverse pH-dependence, reflecting the electrostatic interaction with the adsorbent surface.

Complexation of metals with ligands may increase or decrease the sorption. Benjamin and Leckie[119] suggested three types of interactions:

- Complexes are formed in solution and adsorb weakly or not at all (decreased adsorption in comparison with a system without the ligand)
- Complexes are formed in solution and adsorb strongly (enhanced adsorption in comparison with a system without the ligand)
- Complexing ligand is adsorbed on the surface and enhances the interaction between the surface and the metal ion in solution (bridging effect).

Chloride and sulfate complexes of divalent metals are less strongly adsorbed than the uncomplexed ions. The anionic chloride complex $HgCl_4^{2-}$ that would dominate at high chloride concentrations and low pH (see above) would be poorly adsorbed on negatively charged surfaces (for instance, silicates at neutral pH), but strongly adsorbed on positively charged surfaces (for instance, Al_2O_3 and Fe_2O_3 at low pH). The effects of natural organics on the adsorption given in Figure 10 illustrate all three types of interaction, as well as the importance of pH for the overall distribution between solid phase and solution.

Precipitating amorphous solid phases, notably $Fe(OH)_3$, would act as adsorbents or coprecipitation agents, Figure 11. In this case both $Fe(OH)_3$ and $Al(OH)_3$ exceed their solubility product and precipitate. All the other elements are either adsorbed on the fresh hydroxide precipitate or coprecipitated, and it is not possible to strictly distinguish between adsorption and coprecipitation. Moreover, the removal efficiency would be expected to be identical whether the hydroxide phase is precipitated before contact with the trace metal (adsorption) or afterwards (coprecipitation).[121] Precipitates could form surface coatings or free-flowing particles in a groundwater system, depending on pH, as well as on flow conditions defined by the porosity and tortuosity of the bedrock.

To conclude, adsorption and coprecipitation phenomena involving trace elements must be considered also in groundwater systems, although these processes still have to be verified under field conditions.

IV. SAMPLING AND ANALYSIS

A. SAMPLE HANDLING AND ANALYTICAL CONSIDERATIONS

Accurate determinations of chemical forms of trace elements (speciation) as well as of total concentrations in natural waters are important in the assessment of their mobility and spreading, bioavailability and environmental impacts, background levels, etc. The continuous improvement of analytical methodologies has increased the demand for proper sampling and handling procedures as well as for standardized analytical work and quality control. Several comprehensive texts are dealing with the subject of speciation, handling, and analysis of environmental water samples containing elements at trace levels.[16,113,122-125] Some general precautions and considerations pertinent to work with groundwater systems will be outlined in this chapter.

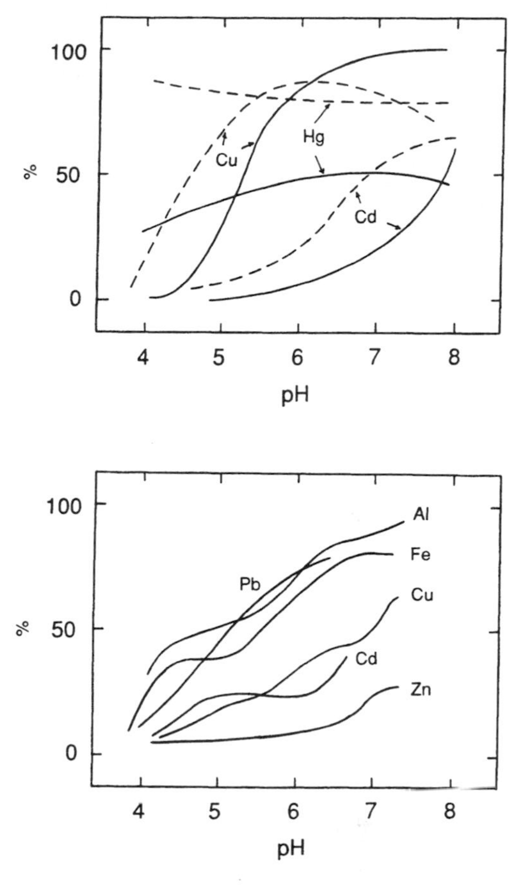

Figure 10 Adsorption of Zn, Cd, and Hg (total concentrations 10^{-6} to 10^{-8} *M*) vs. pH on Al_2O_3. Effects of a fulvic acid, FA (modified after Xu and Allard[101]). Solid lines: No FA; dashed lines: 10 mg l^{-1} FA.

Figure 11 Distribution of trace metals between suspended phase (>0.40 μm) and solution vs. natural variations of pH. Field measurement in a stream receiving metal-enriched surface run-off and alkaline groundwater (modified after Karlsson et al.[120]).

In principle, two major phenomena of importance for the analysis of natural waters with dissolved micro constituents have to be addressed:

- *Increase* of the real concentration by contributions from the environment, equipment, or due to the procedures, etc. (contamination)
- *Reduction* of the real concentration by adsorption, decomposition, evaporation, etc.

Both these aspects have to be considered in the choice of proper procedures.

1. Natural Processes

Certain natural chemical processes will inevitably affect the state and apparent concentration of trace components, particularly in groundwater samples:

- Precipitation or dissolution of particulate matter (colloidal size) due to changes of the hydrochemical conditions, temperature, or pressure
- Release of volatile components
- Adsorption/desorption of trace components on any exposed solid surfaces (particulate matter in the sample, sampling equipment, storage vessels, etc.).

The change in the carbon dioxide pressure when a groundwater sample is taken from a high-pressure subsurface level will affect its total carbonate balance. This will primarily affect pH (and thereby all sorption phenomena that are influenced by pH), as well as the extent of solid carbonate phases in groundwaters that are near calcite saturation. Changes in temperature could have the same principal effects. Changes in the oxygen pressure (exposure to air) lead for most deep groundwaters to a rapid precipitation of the dissolved Fe(II) as amorphous Fe(III) hydroxide, which creates a new, solid trace element scavenging phase in the sample system.[126] Also, other redox-sensitive trace elements would be affected with time.

Adsorption of trace elements onto solids must always be considered as significant at concentrations of 10^{-6} *M* or below, i.e., in the range 50 to 100 μg l^{-1} or below for most heavy metals. An important issue is the operational definition of solid species and species. The solid fraction will serve as an adsorbent phase under proper pH conditions and carry a fraction of the total amount of the trace element. The common choice of 0.45 μm as the proper filter size for separation of particles from solution species is not strictly justified, considering the fact that most natural colloids (hydroxides, silicates, etc.) exist predominantly in the size range 100 to 300 nm.[127]

2. Procedures

The various steps in the total handling of the water sample will be of importance for the apparent concentration of dissolved trace components (see also the discussion by Tschöpel[128]):

- Sampling; choice of technique, equipment, etc.; exposure to air, temperature, and pressure control
- Preservation; the use of chemicals or treatments already in the field to preserve the sample (with respect to the total concentration of dissolved constituents, etc.)
- Transport and storage; choice of equipment, transport conditions, storage conditions (temperature, light, etc.), and storage times
- Handling in the laboratory; analytical procedures (methodology, potential matrix effects, etc.), preconcentration, and enrichment

Of particular importance is the adoption of procedures to avoid contamination from vessels and other equipment and reagents as well as from the laboratory environment. Relevant washing procedures must be adopted to minimize contamination as well as adsorption on exposed equipment.

It should be recognized that the common practice of preserving samples with acid to reduce losses due to adsorption on vessel walls also destroys any natural distribution between the colloidal phase and the solution. A pH reduction will also lead to a partial dissolution of pH-sensitive particulate matter (hydroxides and carbonates) originally present, as well as to a dissociation of metal complexes with organic macromolecular systems (humic and fulvic acids).

Particular aspects on the available techniques for determination of trace elements in natural samples (solid as well as water) are compiled by Stoeppler[16] and outside the scope of this chapter.

B. SPECIATION IN FIELD SAMPLES

There are several techniques for speciation of trace components in natural waters, either based on differences in size (ultrafiltration, ultracentrifugation, hollow-fiber filtration, gel permeation chromatography, etc.), in charge (ion-exchange, electrophoresis, etc.), chemical character (solvent extraction, sequential leaching, etc.), or a combination of these principles.[122–125] There are, however, no really suitable or commonly adopted techniques for combined sampling/speciation of groundwaters, considering the particular requirements (unchanged conditions with respect to hydrochemistry, temperature and pressure, short contact times, high precision in analysis, and phase separations, etc.). Various schemes have been suggested for the characterization of aquatic samples. However, these are rarely applied in studies of deep pristine groundwaters.

Figure 12a and b describes two procedures that have been tested for the speciation of trace metals in shallow as well as deep groundwater, leading to a separation of anionic and cationic species in solution as well as particle fractions of various sizes.[129,130] Organic matter is further characterized by gel filtration (to obtain molecular weight distribution of dissolved organics), and original samples are characterized with respect to the contents of colloidal matter by photon correlation spectroscopy.[131] Solid filter residues are characterized by the sequential leaching procedure given in Table 7,[132] developed from previously suggested procedures.[133–135] Important in these schemes, as well as in other similar procedures, is

- To maintain original hydrochemical conditions; this requires that a separation is performed in a closed system already in the field
- To minimize the separation times (to reduce losses due to adsorption on filters, etc. and contamination as well as changes of equilibrium when phases are separated)

The effect of not controlling the original redox conditions is illustrated in Figure 13, which shows the in-growth of a colloidal Fe(III) hydroxide phase corresponding to almost the total iron content (0.3 mg l^{-1}) in a groundwater sample after only 30 min storage in a closed vessel, but

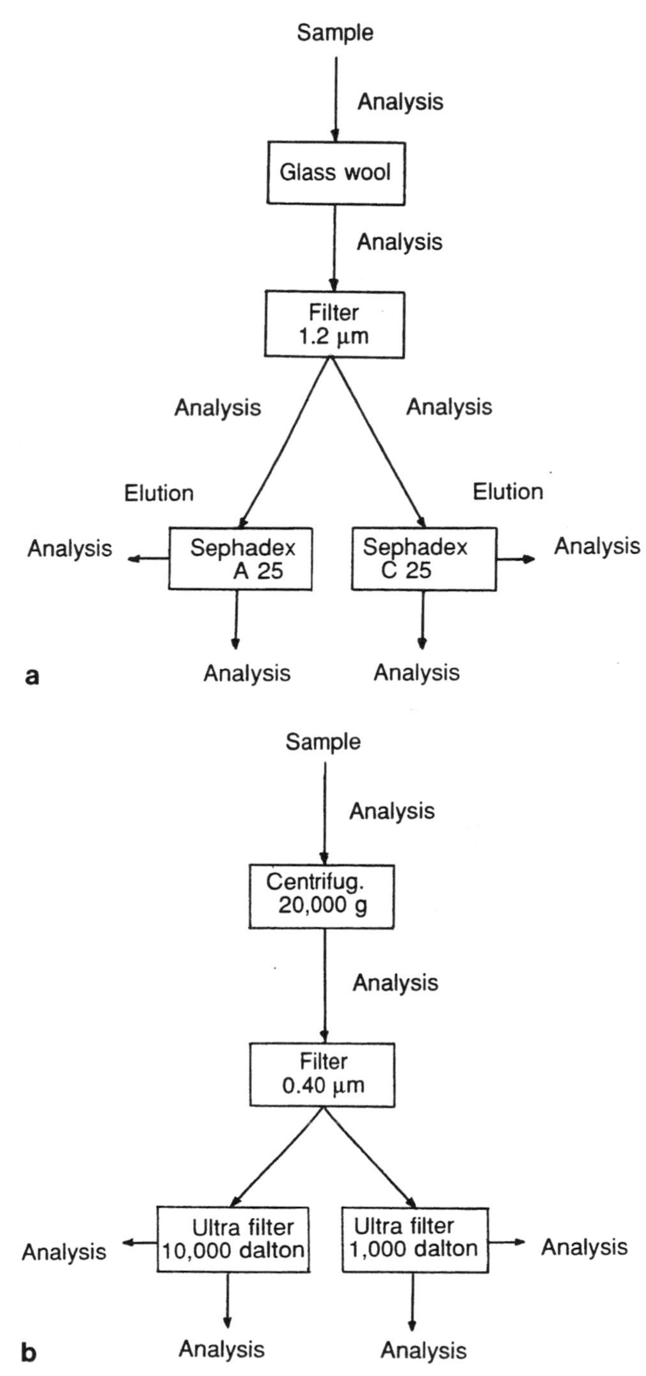

Figure 12 Metal speciation procedure. (a) Separation of anionic and cationic species in solution[96] and (b) of particulate matter and macromolecular species[129].

under air.[126] Filtration without strict redox control would have significantly altered the trace element distribution in this system.

V. CONCENTRATION LEVELS AND SPECIATION

A. TOTAL CONCENTRATIONS AND PHYSICO-CHEMICAL FORMS

An extensive environmental monitoring program for groundwater quality was started in 1978 in Sweden. Groundwater from various geologic environments (wells or pipes in crystalline rock, but also in moraine, gravel, etc.) is sampled from 126 stations with a frequency of 1 to 6 times per year.[136] This program is the basis for several of the background assessments in Table 4,[35,50,56,57] and the data analyzed by Ledin

Table 7 **Sequential leaching of filter residues**[132]

Fraction 1	Ion-exchangeable/carbonate	
	Reagent	1.0 *M* Na acetate at pH 5
	Conditions	5 h at room temperature
Fraction 2	Hydrous oxides	
	Reagent	0.04 *M* NH_2OHHCl in 25% acetic acid
	Conditions	5 h at 85°C
Fraction 3	Organic matter and unconsolidated sulfides	
	Reagent	0.02 *M* HNO_3 + 30% H_2O_2, 1 vol
		3.2 *M* NH_4 acetate in 20% HNO_3, 1 vol
		H_2O, 2 vol
	Conditions	3 h at 90°C
Fraction 4	Residual fraction	
	Reagent	Conc HNO_3, conc HCl
	Conditions	Evaporation to near dryness with HNO_3 at 90°C
		Dissolution in 1 vol HCl + 6 vol H_2O

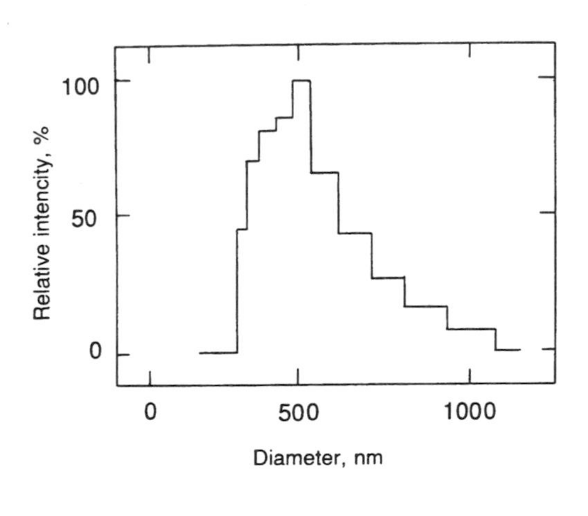

Figure 13 Size distribution histogram (from photon correlation spectroscopy) for $Fe(OH)_3$ colloids precipitated in groundwater after 30 min (sample in sealed bottle but under air; 0.3 mg l^{-1} Fe) (modified after Ledin et al.[126]).

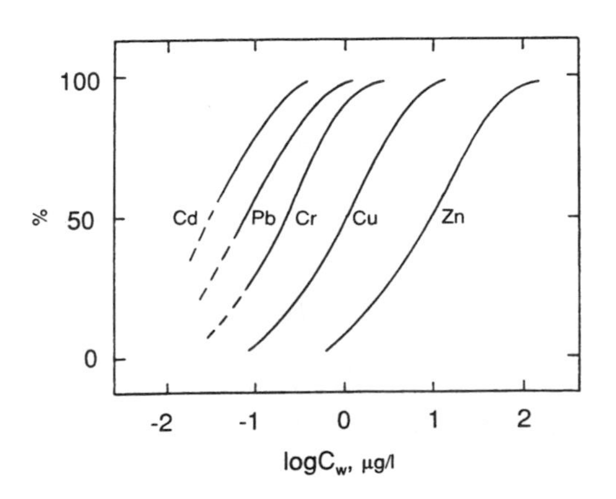

Figure 14 Cumulative curves showing the frequency distribution of the concentrations (C_w, μg l^{-1}) of Cr, Cu, Zn, Cd, and Pb in groundwaters from crystalline bedrock and moraine (modified after Ledin et al.[35]).

et al.[35,58] may serve as a basis for discussion on variations and assessment of background concentrations. Data from 131 (Zn and Pb) and 534 (Cr, Cu, Zn, and Cd) samples are summarized in a cumulative frequency plot (Figure 14). It appears that there is a natural distribution representative of pristine groundwaters similar to the distribution of major cations and anions (Figure 3) related to variations in geologic conditions, etc. When data are divided into groups representing the different pH ranges <5, 5 to 6, 6 to 7, and >7 (Figure 15 [for zinc]), it is evident that the apparent concentrations are dependent on pH. Most of the high concentrations are found in waters of low natural pH, which is further illustrated

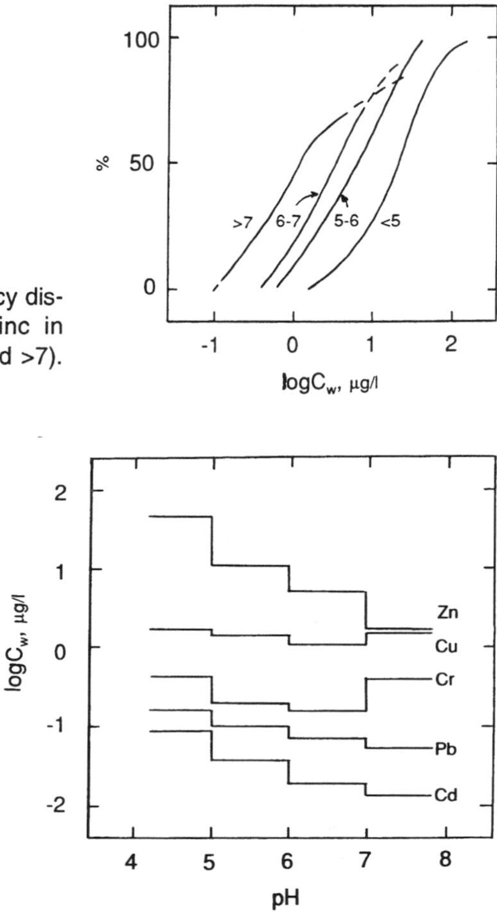

Figure 15 Cumulative curves showing the frequency distribution of the concentrations (C_w, µg l^{-1}) of zinc in groundwaters of different pH (<5, 5 to 6, 6 to 7, and >7). Data from Figure 14.

Figure 16 Median concentrations (C_w, µg l^{-1}) for Cr, Cu, Zn, Cd, and Pb in different pH ranges (<5, 5 to 6, 6 to 7, and >7). Data from Figure 14.

in Figure 16. All of Zn, Pb, and Cd exhibit decreasing median concentrations with increasing pH, while Cr and Cu have higher median concentrations in the high pH range (>7). The pH dependence for Zn, Cd, and Pb could, in fact, indicate adsorption phenomena in the groundwater system, and the enhanced concentration at low pH would reflect desorption rather than dissolution of solid phases. All concentrations are below saturation with respect to solid hydroxy or carbonate phases. The deviations for chromium and copper could indicate that substantial fractions of the elements exist in anionic form, either as complexes with organic acid, or for chromium possibly as CrO_4^{2-} (cf. Figure 8a).

Suggested natural background concentration ranges, mainly from Bowen[31] and Ledin et al.,[35] are compared in Table 8. The data from Bowen, representing measurements before 1979, are generally higher than the data from 1987 to 1988, which can reflect improved analytical procedures, etc. Otherwise, the median values are in reasonable agreement.

B. CONCLUDING REMARKS

The natural variations of trace element background concentrations in groundwater are usually around at least two orders of magnitude. Median values (µg/l) of 0.2 (Cr), 0.5 (Ni, estimated), 1 (Cu), 10 (Zn), 0.9 (As), 0.03 (Cd), 0.001 (Hg, estimated), and 0.1 (Pb) can be assessed in groundwaters from crystalline igneous bedrocks and corresponding soil types. Variations of about 1 order of magnitude to maximum and minimum natural levels, respectively, are reasonable, reflecting the variations in abundance of the trace element in geologic phases as well as natural pH variations in the water. Concentrations up to 2 to 3 orders of magnitude above the median levels can in fact be encountered in areas rich in natural mineralizations of the particular element.

The chemical state of dissolved trace elements in groundwater is largely determined by a few key parameters, notably pH, the presence and concentrations of complexing agents (OH^-, CO_3^{2-} and organic acids), and the redox conditions. Expected chemical speciation can be calculated for most trace elements, however, with some severe limitations:

Table 8 **Suggested background concentration ranges (μg l⁻¹) for Cr, Ni, Cu, Zn, As, Cd, Hg, and Pb in groundwaters**

Element	Min[a]	16th[b]	Median	84th[b]	Max[a]	Source[c]	Ref.
Cr	0.1		1		6	10	31
	0.02	0.04	0.2	0.7	2	543	35
Ni	0.02		0.5		27	8	31
Cu	0.2		3		30	11	31
	0.05	0.3	1	4	9	543	35
Zn	0.2		15		100	16	31
	0.1	2	10	40	100	131	35
As	0.2		0.5		230	11	31
	0.2	0.6	0.9	1.5	40	39[d]	61
Cd	0.01		0.1		3	3	31
	0.002	0.006	0.03	0.1	0.3	543	35
Hg	0.0001		0.1		2.8	13	31
	0.0006				0.052		60
Pb	0.06		0.5		20	6	31
	0.005	0.02	0.08	0.3	1	131	35

[a] 5th and 95th percentiles, respectively;[35] [b] 16th and 84th percentiles; [c] Number of sources[31] and number of samples from 126 locations[35], respectively; [d] Number of locations.

- The effects of complexation with natural organics can not be properly assessed due to a lack of accurate stability constants or functions (reflecting effects of pH, concentrations, and the nature of the organic agent)
- The effects of adsorption and coprecipitation affecting the apparent concentrations are not considered
- The role of natural colloidal matter as a suspended trace element carrying phase is not considered
- Equilibrium is rarely achieved in natural waters; the trace element levels are generally not solubility controlled

There are presently no generally accepted procedures or schemes for the proper sampling and characterization of groundwaters with respect to their trace components (concentrations and chemical state). A rapid separation that does not alter the original hydrochemical conditions is required. Common practices involving long sampling times, sequential filtration procedures using arbitrarily selected pore sizes, etc. can lead to erroneous results due to changes in chemical conditions and phase composition induced by the procedure.

REFERENCES

1. **Garrels, R. M. and Christ, C. L.,** *Solution, Minerals, and Equilibria,* Freeman, Cooper & Co, San Francisco, 1965.
2. **Stumm, W. and Morgan, J. J.,** *Aquatic Chemistry,* John Wiley & Sons, New York, 1981.
3. **Matthess, G.,** *The Properties of Groundwater,* John Wiley & Sons, New York, 1982.
4. **Morel, F. M. M.,** *Principles of Aquatic Chemistry,* John Wiley & Sons, New York, 1983.
5. **Eriksson, E.,** *Principles and Applications of Hydrochemistry,* Chapman and Hall, London, 1985.
6. **Berner, E. K. and Berner, R. A.,** *The Global Water Cycle. Geochemistry and Environment,* Prentice-Hall, Englewood Cliffs, NJ, 1987.
7. **Drever, J. I.,** *The Geochemistry of Natural Waters,* Prentice-Hall, Englewood Cliffs, NJ, 1988.
8. **Voigt, H.-G.,** *Hydrogeochemie,* VEB Deutscher Verlag für Grundstoffindustrie, Leipzig, 1989.
9. **Sigg, L. and Stumm, W.,** *Aquatische Chemie,* Verlag der Fachvereine, Zürich, 1991.
10. **Förstner, U. and Wittmann, G. T. W.,** *Metal Pollution in the Aquatic Environment,* Springer-Verlag, Berlin, 1981.
11. **Salomons, W. and Förstner, U.,** *Metals in the Hydrocycle,* Springer-Verlag, Berlin, 1984.
12. **Hutchinson, T. C. and Meema, K. M., Eds.,** *Lead, Mercury, Cadmium and Arsenic in the Environment,* SCOPE 31, John Wiley & Sons, New York, 1987.

13. **Salomons, W. and Förstner, U., Eds.,** *Environmental Management of Solid Waste,* Springer-Verlag, Berlin, 1988.
14. **Vernet, J.-P., Ed.,** *Heavy Metals in the Environment,* Elsevier, Amsterdam, 1991.
15. **Vernet, J.-P., Ed.,** *Impact of Heavy Metals on the Environment,* Elsevier, Amsterdam, 1992.
16. **Stoeppler, M., Ed.,** *Hazardous Metals in the Environment,* Elsevier, Amsterdam, 1992.
17. **Wedepohl, K. H., Ed.,** *Handbook of Geochemistry,* Springer-Verlag, New York, 1978.
18. **Beall, G. W. and Allard, B.,** *Trans. Am. Nucl. Soc. Annu. Meeting,* 32, 164, 1979.
19. **Thurman, E. M.,** *Organic Geochemistry of Natural Waters,* Martinus Nijhoff/Dr W. Junk Publ., Dordrecht, 1985.
20. **Stevenson, F. J.,** *Humus Chemistry — Genesis, Composition, Reactions,* John Wiley & Sons, New York, 1982.
21. **Christman, R. F. and Gjessing, E., Eds.,** *Aquatic and Terrestrial Humic Materials,* Ann Arbor Science, Ann Arbor, 1983.
22. **Aiken, G. R., McKnight, D. M., Wershaw, R. L., and MacCarthy, P., Eds.,** *Humic Substances in Soil, Sediment and Water,* John Wiley & Sons, New York, 1985.
23. **Allard, B., Larsson, S. Å., Tullborg, E.-L., and Wikberg, P.,** Chemistry of deep groundwater from granitic bedrock, SKBF KBS TR 83–59, Swedish Nuclear Fuel Supply Co., Stockholm, 1983.
24. **Pettersson, C., Ephraim, J., and Allard, B.,** *Org. Geochem.,* 21, 443, 1994.
25. **Allard, B.,** On the buffering effects of the CO_2-CO_3^{2-}-system in deep groundwater, SKBF KBS TR 82–15, Swedish Nuclear Fuel Supply Co., Stockholm, 1982.
26. **Cannon, H. L.,** *Geochem. Environ.,* 3, 17, 1978.
27. National Research Council Canada, Effects of arsenic in the Canadian environment, NRCC No. 15391, Ottawa, 1978.
28. National Research Council Canada, Effects of mercury in the Canadian environment, NRCC No. 16739, Ottawa, 1979.
29. **Fleischer, M., Sarofim, A. F., Fasset, D. W., Hammond, P., Shacklette, H. T., Nisbet, I., and Epstein, S.,** *Environ. Health Perspect.,* 7, 253, 1974.
30. National Research Council of Canada, Effects of chromium in the Canadian environment, NRCC No. 15017, Ottawa, 1976.
31. **Bowen, H. J. M.,** *Environmental Chemistry of the Elements,* Academic Press, New York, 1979.
32. **Cox, D. P.,** in *Copper in the Environment,* Nriagu, J. O., Ed., John Wiley & Sons, New York, 1979, 19.
33. **Swaine, D. J.,** *J. Royal Soc. New South Wales,* 111, 41, 1978.
34. **Shacklette, H. T. and Boerugen, J. G.,** Element concentrations in soils and other surficial materials of the conterminous United States, USGS Prof. Paper 1270, U.S. Government Printing Office, Washington, D.C., 1984.
35. **Ledin, A., Pettersson, C., Allard, B., and Aastrup, M.,** *Water Air Soil Pollut.,* 47, 419, 1989.
36. **Adriano, D. C.,** *Trace Elements in the Terrestrial Environment,* Springer-Verlag, Berlin, 1986.
37. **Kanamori, S. and Sugawara, K.,** *J. Earth Sci. Nagoya Univ.,* 13, 23, 1965.
38. **Ferguson, J. F. and Gavis, J.,** *Water Res.,* 6, 1259, 1972.
39. **Lodemann, C. K. W. and Bukenberger, V.,** *GWF Wasser Abwasser,* 114, 478, 1973.
40. **Brinkmann, F. J.,** *Geol. Mijnbouw,* 53, 117, 1974.
41. **Matthess, G.,** *Geol. Minjbouw,* 53, 149, 1974.
42. **Förstner, U., and Müller, G.,** *GWF Wasser Abwasser,* 116, 74, 1975.
43. **Gibbs, R. J.,** *Geol. Soc. Am. Bull.,* 88, 829, 1977.
44. **Wyttenbach, A., Bajo, S., and Farrenkothenk, K.,** *Gas, Wasser, Abwasser,* 81/82, 653, 1979.
45. National Research Council Canada. Effects of mercury in the Canadian environment, NRCC No. 16739, Ottawa, 1979.
46. **Mayer, R., Heirichs, H., Seekamp, G., and Fassbender, H. W.,** *Z. Pflanzen Ehrnährung Bodenkd.,* 143, 221, 1980.
47. **Tallmann, D. E. and Shaikh, A. U.,** *Anal. Chem.,* 52, 196, 1980.
48. **Kaiser, G. and Tölg, G.,** in *The Handbook of Environmental Chemistry,* Hutzinger, O., Ed., Springer-Verlag, Berlin, 1980.
49. **Crerar, D. A., Means, J. L., Yuretich, R. F., Borcsik, M. P., Amster, J. L., Hastings, D. W., Knox, G. W., Lyon, K. E., and Quiett, R. F.,** *Chem. Geol.,* 33, 23, 1981.
50. **Aastrup, M., Aneblom, T., Henriksson, B., and Persson, G.,** PMK-grundvatten, SGU-Rep. 28, Swedish Geological Survey, Uppsala, 1982.

51. **May, K. and Stoeppler, M.,** in *Proc. Int. Conf. Heavy Metals in the Environment, Heidelberg, 1983,* CEP Consultants Ltd., Edinburgh, 1983, 241.
52. **Ficklin, W. H.,** *Talanta,* 30, 371, 1983.
53. **Borg, H.,** Background levels of trace metals in Swedish freshwaters, SNV PM 1817, The Swedish National Environmental Protection Board, Solna, 1984 (in Swedish).
54. **Subramanian, K. S., Meranger, J. C., and McCurdy, R. F.,** *Atomic Spectrosc.,* 5, 192, 1984.
55. **Aastrup, M.,** in Occurrence of mercury in groundwater, Lindquist, O., Johansson, K., and Timm, B., Eds., Rep. 3265, The Swedish National Environmental Protection Board, Solna, 1986.
56. **Aastruup, M. and Ek, J.,** in *Heavy Metals — Occurrence and Turnover in Nature,* Berne, C., Ed., Monitor 1987, The Swedish National Environmental Protection Board, Solna, 1987 (in Swedish).
57. **Nordberg, L.,** The Swedish National Environmental Protection Board, Solna, 1988 (Personal communication).
58. **Ledin, A., Pettersson, C., Allard, B., and Aastrup, M.,** *Heavy metals in groundwaters,* The Swedish National Environmental Protection Board, Solna, 1988.
59. **Welch, A. H. Lico, M. S., and Hughes, J. L.,** *Groundwater,* 26, 333, 1988.
60. **Aastrup, M., Johnson, J., Bringmark, E., Bringmark, L., and Iverfeldt, Å.,** *Water Air Soil Pollut.,* 56, 155, 1991.
61. **Allard, B., Xu, H., and Grimvall, A.,** Concentration and speciation of arsenic in groundwaters, *Vatten,* in press.
62. **Borg, H.,** Trace metals in forest lakes–Factors influencing the distribution and speciation in water, Dissertation, Uppsala University, 1988.
63. **Runnels, D. D., Shepherd, T. A., and Angino, E. E.,** *Environ. Sci. Technol.,* 26, 2316, 1992.
64. **Hermann, R. and Neumann-Mahlkau, P.,** *Sci. Tot. Environ.,* 43, 1, 1985.
65. **Smith, R. M. and Martell, A. E.,** *Critical Stability Constants,* Vol. 4, *Inorganic Complexes,* Plenum Press, New York, 1976.
66. **Baes, C. F., Jr. and Mesmer, R. E.,** *The Hydrolysis of Cations,* John Wiley & Sons, New York, 1976.
67. **Högfeldt, E.,** *Stability Constants of Metal-Ion Complexes,* Part A, *Inorganic Ligands,* IUPAC Chemical Data Series, No. 21, Pergamon Press, Oxford, 1982.
68. **Ephraim, J. H., Borén, H., Arsenie, I., Pettersson, C., and Allard, B.,** *Sci. Tot. Environ.,* 81/82, 615, 1989.
69. **Randhawa, N. S. and Broadbent, F. E.,** *Soil Sci.,* 99, 362, 1965.
70. **Schnitzer, M. and Hansen, E. H.,** *Soil Sci.,* 109, 333, 1970.
71. **Strohal, P. and Huljev, D.,** in *Proc. Symp. Nucl. Environ. Pollut.,* IAEA, Vienna, 1971, 439.
72. **Cheam, V. and Gamble, D. S.,** *Can. J. Soil Sci.,* 54, 413, 1974.
73. **Millward, G. E. and Burton, J. D.,** *Mar. Sci. Commun.,* 1, 15, 1975.
74. **Ernst, R., Allen, H. E., and Mancy, K. H.,** *Water Res.,* 9, 969, 1975.
75. **Mantoura, R. F. C. and Riley, J. P.,** *Anal. Chim. Acta,* 78, 193, 1975.
76. **Stevenson, F. J.,** *Soil Sci. Soc. Am. J.,* 40, 665, 1976.
77. **Stevenson, F. J.,** *Soil Sci.,* 123, 10, 1977.
78. **Mantoura, R. F. C., Dickson, A., and Riley, J. P.,** *Estuarine Coastal Mar. Sci.,* 6, 387, 1978.
79. **Lu, J. J., Li, C. S., Wang, W. H., and Peng, A.,** in *Proc. Int. Conf. on Heavy Metals in the Environment, Heidelberg, 1983,* CEP Consultants Ltd., Edinburgh, 1983, 780.
80. **Lövgren, L. and Sjöberg, S.,** *Water Res.,* 23, 327, 1989.
81. **Ephraim, J. H. and Xu, H.,** *Sci. Tot. Environ.,* 81/82, 625, 1989.
82. **Ephraim, J. H. and Marinsky, J. A.,** *Anal. Chim. Acta,* 232, 171, 1990.
83. **Ephraim, J.,** *Anal. Chim. Acta,* 267, 39, 1992.
84. **Ephraim, J. H. and Allard, B.,** in *Ion Exchange and Solvent Extraction,* Vol. 11, Marinsky, J. A. and Marcus, Y., Eds., Marcel Dekker, New York, 1993, 235.
85. **Krenkel, P. A., Ed.,** *Heavy Metals in the Aquatic Environment,* Pergamon Press, Oxford, 1975.
86. **Nriagu, J. O., Ed.,** *The Biogeochemistry of Lead in the Environment,* Elsevier, Amsterdam, 1978.
87. **Nriagu, J. O., Ed.,** *The Biogeochemistry of Mercury in the Environment,* Elsevier, Amsterdam, 1979.
88. **Nriagu, N. O., Ed.,** *Copper in the Environment,* John Wiley & Sons, New York, 1979.
89. **Nriagu, J. O., Ed.,** *Cadmium in the Environment,* John Wiley & Sons, New York, 1980.
90. **Nriagu, N. O., Ed.,** *Nickel in the Environment,* John Wiley & Sons, New York, 1980.
91. **Nriagu, N. O. and Sprague, J. B., Eds.,** *Cadmium in the Aquatic Environment,* John Wiley & Sons, Somerset, 1987.

92. **McComish, M. F. and Ong, J. H.,** in *Environmental Inorganic Chemistry,* Bodek, I., Lyman, W. J., Reehl, W. F., and Rosenblatt, D. H., Eds., Pergamon Press, New York, 1988, chap. 7.
93. **Nriagu, N. O. and Nieboer, E., Eds.,** *Chromium in the Natural and Human Environments,* John Wiley & Sons, New York, 1988.
94. **Merian, E., Ed.,** *Metals and Their Compounds in the Environment,* VCH Publishers, Weinheim, 1991.
95. **Pourbaix, M.,** *Atlas of Electrochemical Equilibria in Aqueous Solutions,* Pergamon Press, Elmsford, 1966.
96. **Pettersson, C., Håkansson, K., Karlsson, S., and Allard, B.,** *Water Res.,* 27, 863, 1993.
97. **Braman, R. S.,** in *Biological and Environmental Effects of Arsenic,* Fowler, B. A., Ed., Elsevier, Amsterdam, 1983, 445.
98. **Irgolic, K. J.,** in *Hazardous Metals in the Environment,* Stoeppler, M., Ed., Elsevier, Amsterdam, 1992, 287.
99. **Wood, J. M.,** *Science,* 183, 1049, 1974.
100. **Andreae, M. O,** in *Organometallic Compounds in the Environment,* Craig, P. J., Ed., Longman, Harlow, 1986, 198.
101. **Xu, H. and Allard, B.,** *Water Air Soil Pollut.,* 56, 709, 1991.
102. **Xu, H.,** Effects of humic substances and pH on the speciation and adsorption of cadmium, mercury and arsenic, Dissertation, Linköping University, 1991.
103. **Allard, B. and Arsenie, I.,** *Water Air Soil Pollut.,* 56, 709, 1991.
104. **Dyrssen, D. and Wedborg, M.,** *Water Air Soil Pollut.,* 56, 507, 1991.
105. **Craig, P. J.,** in *Organometallic Compounds in the Environment,* Craig, P. J., Ed., Longman, Harlow, 1986, 65.
106. **Lee, Y. H.,** *Int. J. Environ. Anal. Chem.,* 29, 263, 1987.
107. **Nagase, H., Ose, Y., Sato, T., and Ishikawa, T.,** *Sci. Tot. Environ.,* 32, 147, 1984.
108. **Karlsson, S., Håkansson, K., and Allard, B.,** in *Proc. Int. Conf. Heavy Metals in the Environment, Geneva, 1989,* CEP Consultants Ltd., Edinburgh, 1989, 296.
109. **Karlsson, S.,** Influence of hydrochemical parameters on the mobility and redistribution of metals from a mine waste deposit, Dissertation Linköping University, 1987.
110. **Rai, D. J., Zachara, J., Schwab, A., Smith, R., Girvin, D., and Rogers, J.,** Chemical attenuation rates, coefficients and constants in leachate migration. Vol. 1, Critical review, Report EA-3356, PNL, Richland, 1984.
111. **Wagemann, R.,** *Water Res.,* 12, 139, 1978.
112. **Chizhikov, D. M.,** *Cadmium,* Pergamon Press, New York, 1966.
113. **Benes, P. and Majer, V.,** *Trace Chemistry of Aqueous Solutions,* Elsevier, Amsterdam, 1980.
114. **Tewari, P. H., Ed.,** *Adsorption from Aqueous Solutions,* Plenum Press, New York, 1981.
115. **Muller, A. B., Ed.,** *Sorption. Modelling and Measurement for Nuclear Waste Disposal Studies,* OECD/NEA, Paris, 1983.
116. **Stumm, W., Ed.,** *Aquatic Surface Chemistry,* John Wiley & Sons, New York, 1987.
117. **Leckie, J. O., Benjamin, M. M., Hayes, K., Kaufman, G., and Altman, S.,** Adsorption/coprecipitation of trace elements from water with iron hydroxide, EPRI RP-910, Electric Power Res., Palo Alto, CA, 1980.
118. **Benjamin, M. M. and Leckie, J. O.,** *J. Colloid Interface Sci.,* 79, 209, 1981.
119. **Benjamin, M. M. and Leckie, J. O.,** *Environ. Sci. Technol.,* 162, 1982.
120. **Karlsson, S., Sandén, P., and Allard, B.,** *Nord. Hydrol.,* 18, 313, 1987.
121. **Benjamin, M. M., Hayes, M. M., and Leckie, J. O.,** *J. Water Pollut. Control Fed.,* 54, 1472, 1982.
122. **Leppard, G. G., Ed.,** *Trace Element Speciation in Surface Waters and its Ecological Implications,* Proc. NATO Advanced Research Workshop, Nervi, 1981, Plenum Press, New York, 1983.
123. **Bernhard, M., Brinckman, F. E., and Sadler, P. S., Eds.,** *The Importance of Chemical Speciation in Environmental Processes,* Proc. Dahlem Conf., Springer-Verlag, Berlin, 1986.
124. **Patterson, J. W., Ed.,** *Speciation, Separation and Recovery of Metals,* Proc. Int. Seminar, Lewis Publ., Chelsea, 1986.
125. **Landner, L., Ed.,** *Speciation of Metals in Water, Sediment and Soil Systems,* Lecture Notes in Earth Science 11, Springer-Verlag, Berlin, 1987.
126. **Ledin, A., Karlsson, S., Düker, A., and Allard, B.,** *Water Res.,* 28, 1539, 1994.
127. **Ledin, A.,** Colloidal carrier substances — properties and impact on trace metal distribution in natural waters, Dissertation, Linköping University, 1993.

128. **Tschöpel, P.,** in *Hazardous Metals in the Environment,* Stoeppler, M., Ed., Elsevier, Amsterdam, 1992, 73.
129. **Håkansson, K.,** Metals released from mine waste deposits. Redistribution and fluxes through geological barriers, Dissertation, Linköping University, 1991.
130. **Pettersson, C.,** Properties of humic substances from groundwater and surface waters, Dissertation, Linköping University, 1992.
131. **Ledin, A., Karlsson, S., Düker, A., and Allard, B.,** *Anal. Chim. Acta,* 281, 421, 1993.
132. **Karlsson, S., Allard, B., and Håkansson, K.,** *Appl. Geochem.,* 3, 345, 1988.
133. **Tessier, A., Campbell, P. G. C., and Bisson, M.,** *Anal. Chem.,* 51, 844, 1979.
134. **Förstner, U. and Salomons, W.,** *Trace Element Speciation in Surface Waters and its Ecological Implications,* in Leppard, G. G., Ed., National Water Res. Inst., Burlington, 1983, 245.
135. **Slavek, J. and Pickering, W. F.,** *Water Air Soil Pollut.,* 28, 151, 1986.
136. **Bernes, C.,** The environmental monitoring program in Sweden, SNV PM 1327, The National Environmental Protection Board, Solna, 1980 (in Swedish).

17

Chapter 8

Trace Elements in Lakes

Hans Borg

CONTENTS

I. CONTRIBUTING AND REMOVAL PROCESSES

Water bodies have formed and disappeared many times during different geological eras. The shape, size, and morphological characteristics of lakes and rivers are continuously changing, but the total amount of water on the earth is constant. The amount of fresh water is very small in comparison to the oceans. The distribution on a volumetric basis is concentrated in the large basins of rather few great lakes, with as much as 20% of the world's fresh water in Lake Baikal, Russia. The number of small lake basins concentrated in temperate and subarctic regions is, however, very large.[1] As a consequence of the fast-growing human population, an increasingly large part of the world is suffering from a deficit of fresh water. The habit of using lakes and rivers as recipients for municipal and industrial waste places a further stress on the limited resources.

On today's earth, one of the most important origins of lakes is the massive Pleistocene glaciation in the Northern hemisphere. At the retreat of the last stages of glaciation, thousands of small lakes were created, numerically far exceeding those lakes formed by other processes. Typical examples of such areas are Sweden, Norway, Finland, and the Precambrian Shield in Canada. However, glacial ice movements have also created large lake basins, of which the Great Lakes of the St. Lawrence drainage basin are the most impressive examples. Another important source of lakes is found in the tectonic basins formed by movements of the earth's crust. In such basins, some of the deepest relict lakes in the world are found, such as Lake Tanganyika in the Rift valley in east Africa, and the deepest lake in the world, Lake Baikal, reaching a depth of 1620 m.

The morphometry of the lake basins is important in determining sediment-water interactions and the resulting productivity. Shallow lakes generally show increased biological productivity through the increased percentage exposure of sediment area to water volume and an increased proportion of littoral productivity.[1] Further, the geological formations and the surficial soils of the lake catchment are of major importance in determining the chemical composition of the lake water. Weathering of minerals and flushing of soil organic and inorganic particles are important contributing processes, especially during periods of high discharge, i.e., snow melt. This applies not only to major ions but also to trace elements. These are found in the water of receiving lakes in very low concentrations, from a few nanograms per liter for most elements to some hundred micrograms per liter for more abundant elements such as Fe, Mn, and Al (see Section IV). The concentration in the water depends also on their tendency to adsorb to sedimenting particles, subsequently becoming trapped in the bottom sediment of the lake.

The other way of input to the lake is via direct wet and dry deposition on the lake surface, a pathway which has become increasingly important during recent decades as a result of emissions to the atmosphere from several industrial activities. The increased deposition on the lake surface and the catchment has been clearly documented as vertical profiles in the sediments, with increased metal levels towards the more recent layers. Several metals show this pattern, i.e., Pb, Cd, Hg, and Zn, which are primarily of

0-8493-6304-7/95/$0.00+$.50

anthropogenic origin in the precipitation. Evidence of increased loads of this kind on lakes has been reported, for example, from Sweden,[2,3] Norway,[4] Finland,[5] and North America.[6–9]

A large-scale deposition pattern has been demonstrated in Scandinavia, with higher loads in the southern parts, resulting in higher levels in the sediments compared to more remote lakes in northern Scandinavia.[4,5,10] The same pattern is obvious in the national surveys of land mosses in the Nordic countries[11] as well as in metal concentrations in bulk precipitation in Sweden and Norway.[12–14]

To some extent, the large-scale deposition pattern is also reflected in metal concentrations in lake waters. In regional lake surveys in Norway,[15] Sweden,[16] and Finland,[17] higher levels of Zn, Cd, and Pb were recorded in the southern lakes compared to the northern ones. These concentration differences are partly explained by the fact that the southern parts of the countries receive a higher input of acidifying substances and, consequently, the lakes have a somewhat lower pH than in the northern parts of these countries.

Depending on the ability of the respective element to adsorb to particles and its affinity to organic substances, the acidification of the soil is now an important contributing factor for trace elements in surface waters in areas affected by acidic precipitation.[18] An increased pH-induced mobility of elements such as Cd, Zn, Ni, and Al has been reported from areas with acidified soils.[18–23] Pb and Cu, on the other hand, show a higher terrestrial retention and their export to lakes is primarily coupled to the transport of organic substances.[19,22,23] The run-off of Cu has also been shown to be positively correlated to the watershed area.[21]

The relative contribution of direct deposition on the lake surface and export from the watershed, respectively, depend on the mobility of the element in the soil. For Pb, less than 50% of the total load to lakes originates from terrestrial input.[24] In contrast, Zn and Cd originate to more than 80% from terrestrial input, with the highest leaching in the most acidified areas.[23,24] The runoff of Hg accounts for about 25 to 75% of the total load in Swedish catchments and is primarily correlated to the levels of organic carbon in the water.[24] During periods of high water flow, Hg concentrations increase, especially in humic lakes. Higher Hg to C ratios in lake waters than in surface runoff from forest catchments during summer indicate a significant contribution of direct deposition on the lake surface.[25] These conclusions are consistent with studies in the catchment of Harp Lake, Ontario, where the direct deposition of Hg on the lake surface was estimated to be 55% of the total load on the lake.[26]

Humic substances seem to be the main carrier of Hg in runoff water.[26,27] In contrast to most other elements, the binding of Hg to organic substances has been shown to increase at lower pH.[28–30] Because of the strong association with humic substances, Hg can be considered to be relatively immobile in soil, comparable to Pb.[31]

Cu originates to more than 75% from the drainage area, following the transport of organics, while the long-range airborne anthropogenic input is of minor importance.[23,24] Figure 1 summarizes the atmospheric and terrestrial fluxes of Pb, Hg, Cu, Zn, and Cd to a hypothetical forest lake, based on precipitation and runoff data from Swedish catchments.[24]

The removal processes for trace elements likely to be important in lakes are

- Adsorption to hydrous oxides of Fe, Mn, and possibly Al
- Adsorption to and complexation with organic particulate material
- Adsorption to or uptake by phytoplankton
- Export in dissolved or particulate forms through the outlet

The first three mechanisms supply metals to the sediments, which act as the final or temporal sinks in lakes.

A number of studies have been performed in order to quantify this scavenging of metals from the water column of lakes. Sigg et al.[32] used sediment traps in Lake Zürich to collect settling material. They concluded that the particles consisted mainly of calcium carbonate, biogenic material, Mn and Fe oxides, and silicate minerals. Temporal variations of the composition and fluxes of particles pointed at the importance of sedimentation of biogenic material for removal of trace elements, especially Cu and Zn, but also Pb, Cd, and Cr. Fe oxides contribute to this scavenging and settle together with biological material (cf. review by Murray[33]). Calcium carbonate, however, seems to be inefficient as a carrier material. As is also the case with findings from marine studies, the importance of biogenic material as a carrier has been demonstrated in deep lakes (Lake Washington and Lake Zürich) where well-correlated depth profiles were shown for Pb and Cd.[33]

Fe and Pb show generally short residence times in the water column, about the same as the residence time of particles, while the residence times of Cd, Zn, and Cu are longer.[33] A study of a closed-basin Antarctic lake[34] showed that the residence times for different elements were well correlated to the

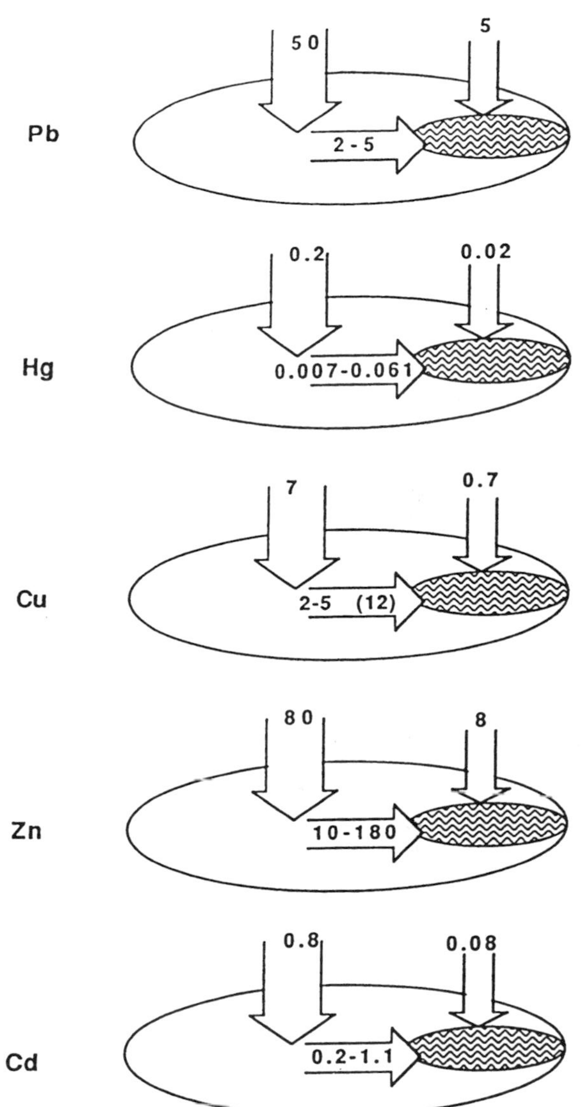

Figure 1 Atmospheric and terrestrial metal fluxes (kg y^{-1}) to a hypothetical lake with an area of 1 km^2 and a drainage area of 10 km^2. (From Borg, H. and Johansson, K., *Water Air Soil Pollut.*, 47, 427, 1989. With permission.)

residence times for the elements in the oceans, but were orders of magnitude lower in the lake. Fe, Co, and Mn showed the shortest residence times (<1 year), while Cd had a residence time of 480 years in this case.

Using lake enclosures, Diamond et al.[35] determined the fluxes of radioisotopes between water and sediment. Model simulations for isotope loss agreed well with experimental results and the sorption to particles decreased in the following order, Co > Sn > Fe > Hg > As > Zn > Cs. In agreement with earlier results from radiotracer experiments performed in the experimental lakes area, Ontario,[28,38] the losses from the water column were exponential during the first 2 weeks, with ^{57}Co showing a half-time of only 3.6 days and ^{134}Cs, 23 days. After about 25 days, a migration back to the water column could be observed.[35]

Table 1 gives some examples of flux rates of metals to sediments in lakes. With some exceptions, the uniformity of the deposition figures is good, considering the fact that they represent different lake types with different hydrology, morphology, redox conditions, etc. Further, the data have been obtained by different methods: sediment traps as well as calculation from vertical concentration profiles in the bottom sediments. The three acidified lakes in South Sweden show lower deposition rates than the more circumneutral ones. Lake number 2013 is also a very humic lake, and the higher figures are probably a result of an increased scavenging of trace metals by Fe oxide-humic aggregates (Table 1).

The redox potential of the lake water typically influences the fluxes of Fe and Mn, which secondarily might influence other trace elements as well. Evidence of a redox coupling of metals has been reported

Table 1 **Sedimentation rate of metals in lakes (mg $m^{-2}yr^{-1}$), measured in sediment traps or recent sediment layers**

	Al	Fe	Zn	Cu	Pb	Cd	Hg	Method	Ref.
5 lakes in S. Sweden (mean of 4 years) Lake number:									
2011	1,588	11,310	33	3.3	22	0.27		Sediment traps	94
2012	3,455	11,730	82	7.3	45	0.82			
2013 acidified	4,840	25,800	35	7.4	36	0.33			
2014 acidified	480	2,630	6.2	1.4	9.8	0.08			
2015 acidified	370	1,220	7.6	1.2	4.9	0.07			
5 lakes in N. Sweden					5–15	0.15–0.30	0.01–0.02	Sediment cores	36
4 lakes in S. Finland (mean)			14.4	2.3	13.4	0.19	0.04	Sediment cores	5
7 lakes in N. Finland (mean)			7.4	1.2	6.1	0.08	0.02	Sediment cores	5
Found Lake, Ontario (2 cores)					15–35			Sediment cores	6
Turkey Lakes, Ontario (mean of 4 lakes)	3,150	1,210	18.8	3.5	8.9	0.20	0.03	Sediment cores	8
Adirondacks, U.S. 10 lakes			20–100		20–55			Sediment cores	9
Kejimkujik National Park			55	20	8.6			Sediment traps	37
Nova Scotia, 2 lakes			73	29	128				
Lake Zürich, Switzerland									
April		4,161	128	19	40	1.2		Sediment traps	32
August		10,403	168	33	84	2.3			
Little Rock Lake, Wisconsin							0.01	Sediment traps	89

from the hypolimnion of Lake Zürich, where the concentrations of Cu, Ni, and Cd decrease when the anoxic interface moves upwards into the water column.[32,33] In contrast, the large seasonal variations in the concentrations of Fe and Mn in the periodically anoxic Esthwaite Water were not reflected in simultaneous changes in the concentrations of trace metals.[39]

The lake sediments not only act as a sink for metals but may also be a source of metals entering the water column. That is especially the case for elements with more mobile reduced forms such as Fe and Mn, which are released from the sediments when the oxygen concentration decreases in the near-bottom water. Reduced pH in the hypolimnion of lakes also influences the mobility of metals in the surface sediment. In enclosure experiments in Lake 223, Experimental Lakes Area in Ontario, where the pH was lowered to 5.1, a release of Al, Mn, Zn, and Fe from the sediment was recorded at pH 5.7 and, more pronounced, at 5.1 (Cu, Cd, Co, Cr, and Pb were mostly below the detection limit of the determination method used). By addition of radioisotopes it was also concluded that acidification made some elements more soluble (Fe, Co, Mn, and Zn) and slowed the loss from water column to sediments (Mn and Zn).[28] Later, laboratory experiments using highly contaminated sediments from Ramsey Lake in the Sudbury area, Ontario, showed that metals are released from sediments below a threshold pH of about 4.0 for Cu, Ni, Zn, Cd, and about 3.0 for Fe.[40] However, the quantitative importance of a net release of sediment-bound metals for the total metal fluxes in lakes of different pH and contamination level still remains to be evaluated.

II. SAMPLING AND PREANALYSIS HANDLING

The sampling of lake waters generally requires the same equipment and precautions as sampling of seawater, especially when sampling unpolluted waters. The problems with contamination during sampling and analysis are clearly demonstrated by the reported "background" levels over time in seawater as well as fresh water. The reported data for most trace metals show a drastically decreasing trend with time during recent decades (further discussed in Section IV).

Samples of surface waters of lakes are preferably taken from a small plastic boat by hand (wearing shoulder-length plastic gloves) from the bow while the boat is slowly rowed upwind. However, if the boat is drifting with the wind, the samples should be collected downwind, where the virgin water is. The surface microlayer should not be included as it is often enriched in trace elements, which could cause the sample to be less representative of the epilimnetic water column of the lake.[41,42] During early summer, the surface of forest lakes is sometimes covered with a thick layer of pollen from, e.g., spruce and pine, which can contribute significantly to the metal concentrations of a water sample.

If a sampler must be used, it should be appropriate for trace element work, i.e., no metal or rubber parts, and should withstand cleaning in dilute acids before use. A number of samplers suitable for trace element sampling from small boats in lakes are available on the market. PVC samplers with a Teflon coating inside ("Go-Flo" and modified "Niskin" bottles) have successfully been used primarily in seawater[43,44] as well as samplers made entirely of Teflon ("MERCOS",[45] "WATES"[46]). In our laboratory, we have used a modified, all-plastic "Ruttner" sampler (plexiglass, PVC) for sampling of unpolluted lake waters with satisfactory results.[16,47] Some authors have described depth-integrated systems[48] and pumping systems essentially based on Teflon pumps and PFA or polyethylene tubing.[49,50] For a detailed review of sampling techniques and equipment, see Reference 44.

A comparison of metal concentrations obtained after sampling by hand and by a modified Go-Flo bottle is presented in Table 2. The samples were collected in 250-ml polypropylene bottles from a fiberglass boat at 4 sites in the epilimnion of a small acidified forest lake in Sweden. The direct sampling was performed at arms-length depth wearing shoulder-length plastic gloves. The bottles were uncapped and recapped while submerged so as not to include the surface microlayer. The 1.7-liter Go-Flo bottle had a Teflon-coated inside and Teflon spigots, and it was triggered by a PVC-coated weight at a depth of 1 m. The results showed that the levels of Co, Cu, and Pb were higher in the hand-collected samples and that the variation in Co and Pb results were somewhat higher in the Go-Flo samples. However, the overall results compared well, and no significant contamination could be detected from the Go-Flo bottle.[51] Several other comparisons performed in seawater have also showed good agreement between Go-Flo bottles and other sampling techniques (reviewed by Sturgeon and Berman, 1987).[44]

Sampling under the ice cover during winter generally requires some kind of messenger-triggered sampler or a pump system. A special risk of contamination during winter sampling in lakes is constituted by polluted snow or melt water, which drains into the bore hole and might contaminate the sampler on the way up.

Table 2 **Comparison of metal concentrations (μg l⁻¹) in water samples from an acidified forest lake (Lake Årsjön) in Sweden, collected by hand and by using a 1.7-l Go-Flo bottle[51]**

Sampling Method	Al	Cd	Co	Cr	Cu	Fe	Mn	Ni	Pb	Zn
Hand-collected	250	0.105	0.37	0.23	0.34	180	75	0.8	0.71	14
samples,	242	0.111	0.38	0.18	0.74	170	70	1.1	0.77	18
0.5 m depth	245	0.106	0.44	0.25	0.27	100	75	0.9	0.65	16
	242	0.114	0.40	0.29	0.45	150	75	1.0	0.65	18
Mean	245	0.109	0.40	0.24	0.45	150	74	1.0	0.70	16.5
SD (%)	1.5	3.7	7.5	19	47	24	3.4	13	8.6	12
Go-Flo	239	0.111	0.35	0.24	0.29	150	75	0.7	0.43	20
bottle,	245	0.109	0.34	0.22	0.27	110	75	0.8	0.43	17
1 m depth	239	0.111	0.31	0.25	0.29	100	75	0.9	0.39	17
	239	0.109	0.24	0.30	0.32	92	75	1.0	0.77	18
Mean	241	0.110	0.31	0.25	0.29	113	75	0.9	0.54	17.7
SD (%)	1.2	0.9	16	14	7.2	22	0	16	35	7.9

The choice of suitable materials for sampling bottles, filtration equipment, and other laboratory wares in contact with the samples, as well as the need for washing procedures, is discussed in Chapter 3. As these requirements are essentially the same, regardless of which part of the hydrological cycle samples are collected from, they will not be discussed in detail here. Sturgeon and Berman[44] comprehensively summarized data on trace metal content of various materials commonly used for sampling and storage of waters for trace metal analyses.

The materials most widely used for storage bottles include LD-polyethylene, polypropylene, or FEP-Teflon. For determination of low level Hg concentrations, borosilicate glass or quartz bottles have proved to be suitable. The plastic bottles mentioned above are relatively clean from the start, but to be acceptable for low-level trace metal sampling, they still need a thorough cleaning in acid. As nitric and hydrochloric acid appear to leach various elements with different efficiencies, Moody and Lindstrom[52] and Patterson and Settle[53] recommended the use of both acids in sequence. A thorough acid leaching is, of course, also necessary for all other equipment in contact with the sample, such as filters and filter-holders, etc. Appropriate cleaning procedures for different materials are reviewed by Sturgeon and Berman.[44]

In spite of the improvement in analytical sensitivity during recent years, many elements are present in concentrations still not making them directly determinable. A preconcentration or separation step may, therefore, be necessary. The techniques which are potentially available for preconcentration of trace elements from lake waters are generally identical to those applied to seawater. The most widely used techniques include solvent extraction, freeze concentration, carbon adsorption, evaporation, electro-deposition, and ion exchange (reviewed by Orpwood;[54] Leyden and Wegscheider[55]). However, the higher concentration of humic substances and associated Fe hydrous oxides in fresh water compared to seawater limits the applicability of some techniques. Chelating ion-exchange resins, such as Chelex-100, have proved capable of extracting a wide range of elements from seawater with good recovery.[56,57] However, the complexing capacity of humic fresh water for some elements might cause an incomplete and varying recovery on the chelating resin, as the organically complexed metal forms may not be quantitatively retained by the resin.[58]

Solvent extraction techniques using chelating agents such as 8-hydroxyquinoline (oxine), ammonium pyrrolidine dithiocarbamate (APDC), and diethyl ammonium-diethyl dithiocarbamate (DDOC) have been widely used, as these agents have the ability to extract a wide variety of metal ions.[54,55,59] To minimize the contamination problems and increase the sample throughput, some authors have presented automated extraction techniques in flow systems with a reduced number of working steps.[60,61] van Geen and Boyle[62] presented an automated procedure which is a combination of organic complexation of metals and then adsorption onto a resin column. The relatively high content of organic substances and Fe in fresh water may, however, limit the direct use of extraction procedures in fresh water. Some preceding oxidation steps might be necessary to obtain a good recovery.

On the other hand, the lower concentrations of major ions, e.g., Ca, Na, Cl, and SO_4 in fresh water compared to sea water, simplifies the determination of trace elements with methods susceptible to interference from the salt matrix. Evaporation and freeze concentration techniques offer possibilities to concentrate all forms of the elements.[54] The sample matrix is concentrated, as well as the trace metals, which may reduce the maximum concentration factor. Evaporation by heat suffers from contamination problems and also from the risk of losses of volatile elements. Görlach and Boutron[63] described a nonboiling evaporation procedure performed in a laminar flow clean hood, which permitted the determination of Pb, Cd, Cu, and Zn at the pg g^{-1} level.

If the evaporation is performed by freeze-drying directly in the sampling containers, the contamination problems are much reduced. Hall and Godinho[64] compared freeze-drying of fresh water with solvent extraction and chelating ion-exchange and found a good agreement and recovery. Pb, however, was not completely recovered from the freeze-drying residue of natural water. They allowed the water to dry completely during the freeze-drying and dissolved the residue in hydrochloric acid. If the freeze-drying is not allowed to continue to complete dryness, but is stopped when the water has been concentrated about ten times in the bottles, which is sufficient, e.g., for determination by graphite-furnace AAS, a complete recovery can be achieved for Fe, Pb, and Cu.[51,65] The preconcentration is performed on the acid-preserved sample (containing nitric acid, 2 ml/1) which probably also improves the recovery of elements with a strong affinity to particles in the water.

Modern graphite-furnace AAS systems with computer-controlled autosamplers offer possibilities to perform preconcentration of the water automatically in the furnace. The autosampler is programmed to make replicate injections before the atomization and measurement of the signal. It is then possible to get a reproducible concentration factor adjusted to the need for every single element of interest in the sample. We generally use an injection volume of 80 μl injected 8 times for Co and Ni, 7 times for Pb, and 3 times for Cd. Figures 2 and 3 show a comparison of results of ETAAS analyses for Cd, Pb, Co, and Ni in Swedish lake waters after preconcentration by freeze-drying and replicate injection, respectively.[65] However, even in these dilute waters with low concentrations of electrolytes, some interferences are recognized at ETAAS determination of Pb and Cd after preconcentration of the water. To obtain accurate results, it is, therefore, necessary to use a chemical matrix modifier. Ammonium hydrogenphosphate and magnesium nitrate have been widely used in order to compensate for interferences in the determination of Pb and Cd.[65] Several studies have shown that palladium also has a capability to act as a universal matrix modifier in ETAAS determinations.[66,67]

III. ANALYTICAL CONSIDERATIONS

Accurate determination of trace metals in water at background levels has traditionally been hampered by the low concentration levels in the water column. It is only during the last 15 to 20 years or so that instrumental analytical techniques have been available, at reasonable cost, for monitoring of trace metals in fresh waters. The general analytical considerations of trace metal determinations are discussed in Chapter 3. Suitable methods for lake waters are similar as for other fresh waters, even if the sometimes higher content of organic substances requires additional pretreatment steps for the determination of total concentrations of some elements (e.g., Hg, As, and Se). Relatively few instrumental techniques are sensitive enough to determine more than a few elements in lake waters directly without preconcentration. The most widely used technique for determination of low concentrations of trace metals in water is probably electrothermal atomic absorption spectrometry (ETAAS). The modern ETAAS systems presently available offer excellent possibilities to determine low levels of metals with high accuracy and a precision of a few percent. Automated preconcentration by replicate injections can be used in order to further increase sensitivity and precision (cf. Section II). A preconcentration is generally necessary in spite of the high sensitivity of ETAAS, as natural background levels of most metals in fresh waters normally are below the detection limit for direct analysis.

However, the high sensitivity obtained when atomizing a sample in the graphite furnace also results in susceptibility to various interferences, e.g., molecular absorption, particles, and reactions with salts during the atomization. In relatively dilute samples such as lake waters, at least some of the interferences are reduced by continuum source background correction or line splitting by Zeeman effect systems.[65,68] However, to obtain accurate results for Cd and Pb even in a dilute fresh water, preconcentrated about ten times or more, it is necessary to use chemical matrix modifiers (cf. Section II) and possibly stabilized temperature atomization by the use of the L'vov platform.[65,68]

One limitation of the ETAAS technique is the relatively slow sample throughput, especially when using replicate injections. Another limitation, as for all AAS applications, is that only one element can be determined at a time because of the single element light sources used and the different optimum conditions for atomization and detection of the respective elements.

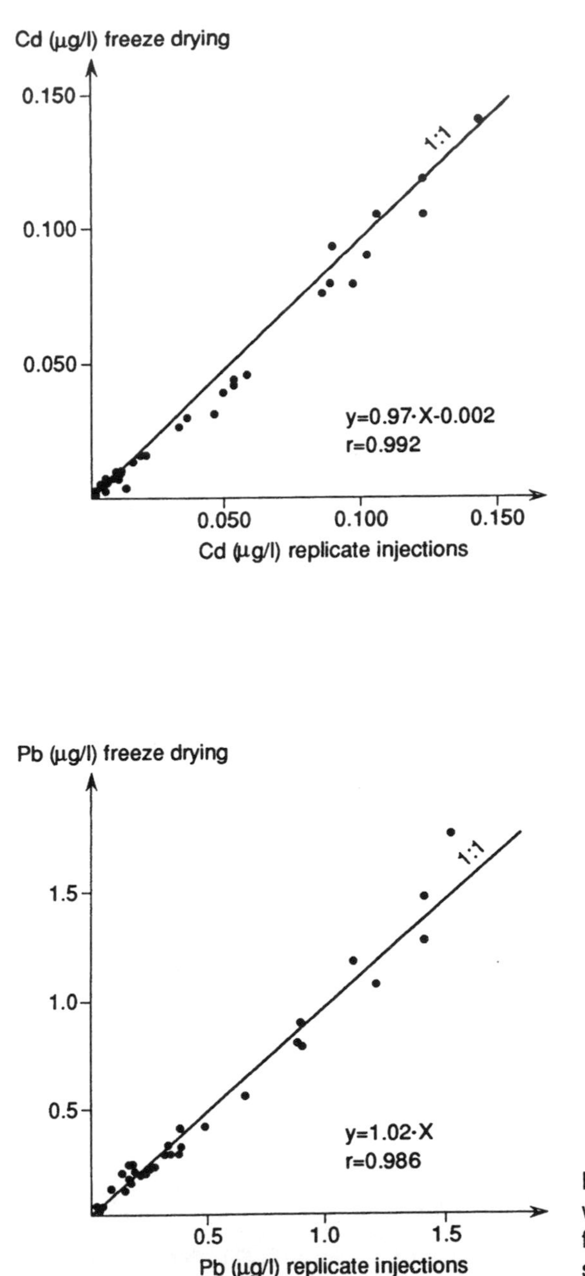

Figure 2 Determination of Pb and Cd in lake waters by ETAAS after preconcentration by freeze-drying and by replicate injections, respectively (Holm et al.[65]).

A new technique, which has the sensitivity of ETAAS together with multielement capability, is inductively coupled plasma mass spectrometry (ICP-MS). As in ICP-AES (ICP-atomic emission spectrometry), the sample solution is aspirated through a nebulizer to an argon plasma which serves as an ion source to a quadrupole mass spectrometer.[69–71] Commercial instruments which utilize this technique became available as recently as 1983, and, since then, there has been a fast increase in the number of applications for environmental analyses.[72–77] Besides the multielement capacity and high sensitivity, a further advantage is the possibility to perform isotope dilution analyses with ICP-MS, a method to increase the accuracy and quality assurance in environmental monitoring.[78]

Henshaw et al.[77] used ICP-MS for the determination of 49 elements in waters from a lake survey in the U.S. They concluded that, by using a multielement mode without any preconcentration, about 80% of the elements were present at concentrations above the system detection limit. The interferences in lake waters are moderate and can generally be completely corrected. In Table 3, some examples of detection limits are presented for ETAAS and ICP-MS. The data for ETAAS are the levels normally reached during

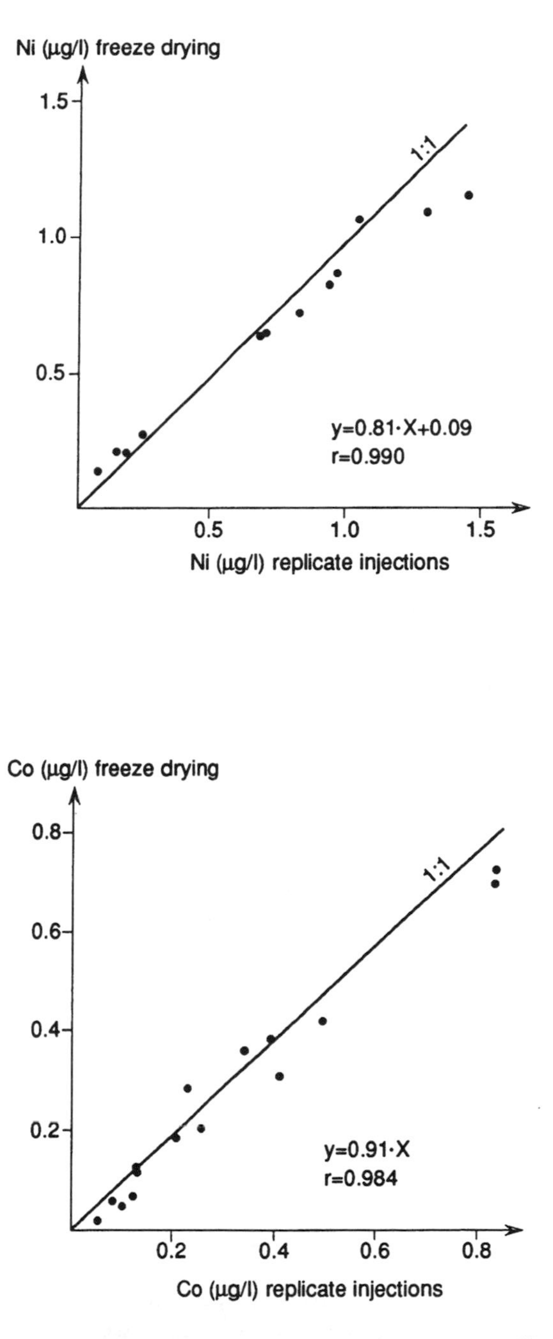

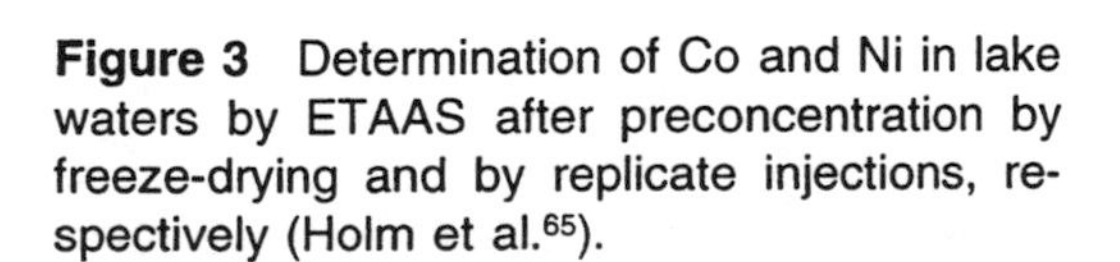

Figure 3 Determination of Co and Ni in lake waters by ETAAS after preconcentration by freeze-drying and by replicate injections, respectively (Holm et al.[65]).

routine analyses of Swedish soft water forest lakes using a Perkin-Elmer Zeeman 3030 instrument.[65] Arsenic is determined with an automated hydride generation AAS system with an electrically heated quartz cuvette.[79] The ICP-MS data represent quantitative determinations of the listed elements on a Perkin-Elmer Sciex ELAN 5000 (K. Holm, work in progress). As the detection limits are largely dependent on the blank values as well as the sample composition, etc., and might vary from day to day, such figures should never be taken at face value but rather as approximate guidelines for the capacity of the procedures. It is evident from the data in Table 3 that both techniques provide excellent possibilities for monitoring of trace elements in fresh waters. ICP-MS offers a somewhat wider variety of elements and is superior for the heavier elements, which are difficult to determine with ETAAS, such as the rare earth elements.

The determination of Hg in natural waters has traditionally been difficult because of the low levels, a few ng per liter. By using ultraclean techniques throughout sampling and analysis, an increasing number of data on background concentrations have been produced during recent years. Suitable procedures

Table 3 **Examples of detection limits for ICP-MS and ETAAS (μg l⁻¹)**

Element	ICP-MS[a]	ETAAS:[b]	Injection Volume (μl)	Number of injections
As	0.02	0.05[c]		
Se	0.2	0.1[c]		
Mo	0.003			
Ag	0.003			
Cd	0.005	0.003	80	3
Cs	0.001			
Be	0.014			
Al	0.025	0.3	50	1
V	0.009			
Cr	0.038	0.03	50	2
Fe	0.5	1.0	50	1
Mn	0.005	0.3	10	1
Ni	0.047	0.2	80	8
Co	0.005	0.1	80	8
Cu	0.015	0.02	80	5
Zn	0.046	0.1	10	1
W	0.002			
Au	0.012			
Hg	0.005	0.1[d]		
Tl	0.002			
Pb	0.006	0.03	80	7
Th	0.001			
U	0.001			

[a] ICP-MS: inductively coupled plasma mass spectrometry (Perkin-Elmer Sciex Elan 5000), 1% HNO_3 solution, 3 · SD, 1 second integration time; [b] ETAAS: graphite furnace atomic absorption spectrometry (Perkin-Elmer Zeeman 3030), lake waters preserved with 0.2% HNO_3, 3 · SD; [c] Hydride generation AAS; [d] Cold vapor AAS

include gold amalgamation traps with AAS or the more sensitive AES detection. The recently applied ICP-MS and especially atomic fluorescence (AFS) detection have further increased the possibilities to determine low Hg levels accurately.[80–82]

IV. CONCENTRATION LEVELS

The toxicity of metals is generally negatively correlated to their abundance in the earth's crust and to the natural background concentrations in water. Thus, the "lowest known effect level" (LKE) for aquatic organisms is directly proportional to the background concentration in fresh water.[83] The LKE is only about three to five times higher than the background concentration for many elements. These relationships clearly demonstrate the importance of accurate measurements of background concentrations, considering monitoring, emission control, and hazard assessment of various elements in the aquatic environment.

The documentation of the distribution of trace metals in water at natural or close to natural levels has been seriously obstructed by contamination at sampling and analyses and by insufficient sensitivity of the analytical procedures. The reported "background" levels show a typical declining trend during recent decades. Some examples of historical data for Pb in fresh water are presented in Figure 4, but the trend is similar for most trace elements in both seawater and fresh water (Sturgeon and Berman[44]). Such an example is the drastic decrease of about 2 orders of magnitude during only 10 years in Lake Huron. This decrease in concentrations does not reflect a real decline but is primarily a result of increasing control of contamination sources at sampling and analysis, as well as more accurate and sensitive analytical methods.

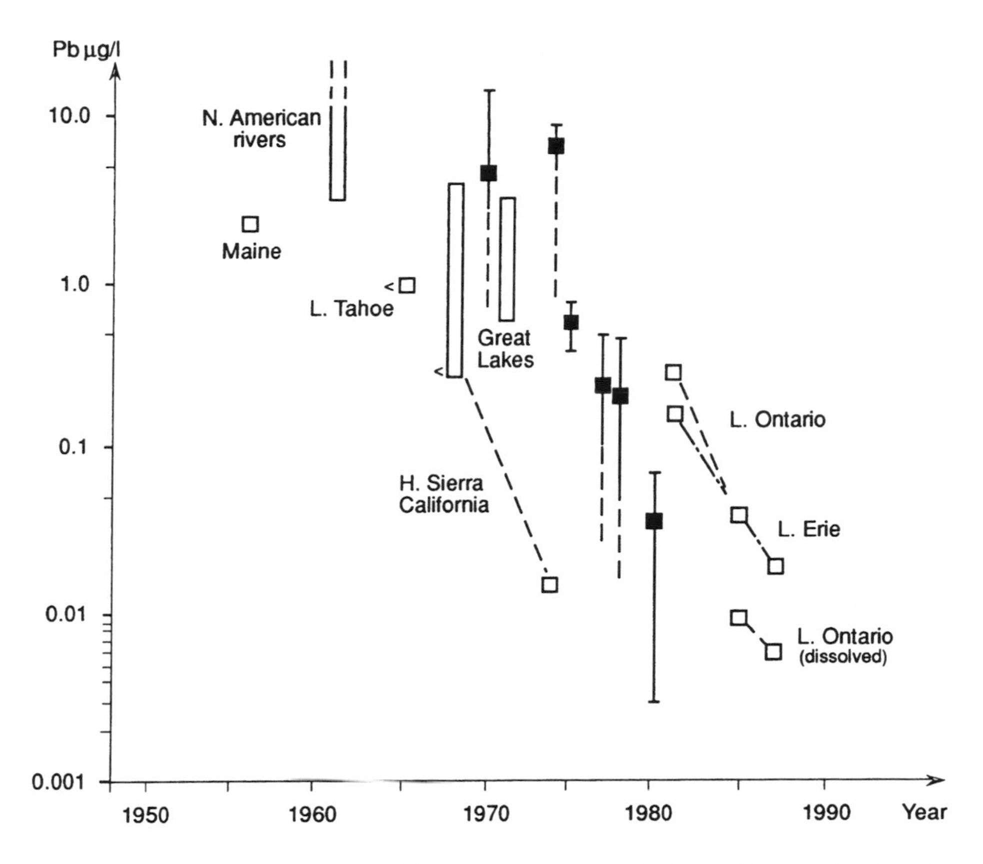

Figure 4 Historical changes in reported "background" concentrations of lead in fresh waters. The data from Lake Huron (filled symbols) show a decline over 2 orders of magnitude during only 10 years, following the application of cleaner sampling and analytical techniques (revised by Borg,[84] Rossmann and Barres,[50] Coale and Flegal[88]).

In spite of the improvements in analytical sensitivity and contamination control during recent years, the documentation of concentrations of most trace elements in water is still very sparse, especially in remote lakes with little anthropogenic influence. In regional lake surveys performed during the winter of 1980 in southern and northern Sweden, about 15 elements were determined using NAA, ETAAS and ASV (Table 4). Only small differences were found between the two areas investigated, but Co, Zn, Pb, and Mn were slightly higher in the southern area. The levels of Al, Cd, and Pb were elevated in the acidified lakes included in the study. If these lakes were excluded, the mean values for the metals concerned were about half of the levels for the total database.[47,84] In a survey of 59 lakes along a section of about 1000 km in length from central to northern Sweden, performed during the summer of 1980, a decrease to the north was found for Zn, Cd, and Pb, being most pronounced in the southern half of the material.[16] This regional distribution pattern, with higher levels in the south, is consistent with results of lake surveys in Finland[17] and Norway.[85] It reflects the large-scale deposition pattern over Scandinavia, influenced by long-range airborne pollutants (cf. Section I). The data from the recent lake surveys in Scandinavia show good agreement, but some of the earlier Cd data seem to be influenced by contamination (Table 5).

All the data presented in Tables 4 and 5 represent lakes with a varying degree of long-range airborne pollution load, resulting in a decreased pH value in some of the lakes. The reported maximum levels of Zn, Cd, and Ni might, consequently, be caused by acidification and not exclusively a result of methodological errors. Not all of the lakes in these studies can, therefore, be considered to represent true background levels of metals.

Table 6 presents data from mountain lakes in northern Sweden, in an area with low atmospheric deposition of metals as well as sulfur and nitrogen.[13] These data should represent a situation at least very close to a true natural background. The samples were collected in the outlet streams directly into polypropylene bottles, which were cleaned according to the procedure described in Section II. It can be

Table 4 **Metal concentrations in small headwater forest lakes in Sweden, sampled in the winter of 1980, $\mu g\ l^{-1}$ [47,84]**

Element	S. Sweden			N. Sweden			Determination
	Mv	Range	n	Mv	Range	n	Method
Al	145	44–306	20	134	9.0–361	18	NAA
As	0.19	0.10–0.36	20	0.22	0.11–0.40	17	HAAS
Cd	0.042	0.005–0.12	19	0.038	0.004–0.09	16	ASV
Co	0.55	<0.2–1.2	16	0.21	0.1–0.7	18	NAA
Cr	0.21	0.11–0.35	7	0.18	0.09–0.33	10	ETAAS
Cs	<0.05		2	<0.05		10	NAA
Cu	0.68	0.3–1.0	20	0.90	0.25–2.66	17	AVS
Fe	616	50–2200	20	416	40–2600	18	NAA
Hg	<0.01		2	<0.01		18	NAA
La	0.38	<0.1–1.0	16	0.39	<0.1– 1.1	18	NAA
Mn	195	8.0–553	16	42	2.0–149	17	NAA
Mo	<0.8	<0.1–2.4	16	<0.8	<0.1– 1.5	17	NAA
Ni	<1.0	<0.9–1.2	6	<0.9	—	10	ETAAS
Pb	0.67	0.3–1.1	7	0.42	0.2–1.1	10	ETAAS
Rb	2.0	0.6–5	17	1.9	<1.0–5	10	NAA
Sb	0.14	0.08–0.19	16	0.13	0.06–0.17	18	NAA
Sc	0.025	0.01–0.05	16	0.059	<0.005–0.46	18	NAA
Se	<0.3	<0.1–0.4	16	<0.2	<0.1–1.0	18	NAA
Sm	0.07	0.03–0.14	16	0.08	0.02–0.26	18	NAA
Ti	<10	—	20	<10		18	NAA
U	—	0.01–0.06	2	0.21	0.05–0.56	9	NAA
V	—	<0.1–1.0	20	—	<0.2–1.5	18	NAA
W	<0.05	—	2	<0.05	—	18	NAA
Zn	11	4.0–25	15	7	<2.0–20	18	NAA

Note: NAA, neutron activation analysis; ETAAS, Electrothermal atomic absorption spectrometry; HAAS, hydride generation atomic absorption spectrometry; ASV, anodic stripping voltammetry.

Table 5 **Mean values and ranges of trace metal concentrations in water ($\mu g\ l^{-1}$) from Scandinavian surveys of small headwater lakes**

		Cd	Cu	Pb	Zn	Ni	Ref.
S. Sweden		0.042	0.68	0.67	11.0	<1.0	47
1980 winter		0.005–0.12	0.3–1.0	0.3–1.1	4.0–25	<0.9–1.2	
N. Sweden		0.038	0.90	0.42	7.0	<0.9	47
1980 winter		0.004–0.09	0.90	0.25–2.66	<2.0–20	—	
N. Sweden		0.014	0.51	0.27	2.2	<0.42	16
1980 summer		0.007–0.036	0.1–2.0	0.1–0.8	<0.4–8.5	<0.2–1.0	
Norway		—	1.5	0.9	15.1		85
1986		<0.1–0.54	0.4–9.1	<0.5–4.5	0.4–39	—	
Norway		0.1–0.5	<0.5–2	<0.5–2	0.5–12	—	15
1974							
S. Finland	Mean	0.031	0.38	0.13	5.6	0.40	17
1987	Max	0.13	2.6	0.9	28	5.3	
N. Finland	Mean	0.022	0.43	0.12	2.5	0.25	17
1987	Max	0.13	3.0	0.7	20	2.6	

Table 6 **pH, water color, and trace metal concentrations (μg l⁻¹) in mountain lakes in the upper catchment of R. Ångermanälven, province of Västerbotten, N. Sweden (samples taken in the outlet streams in July and August, 1983 to 1990)**

		pH	Color MgPt l^{-1}	Al	Fe	Mn	Zn	Cu	Pb	Cd	As	Cr	Ni	Co
Lake Ransaren,	Mv	7.41	10	29	29	4.0	0.60	0.39	0.08	0.008	0.18	0.10	0.56	0.30
catchment above	SD	±0.12	±3.5	±6	±8	±2.2	±0.32	±0.07	±0.016	±0.007				
tree line	n	5	5	5	4	5	5	5	5	5	2	1	1	1
Lake Ö.Marssjön,	Mv	7.13	18	26	89	6.8	0.75	0.30	0.10	0.006	0.13	0.22	0.74	0.19
catchment mainly	SD	0.34	±13	±11	±32	±1.2	±0.36	±0.13	±0.06	±0.003				
forested	n	6	6	4	6	6	6	6	6	6	2	2	2	2
Lake Kultsjön,	Mv	7.29	12	21	36	4.3	2.3	0.45	0.10	0.008	0.15	0.11	0.59	0.20
catchment partly	SD	±0.15	±6	±7	±11	±1.0	±1.1	±0.09	±0.03	±0.005				
forested	n	5	5	4	5	5	5	5	5	5	2	2	2	2
5 headwater	Mv	6.58	30	—	104	10	1.3	0.20	0.04	0.005	—	—	0.41	—
lakes on the	SD	±0.33	±9	—	±38	±10	±0.4	±0.06	±0.006	0.001			±0.01	
Njakafjäll area (sampled 4 times in autumn 1989), catchments covered with bogs and coniferous forest	n	5	5		5	5	5	5	5	5			5	

Note: Mv, mean value; SD, standard deviation; n, number.

concluded that the levels of some elements are lower than in the surveys mentioned above. The concentration of zinc is, for example, generally below or around 1 μg l^{-1} in these lakes. However, for most metals there is not a large difference in the mean concentrations of metals between the former studies in Scandinavia and the lakes presented in Table 6. The earlier surveys therefore may be acceptable to use as background data for most trace metals in lake waters. There is also a reasonable consistency between these data and results from other unpolluted and moderately polluted areas. The low Cd concentrations from the surveys in the 1980s is similar to those reported from studies in Ontario[86] and northwestern England.[87] The concentrations of Cd, Cu, Zn, As, and Cr are in agreement with median values from Lake Huron and Lake Superior,[50] while the concentration of Pb is generally higher than that found in Lake Superior (0.029 μg l^{-1}). Coale and Flegal[88] found lower concentrations of Zn, Pb, and Cd in Lake Ontario and Lake Erie than in earlier investigations published, which indicates that even fairly recent data might be significantly influenced by contamination (cf. Figure 4).

The concentration of Hg in natural waters has been difficult to measure because of the relatively low levels. The development of sensitive routines for sample handling and analysis during recent years has made it possible to determine natural Hg levels with sufficient accuracy. An example of that development is given in Figure 5, showing results from three different investigations performed in Wandercook Lake, Wisconsin.[89] The concentration decreased considerably with time as cleaner procedures were adopted (cf. Figure 4). In Table 7 some data on Hg concentrations in lake waters are summarized. It can be concluded that the total concentrations are in the range of 1 to 10 ng l^{-1} in most areas, a level not too far above the concentrations found in open ocean water (0.2 to 2 ng l^{-1}, Fitzgerald and Watras[89]). In humic lake waters, the concentrations are sometimes around 15 ng l^{-1}.[90]

A. SEASONAL VARIATIONS

The time of sampling of lake waters influences the concentration levels of most trace metals significantly, as a temporal variation can be found in most lakes. The concentrations of Pb, Cd, and Zn were higher in May than in the autumn in Finnish forest lakes,[17] and Swedish forest lakes showed a similar trend. The ratios winter to summer for Al, Pb, Cd, Fe, and Cu in these Swedish forest lakes were in the range 2 to 3 as a mean.[16] In three lakes in northern Sweden, the seasonal variations of Zn, Cd, Pb, and Cu showed concentration maxima during March/April and minima during July/August.[91] Similar seasonal variations were found for Pb in Darts Lake, New York,[92] as well as in California,[93] and for Zn, Cd, Pb, and Cu in forest lakes in southern Sweden.[94] An example of typical temporal variation curves from the latter study is given in Figure 6. Generally, the variation in metal concentrations of the lake water is the opposite of the variation in pH. The concentration of Hg in small forest lakes in Sweden showed similar trends with minima during July/August and increasing values in the autumn. The variation was greater in the northern lakes than in the south.[95]

The reasons for these temporal variations are connected to hydrological conditions in the watersheds as well as biological productivity in the lakes. The peak values recorded in spring are mostly caused by the increased inflow of meltwater, which often is more acid and contaminated than the lake water. Simultaneously, an increase in the runoff of organic substances from the watershed often occurs, favoring the transport of complexed metal forms (e.g., Hg and Pb). The increased plankton production during summer favors an increased sedimentation of metals bound to particles and a subsequent decrease in the content of most elements in the water column. The peak values sometimes recorded in autumn are generally caused by an increased runoff of metals from the watershed during rainy periods, similar to the conditions during spring snowmelt.

In conclusion, the time of sampling is important, and the seasonal variations should be considered when performing lake surveys, especially when investigating small lakes with a relatively short turnover time.

V. SPECIATION

Trace metals are present in natural waters in a number of physico-chemical forms including:

1. Free metal ions
2. Inorganic complexes (CO_3^{2-}, Cl^-)
3. Organic complexes with fulvic and humic acids
4. Associated with colloidal and particulate materials such as clay minerals, hydrous oxides of Fe and Mn, and biogenic material as living algae and detritus

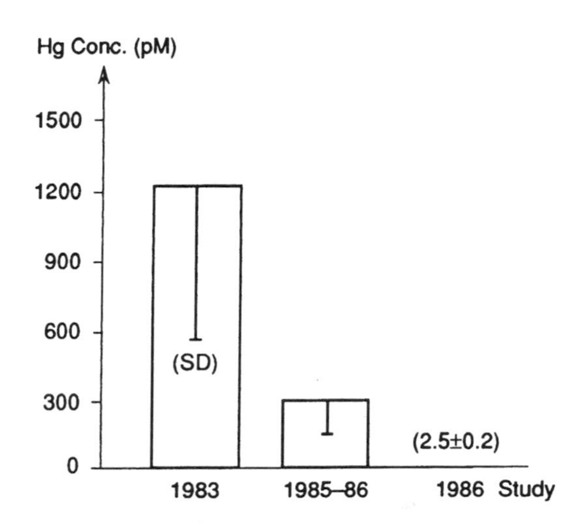

Figure 5 Repeated studies of mercury concentrations in surface waters of Vandercook Lake, U.S. Progressively cleaner sample collection and handling techniques have lowered the reported Hg concentrations considerably. (From Fitzgerald, W. F. and Watras, C. J. *Sci. Tot. Environ.*, 87/88, 223, 1989. With permission.)

Table 7 **Concentration of Hg in lake waters (ng l^{-1})**

	Hg Total	Hg Reactive[a]	Reference
25 forest lakes in Sweden, median			
spring (April to June)	3.7	1.5	90
summer (July to August)	3.1	1.3	
autumn (September to November)	6.2	1.1	
4 lakes in northern Wisconsin, autumn	0.9–2.0	0.1–0.6	89
Little Rock Lake, Wisconsin	0.37	0.17	145
Lake Huron	11		50
Lake Superior	2		
Lake Ontario	10		
Lake Ontario	0.9		
	0.7[b]		135
Lakes in California			
Lake Beryessa	8.8	6.4	135
Lake Nacimiento	1.5	0.2	135
Lake San Antonio	1.2	0.6	135
Silver Lake	0.6	0.4	135
Huntington Lake	0.5	0.3	135

[a] Acid labile, easily reducible (HCl–$SnCl_2$); [b] Dissolved.

Metal ions in lake waters interact in different ways with dissolved inorganic and organic substances as well as particles. The relative importance of these interactions is dependent on a number of biological, chemical, physical, and hydrological factors. The form in which a metal is found in natural water is not always similar to what is expected from laboratory experiments, as environmental processes often do not achieve thermodynamic equilibrium. Frequently, anticipated calculated chemical equilibria, including certain forms and concentrations of molecular species, do not correspond to the actual conditions found in the natural environment.[96] Different forms of biological activity are often responsible in such cases.

As the bioavailability and toxicity of metals is dependent on the physico-chemical form in which they occur in the water, there is a need for analytical methods capable of distinguishing between these forms. Determining the nature of all the coordination compounds formed by metal ions in the environment would demand enormous efforts, considering the competition for a very large number of ligands with largely

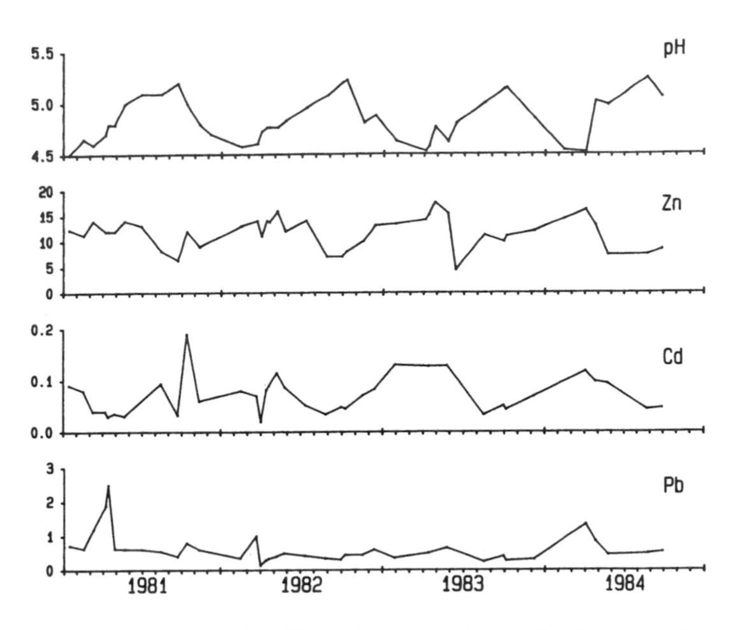

Figure 6 Seasonal variation of pH and trace metals (μg l^{-1}) in Lake Gyslättasjön, an acidified forest lake in S. Sweden. (From Borg, H., Andersson, P., and Johansson, K., *Sci. Tot. Environ.,* 87/88, 241, 1989. With permission.)

unknown characteristics. In environmental investigations, much work has been devoted to the determination of different molecular weight fractions of complexed and particle-bound metals in water.[97,98] As it is generally assumed that free metal ions are more toxic to aquatic biota than ions bound to large organic molecules (i.e., humic substances), it is especially important to measure the proportion of uncomplexed forms in water. Different size separation techniques have been used for this purpose. Filtration and ultrafiltration procedures provide possibilities to separate particulate from colloidal and low molecular weight species,[99–101] while dialysis and hollow fiber techniques offer possibilities to separate truly dissolved forms in laboratory as well as *in situ* experiments.[99,102–104] It should be remembered that the often-applied definition of "dissolved" metals as being equal to the fraction passing a membrane filter with a pore size of 0.45 mμ is not correct, as the filtrate also includes fine particulate and colloidal forms. Generally, the *in situ* applications are less susceptible to contamination and loss of elements by adsorption on membranes and container walls.

Iron is generally found in the Fe(III) state in oxygenated waters. The Fe(II) ion is only stable in very acidic and strongly anaerobic waters, but dissolved organic matter may influence the oxidation rate of Fe(II) and increase the level found in the water column of lakes.[105] The concentration of filterable Fe in natural waters is generally much higher than what would be predicted from the solubility of Fe (III) in aqueous solution, especially in humic waters. This is explained by the occurrence of dissolved and colloidal Fe-humic complexes and colloidal particles of hydrous Fe oxides, possibly associated with organic substances.[105] Consequently, Fe is generally found in a particulate, nondialyzable fraction even in relatively acidic lake waters.[99,101,106] Only in clear waters (TOC < 5 mg/l) can more than 30% of Fe be found in the dialysable fraction.[106]

In contrast to Fe, Mn occurs to a large degree in dissolved form in lake waters. Studies in Swedish lakes showed that the dialyzable form is dominating, especially in clear-water lakes (90 to 100% of the total concentration), and decreases to about 30 to 50% in lakes with TOC values of 25 mg/l.[106] Similarly, LaZerte and Burling[107] demonstrated Mn to be in a dissolved state (partly as Mn(II)) in acidic dilute lakes in Ontario with an increasing portion as colloidal oxyhydroxides at higher pH. These results are consistent with results from English lakes where Mn was present mainly in a soluble form, possibly as manganous ions.[101]

The oxyhydroxides of Fe and Mn have been shown to adsorb other trace metals effectively. The degree of adsorption generally increases with pH.[108] The pH at which 50% of the metal concentration is adsorbed on $Fe(OH)_3$ increases in the following order: pH 3.0(Pb), pH 4.3(Cu), pH 5.3(Zn), pH 5.7(Ni), pH 5.9(Cd,Co).[31] Further, positive correlations have been demonstrated between pH and the equilibrium constant for the adsorption of Zn, Cu, Cd, and Pb on natural Fe oxyhydroxide.[109] Strong complexes of hydrous Mn oxides and Pb, Cu, Zn, and Cd have been shown experimentally to occur at low metal-to-ligand ratios. The degree of complexation increases with pH in the range of 5.5 to 6.5.[110]

Aluminum is the most abundant metallic element in the earth's crust, but the details of its fairly complicated chemistry and transformation processes in the aquatic environment are still poorly understood.

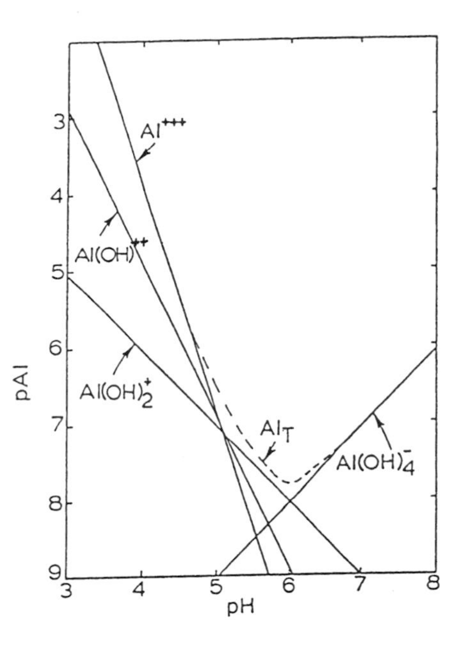

Figure 7 The solubility of Al in equilibrium with $Al(OH)_3$. (From Driscoll, C. T. and Schecher, W. D., *Environ. Geochem. Health,* 12, 28, 1990. With permission.)

It is highly insoluble in natural waters, but under very acidic or alkaline conditions, or in the presence of complexing ligands, it may be found in elevated concentrations in lake waters. During recent years there has been growing concern about the acid-induced mobility of Al in soils and surface waters, and increasing numbers of published articles focus on the question of Al mobility, speciation, and toxicity in acidified fresh waters.[111–117]

Al shows minimum solubility at around pH 6.0. The Al concentration increases drastically when pH decreases below 6.0. Monomeric Al occurs as complexes including H_2O, OH^-, F^-, SO_4^{2-}, HCO_3^-, and organic ligands. Lower pH generally induces an increase of monomeric inorganic Al as aquo ions and positively charged OH-complexes (Figure 7), resulting in increased concentration in acidified lakes.[111,117,118] However, a negative correlation with pH also for organic Al has been demonstrated in lakes in the northeastern U.S., suggesting that leaching of alumino-organic solutes from soil is enhanced under acidic conditions.[117]

The water temperature may have an influence on the distribution of particulate and dissolved Al. The formation and precipitation of particulate $AlOH_3$ has been shown to be faster at 25°C than at 2°C. There is also a shift to higher molecular weight species and a more advanced polymerization at the higher temperature, while the high molecular weight species exist mainly as colloids at 2°C.[119]

Size separation of trace metals in Swedish lake waters has shown that Pb occurs almost exclusively in a nondialyzable fraction (>2 nm), of which a large part is also nonfilterable (>400 nm). Cu is largely found in a fine particulate, nonfilterable fraction. Zn and Cd, on the other hand, are generally found to more than 50% in the dialyzable fraction. Also, As and Co occur mostly in the dialyzable fraction, while Cr is present more in the particulate fraction.[94,99,103,120]

One of the most important factors influencing the distribution and speciation of trace metals in lake waters is the binding to organic substances. In forest lakes, humic and fulvic acids are the main contributors to metal complexation. A number of studies during recent years have addressed the question of determining the degree of complexation of different elements as well as the determination of conditional stability constants for metal organic complexes (see, for example, the books edited by Christman and Gjessing,[121] Kramer and Duinker,[122] Broekaert et al.,[123] and the review by Livens[124]).

The stability of metal-humic complexes in natural water is generally higher than for the corresponding inorganic metal complexes, and major cations, e.g., Ca, are not effective competitive ions against trace metals for organic binding sites.[98] Many different approaches have been used to determine the binding capacity of humic and fulvic acids and the conditional stability constants for different elements (reviewed by Neubecker and Allen[125]). The reported log K_1 values for Cu generally fall within the range 5.5 to 7.0 and the log K_2 values in the range 3.8 to 5.4 for natural waters.[120,126–130] Lövgren and Sjöberg[131] reported log K values of 3.6 to 4.1 for Cu in concentrated bog water. The stability constants for Pb fall within the same range as those for Cu, while Cd is more weakly bound with log K values in the range 2 to 4.[126,128,129,131,132]

Mercury is strongly associated to humic substances in water and has been shown to effectively compete with other metal ions for organic binding sites.[30,131,133,134] The association to organic matter is reflected in the positive correlations found between color values and Hg species in Swedish lake waters. The species associated with dissolved organic matter constitute about 55 to 90% of the total Hg in these lakes[90] and up to 89% of the dissolved Hg in lakes in California.[135]

It should be pointed out that the data on complexing capacity of natural organic substances are difficult to compare as the molecular weight and structure of humic compounds are not very well defined. Further, the calculated equilibrium constants are highly conditional, i.e., very dependent upon the experimental conditions, such as pH, temperature, ionic strength, etc.[136]

The pH of the lake water is an important modifier of the complexation of most elements. Acidification of the water causes not only an increased total concentration of many elements by increased leaching from the watershed (cf. Sections I and IV), but also a change of speciation in the lake water towards more dissolved uncomplexed metal forms. A marked increase of the monomeric inorganic Al forms has been documented from acidified lakes and streams (discussed above). By using *in situ* dialysis, it was possible to show a negative correlation with pH for dialyzable Cd and Al in forest lakes. The proportion of dialyzable Cd increased from around 50% at pH 7.0 to almost 100% in lakes with pH 4.5 to 5.0, and Al increased from about 10 to about 60% in the same lakes.[103]

The dialyzable proportion of Cu did not show any dependence on pH but a negative correlation with organic carbon (TOC) values, illustrating the association to humic substances. The dialyzable Pb fraction showed a similar correlation, with the highest values in the lakes with the lowest TOC values. In these clear-water lakes there was also a negative correlation between pH and dialyzable Pb, which was not found in the more humic waters.[103] Similarly, particulate Pb decreased in the experimentally acidified basin of Little Rock Lake, Wisconsin, compared to the reference basin.[118] These results demonstrate the tendency for Pb to bind in high-molecular nondialyzable compounds in humic lake waters. The same distribution of Cd, Cu, and Pb fractions in lake waters has also been demonstrated in several other studies using ASV measurements,[137,138] ion exchange, and dialysis techniques.[139–141]

Another element which has been shown to increase dramatically in acidified waters is beryllium. The Be concentration at pH 4.0 was about 100 times higher than in weakly alkaline waters in Czechoslovakia. Beryllium occurred in acid waters primarily as free ions and low-molecular fulvates. A positive correlation to fluoride and iron was also demonstrated and the binding to particles increased at higher pH.[142] The authors propose, in view of the analogous behavior of Al and Be in acidified waters, that some of the negative effects on organisms, which so far have been attributed to Al, might have been caused by Be.

The binding of Hg to organic matter does not seem to be influenced by pH in the way described for the elements mentioned above. The adsorption of Hg to organic matter seems instead to increase at lower pH.[28–30,143] In Swedish lake waters, weak negative correlations between pH and Hg fractions were found, owing to pH and water color being strongly intercorrelated. Multiple regression analysis indicated a much stronger influence by water color ($p < 0.0001$) than by pH ($p < 0.02$ to 0.8).[90]

The organic Hg forms detected in lake waters include methyl-Hg.[90,135] The concentration of Me-Hg in lake waters is very low. Some data from Swedish lakes indicate a level of 0.1 to 0.4 ng l^{-1},[144] and 0.04 to 0.06 ng l^{-1} was reported from lakes in the U.S.[145] Nevertheless, the concentration of Me-Hg in water probably has a great influence on the Hg levels in fish, as shown by the significant positive correlation between dissolved organo Hg in water and Hg in fish tissue.[135]

Besides Hg, other elements forming metal-carbon bonds which are stable enough to allow methylmetal species to exist in the aqueous environment include Pb, As, Se, Sn, Ge, Te, and Sb. Methylation proceeds in biological systems but can also take place abiotically. Metal organic compounds with more complex organic groups are also found in nature (e.g., As and Se).[146]

The documentation of the distribution of organo-metal compounds in lake waters is still very sparse, but the recent development of analytical techniques capable of detecting these compounds at natural levels has led to an increased knowledge of at least some of the substances. The determination methods generally utilize a combination of gas or liquid chromatography with furnace atomic absorption spectrometry or ICP-mass spectrometry as detectors.[147–150]

In contrast to mercury, the formation of methyl-element compounds does not always increase the toxicity of the element. Arsenite, for example, is converted through methylation to the less toxic methylarsonic acid and dimethylarsinic acid.[146]

Alkylated Pb compounds are widespread in the environment because of their use as additives in gasoline. The alkyl-Pb species dominating in lake waters seems to be trimethyl-Pb, followed by triethyl-Pb and with other alkylated species accounting for minor parts of the Pb concentration.[151] Whether Pb is actually biomethylated in natural waters is still open to question. The methyl-Pb compounds probably produced (mono- and dimethyl lead(IV) species) generally show poor stability in water and decompose rapidly.[96] The occurrence of methyl-Pb even in remote waters might be a result of long-range airborne transport and degradation of tetraalkyl-Pb from vehicle exhausts.[96,151]

An element occurring in stable organometallic forms and considered to be biomethylated is tin (Sn).[96] It has been widely used as an antifouling agent on boats, in the form of fairly toxic butyl-Sn compounds. The butyl-Sn compounds have been determined in seawater and lake water and show enrichment in sites with heavy shipping traffic, such as harbors and marinas. There is also a considerable enrichment in the surface microlayer of lakes.[152] Also methyl-Sn compounds have been detected in lake waters in concentrations generally below 0.3 $\mu g\ l^{-1}$ with monomethyl-Sn dominating. No direct correlation has been demonstrated between butyl-Sn and methyl-Sn, but both compounds occur in higher concentrations at sites with chemical plants or heavy shipping traffic. The presence of methyl-Sn in remote lakes might be caused by atmospheric transport, but abiotic or biotic methylation of Sn of a natural origin could also be a reason.[152]

Considering the mobility (cf. Section I) of different elements, as well as their ability to form complexes and to adsorb to particles, the following rough classification of the performance of the most studied elements in natural lake waters could be proposed:

1. Elements mainly present in dissolved (dialyzable) form: Zn, Cd, Ni, Mn, and As.
2. Elements mainly present in particulate-bound form: Al, Fe, Pb, and Cr.
3. Elements mainly present as dissolved or colloidal organic complexes: Cu, Pb, Hg, Al, and Fe.
4. Elements very susceptible to pH decrease, concerning mobility, sedimentation, concentration, and speciation: Al, Mn, Zn, Cd, Ni, and Be.
5. Elements occurring in organometal compounds formed by biomethylation: Hg, As, Sn, Se, Pb (?), Tl (?), and Cd (?).

This classification places some elements in more than one group, an example being Al, which occurs largely bound to particulate material in circum-neutral waters, but changes towards dissolved ionic species when the water becomes more acidic. Al, Pb, and Fe are placed in both groups two and three owing to the rather diffuse limit between "dissolved organic complexed" forms and "particulate" forms in lake waters. The organic substances are partly colloidal and particulate, often associated with Fe oxyhydroxides, forming aggregates with humic substances. Consequently, some of the particle-associated metal forms may also include dissolved complexes of metals with humic and fulvic acids.

ACKNOWLEDGMENTS

I am very grateful to Margareta Wigh and Karin Holm, Institute of Applied Environmental Research, for editing the manuscript including the list of references, and for valuable comments, respectively.

REFERENCES

1. **Wetzel, R. G.,** *Limnology,* W. B. Saunders, Philadelphia, 1975, chap. 3.
2. **Johansson, K.,** Heavy metals in acid forest lakes, Report 1359, Swedish Environmental Protection Agency, 1980 (in Swedish).
3. **Johansson, K.,** Heavy metals in Swedish forest lakes — factors influencing the distribution in sediments, Ph.D. thesis, Institute of Limnology, University of Uppsala; *Acta Univ. Ups.,* 1988. 144.
4. **Rognerud, S. and Fjeld, E.,** Landsomfattende undersøkelse av tungmetaller i innsjøsedimenter og kvikksølv i fisk, Report 426/90, Statens forurensningstilsyn, 1990 (in Norwegian).
5. **Verta, M., Tolonen, K., and Simola, H.,** History of heavy metal pollution in Finland as recorded by lake sediments, *Sci. Tot. Environ.,* 87/88, 1, 1989.
6. **Evans, R. D. and Dillon, P. J.,** Historical changes in anthropogenic lead fallout in southern Ontario, Canada, 2nd Int. Sediment/Freshwater Symp., Kingston, Ontario, Nijhoff/Junk Publ., The Hague, 1982.
7. **Evans, H. E., Smith, P. J., and Dillon, P. J.,** Anthropogenic zinc and cadmium burdens in sediments of selected Southern Ontario lakes, *Can. J. Fish. Aquat. Sci.,* 40, 570, 1983.

8. **Johnson, M. G., Culp, L. R., and George, S. E.,** Temporal and spatial trends in metal loadings to sediments of the Turkey Lakes, Ontario, *Can. J. Fish. Aquat. Sci.,* 43, 754, 1986.
9. **Norton, S. A.,** A review of the chemical record in lake sediment of energy related air pollution and its effects on lakes, *Water Air Soil Pollut.,* 30, 331, 1986.
10. **Johansson, K.,** Metals in sediment in lakes in northern Sweden, *Water Air Soil Pollut.,* 47, 441, 1989.
11. **Rühling, Å., Rasmussen, L., Pilegaard, K., Mäkinen, A., and Steinnes, E.,** Survey of atmospheric heavy metal deposition — monitored by moss analyses, The Nordic Council of Ministers, NORD 1987:21, 1987.
12. **Bernes, C. (Ed.),** Tungmetaller—förekomst och omsättning i naturen, Monitor 1987, Swedish Environmental Protection Agency, 1987 (in Swedish).
13. **Ross, H. B. and Granat, L.,** Deposition of atmospheric trace metals in northern Sweden as measured in the snowpack, *Tellus,* 38B, 27, 1986.
14. **Hanssen, J. E., Rambaek, J. P., Semb, A., and Steinnes, E.,** Atmospheric deposition of trace elements in Norway, in *Ecological Impact of Acid Precipitation,* Drablös, D. and Tollan, A., Eds., SNSF-project, Norway, 1980, 116.
15. **Henriksen, A. and Wright, R. F.,** Concentrations of heavy metals in small Norwegian lakes, *Water Res.,* 12, 101, 1978.
16. **Borg, H.,** Trace metals and water chemistry of forest lakes in Northern Sweden, *Water Res.,* 21, 65, 1987.
17. **Verta, M., Mannio, J., Iivonen, P., Hirvi, J-P., Järvinen, O., and Piepponen, S.,** Trace metals in Finnish headwater lakes — Effects of acidification and airborne load, in *Acidification in Finland,* Kauppi, P., et al., Eds., 1990, 883.
18. **Berggren, D., Bergkvist, B., Falkengren-Grerup, U., Folkeson, L., and Tyler, G.,** Metal solubility and pathways in acidified forest ecosystems of South Sweden, *Sci. Tot. Environ.,* 96, 103, 1990.
19. **Tyler, G.,** Leaching of metals from the A-horizon of a spruce forest soil, *Water Air Soil Pollut.,* 15, 353, 1981.
20. **Esser, J. and El Bassam, N.,** On the mobility of cadmium under aerobic soil conditions, *Environ. Pollut. (Series A),* 26, 15, 1981.
21. **Schut, P. H., Evans, D., and Scheider, W. A.,** Variation in trace metal exports from small Canadian shield watersheds, *Water Air Soil Pollut.,* 28, 225, 1986.
22. **Bergkvist, B.,** Soil solution chemistry and metal budgets of spruce forest ecosystems in S. Sweden, *Water Air Soil Pollut.,* 33, 131, 1987.
23. **Lazerte, B., Evans, D., and Grauds, P.,** Deposition and transport of trace metals in an acidified catchment of central Ontario, *Sci. Tot. Environ.,* 87/88, 209, 1989.
24. **Borg, H. and Johansson, K.,** Metal fluxes to Swedish forest lakes, *Water Air Soil Pollut.,* 47, 427, 1989.
25. **Meili, M.,** Mercury in boreal forest lake ecosystems, Ph.D. thesis, Institute of Limnology, University of Uppsala; *Acta Univ. Ups.,* 1991, 336.
26. **Mierle, G.,** Aqueous inputs of mercury to Precambrian shield lakes in Ontario, *Environ. Toxicol. Chem.,* 9, 843, 1990.
27. **Iverfelt, Å. and Johansson, K.,** Mercury in run-off water from small watersheds, *Verh. Int. Verein. Limnol.,* 23, 1626, 1988.
28. **Schindler, D. W., Hesslein, R. H., Wagemann, R., and Broecker, W. S.,** Effects of acidification on mobilization of heavy metals and radionuclides from the sediments of a freshwater lake, *Can. J. Fish. Aquat. Sci.,* 37, 373, 1980.
29. **Jackson, T. A., Kipphut, G., Hesslein, R. H., and Schindler, D. W.,** Experimental study of trace metal chemistry in soft-water lakes at different pH levels, *Can. J. Fish. Aquat. Sci.,* 37, 387, 1980.
30. **Lodenius, M., Seppänen, A., and Uusi-Rauva, A.,** Sorption and mobilization of mercury in peat soil, *Chemosphere,* 12, 1575, 1983.
31. **Nelson, W. O. and Campbell, P. G. C.,** The effects of acidification on the geochemistry of Al, Cd, Pb and Hg in freshwater environments: a literature review, *Environ. Pollut.,* 71, 91, 1991.
32. **Sigg, L., Sturm, M., and Kistler, D.,** Vertical transport of heavy metals by settling particles in Lake Zürich, *Limnol. Oceanogr.,* 32 (1), 112, 1987.
33. **Murray, J. W.,** Mechanisms controlling the distribution of trace elements in oceans and lakes, in *Sources and fates of aquatic pollutants,* American Chemical Society, Washington, D.C., 1987, chap. 6.
34. **Green, W. J., Ferdelman, T. G., Gardner, T. J., Lawrence, C. V., and Angle, M. P.,** The residence times of eight trace metals in a closed-basin Antarctic lake: Lake Hoare, *Hydrobiologia,* 134, 249, 1986.

35. **Diamond, M. L., Mackay, D., Cornett, R. J., and Chant, L. A.,** A model of the exchange of inorganic chemicals between water and sediments, *Environ. Sci. Technol.,* 24, 713, 1990.
36. **Renberg, I.,** Concentration and annual accumulation values of heavy metals in lake sediments: their significance in studies of the history of heavy metal pollution, *Hydrobiologia,* 143, 379, 1986.
37. **Nriagu, J. O. and Wong, H. K. T.,** Dynamics of particulate trace metals in the lakes of Kejimkujik National Park, Nova Scotia, Canada, *Sci. Tot. Environ.,* 87/88, 315, 1989.
38. **Hesslein, R. H. and Broecker, W. S.,** Fates of radiotracers added to a whole lake: sediment-water interactions, *Can. J. Fish. Aquat. Sci.,* 37, 378, 1980.
39. **Morfett, K., Davison, W., and Hamilton-Tylor, J.,** Trace metal dynamics in a seasonally anoxic lake, *Environ. Geol. Water Sci.,* 11, 107, 1988.
40. **Arafat, N. and Nriagu, J. O.,** Simulated mobilization of metals from sediments in response to lake acidification, *Water Air Soil Pollut.,* 31, 991, 1986.
41. **Elzerman, A. W. and Armstrong, D. E.,** Enrichment of Zn, Cd, Pb and Cu in the surface microlayer of Lakes Michigan, Ontario, and Mendota, *Limnol. Oceanogr.,* 24(1), 133, 1979.
42. **Elzerman, A. W., Armstrong, D. E., and Andren, A. W.,** Particulate zinc, cadmium, lead, and copper in the surface microlayer of Southern Lake Michigan, *Environ. Sci. Technol.,* 13, 720, 1979.
43. **Bewers, J. M. and Windom, H. L.,** Comparison of sampling devices for trace metal determinations in seawater, *Mar. Chem.,* 11, 71, 1982.
44. **Sturgeon, R. E. and Berman, S. S.,** Sampling and storage of natural water for trace metals, *CRC Crit. Rev. Anal. Chem.,* 18, 209, 1987.
45. **Freimann, P., Schmidt, D., and Schomaker, K., Mercos** — A simple teflon sampler for ultratrace metal analysis in seawater, *Mar. Chem.,* 14, 43, 1983.
46. **Brügmann, L., Geyer, E., and Kay, R.,** A new teflon sampler for trace metal studies in seawater — 'Wates', *Mar. Chem.,* 21, 91, 1987.
47. **Borg, H.,** Trace metals in Swedish natural fresh waters, *Hydrobiologia,* 101, 27, 1983.
48. **Fabris, J. G., Smith, K. A., Atack, J. E., Hefter, G., and Kilpatrick, A. L.,** Submersible integrating water sampler for heavy metals, *Water Res.,* 20, 1393, 1986.
49. **Harper, D. J.,** A new trace metal-free surface water sampling device, *Mar. Chem.,* 21, 183, 1987.
50. **Rossmann, R. and Barres, J.,** Trace element concentrations in near-surface waters of the Great lakes and methods for collection, storage, and analysis, *J. Great Lakes Res.,* 14(2), 188, 1988.
51. **Borg, H.,** Trace metals in Swedish forest lakes — factors influencing the distribution and speciation in water, Ph.D. thesis, Institute of Limnology, University of Uppsala; *Acta Univ. Ups.,* 1988, 145.
52. **Moody, J. R. and Lindstrom, R. M.,** Selection and cleaning of plastic containers for storage of trace element samples, *Anal. Chem.,* 49, 2264, 1977.
53. **Patterson, C. C. and Settle, D. M.,** The reduction of orders of magnitude errors in lead analysis of biological materials and natural waters by evaluating and controlling the extent and sources of industrial lead contamination introduced during sample collecting and analysis, National Bureau of Standards, Special Publication 422, La Fleur, P.D., Ed., Washington, D.C., 1976, 321.
54. **Orpwood, B.,** Concentration techniques for trace elements: a review, Water Research Centre, Technical report 102, 1979.
55. **Leyden, D. E. and Wegscheider, W.,** Preconcentration for trace element determination in aqueous samples, *Anal. Chem.,* 53, 1059, 1981.
56. **Kingston, H. M., Barnes, I. L., Brady, T. J., Rains, T. C., and Champ, M. A.,** Separation of eight transition elements from alkali and alkaline earth elements in estuarine and seawater with chelating resin and their determination by graphite furnace atomic absorption spectrometry, *Anal. Chem.,* 50, 2064, 1978.
57. **Greenberg, R. R. and Kingston, H. M.,** Trace element analysis of natural water samples by neutron activation analysis with chelating resin, *Anal. Chem.,* 55, 1160, 1983.
58. **Cox, J. A., Slonawska, K., and Gatchell, D. K.,** Metal speciation by Donnan dialysis, *Anal. Chem.,* 56, 650, 1984.
59. **Danielsson, L. G., Magnusson, B., and Westerlund, S.,** An improved metal extraction procedure for the determination of trace metals in sea water by atomic absorption spectrometry with electrothermal atomization, *Anal. Chim. Acta,* 98, 47, 1978.
60. **Bäckström, K.,** Continuous flow extraction for trace metal analysis, thesis, The Royal Institute of Technology, Stockholm, 1990.
61. **Nakashima, S., Sturgeon, R. E., Willie, S. N., and Berman, S. S.,** Determination of trace metals in seawater by graphite furnace atomic absorption spectrometry with preconcentration on silica-immobilized 8-hydroxyquinoline in a flow-system, *Fresenius Z. Anal. Chem.,* 330, 592, 1988.

62. **van Geen, A. and Boyle, E.,** Automated preconcentration of trace metals from seawater and freshwater, *Anal. Chem.,* 62, 1705, 1990.
63. **Görlach, U. and Boutron, C. F.,** Preconcentration of lead, cadmium, copper and zinc in water at the pg g^{-1} level by non-boiling evaporation, *Anal. Chim. Acta,* 236, 391, 1990.
64. **Hall, A. and Godinho, M. C.,** Concentration of trace metals from natural waters by freeze-drying prior to flame atomic absorption spectrometry, *Anal. Chim. Acta,* 113, 369, 1980.
65. **Holm, K., Borg, H., and Korhonen, M.,** Determination of trace metals in natural fresh waters, Report 3629, Swedish Environmental Protection Agency, 1989, (in Swedish, English summary).
66. **Schlemmer, G. and Welz, B.,** Palladium and magnesium nitrates, a more universal modifier for graphite furnace atomic absorption spectrometry, *Spectrochim. Acta,* 41B, 1157, 1986.
67. **Zhe-Ming, N. and Xiao-Quan, S.,** The reduction of matrix interferences in graphite furnace atomic absorption spectrometry, *Spectrochim. Acta,* 42B, 937, 1987.
68. **Fishman, M. J., Perryman, G. R., Schroder, L. J., and Matthews, E. W.,** Determination of trace metals in low ionic strength water using Zeeman and deuterium background correction for graphite furnace atomic absorption spectrometry, *J. Assoc. Off. Anal. Chem.,* 69, 704, 1986.
69. **Houk, R. S.,** Mass spectrometry of inductively coupled plasmas, *Anal. Chem.,* 58, 97, 1986.
70. **Herzog, R. and Dietz, F.,** Anwendung der ICP-MS in der Wasseranalytik, *Gewässerschutz Wasser Abwasser,* 92, 109, 1987.
71. **Gurka, D. F., Betowski, L. D., Hinners, T. A., Heithmar, E. M., Titus, R., and Henshaw, J. M.,** Environmental applications of hyphenated quadrupole techniques, *Anal. Chem.,* 60, 454A, 1988.
72. **McLaren, J. W., Mykytiuk, A. P., Willie, S. N., and Berman, S. S.,** Determination of trace metals in seawater by inductively coupled plasma mass spectrometry with preconcentration on silica-immobilized 8-hydroxyquinoline, *Anal. Chem.,* 57, 2907, 1985.
73. **Beauchemin, D., McLaren, J. W., Mykytiuk, A. P., and Berman, S. S.,** Determination of trace metals in a river water reference material by inductively coupled plasma mass spectrometry, *Anal. Chem.,* 59, 778, 1987.
74. **McLaren, J. W., Beauchemin, D., and Berman, S. S.,** From lithium to uranium, picograms to per cent — the versatility of ICP-MS, Third Chem. Cong. North America, Toronto, Canada, 1988, 125.
75. **Boomer, D. and Bennet, R.,** Development of an ICP/MS based method for analysis of potable water samples, Third Chem. Cong. North America, Toronto, Canada, 1988, 126.
76. **Beauchemin, D. and Berman, S. S.,** Determination of trace metals in reference water standards by inductively coupled plasma mass spectrometry with on-line preconcentration, *Anal. Chem.,* 61, 1857, 1989.
77. **Henshaw, J. M., Heithmar, E. M., and Hinners, T. A.,** Inductively coupled plasma mass spectrometric determination of trace elements in surface water subject to acidic deposition, *Anal. Chem.,* 61, 335, 1989.
78. **Garbarino, J. R. and Taylor, H. E.,** Stable isotope dilution analysis of hydrologic samples by inductively coupled plasma mass spectrometry, *Anal. Chem.,* 59, 1568, 1987.
79. **Borg, H., Holm, K., and Holmgren, K.,** Determination of arsenic with automated hydride generation and flameless atomic absorption spectrometry, Report 1985, Swedish Environmental Protection Agency, 1985 (in Swedish, English summary).
80. **Iverfeldt, Å.,** Structural, thermodynamic and kinetic studies of mercury compounds; applications within the environmental cycle, Ph.D. thesis, Dept. of Inorganic Chem., CTH, Göteborg, Sweden, 1984.
81. **Bloom, N. S. and Fitzgerald, W. F.,** Determination of volatile mercury species at the picogram level by low-temperature gas chromatography with cold-vapor atomic fluorescence detection, *Anal. Chim. Acta,* 208, 151, 1988.
82. **Haraldsson, C., Westerlund, S., and Öhman, P.,** Determination of mercury in natural samples in the sub-nanogram level using inductively coupled plasma/mass spectrometry after reduction to elemental mercury, *Anal. Chim. Acta,* 221, 77, 1989.
83. **Lithner, G.,** Some fundamental relationships between metal toxicity in freshwater, physico-chemical properties and background levels, *Sci. Tot. Environ.,* 87/88, 365, 1989.
84. **Borg, H.,** Background levels of trace metals in Swedish freshwaters, Report 1817, Swedish Environmental Protection Agency, 1984 (in Swedish, English summary).
85. **Steinnes, E., Hovind, H., and Henriksen, A.,** Heavy metals in Norwegian surface waters, with emphasis on acidification and atmospheric deposition, Int. Conf. Heavy Metals in the Environment, Geneva, 1, 36, 1989.

86. **Stephenson, M. and Mackie, G. L.,** Total cadmium concentrations in the water and littoral sediments of central Ontario lakes, *Water Air Soil Pollut.,* 38, 121, 1988.
87. **Laxen, D. P. H.,** Cadmium in freshwaters: concentrations and chemistry, *Freshwater Biol.,* 14, 577, 1984.
88. **Coale, K. H. and Flegal, A. R.,** Copper, zinc, cadmium and lead in surface waters of Lakes Erie and Ontario, *Sci. Tot. Environ.,* 87/88, 297, 1989.
89. **Fitzgerald, W. F. and Watras, C. J.,** Mercury in surficial waters of rural Wisconsin lakes, *Sci. Tot. Environ.,* 87/88, 223, 1989.
90. **Meili, M., Iverfeldt, Å., and Håkansson, L.,** Mercury in the surface water of Swedish forest lakes — concentrations, speciation and controlling factors, in Mercury in Boreal Forest Lake Ecosystems, Meili, M., Ed., Ph.D. thesis, Institute of Limnology, University of Uppsala; *Acta Univ. Ups.,* 1991, 336.
91. **Lithner, G.,** Assessment criteria for lakes and watercourses, Background Document 2, Metals, Report 3628, Swedish Environmental Protection Agency, 1989, (in Swedish, English summary).
92. **White, J. R. and Driscoll, C. T.,** Lead cycling in an acidic Adirondack lake, *Environ. Sci. Technol.,* 19, 1182, 1985.
93. **Byrd, J. E. and Perona, M. J.,** The temporal variations of lead concentration in a freshwater lake, *Water Air Soil Pollut.,* 13, 207, 1980.
94. **Borg, H., Andersson, P., and Johansson, K.,** Influence of acidification on metal fluxes in Swedish forest lakes, *Sci. Tot. Environ.,* 87/88, 241, 1989.
95. **Lindqvist, O.,** Ed., Mercury in the Swedish environment, *Water Air Soil Pollut.,* 55, 1, 1991.
96. **Brinkman, F. E.,** Environmental inorganic chemistry of main group elements with special emphasis on their occurrence as methyl derivatives, in *Environmental Inorganic Chemistry,* Irgolic, K. J. and Martell, A. E., Eds., VCH Publishers, Deerfield Beach, FL, 1985, chap. 9.
97. **Salbu, B.,** Analytical techniques in speciation studies, in *Speciation of Metals in Water, Sediment and Soil Systems,* Landner, L., Ed., Springer-Verlag, Berlin, 1987, 43.
98. **Giesy, J. P., Jr.,** Metal binding capacity of soft, acid, organic-rich waters, *Toxicol. Environ. Chem.,* 6, 203, 1983.
99. **Bene, P. and Steinnes, E.,** *In situ* dialysis for the determination of the state of trace elements in natural waters, *Water Res.,* 8, 947, 1974.
100. **Laxen, D. P. H. and Harrison, R. M.,** A scheme for the physico-chemical speciation of trace metals in freshwater samples, *Sci. Tot. Environ.,* 19, 59, 1981.
101. **Laxen, D. P. H. and Chandler, I. M.,** Size distribution of iron and manganese species in freshwaters, *Geochim. Cosmochim. Acta,* 47, 731, 1983.
102. **Lydersen, E., Bjørnstad, H. E., Salbu, B., and Pappas, A. C.,** Trace element speciation in natural waters using hollow-fiber ultrafiltration, in *Speciation of Metals in Water, Sediment and Soil Systems,* Landner, L., Ed., Springer-Verlag, Berlin, 1987, 85.
103. **Borg, H. and Andersson, P.,** Fractionation of trace metals in acidified fresh waters by *in situ* dialysis, *Verh. Int. Verein. Limnol.,* 22, 725, 1984.
104. **Morrison, G. M. P.,** Approaches to metal speciation analysis in natural waters, in *Speciation of Metals in Water, Sediment and Soil Systems,* Landner, L., Ed., Springer-Verlag, Berlin, 1987, 55.
105. **Hart, B. T.,** A study of the physico-chemical forms of trace metals in natural waters and wastewaters, Australian Water Resources Council, Report 35, 1978.
106. **Borg, H.,** Metal fractionation by dialysis — Problems and possibilities, in *Speciation of Metals in Water, Sediment and Soil Systems,* Landner, L., Ed., Springer-Verlag, Berlin, 1987, 75.
107. **LaZerte, B. D. and Burling, K.,** Manganese speciation in dilute waters of the Precambrian shield, Canada, *Water Res.,* 24, 1097, 1990.
108. **Allard, B., Håkansson, K., and Karlsson, S.,** The importance of sorption phenomena in relation to trace element speciation and mobility, in *Speciation of Metals in Water, Sediment and Soil Systems,* Landner, L., Ed., Springer-Verlag, Berlin, 1987, 99.
109. **Tessier, A. and Campbell, P. G. C.,** Partitioning of trace metals in sediments and its relationship to their accumulation in benthic organisms, in *Metal Speciation in the Environment,* NATO ASI Series G: Ecological Sciences, Vol. 23, Broekart, J. A. C., Gücer, S., and Adams, F., Eds., Springer-Verlag, Berlin, 1990, 545.
110. **Balikungeri, A. and Haerdi, W.,** Complexing abilities of hydrous manganese oxide surfaces and their role in the speciation of heavy metals, *Int. J. Environ. Anal. Chem.,* 34, 215, 1988.
111. **Dickson, W.,** Properties of acidified waters, in *Ecological Impact of Acid Precipitation,* Drabløs, D. and Tollan, A., Eds., SNSF-project, Oslo-Ås, Norway, 1980.

112. **Driscoll, C. T.,** Chemical characterization of some dilute acidified lakes and streams in the Adirondack region of New York State, Ph. D. thesis, Cornell University, 1980.
113. **Baker, J. P. and Schofield, C. L.,** Aluminium toxicity to fish as related to acid precipitation and Adirondack surface water quality, in *Ecological Impact of Acid Precipitation,* Drabløs, D. and Tollan, A., Eds., SNSF-project, Oslo-Ås, Norway, 1980, 292.
114. **Muniz, I. P. and Leivestad, H.,** Toxic effects of aluminium on the brown trout, Salmo trutta L., in *Ecological Impact of Acid Precipitation,* Drabløs, D. and Tollan, A., Eds., SNSF-project, Oslo-Ås, Norway, 1980, 320.
115. **LaZerte, B.,** Forms of aqueous aluminium in acidified catchments of central Ontario: a methodologic analysis, *Can. J. Fish. Aquat. Sci.,* 41, 766, 1984.
116. **Borg, H.,** Metal speciation in acidified mountain streams in central Sweden, *Water Air Soil Pollut.,* 30, 1007, 1986.
117. **Driscoll, C. T. and Schecher, W. D.,** The chemistry of aluminium in the environment, *Environ. Geochem. Health,* 12, 28, 1990.
118. **Mach, C. E. and Brezonik, P. I.,** Trace metal research at Little Rock Lake, Wisconsin: background data, enclosure experiments, and the first three years of acidification, *Sci. Tot. Environ.,* 87/88, 269, 1989.
119. **Lydersen, E., Salbu, B., Poleo, A. B. S., and Muniz, I. P.,** The influences of temperature on aqueous aluminium chemistry, *Water Air Soil Pollut.,* 51, 203, 1990.
120. **Borg, H.,** Speciation of metals in lake waters in the Rönnskär area, Report 3124, Swedish Environmental Protection Agency, 1986, (in Swedish).
121. **Christman, R. F. and Gjessing, E. T., Eds.,** *Aquatic and Terrestrial Humic Materials,* Ann Arbor Science, Ann Arbor, MI, 1983.
122. **Kramer, C. J. M. and Duinker, J. C., Eds.,** *Complexation of Trace Metals in Natural Waters,* Nijhoff/Junk Publ., The Hague, 1984.
123. **Broekaert, J. A. C., Gücer, S., and Adams, F., Eds.,** *Metal Speciation in the Environment,* NATO ASI Series G: Ecological Sciences, 23, Springer-Verlag, Berlin, 1990.
124. **Livens, F. R.,** Chemical reactions of metals with humic material, *Environ. Pollut.,* 70, 183, 1991.
125. **Neubecker, T. A. and Allen, H. E.,** The measurement of complexation capacity and conditional stability constants for ligands in natural waters, *Water Res.,* 17, 1, 1983.
126. **Saar, R. A. and Weber, J. H.,** Fulvic acid: modifier of metal-ion chemistry, *Environ. Sci. Technol.,* 16, 510A, 1982.
127. **Giesy, J. P., Newell, A., and Leversee, G. J.,** Copper speciation in soft, acid, humic waters: effects on copper bioaccumulation by and toxicity to Simocephalus cerrulatus (Daphnidae), *Sci. Tot. Environ.,* 28, 23, 1983.
128. **Steritt, R. M. and Lester, J. N.,** Comparison of methods for the determination of conditional stability constants of heavy metal-fulvic acid complexes, *Water Res.,* 18, 1149, 1984.
129. **Varney, M. S., Turner, D. R., Whitfield, M., and Mantoura, R. F. C.,** The use of electrochemical techniques to monitor complexation capacity titrations in natural waters, in *Complexation of Trace Metals in Natural Waters,* Kramer, C. J. M. and Duinker, J. C., Eds., Nijhoff/Junk Publ., The Hague, 1984.
130. **Giesey, J. P., Alberts, J. J., and Evans, D. W.,** Conditional stability constants and binding capacities for copper(II) by dissolved organic carbon isolated from surface waters of the southeastern United States, *Environ. Toxicol. Chem.,* 5, 139, 1986.
131. **Lövgren, L. and Sjöberg, S.,** Equilibrium approaches to natural water systems. VII. Complexation reactions of copper(II), cadmium(II) and mercury(II) with dissolved organic matter in a concentrated bog-water, *Water Res.,* 23, 327, 1989.
132. **Saar, R. A. and Weber, J. H.,** Complexation of cadmium(II) with water- and soil derived fulvic acides: effect of pH and fulvic concentration, *Can. J. Chem.,* 57, 1263, 1979.
133. **Beneš, P., Gjessing, E. T., and Steinnes, E.,** Interactions between humus and trace elements in fresh water, *Water Res.,* 10, 711, 1976.
134. **Schnitzer, M. and Kerndorf, H.,** Reactions of fulvic acid with metal ions, *Water Air Soil Pollut.,* 15, 97, 1981.
135. **Gill, G. A. and Bruland, K. W.,** Mercury speciation in surface freshwater systems in California and other areas, *Environ. Sci. Technol.,* 24, 1392, 1990.

136. **Perdue, E. M.,** Effects of humic substances on metal speciation, Report EPA/600/01, U.S. Environmental Protection Agency, Athens, GA, 1989.
137. **Chau, Y. K. and Lum-Shue-Chan, K.,** Determination of labile and strongly bound metals in water, *Water Res.,* 8, 383, 1974.
138. **Florence, T. M.,** Trace metal species in fresh waters, *Water Res.,* 11, 681, 1977.
139. **Hart, B. T. and Davies, S. H. R.,** Trace metal speciation in three Victorian lakes, *Aust. J. Mar. Freshwater Res.,* 32, 175, 1981.
140. **Truitt, R. E. and Weber, J. H.,** Copper(II) and cadmium(II)-binding abilities of some New Hampshire freshwaters determined by dialysis titration, *Environ. Sci. Technol.,* 15, 1204, 1981.
141. **Campbell, J. H. and Evans, R. D.,** Inorganic and organic ligand binding of lead and cadmium and resultant implications for bioavailability, *Sci. Tot. Environ.,* 62, 219, 1987.
142. **Vesely, J., Beneš, P., and Sevicik, K.,** Occurrence and speciation of beryllium in acidified freshwaters, *Water Res.,* 23, 711, 1989.
143. **Lodenius, M.,** Factors affecting the mobilization of mercury from soil, in *Heavy Metals in the Environment,* Lindberg, S. E. and Hutchinson, T. C., Eds., Proc. Internat. Conf., New Orleans, 1987.
144. **Lee, Y. H. and Hultberg, H.,** Methylmercury in some Swedish surface waters, *Environ. Toxicol. Chem.,* 9, 833, 1990.
145. **Bloom, N. S. and Watras, C. J.,** Observations of methylmercury in precipitation, *Sci. Tot. Environ.,* 87/88, 199, 1989.
146. **Irgolic, K. J. and Martell, A. E., Eds.,** *Environmental Inorganic Chemistry,* VCH Publishers, Deerfield Beach, FL, 1985.
147. **Chau, Y. K., Wong, P. T. S., and Bengert, G. A.,** Determination of methyltin (IV) and tin (IV) species in water by gas chromatography/atomic absorption spectrophotometry, *Anal. Chem.,* 54, 246, 1982.
148. **Chau, Y. K., Wong, P. T. S., and Kramar, O.,** The determination of dialkyllead, trialkyllead, tetraalkyllead and lead(II) ions in water by chelation/extraction and gas chromatography/atomic absorption spectrometry, *Anal. Chim. Acta,* 146, 211, 1983.
149. **Parks, E. J., Brinckman, F. E., Jewett, K. L., Blair, W. R., and Weiss, C. S.,** Trace speciation by HPLC-GF AA for tin- and leadbearing organometallic compounds, with signal increases induced by transition-metal ions, *Appl. Organometallic Chem.,* 2, 441, 1988.
150. **Marshall, W. D., Blais, J. S., and Adams, F. C.,** HPLC-AAS interfaces for the determination of ionic alkyllead, alkyltin, arsonium and selenonium compounds, in *Metal Speciation in the Environment,* Broekaert, J. A. C., Gücer, S., and Adams, F., Eds., NATO ASI Series G: Ecological Sciences, 23, Springer-Verlag, Berlin, 253, 1990.
151. **van Cleuvenbergen, R. J. A., Marshall, W. D., and Adams, F. C.,** Speciation of organolead compounds by GC-AAS, in *Metal speciation in the Environment,* Broekaert, J. A. C., Gücer, S., and Adams, F., NATO ASI Series G: Ecological Sciences, 23, Springer-Verlag, Berlin, 307, 1990.
152. **Maguire, R. J., Chau, Y. K., Bengert, G. A., and Hale, E. J.,** Occurrence of organotin compounds in Ontario lakes and rivers, *Environ. Sci. Technol.,* 16, 698, 1982.

Chapter 9

Trace Elements in Rivers

Barry T. Hart and Tina Hines

CONTENTS

I. INTRODUCTION

Thc dynamics of trace elements, such as heavy metals and nutrients, are often more complicated in rivers and streams (running water or lotic systems) than in lakes and the oceans. In all aquatic systems, the behavior of materials is coupled with the physical movement of the water; however, in rivers and streams these hydrologic and hydrodynamic factors have a much greater influence on behavior than in other aquatic systems. Rivers and streams are essentially open systems, with allochthonous materials (dissolved and in suspension) transported through them, resulting in a net flux generally in the downstream direction.

Elements or materials in streams are commonly categorized according to their biological or chemical reactivity. Substances used by organisms or otherwise changed biologically or abiotically are called nonconservative (or reactive).[1] Conservative (nonreactive) materials are not altered substantially by either biological demand or chemical reactions. It is important to note that at a particular time or place an element may be reactive, while at another point the same element may be conservative; this is particularly so with some biologically essential nutrients, which may act conservatively at higher concentrations because they are so abundant their concentration is not appreciably changed by biological uptake.

For the purpose of understanding the dynamics of trace elements, a river or stream can be conceptualized as consisting of a number of *compartments,* with transfers of elements and materials between these compartments being possible via a number of *transfer pathways.* This model is known as a *biogeochemical cycle.* In rivers and streams, however, the concept of *material spiraling* has been introduced to indicate that in addition to the abiotic/biotic cycling, materials are transported through the system by the water flow.[1-9]

Figure 1 provides a model of a typical flowing water biogeochemical cycle (or spiral). In this rather simple model, three compartments are assumed to exist. The *water compartment* will consist of elements present in "dissolved" and colloidal forms, with the possibility that there can be interactions between these two forms. The presence of natural dissolved organic matter (DOM) will play an important role in these interactions. For example, the final speciation of copper in the presence of DOM and colloidal matter will depend upon the relative strength of the Cu-DOM and Cu-colloid formation constants and the concentration of binding sites associated with the DOM and the colloidal matter. The *seston compartment* will consist of two components, abiotic SPM transported by the river, and biological material such as bacteria and phytoplankton. Phytoplankton can make up a large proportion of the seston in large, slow moving rivers and regulated rivers (e.g., in the River Murray, Australia, phytoplankton make up approximately 10 to 20% of the total seston concentration [typically 30 to 40 mg/l]), but will normally be a minor

0-8493-6304-7/95/$0.00+$.50

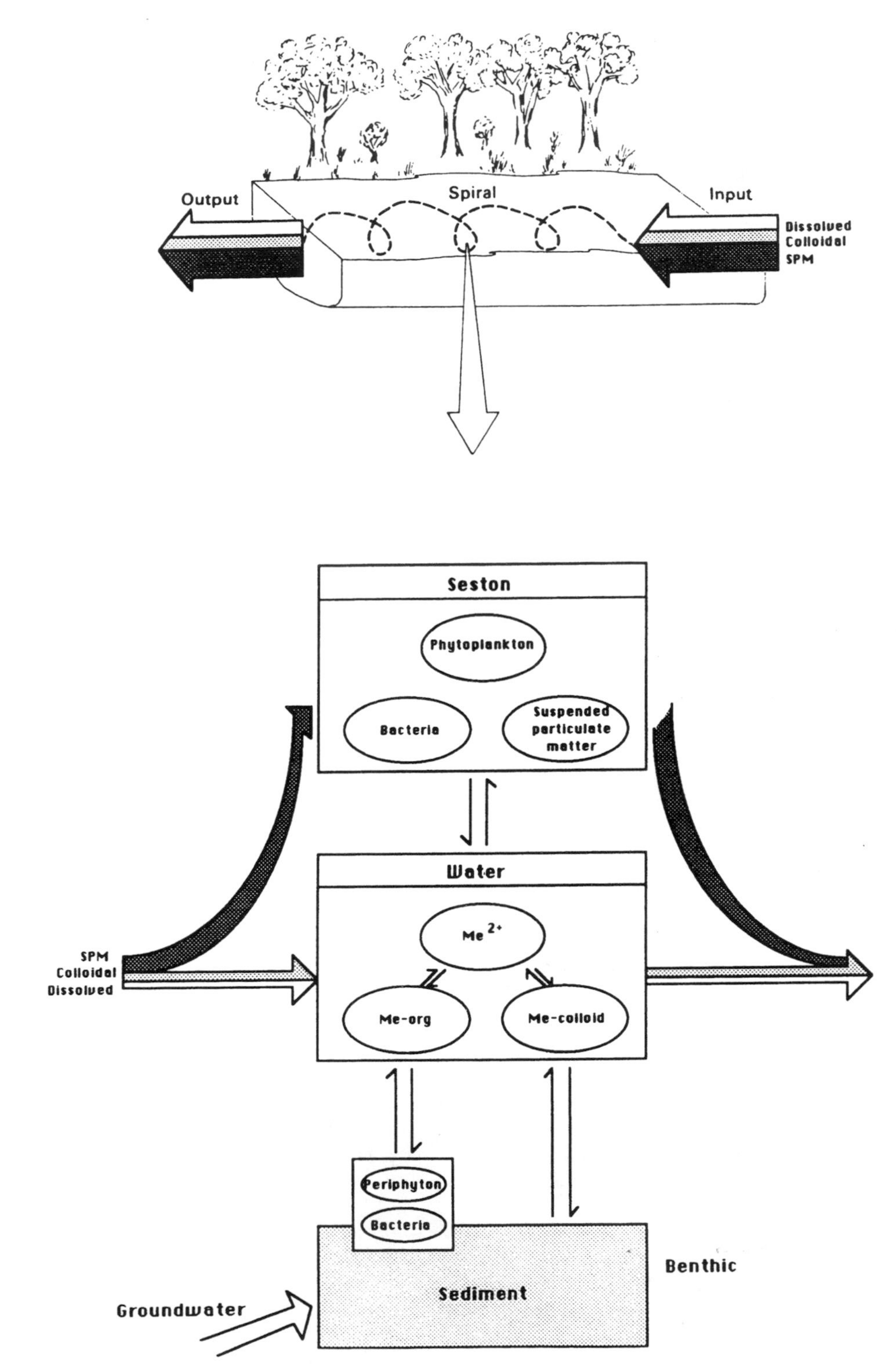

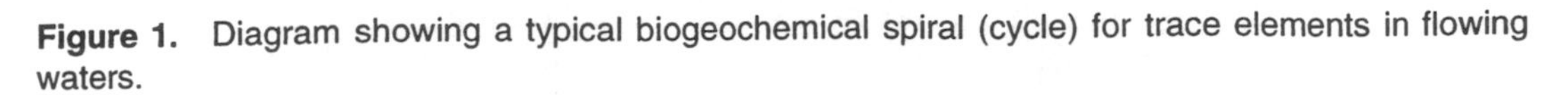

Figure 1. Diagram showing a typical biogeochemical spiral (cycle) for trace elements in flowing waters.

component in most upland streams and in highly turbid rivers. The *benthic compartment* consists of two components, the sediments and the attached algae and microbial films. This biological component is often a dominant part of the benthic compartment in small, rocky upland streams, but will be essentially absent in larger, deep rivers. The actual composition of each of the three compartments, particularly the amount and type of biological material, will depend upon the type of river. For example, in small, fast flowing upland streams, the biota is predominantly associated with the benthic compartment, while in larger rivers the biota will mostly be in the water column in the seston compartment.

In this model, trace elements are assumed to enter and leave each stream segment in three forms — soluble, colloidal, and particulate. Of course, these are quite arbitrary subdivisions, but they do permit us to emphasize some of the differences in reactivity and behavior of trace elements associated with these three fractions. Most discussions of the dynamics of trace elements in rivers to date have assumed that trace elements are present in only two forms, particulate and dissolved. However, as discussed later, there is now good evidence to suggest that an improved understanding of trace element dynamics must also consider the very active colloidal form. The downstream transport processes deliver the trace elements to reactive sites, where exchanges between sites can occur. Exchanges may include chemical transformations (e.g., change in chemical species), sorption and desorption, and biological processes such as algal or microbial uptake, microbial oxidation and reduction, and invertebrate consumption of algae. In small rivers and streams, most of these transfers occur between the water compartment and the benthic compartment; however, in larger systems it is also possible that transfers between the water compartment and the seston compartment will also occur. Transfers in the other direction, from the streambed (or phytoplankton) to the water column, result in release of trace elements to transport. Retention is the net difference between uptake and release and will be flow dependent.[1]

It is possible to broadly classify trace elements in rivers and streams on the basis of the primary factor(s) — biological or physico-chemical — likely to control their behavior. Three broad categories can be considered:

1. *Elements primarily influenced by biological processes* — for example, the behavior of nitrate in natural waters is dominated by biological uptake (by algae and bacteria) which, depending upon the redox conditions, may lead to either increased biomass or loss of nitrogen from the system by denitrification.[10] The behavior of carbon is also mainly controlled by biological processes in aquatic systems. In fact, organic matter (allochthonous and autochthonous, dissolved and particulate) is one of the main driving forces in structuring aquatic ecosystems.[11,12]
2. *Elements primarily controlled by abiotic or geochemical processes* — Cu, Pb, Zn, and Cd are examples; their behavior is highly dependent upon the balance between complexation with dissolved organic matter and association with the surfaces of suspended particulate and colloidal matter; biological uptakes has only a minor influence on their behavior.
3. *Elements that may be influenced by both biological and geochemical processes depending upon the particular aquatic system* — P, Fe, and Mn are examples. Phosphorus can be controlled by both chemical processes (particularly adsorption to iron oxide colloids and particles)[13–15] and biological processes (uptake by bacteria and algae), the final balance being dependent upon the particular characteristics of the stream system.[7] The behaviour of Fe and Mn is very dependent upon the redox conditions existing in the particular system.[16] At low redox potential (anoxic condition), these trace metals are generally present in the soluble, reduced form (Fe(II) and Mn(II), respectively), while under oxic conditions the most stable forms are insoluble hydroxyoxides. Direct microbial[17,18] and photochemical[19,20] reactions can also influence the form of Fe and Mn present.

In this chapter the dynamics of trace elements in rivers and streams is reviewed. The trace elements considered include Fe, Mn, Cd, Cu, Cr, Pb, Zn, As, and Hg, but not the nutrients P, N, and Si, despite their obvious importance in rivers and streams. The review covers the sources, concentrations, and fluxes of trace elements, with a focus on the factors influencing the biogeochemical cycling of trace elements in rivers. The importance of a detailed understanding of the physico-chemical forms or speciation of trace elements is stressed. A number of reviews cover earlier work on the dynamics of trace elements in rivers and streams.[1,7,22–24]

II. CLASSIFICATION OF RIVERS AND STREAMS

The broad term "rivers and streams" covers a very wide range of systems, with essentially one feature in common: they are all flowing water systems. However, rivers and streams can vary considerably in size, altitude, gradient, flow regime, turbulence, temperature, substrate type, turbidity, and water chemistry, being influenced in these features particularly by geographic location and climatic regime. Additionally, a range of activities in the catchment can also influence the associated rivers and streams.[25–27] For example, rivers and streams receive water, solutes, and energy (particularly allochthonous carbon) from their catchment and, also, increasingly large quantities of unwanted contaminants from waste discharges and diffuse runoff.

Obviously, a large number of physical, chemical, and biological features can potentially influence the behavior of materials within a particular river or stream system. For this reason, it would assist the discussion of trace element dynamics if these flowing systems could be classified into a broad set of categories based on their major physical, chemical, and biological features.

Earlier classification methods were based primarily on hydrological considerations. The concept of stream order, developed by Strahler,[28] has been widely used to relate stream characteristics by providing some broad understanding of stream size, watershed size, and in some instances even the quantity of water. However, as pointed out by Hughes and Omernik,[29] there are considerable problems with reliance on stream order only; they recommended using mean annual discharge per unit area and watershed area. Hughes and James[30] used 16 hydrological variables (coefficient of variation of annual, monthly, low, and peak flows) to classify the streams of Victoria, Australia into 5 distinctive and spatially significant regions. These workers noted that this stream regionalization procedure should be most useful for stream ecologists.

Rivers and streams have also been classified on the basis of their chemistry[31,32] and geomorphology.[33]

Consistent with the increasing emphasis worldwide on the protection of aquatic ecosystems, there has been considerable interest in developing more ecologically based classification of rivers and streams. One of note is the river continuum concept (RRC)[12,34] where streams are seen as longitudinally linked systems with processes occurring in the headwaters affecting the downstream system. Edwards et al.[35] has used a geographic information system to classify subcatchments within the Acheron River basin on the basis of areal coverage of catchment attributes known or hypothesized to influence stream processes (e.g., vegetation, slope, geology, rainfall). This multivariate categorization has highlighted the drawbacks of stream order and should allow more objective and accurate selection of sampling sites for studies intended to characterize longitudinal trends.

Obviously, at a broad level the biogeochemical spiralling of trace elements will be influenced by the type of river or stream being studied. The present methods for classifying rivers and streams are generally based on either stream order, hydrological features, or geomorphology, and there is a need for a new classification scheme which includes all these features as well as the biological features.

III. SPECIATION IN RIVERS

Trace elements can exist in a number of physico-chemical forms in natural waters. These include free aquated ionic forms (e.g., Cu^{2+}, $Fe(OH)^{2+}$), "dissolved" inorganic or organic complexes, complexes with colloidal or particulate matter (may be organic and/or organic in nature), and complexes associated with the biota (e.g., phytoplankton, periphyton, bacteria).[22,36–38] The three main approaches used to determine the speciation of trace elements in natural water samples, size separation, electrochemical techniques, and thermodynamic equilibrium calculations,[39] are discussed more extensively in Chapter 2.

The speciation can have a major influence on the behavior, bioavailability, and toxicity of a trace element.[40,41] For example, a trace element in its free ionic form is more likely to become associated with sediments, suspended particulate matter, or colloids than if it is already complexed to natural organic matter.[37] Also, the free ionic form of a number of trace elements (e.g., Cu, Zn, Cd, Pb) is considerably more toxic to aquatic biota than when complexed either to dissolved organic matter or to colloidal or particulate matter.[40] This, however, is not always the case as is illustrated by Hg and Sn, where the methylated forms are considerably more toxic than the ionic forms.[41]

Despite the knowledge that speciation can significantly affect the fate and effects of trace elements, most studies of rivers still use a simple size separation method (based on membrane filtration using 0.4-μm or 0.45-μm filters) to separate the total element concentration into two fractions: the "particulate" fraction that is retained by the filter, and the "dissolved" or "filterable" fraction that passes through the filter. There is, however, increasing evidence that in many river systems a considerable proportion of the fraction passing through a 0.4-μm filter should strictly be regarded as colloidal matter and not in truly "dissolved" forms, and that this colloidal matter is a major transport mechanism for numerous anthropogenic contaminants in surface and groundwater systems.[42–46]

Colloids are small particles (typically 0.001 to 1 μm in diameter) in which the surface free energy dominates the bulk free energy.[42,46] They do not settle under gravity, but remain dispersed for long periods of time. Recent work summarized by Ongley et al.[47] suggests that at least some of this very fine colloidal material may be present *in situ* as large (perhaps tens of μms in diameter), loosely associated flocs; most methods used to sample this material have been shown to break up these flocs.[48] There is now also good evidence that all colloids in aquatic systems are coated with natural organic matter[42] and that this organic coating is the main reason why these natural colloids are negatively charged and very reactive to trace elements and many toxic hydrophobic organic compounds.

The few studies of colloidal matter in rivers have shown that the amount can be significant both in terms of absolute concentration and as a fraction of the total suspended matter. In a study of the Mississippi River, Rees and Ranville[46] found the colloidal matter (size 0.005 to 3 μm) ranged from 0.8 to 348 mg/l or 1 to 50% of the total particulate matter. In a study of a small flood event in Magela Creek in northern Australia, Hart et al.[49] found that the concentration of suspended particulate matter (>1 μm) was highest in the early part of the flood (7.9 to 26 mg/l) and was considerably reduced (to 2.4 mg/l) after the flood peak had passed. The concentration of colloidal matter (size 0.015 to 1 μm) was relatively high and constant during the flood event (8.0 ± 1.2 mg/l; n = 8), and made up between 26 and 74% of the total particulate matter. Similarly, a study of the Murray-Darling River system in Australia showed that the colloidal matter (size 0.003 to 1 μm) ranged from 14 to 17 mg/l (32–42% of the total particulate matter) in the Murray and Ovens Rivers and was 158 mg/l (67%) in the more turbid Darling River (Douglas, unpublished data). Clearly, the concentrations of colloidal matter reported in rivers (1 to 350 mg/l) are of the same order of magnitude as the concentrations reported to occur in marine interstitial waters (1 to 100 mg/l[50]) and groundwaters (up to 63 mg/l[51]).

A number of methods have been employed to separate the colloidal matter from natural waters, with most success being obtained with hollow fiber filtration and tangential flow filtration (TFF; also known as cross-flow filtration).[52–54] In TFF at any one time only a small amount of the sample passes through the filter membrane, with the majority of the sample passing over the membrane and returning to the sample container. Since the sample is continually flushed over the membrane surface, clogging of the membrane is decreased and large volumes of the sample can be filtered. The process results in a colloid or particle concentrate in the sample container and particle-free water which has passed through the membrane. Different sized membranes can be used sequentially to permit several particulate and colloidal fractions to be separated. For example, Douglas et al.[54] recently reported a new method in which continuous flow centrifugation (CFC) and TFF were used sequentially to separate natural waters from the Murray-Darling River system in Australia into 5 fractions; 1 particulate matter fraction (>1 μm, CFC); 3 colloidal fractions (1 μm to 0.2 μm; 0.2 μm to 0.006 μm (100,000 Da); 0.006 μm to 0.003 μm (10,000 Da); TFF), and 1 "dissolved" fraction (<10,000 Dalton).

Aquatic colloidal matter has been shown to complex a range of trace elements and hydrophobic organic compounds[42–46,55,56] and in this way to enhance the transport of contaminants, particularly in groundwaters. In one study, Puls and Powell[51] showed that the rate of colloid-associated arsenate transport was over 20 times greater than that of dissolved arsenate.

An example of the very large concentrations of trace elements that can be associated with the colloidal matter is provided by a study undertaken by the authors into the speciation of Cu, Zn, and Fe in a polluted section of the Tambo River, situated in the Australian Highlands. Hart et al.[57] have shown that this river receives input from a metal-enriched spring through the base of the river over a distance of around 200 m (between –50 and 150 m). Water samples were collected from the river in December 1991 at approximately 50-m intervals over the input region, and separated into 3 operationally defined fractions (particulate, >0.2 μm; colloidal, 0.2 to 0.003 μm; dissolved, <0.003 μm) using TFF; a flow diagram of the separation procedure is shown in Figure 2. There were a number of obvious differences in the speciation of these three elements (Figure 3), in addition to the rather marked increase and then decrease in the absolute concentration of all three elements over the study region. For example, colloidal forms dominated the speciation of Fe and Cu, while soluble forms dominated the speciation of Zn. If the commonly used separation into particulate and dissolved fractions (based on size separation at 0.2 μm) were undertaken for Cu, most would have been assessed to be present as "dissolved" forms, despite the fact that on average 85% of the so-called "dissolved" Cu was actually associated with colloidal forms. The actual form of the Cu and Zn in this colloidal fraction is not known, although it seems likely that the metals would be largely present adsorbed to the amorphous iron oxyhydroxide which makes up a large proportion of the colloidal fraction (ca. 22% [as FeOOH] compared with ca. 8.6% in particulate matter). However, the average surface densities of Cu (0.26 mol Cu/mol Fe) and Zn (0.22 mol/mol Fe) in the colloidal fraction are considerably higher than the maximum values reported in the literature (Cu: 0.012 to 0.03 mol/mol Fe; Zn: 0.020 to 0.03 mol/mol Fe).[58–60] This may mean that some of the Cu and/or Zn is precipitated onto the surface of colloidal particles.

Obviously, the simple classification of trace elements into particulate and dissolved fractions, without consideration of the colloidal fraction, is totally inadequate if a full and detailed understanding of the

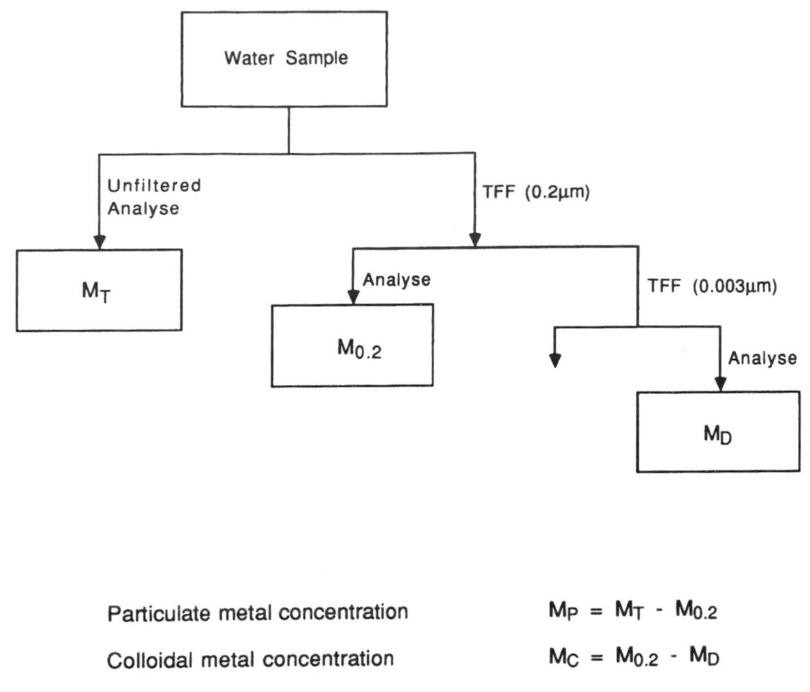

Figure 2. Scheme used to separate particulate, colloidal, and dissolved fractions from natural water samples for trace element analysis.

biogeochemistry of trace elements is required. We have shown that tangential flow filtration is a potentially useful technique for the separation of colloidal matter from natural waters. A disadvantage of this TFF-based separation procedure is that it does not provide information on the detailed speciation of the trace elements in each particular fraction; for example, the proportion of the dissolved fraction present as organic complexes will not be known without further analysis. However, despite this disadvantage, we believe that considerable advances can be made by including an additional separation step to obtain a measure of the trace elements in colloidal forms. A suggested separation scheme is given in Figure 2, in which the broad fractions are designated "dissolved", colloidal, and particulate. A convenient size cutoff between colloidal and particulate is 0.2 μm or 0.4 μm, and between "dissolved" and colloidal around 10,000 Da (or around 3 nm). This speciation scheme is a compromise between being simple enough to be undertaken by a large number of laboratories and yet providing additional information on the speciation of trace elements in rivers.

IV. CONCENTRATIONS, SOURCES, AND FLUXES

A. CONCENTRATIONS

1. Dissolved Trace Elements

Reliable measurements of the concentrations of dissolved trace elements in rivers are needed for calculating oceanic chemical mass balances, understanding continental weathering processes, and evaluating the scale and importance of anthropogenic chemical pollution. However, as Ahlers et al.[61,62] have pointed out, obtaining reliable dissolved trace element concentrations in unpolluted river systems is very difficult to achieve because of problems of sample contamination during collection, handling, and analysis.

Recent data on the dissolved trace element concentrations in a number of uncontaminated river systems around the world are collated in Table 1. Most of these studies used filtration through a 0.4-μm filter to obtain the dissolved fraction. Also shown in this table are the generally accepted background concentrations obtained from older studies. The most striking feature of these data is the very much lower dissolved concentrations in uncontaminated rivers and streams for the most recent studies. For example, Erel et al.[63] found dissolved Pb

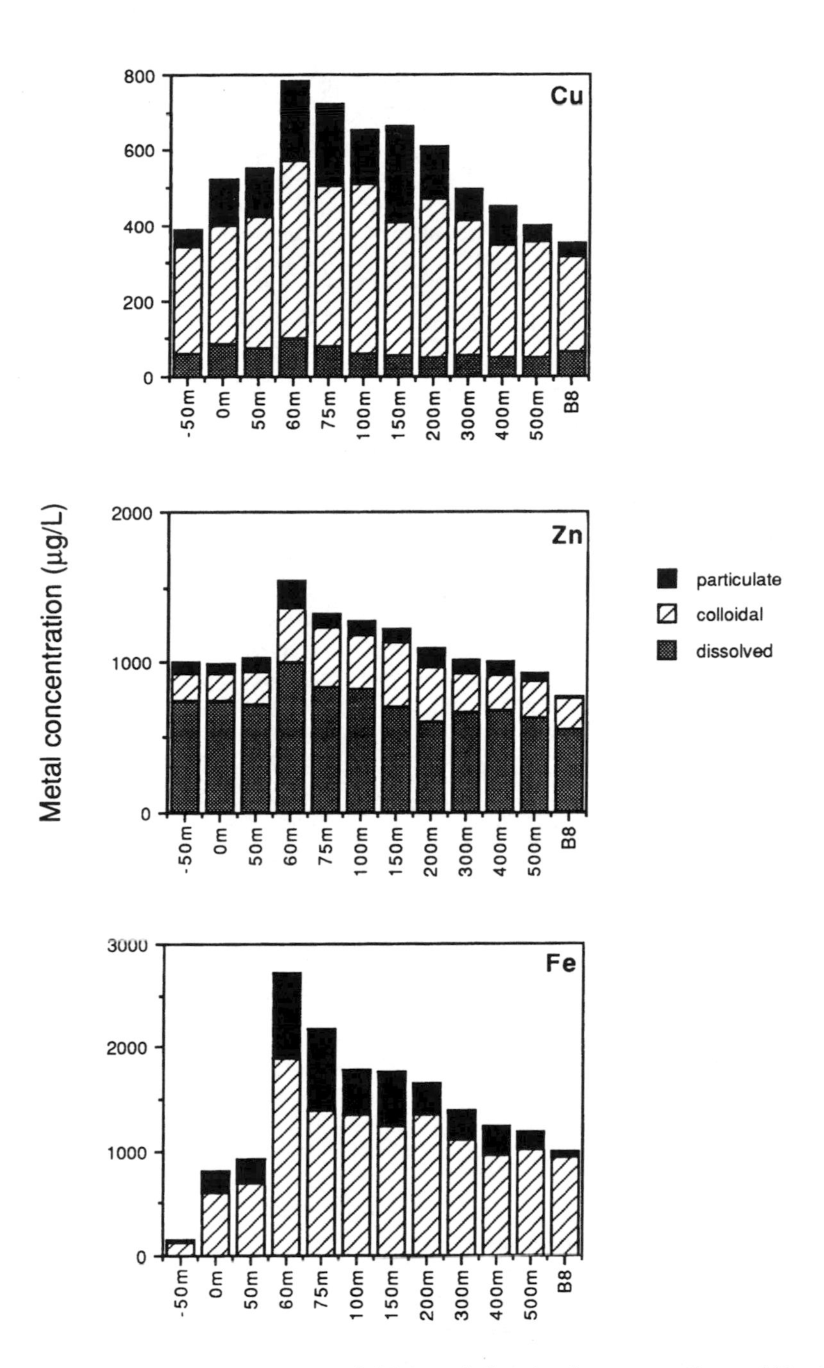

Figure 3. Concentrations of particulate, colloidal, and dissolved concentrations of Fe, Cu, and Zn in a polluted section of the Tambo River, Australia. See Figure 2 for the separation scheme used.

concentrations in a remote mountain stream were over 2 orders of magnitude (9 ng/l cf. 200 ng/l) lower than the value considered by Salomons and Förstner[38] to be the background level, and over 3 orders less than the older value of 1000 ng/l quoted by Martin and Meybeck.[64] Similarly, the most recent concentrations for zinc (ca. 200 ng/l) reported by Shiller and Boyle[65] for uncontaminated rivers from the Amazon and Orinoco basins and the Yangtze river were also considerably lower than background concentrations quoted by Salomons and Förstner.[38]

Concern has previously been expressed[66] that much of the trace elements data published for world rivers and streams prior to the 1980s may be incorrect, with the true concentrations of dissolved trace elements being up to 2 orders of magnitude lower. On the basis of the more recent data, where very great care was taken to avoid contamination, this appears to be true.[61,63,67] This same situation has been found

Table 1 **Dissolved trace element concentrations in uncontaminated rivers**

Element	System	Form	Concentration ng/l	Ref.
Cd	Mountain stream, California	NR[a]	0.3–2.1 (1.2)[b]	63
	Manuherikia River, New Zealand	<0.4 µm	8	62
	Amazon River	<0.4 µm	10–90	71
	Background	NR	20	38
Cu	Manuherikia River, New Zealand	<0.4 µm	150	62
	Niger River	<0.4 µm	140	70
	Negros River	<0.4 µm	320	71
	Amazon River	<0.4 µm	1,200–1,400	70
	Amazon River	<0.4 µm	2,200	71
	Orinoco River	<0.4 µm	1,200	70
	Background	NR	1,000	38
	"World average"	NR	10,000	64
Pb	Mountain stream, California	<0.4 µm	1–24(9)[b]	63
	Manuherikia River, New Zealand	NR	20–30	62
	Southeastern U.S. rivers	<0.4 µm	20–500	72
	Background	NR	200	38
	"World average"	NR	1,000	64
Zn	Magela Creek, Australia	<0.4 µm	87–121	67
	Manuherikia River, New Zealand	<0.4 µm	150–200	62
	Amazon basin, Orinoco basin, Yangtze	<0.4 µm	20–1,800 (~200)[b]	65
	Background	NR	500–10,000	38
	"World average"	NR	30,000	64
Ni	Manuherikia River, New Zealand	<0.4 µm	100–150	62
	Yangtze River	<0.4 µm	130	70
	Orinoco River	<0.4 µm	290	70
	Amazon River	<0.4 µm	320	70
	Background	NR	300	38
	"World average"	NR	2,200	64
Hg	Manuherikia River, New Zealand	[c]	0.3	62
	Eight rural lakes, Wisconsin	[c]	0.2–0.5	134
	Background	NR	10	38

[a] NR, not recorded; [b] Range (mean); [c] Sample not filtered, total Hg released by borohydride reduction.

in marine chemistry where the accepted dissolved trace element concentrations in ocean waters have decreased by orders of magnitude over the past 15 years or so as improved sampling and analytical methods have been introduced.[61]

The need for laboratories analyzing trace elements in waters, sediments, and biota to employ adequate quality assurance procedures is continually being stressed.[61,68,69] However, it now seems clear that greater care needs to be taken, and perhaps improved techniques introduced, generally in trace element studies to ensure that contamination at the point of sampling does not occur.

2. Particulate Trace Elements

The role of SPM in the biogeochemical cycling of trace elements in fluvial systems is well established.[38,73] On the basis of rather few data a number of "truths" have become fashionable, in particular: (1) it is only the smaller fraction (<63 µm) of the SPM that is geochemically important, and (2) the trace metal concentrations in SPM are primarily controlled by grain size (as grain size decreases [and surface area increases] trace element concentration normally increases) and the bulk chemical composition of the SPM. The most important SPM-trace element concentrators are Fe and Mn oxides and hydroxides and organic matter (and in some cases clay minerals).[73–75,77,78]

The "particulate" trace element concentrations reported in the literature are considerably more variable than the dissolved concentrations for a number of reasons:

1. Sampling: There is still considerable controversy about the "best" method to sample fluvial cross sections to determine the SPM concentrations. Infrequent sampling, poor selection of cross-sectional sampling sites, and/or missed high flow events can lead to substantial over- or underestimation of SPM concentrations. In a study of six U.S. rivers using depth- and width-integrated point and pump sampling, Horowitz et al.[76] concluded that in sampling rivers where there is a lack of detailed knowledge regarding SPM concentrations, grain size distributions and cross-sectional heterogeniety, it would be best to use depth- and width-integrated sampling.
2. SPM fraction: A number of workers have concluded that only the <63 μm fraction needs to be considered in determining the role of the SPM. This will, of course, depend upon the objective of the survey. If, for example, the annual SPM-associated trace element transport in a particular river is required, recent work by Horowitz et al.[76] shows that one cannot ignore the contribution from the >63 μm fraction. However, if the objective was to assess the trend in an anthropogenically enriched river system, one would expect that most of the increase over background would occur in the finer <63 μm fraction, in which case it would be sensible to concentrate on this fraction exclusively. Consideration also needs to be given to the lower size limit used to define SPM. A wide range of size limits have been used, although generally the cut is considered to be that achieved with a glass fiber filter (such as that used to determine suspended solids or nonfiltrable residue, size ca. 0.8 to 1.0 μm) or a membrane filter (generally 0.4 μm or 0.2 μm). As was discussed earlier, these separation methods will result in a (major) proportion of the colloidal matter being included in the filtrate and therefore considered as "dissolved".
3. Variations with flow: Walling and Moorehead[79] have shown that both the amount and size characteristics of the SPM transported by rivers can vary widely with flow. In broad terms, the transport and cross-sectional distribution of the coarse SPM (sand-sized) is controlled by discharge, while the transport and vertical and horizontal distribution of the finer SPM (salt/clay) is homogeneous with transport controlled by supply to the river.[80] However, in their review, Walling and Moorehead[79] clearly show that the factors controlling the delivery and transport of this SPM are complex and generally not well understood. The flow dependence of both the amount and characteristics of the SPM has implications for the "particulate" trace element fraction. For example, if the amount of SPM increases dramatically with flow, as is often observed to occur in many small rivers and streams during the very early part of a flood event, and this is missed because of the sampling frequency, the flux of "particulate" trace elements will be underestimated. It is possible that, depending upon the source material, the composition of the SPM, and particularly the trace element concentrations, will also vary during flood events. As is emphasized below, the flux of the trace element transported by the river may be considerably underestimated if the high flow events are not adequately sampled.
4. Analytical method used: The amount of a trace element associated with the SPM will vary depending on the analytical method used. For example, different amounts of Cu, Pb, and Zn can be liberated by HNO_3 acidification compared with more complete digestion using $HNO_3/HClO_4/HF$. Again, the selection of the digestion procedure will depend upon the purpose of the investigation. Complete digestion will be needed if the main purpose is to determine the geochemistry of particular trace elements. However, if the purpose of the investigation is to assess the transport of biologically available (or potentially biologically available) elements, less rigorous digestion procedures would be desirable.

B. SOURCES

Trace elements can enter rivers and streams from a number of sources, including natural and anthropogenic sources within the catchment and directly from effluent discharges.

Natural sources of trace elements include the general weathering of soils and base rocks in the catchment[31,81,82] and atmospheric inputs. In most cases, weathering is a relatively slow process, being influenced by rainwater made slightly acidic through the solution of CO_2 from the atmosphere.[16] As yet there is no comprehensive rationale for predicting the dissolution rates of the common rock- and soil-forming minerals (but see Reference 83).

There have been few studies of the origins of trace elements in natural catchments. Stallard and Edmonds[84–86] studied the geochemistry of the Amazon Basin and found that substrate lithology and erosional regime (seen in terms of transport-limited and weathering-limited denudation) exerted the most fundamental control on the elemental concentrations in this system. The effects of atmospheric salt inputs and biological uptake and release were secondary in this system, but this may not be the case in all systems. However, this and most other studies on the natural sources of dissolved elements in rivers and streams focus mainly on the major cations and anions (e.g., Na, K, Ca, Mg, Cl, SO_4, HCO_3) and not on the trace elements. In a recent study, Casey and Westrich[87] provided evidence for the hypothesis that the mechanisms of mineral dissolution are linked to the mechanisms by which dissolved elements exchange ligands. For the first row transition metals, dissolution rates of orthosilicate minerals correlate with the ligand-field stabilization energies (e.g., $Zn(H_2O)_6^{2+} = Mn(H_2O)_6^{2+} > Fe(H_2(H_2O)_6^{2+} > Co(H_2O)_6^{2+} > Ni(H_2O)_6^{2+}$), suggesting that the dissolution rate is controlled by the character of the bonds between the element and neighboring oxygen atoms.

The other main natural sources of trace elements to the catchment will be that from the atmosphere, in both wet and dry precipitation. This is covered fully in Chapter 5 and will not be repeated here, except to note that with the increased pollution of the atmosphere two effects are apparent. First, there is now an increased input of trace elements from the atmosphere to almost all catchments. Obviously, the absolute amount deposited is greatest in those catchments close to urban and industrial centers. However, since atmospheric transport can occur over great distances, it seems likely that all catchments around the world are now experiencing greater inputs of trace elements than occurred in the past.

The second effect is the generally increased acidity of rain in many parts of the world.[88,89] Acidic waters are defined as those having acid neutralizing capacity (ANC) ≤ 0 equiv/l.[89] Operationally, rain with a pH of less than 5.6 (lowest value for pure water in equilibrium with atmospheric CO_2) is defined as "acid rain"; recent work by Likens et al.[90] suggests that for unpolluted areas in the southern hemisphere, a reference value of around 5.0 is more appropriate. There is little question that the acidic deposition — generally caused by the increased emissions of sulfur and nitrogen oxides — has caused acidification of many surface waters in the northern hemisphere.[91] A number of studies have also shown that this increased acidic deposition has accelerated the "natural" weathering process in a number of catchments. This effect has been particularly noticeable for Al and in poorly buffered catchments.[92,93]

The major human source of trace elements, apart from that in effluents, is that added to agricultural land. This source is particularly important for nutrients and toxic organic compounds, but can also be important for trace elements.[38,73,94]

Trace elements may also be directly discharged to rivers and streams in a variety of effluents. In many parts of the world, anthropogenic discharges are the major contributors of trace elements to rivers, and, for this reason, environmental protection authorities in most countries are placing increasingly stringent requirements on the trace element concentrations permitted in effluents discharged directly into rivers and streams. It is not the purpose of this chapter to review recent water quality guidelines; however, the interested reader is referred to USEPA,[95] CCREM,[96] and Hart et al.[97] for more details on this subject. Additionally, a number of case studies of rivers polluted by trace elements resulting from waste discharges are reviewed by Salomons and Förstner.[38]

C. TRACE ELEMENT FLUXES IN FLUVIAL SYSTEMS

The paths followed by rainfall from the time it strikes the land surface until it appears as streamflow have a major influence on the chemical composition of the stream water, since the detailed water pathways may exert a controlling influence on weathering reactions and drainage water chemistry.[99] There are a variety of ways in which trace elements may be transported through the catchment to the stream, and as was noted in Section III, these transport pathways will be influenced by and will influence the physico-chemical form (speciation) of the element.

Hydrologists generally identify three sources of water entering a stream: overland flow, groundwater, and subsurface stormflow (i.e., streamflow derived from subsurface sources that arrives quickly enough to become part of the storm hydrograph). The contributions from these different sources will vary with a number of factors, including time, topography, soil type, plant cover, climate, hydrologic conditions, storm intensity, and antecedent conditions.[99] When the streamflow is low, the major source of water will be groundwater, and at higher streamflows, a greater proportion will be from subsurface stormflow and possibly overland flow. In a recent review of streamflow generation processes, Pearce[100] concluded that although there have been many studies that "have greatly increased the conceptual understanding of flow generation, there has not been a corresponding increase in the ability to model or predict flow processes."

The physico-chemical forms of the elements in these three source waters can be quite different. The surface runoff will contain much higher concentrations of eroded SPM and colloidal matter. The predominant elements in subsurface stormflow will be those that can be rapidly transferred into the aqueous phase by processes such as ion exchange and leaching, whereas in groundwaters the weathering of bedrock will be a more important source of trace elements because of the longer residence times during which these reactions can occur. Two other mechanisms can influence the concentrations and fluxes of stream solutes during small storms: (1) leaching from riparian vegetation and from organic debris that accumulates in parts of the channel that are dry during low flow, and (2) flushing of solutes concentrated in soil water in the riparian zone.[101]

Recent work has also shown that subsurface stormflow and groundwater, as well as overland flow, can contain colloidal matter and dissolved organic matter, and that this material can significantly increase the amounts of trace elements transported to the river or stream.[43–45,51,55] This latter point has been fully covered in Section III.

The role of flood events in transporting trace elements in rivers has probably been underestimated in many areas of the world, resulting in a corresponding underestimate of the global flux of trace elements from world rivers to the oceans.[49] For example, Cullen et al.[102] found that almost 70% of the total phosphorus transported to Lake Burley Griffin in Canberra, Australia during an 18-month study occurred during flood events which made up only 9% of the time. Similarly, Cosser[103] found that the 86% of the total phosphorus transported by the South Pine River in Brisbane, Australia over a 24-month study period occurred during a small number of flood events (which took up ca. 3% [20 days] of the time). In a recent study of herbicide transport in the Cedar River, Illinois, Squillace and Thurman[104] reported that 94% of the annual load of atrazine was transported with overland flow and only 6% with the groundwater component of the flow.

There are fewer studies devoted to the transport of trace elements. However, it seems unlikely that the situation would be significantly different from that illustrated above for phosphorus and herbicide transport. Hart et al.[105] in a study of a major flood event in the Annan River in North Queensland, Australia showed that the major amount of the heavy metals were transported in particulate forms (Fe 99%; Mn 95%; Pb, Zn, Sn ca. 80%; Cu ca. 60%); filterable metal concentrations were low and changed little with flow. Since the flux of particulate matter generally increases markedly during flood events, it follows that a major increase in the flux of trace elements will also occur during higher flows. Others have also observed higher concentrations of hydrogen ions and dissolved aluminum during storm events in acid-rain affected streams in north Europe and North America (see Reference 106 for summary).

We are not aware of any studies where the flux of colloid-associated trace elements has been determined for a river or stream system. Although two studies[46,49] have found that the concentration of colloidal matter was not correlated with river flow, it has yet to be determined whether this is a general feature of the transport of colloidal matter.

The evidence suggests that in many rivers and streams, a high proportion (perhaps most) of the trace elements transported annually will occur during a small number of high flow events, mostly associated with SPM (and colloidal matter) and not in dissolved forms. Clearly, if meaningful trace element budgets are to be obtained, it is essential that considerable effort be expended in sampling during these high flow events. This is not a trivial exercise since it is difficult to predict and sample flood events.

V. PROCESSES CONTROLLING ELEMENT DYNAMICS IN RIVERS

As was noted in the introduction, the behavior of trace elements in aquatic systems is complicated and difficult to predict,[38,107] and this is particularly so in rivers and streams where physical (hydrological), chemical, and biological interactions must be coupled if the behavior of trace elements is to be understood and eventually predicted. Figure 1 shows the major interactions that can occur between the three main compartments containing trace elements in a river system.

The scientific literature contains a large amount of information on the behavior of trace elements in the aquatic environment. These, however, are dominated by laboratory-based studies of trace element speciation and the interactions of elements with a variety of substrates (e.g., ligands, colloids, SPM), by reports of the levels of trace elements in waters, sediments, and biota, and by laboratory studies of metal toxicity to aquatic organisms.[108] There are relatively few field studies that have investigated specifically the changes in the biogeochemistry of reactive trace elements that can occur as a result of the wide range of natural variations that are characteristic of stream ecosystems.

There have, however, been a number of studies reported which have concentrated on the main processes controlling the dynamics of nutrients in streams. The main emphasis has been on phosphorus because it appears to be the limiting nutrient in many streams.[3,4,6,109,110] The dynamics of nitrogen has also been well studied,[111–113] particularly in arid zone streams where it appears that nitrogen rather than phosphorus is the nutrient limiting primary production.[114] The behavior of carbon in streams has also been well studied, with the input and processing of particulate organic carbon having received the most attention because of its key role in the river continuum concept,[12,34] where streams are seen as longitudinally linked systems with processes occurring in the headwaters having an impact downstream. The application of the river continuum concept to hydrologically diverse rivers, such as exist in many regions of Australia, has been questioned.[30,65] More recently, the lateral linkages between the stream and its riparian zone (taken to include the floodplain) have been shown to be as important as the longitudinal linkages emphasized in the river continuum concept. This is particularly so in blackwater rivers, such as the Ogeechee River in Georgia, where Meyer[116] has shown the importance of dissolved organic matter input to the river from floodplain swamps on a range of ecosystem processes.

There have been few studies of the processes controlling the dynamics of trace metals in river systems,[117–122] and those that have been undertaken have mostly been associated with acid mine drainage. It is questionable as to how helpful these studies will be to improving the understanding of trace element behavior in unpolluted systems.

Most studies of trace elements in aquatic systems assume that chemical equilibrium exists.[38] However, as Morgan and Stone[123] point out, the rates (kinetics) of transfer of chemical species can vary enormously depending upon whether the reaction is homogeneous or heterogenous and whether it is catalyzed by organisms, other aqueous species, light, or surfaces. For biochemically essential elements (e.g., Cu, Zn, Fe, and Mn), the kinetics of removal from and return to the water column are strongly influenced by rates of biological processes such as photosynthesis, respiration, nitrate reduction, and sulfate reduction. McKnight and Bencala,[122] in discussing the modeling of the responses of iron and other heavy metals in acidic streams, suggest that the spatial variability in the physical properties of different reaches of the stream and the spatial and seasonal variability in the nature of the reacting hydrous iron oxides in the stream bed may represent greater challenges than the uncertainties in the kinetics of different chemical processes.

There is now considerable evidence showing that trace element behavior in aquatic systems is significantly influenced by sorption reactions involving colloidal matter, SPM, and sediments. Fox[13–15] has recently shown that the dissolved phosphorus concentration in large, turbid rivers is largely controlled by sorption to iron hydroxyoxide colloidal and suspended particulate matter. In these turbid systems, biological processes, such as uptake by bacteria and algae, would be minimal.

However, despite the obvious importance of sorption reactions in controlling the behavior of trace elements in natural waters, Honeyman and Santschi[107] have pointed out that our ability to model trace element transport is still very limited. This arises largely from a lack of detailed quantitative understanding of the factors that influence the distribution of trace elements between dissolved and sorbed phases in real systems, factors that will include pH, nature and concentration of sorption sites, site heterogeneity, kinetic effects, and particle-particle interactions. Although the theory relating the sorption of trace elements onto model oxide surfaces is now generally quite well developed, there are still many unknowns that prevent this theory from being used to explain metal sorption reactions in natural systems. For example, sorption sites in "real" particles are heterogeneous; sorption kinetics are often very much slower than expected (days to weeks cf. hours for model systems) and there are anomalous effects of particle concentrations (e.g., K_ds can be inversely related to particle concentrations).[107] Honeyman and Santschi[124,125] have recently postulated a new model for the sorption process which explains some of these effects. This model (called "Brownian pumping") assumes that truly dissolved metal species are transferred to the particulate phase through a colloidal intermediate; the process involves two steps: (1) rapid formation of metal-colloid surface site complexes (i.e., adsorption), and (2) slow agglomeration and coagulation of these colloids to larger particles.

It is clear that our understanding of the sorption process in natural systems requires a more complete knowledge of the surface chemistry of the natural colloidal and particulate matter, the particle size distribution of the entire colloidal and particulate range, and the interactions between particles (e.g., agglomeration). A relatively new separation technique of field flow fractionation (FFF) shows considerable promise for providing this information.[126,127]

Field studies that have shed light on the mechanisms involved in trace element dynamics in flowing waters have generally employed one of two techniques. The first approach involves a controlled perturbation of the stream by the addition of a single heavy metal or mixtures of metals, subsequent sampling over time to measure the changes in concentration, and often the use of an hydrological/chemical model to simulate the behavior of the metals. For example, Kuwabara et al.[117] added Cu to Convict Creek, a small stream in the Sierra Nevada in California, and showed that the subsequent behavior of the Cu was controlled by hydrological (advection, dispersion), biological (periphyton uptake), and chemical (sediment sorption) processes. A major problem with this approach is the difficulty in obtaining approval from the relevant authorities to add toxic metals to the stream.

The other approach employs intensive *in situ* studies of streams that receive either acid-mine effluent or acidic runoff from naturally mineralized areas.[58,119,121,122,128–131] Chapman et al.[119] reported an interesting study of a small acidic (pH 2.5 to 3.5) stream in New South Wales, Australia that received acid mine drainage. Biologically mediated oxidation of Fe(II) reduced its concentration by over an order of magnitude (from ca. 1.5 m*M*), and resulted in precipitation of amorphous iron hydroxide along with small quantities of potassium and lead jarosites ($M^{n+}_{1/n}Fe_3(SO_4)_2(OH)_6$). The majority of the Al, As, and sulfate lost from the water column was probably sorbed onto the amorphous iron hydroxide. Not surprisingly given the low pH, there was little change in the concentrations of Cu and Zn over the stream section studied. Chapman et al.[119] attributed the observed changes in pH and Cu and Zn concentrations mainly to dilution by seeps and tributaries. Filipek et al.[131] studied the input of acid mine drainage to the West Squaw Creek, a California stream draining igneous rocks of low acid-neutralizing capacity. Above the acidic inputs, the creek had a pH of 6.2 and low concentrations of dissolved metals. Over a distance of approximately 5 km, an almost equal volume of acid-sulfate drainage water entered the creek, reducing the pH to around 2. Although the stream contained an ubiquitous orange X-ray amorphous precipitate, probably a mixture of ferric hydroxides and jarosites or ferric hydroxides with adsorbed sulfate, Cu, Zn, Mn, and Al were all found to remain in solution rather than precipitating or sorbing onto the solid phase.

In a study of the Cannon River, a river in southwest England, exhibiting a wide pH range from 7.5 above to around 3.1 below where acidic mine water enters it, Johnson[58] found that sorption by amorphous iron hydroxyoxide particles (taken as the particulate iron concentration) controlled the concentration of dissolved copper and zinc.

McKnight and co-workers[121,122,128] have also used acidic streams in the Colorado Rocky Mountains to study the behavior of dissolved and particulate iron. In the Snake River, a stream receiving acidic inputs from the weathering of disseminated pyrite, McKnight and Bencala[121] showed that the diel variation in dissolved iron concentration was due to the photoreduction of hydrous iron oxides which are abundant in the fine sediments and as coatings on rocks in the stream. In another acidic stream in Colorado, McKnight et al.[128] showed that the daytime production of ferrous iron by photoreduction was almost four times as great as the nighttime oxidation of ferrous iron. In another experiment in the Snake River, Colorado, McKnight and Bencala[122] lowered the pH from 4.2 to 3.2 for a period of 3 h and observed a simultaneous increase in ferrous iron concentration. The chemical processes involved were probably rapid increases in dissolved ferric iron by dissolution of hydrous iron oxide and then rapid reduction to ferrous iron by a solution phase photochemical reaction. At least two different hydrous iron oxide compartments were implicated; one was more readily dissolved but was in trace quantities, and the other was relatively abundant but was less readily dissolved.

The few attempts to model the transport of trace elements in rivers have been concerned with rather small rivers or creeks.[1,24,119,132] Solute transport models attempt to couple the physical transport processes (e.g., advection, dispersion, groundwater inputs) with the chemical (e.g., precipitation, sorption, complexation) and biological (e.g., microbial uptake, algal uptake) transformations that can take place in the system. Generally, the chemical and biological transformation processes are poorly known and need to be approximated with rather gross simplifications. A more detailed discussion of the current models can be found in Bencala and Walters,[132] Kuwabara and Helliker,[24] and the Stream Solute Workshop.[1] However, despite the often gross simplifications made, these simulations can assist considerably in attempts to assess the major processes controlling trace elements in rivers and streams. Even rather coarse information on the relative importance of chemical and biological uptake and release processes under different conditions can be of considerable help in improving the understanding of the controlling processes and in establishing better management guidelines.

In the future, the most promising approaches to explaining the behavior of trace elements in rivers will be those that more completely couple the physical, chemical, and biological components occurring in the system.[133]

VI. CONCLUSIONS

The dynamics of trace elements in flowing waters (rivers and streams) is complicated. In addition to chemical and biological uptake and release processes, the trace elements will also be physically transported downstream by the water flow. This behavior can be thought of as an extended biogeochemical cycle, or better spiral, in which the trace elements reside in three major compartments (water, seston, and benthic) with transfers occurring between these compartments depending upon the physical, chemical, and biological conditions at the time. It is, therefore, not surprising that the detailed understanding of the dynamics of trace elements in such systems is still in its infancy.

Obviously, at a broad level the biogeochemical spiralling of trace elements will be influenced by the type of river or stream being studied. For example, in a large turbid river the spiralling would be most influenced by interactions with particulate and colloidal matter, whereas in a stony upland stream, interactions with attached algae and microbial films attached to rocks would dominate. The present methods available for classifying rivers and streams are generally based on either stream order, hydrological features, or geomorphology. There is a need for a new classification scheme which includes all these features as well as the biological features.

Trace elements can exist in a number of physico-chemical forms in natural waters, and it is well known that the form or speciation can have a marked influence on the behavior and effect of the element. Evidence is provided showing that the colloidal fraction (size range approximately 1 μm to 3 nm), which is often poorly characterized, can complex a range of trace elements and hydrophobic organic compounds and enhance their transport. However, despite this evidence it is still common for trace element studies to only determine two fractions, particulate and "dissolved", based on filtration through a 0.4-μm membrane filter. Reliable and easy-to-use methods based on tangential flow filtration are now available to obtain the colloidal fraction, and it is recommended that future trace element studies should collect three fractions, namely particulate, colloidal, and dissolved. This will considerably assist attempts to better understand the biogeochemical spiraling of trace elements in flowing waters.

There is now convincing evidence that most of the trace element transport in rivers occurs during the high flow events. However, these events are often either not sampled or poorly sampled so that many of the trace element budgets quoted in the literature are probably seriously underestimated. This calls into question the present global fluxes of trace elements by the world rivers. On the other hand, there is increasing evidence that suggests that the concentrations of many trace elements in uncontaminated rivers may be considerably lower than originally thought. In fact, these background concentrations may be up to two orders of magnitude lower than presently accepted. This work reemphasizes the need for rigorous quality assurance procedures in trace element studies, not only in the laboratory but also in the sampling to ensure that unwanted contamination does not occur.

Conceptually, the main processes likely to influence the behavior of trace elements in rivers are well known from the many laboratory studies that have been undertaken. However, this laboratory-based information is not easily transferable to explain (quantitatively) the actual behavior in real systems. The development of a number of mathematical models, which attempt to couple the hydrology with the chemical and biological uptake processes, have led to some advances in understanding the relative importance of the different processes. However, most of the field studies on trace elements have been on river systems polluted with acid mine drainage, because of the difficulties in undertaking whole-stream release experiments. In the future, the most promising approaches to explaining the behavior of trace elements in rivers will be those that more completely couple the physical, chemical, and biological components occurring in the system.

ACKNOWLEDGMENTS

We are grateful to Macquarie Resources Ltd for providing funding for the Tambo River investigation and for the postgraduate scholarship awarded to TH. Dr. Ron Beckett provided constructive comments on an earlier draft.

REFERENCES

1. **Stream Solute Workshop,** Concepts and methods for assessing solute dynamics in stream ecosystems, *J. N. Am. Benthol. Assoc.*, 9, 95, 1990.
2. **Elwood, J. W., Newbold, J. D., O'Neill, R. V., and Van Winkle, W.,** Resource spiralling: an operational paradigm for analysing lotic ecosystems, in *Dynamics of Lotic Ecosystems,* Fontine, T. D. and Bartell, S. M., Eds., Ann Arbor, Sci. Publ., Ann Arbor, MI, 1983, 3.
3. **Hart, B. T., Freeman, P., McKelvie, I. D., Pearse, S., and Ross, G. R.,** Phosphorus spiralling in Myrtle Creek, Victoria, Australia, *Verh. Int. Verein. Limnol.,* 24, 2065, 1990.
4. **Hart, B. T., Freeman, P., and McKelvie, I. D.,** Whole-stream phosphorus release studies: variation in uptake length with initial phosphorus concentration, *Hydrobiology,* 235/236, 573, 1992.

5. **Mulholland, P. J., Newbold, J. D., Elwood, J. W., and Ferren, L. A.,** Phosphorus spiralling in a woodland stream: seasonal variations, *Ecology,* 66, 1012, 1985.
6. **Mulholland, P. J., Steinman, A. D., and Elwood, J. W.,** Measurement of phosphorus uptake length in streams: comparisons of radiotracer and stable PO_4 releases, *Can. J. Fish. Aquatic Sci.,* 47, 235, 1990.
7. **Meyer, J. L., McDowell, W. H., Bott, T. L., Elwood, J. W., Ishizaki, C., Melack, J. M., Peckarsky, B. L., Peterson, B. J., and Rublee, P. A.,** Elemental dynamics in streams, *J. N. Am. Benthol. Soc.,* 7, 410, 1988.
8. **Newbold, J. D., Elwood, J. W., O'Neill, R. V., and Van Winkle, W.,** Measuring nutrient spiralling in streams, *Can. J. Fish. Aquatic Sci.,* 38, 860, 1981.
9. **Newbold, J. D., O'Neill, R. V., Elwood, J. W., and Van Winkle, W.,** Nutrient spiralling in streams: implications for nutrient limitation and invertebrate activity, *Am. Nat.,* 120, 628, 1982.
10. **Nielsen, L. P., Christensen, P. B., and Revsbech, N. P.,** Denitrification and photosynthesis in stream sediment studied with microsensor and whole-core techniques, *Limnol. Oceanogr.,* 35, 1135, 1990.
11. **Cummins, K. W.,** Structure and function of stream ecosystems, *BioScience,* 24, 631, 1974.
12. **Minshall, G. W., Cummins, K. W., Peterson, R. C., Cushing, C. E., Burns, D. A., Sedell, J. R., and Vannote, R. L.,** Developments in stream ecosystem theory, *Can. J. Fish. Aquatic Sci.,* 42, 1045, 1985.
13. **Fox, L. E.,** A model for inorganic control of phosphate concentrations in river waters, *Geochim. Cosmochim. Acta,* 53, 417, 1989.
14. **Fox, L. E.,** Geochemistry of dissolved phosphate in the Sepik River and Estuary, Papue, New Guinea, *Geochim. Cosmochim. Acta,* 54, 1019, 1990.
15. **Fox, L. E.,** Phosphorus chemistry in the tidal Hudson River, *Geochim. Cosmochim. Acta,* 55, 1529, 1991.
16. **Stumm, W. and Morgan, J. J.,** *Aquatic Chemistry: An Introduction Emphasizing Chemical Equilibria in Natural Waters,* John Wiley & Sons, New York, 1981.
17. **Chapnick, S. D., Moore, W. S., and Nealson, K. H.,** Microbially mediated manganese oxidation in a freshwater lake, *Limnol. Oceanogr.,* 27, 1004, 1982.
18. **Sunda, W. G. and Huntsman, S. A.,** Microbial oxidation of manganese in a North Carolina estuary, *Limnol. Oceanogr.,* 32, 552, 1987.
19. **Bertino, D. J. and Zepp, R. G.,** Effects of solar radiation on manganese oxide reactions with selected organic compounds, *Environ. Sci. Technol.,* 25, 1267, 1991.
20. **Waite, T. D. and Morel, F. M. M.,** Photoreductive dissolution of colloidal iron oxides in natural waters, *Environ. Sci. Technol.,* 18, 860, 1984.
21. **Waite, T. D.,** Photochemistry of colloids and surfaces, in *Surface and Colloid Chemistry in Natural Waters and Water Treatment,* Beckett, R., Ed., Plenum Press, Melbourne, 1990, 27.
22. **Hart, B. T.,** Uptake of trace metals by sediments and suspended particulates: a review, *Hydrobiology,* 91, 299, 1982.
23. **Fontaine, T. D. and Bartell, S. M.,** *Dynamics of Lotic Ecosystems,* Ann Arbor Sci. Publ., Ann Arbor, MI, 1983.
24. **Kuwabara, J. S. and Helliker, P.,** Trace contaminants in streams, in *Civil Engineering Practice,* Vol. 5, *Water Resources/Environmental,* Cheremisinoff, P. N., Cheremisinoff, N. P., and Cheng, S. L., Eds., Technomic Publishing Co., Lancaster, PA, 1988, chap. 26.
25. **Hynes, H. B. N.,** The stream and its valley, *Verh. Int. Verin. Theoret. Angew. Limnol.,* 19, 1, 1975.
26. **Likens, G. E.,** Beyond the shoreline: a watershed ecosystem approach, *Verh. Int. Verein. Limnol.,* 22, 1, 1984.
27. **Wetzel, R. G.,** Land-water interfaces: metabolic and limnological regulators, *Verh. Int. Verein. Limnol.,* 24, 6, 1990.
28. **Strahler, A. N.,** Quantitative analysis of watershed geomorphology, *Trans. Am. Geophys. Union,* 38, 913, 1957.
29. **Hughes, R. M. and Omernik, J. M.,** An alternative for characterising stream size, in *Dynamics of Lotic Ecosystems,* Fontaine, T. D. and Bartell, S. M., Eds., Ann Arbor Sci. Publ., Ann Arbor, MI, 1983, 87.
30. **Hughes, J. M. R. and James, B.,** A hydrological regionalisation of streams in Victoria, Australia, with implications for stream ecology, *Aust. J. Mar. Freshwater Res.,* 40, 303, 1989.
31. **Gibbs, R. J.,** Mechanisms controlling world water chemistry, *Science,* 170, 1088, 1970.
32. **Hart, B. T. and McKelvie, I. D.,** Chemical limnology in Australia, in *Limnology in Australia,* De Deckker, P. and Williams, W. D., Eds., CSIRO, Melbourne, Australia, 1986.

33. **Selby, M. J.,** *Earth's Changing Surface: An Introduction to Geomorphology,* Oxford University Press, Oxford, 1985.
34. **Vannote, R. L., Minshall, G. W., Cummins, K. W., Sedell, J. R., and Cushing, C. E.,** The river continuum concept, *Can. J. Fish. Aquatic Sci.,* 37, 130, 1980.
35. **Edwards, R. E., Poulson S., and Hart, B. T.,** Use of GIS for stream ecology and river management, in preparation.
36. **Florence, T. M.,** The speciation of trace elements in waters, *Talanta,* 29, 345, 1982.
37. **Pickering, W. F.,** Metal ion speciation — soil and sediments (a review), *Ore Geol. Rev.,* 1, 83, 1986.
38. **Salomons, W. and Förstner, U.,** *Metals in the Hydrosphere,* Springer-Verlag, Heidelberg, 1984.
39. **Batley, G. E.,** *Trace Element Speciation: Analytical Methods and Problems,* CRC Press, Boca Raton, FL, 1989.
40. **Florence, T. M.,** Trace element speciation and aquatic toxicology, *Trends Anal. Chem.,* 2, 162, 1983.
41. **Hart, B. T.,** *Australian Water Quality Criteria for Heavy Metals,* A.W.R.C. Tech. Paper 77, Australian Government Publishing Service, Canberra, 1982.
42. **Beckett, R.,** *Surface and Colloid Chemistry in Natural Waters and Water Treatment,* Plenum Press, Melbourne, 1990.
43. **Dunnivant, F. M., Jardine, P. M., Taylor, D. L., and McCarthy, J. F.,** Cotransport of cadmium and hexachlorobiphenyl by dissolved organic carbon through columns containing aquifer material, *Environ. Sci. Technol.,* 26, 360, 1992.
44. **McCarthy, J. F. and Zachara, J. M.,** Subsurface transport of contaminants, mobile colloids in the subsurface environment may alter the transport of contaminants, *Environ. Sci. Technol.,* 23, 496, 1989.
45. **McDowell-Boyer, L. M.,** Chemical mobilization of micron-sized particles in saturated porous media under steady state, *Environ. Sci. Technol.,* 26, 586, 1992.
46. **Rees, T. F. and Ranville, J. F.,** Comparison of colloidal materials transported in the Mississippi River system during low-flow and high-flow conditions, U.S. Geological Survey, 1990.
47. **Ongley, E. D., Krishnappan, B. G., Droppo, I. G., Rao, S. S., and Maguire, R. J.,** Cohesive sediment transport: emerging issues for toxic chemical management, in *Sediment-Water Interactions* vol. 5, Hart, B. T. and Sly, P. G., Eds., Kluwer Academic Publishers, Dordrecht, 1992.
48. **Droppo, I. G. and Ongley, E. D.,** The state of suspended sediment in the freshwater fluvial environment: a method of analysis, *Water Res.,* 26, 65, 1992.
49. **Hart, B. T., Douglas, G. B., Beckett, R., Van Put, A., and Van Grieken, R. E.,** Characterisation of colloidal and particulate matter in the Magela Creek system, northern Australia, *Hydrol. Process.,* 7, 105, 1993.
50. **Chin, T.-P. and Gschwend, P. M.,** The abundance, distribution, and configuration of porewater organic colloids in recent sediments, *Geochim. Cosmochim. Acta,* 55, 1309, 1991.
51. **Puls, R. W. and Powell, R. M.,** Transport of inorganic colloids through natural aquifer material: implications for contaminant, *Environ. Sci. Technol.,* 26, 614, 1992.
52. **Horowitz, A. J., Elrick, K. A., and Hooper, R. C.,** A comparison of dewatering methods for the separation and concentration of suspended sediment for sequential trace element analysis, *Hydrol. Process.,* 2, 163, 1988.
53. **Kuwabara, J. S. and Harvey, R. W.,** Application of a hollow fibre, tangential flow device for sampling suspended bacteria and particles from natural waters, *J. Environ. Qual.,* 19, 625, 1990.
54. **Douglas, G. B., Beckett, R., and Hart, B. T.,** Separation and fractionation of suspended particulate matter from rivers, *Hydrol. Process.,* 7, 177, 1993.
55. **Chiou, G. T., Kile, D. E., Brinton, T. I., Malcolm, R. L., Leenheer, J. A., and MacCarthy, P.,** A comparison of water solubility enhancements of organic solutes by aquatic humic materials and commercial humic acids, *Environ. Sci. Technol.,* 21, 1231, 1987.
56. **Backhus, D. A. and Gschwend, P. M.,** Fluorescent polycyclic aromaic hydrocarbons as probes for studying the impact of colloids on pollutant transport in groundwater, *Environ. Sci. Technol.,* 25, 1214, 1990.
57. **Hart, B. T., Sdraulig, S., and Jones, M. J.,** Behaviour of copper and zinc added to the Tambo River, Australia by a metal-enriched spring, *Aust. J. Mar. Freshwater Res.,* 43, 457, 1992.
58. **Johnson, C. A.,** The regulation of trace element concentrations in river and estuarine waters contaminated with acid mine drainage: the adsorption of Cu and Zn on amorphous Fe oxyhydroxides, *Geochim. Cosmochim. Acta,* 50, 2433, 1986.
59. **Millward, G. E. and Moore, R. M.,** The adsorption of Cu, Mn and Zn by iron oxyhydroxide in model estuarine systems, *Water Res.,* 16, 981, 1982.

60. **Tessier, A., Rapin, F., and Carignan R.,** Trace metals in oxic lake sediments: possible adsorption onto iron oxyhydroxides, *Geochim. Cosmochim. Acta,* 49, 183, 1985.
61. **Ahlers, W. W., Reid, M. R., Kim, J. P., and Hunter, K. A.,** Contamination-free sample collection and handling protocols for trace elements in natural fresh waters, *Aust. J. Mar. Freshwater Res.,* 41, 713, 1990.
62. **Ahlers, W. W., Kim, J. P., and Hunter, K. A.,** Dissolved trace metals and their relationship to major elements in the Manuherikia River, a pristine subalpine catchment in central Otago, New Zealand, *Aust. J. Mar. Freshwater Res.,* 42, 409, 1991.
63. **Erel, Y., Morgan, J. J., and Patterson, C. C.,** Natural levels of lead and cadmium in a remote mountain stream, *Geochim. Cosmochim. Acta,* 55, 707, 1991.
64. **Martin, J. M. and Meybeck, N.,** Elemental mass-balance of material carried by major world rivers, *Mar. Chem.,* 7, 173, 1979.
65. **Shiller, A. M. and Boyle, E.,** Dissolved zinc in rivers, *Nature,* 317, 49, 1985.
66. **Martin, J. M., Knauer, G. A., and Flegal, A. R.,** in *Zinc in the Environment,* Nriagu, J. O., Ed., John Wiley & Sons, New York, 1980, 193.
67. **LeGras, C. A. A. and Noller, B. N.,** *The Determination of Zinc in Magela Creek,* ARRRI Technical Memorandum, 24, Australian Government Publishing Service, Canberra, 1989.
68. **Adeloju, S. B. and Bond, A. M.,** Influence of laboratory environment on the precision and accuracy of trace analysis, *Anal. Chem.,* 57, 1728, 1985.
69. **Bloom, N. S.,** On the chemical form of mercury in edible fish and marine invertebrate tissue, *Can. J. Fish. Aquat. Sci.,* 49(15), 1010, 1992.
70. **Grant, B., Ming-Hu, H., Boyle, E. A., and Edmonds, J. M.,** Comparison of the trace metal chemistry in the Amazon, Orinoco and Yangtze plumes, *EOS: Trans. Am. Geophys. Union,* 63, 48, 1982.
71. **Boyle, E. A., Huested, S. S., and Grant, B.,** The chemical mass-balance of the Amazon plume, II, copper, nickel and cadmium, *Deep-Sea Res.,* 29, 1355, 1982.
72. **Windom, H. L. and Smith, R. G.,** The geochemistry of lead in rivers, estuaries and the continental shelf of the southeastern United States, *Mar. Chem.,* 17, 43, 1985.
73. **Föstner, U. and Wittmann, G. T. W.,** *Metal Pollut. Aquatic Environ.,* 2nd ed., Springer-Verlag, Heidelberg, 1981.
74. **Horowitz, A. J. and Elrick, K. A.,** The relation of stream sediment surface area, grain size and composition to trace element chemistry, *Appl. Geochem.,* 2, 437, 1987.
75. **Horowitz, A. J., Elrick, K. A., and Hooper, R. P.,** The prediction of aquatic sediment-associated trace element concentrations using selected geochemical factors, *Hydrol. Process.,* 3, 347, 1989.
76. **Horowitz, A. J., Rinella, F. A., Lamothe, P., Miller, T. L., Edwards, T. K., Roche, R. L., and Rickert, D. A.,** Variations in suspended sediment and associated trace element concentrations in selected riverine cross sections, *Environ. Sci. Technol.,* 24, 1313, 1990.
77. **Jenne, E.,** Controls of Mn, Fe, Co, Ni, Cu and Zn concentrations in soils and water: the significance of hydrous Fe and Mn oxides, *ACS Adv. Chem. Series,* 73, 337, 1968.
78. **Smith, R. W. and Jenne, E. A.,** Recalculation, evaluation, and prediction of surface complexation constants for metal adsorption on iron and manganese oxides, *Environ. Sci. Technol.,* 25, 525, 1991.
79. **Walling, D. E. and Moorehead, P. W.,** The particle size characteristics of fluvial suspended sediments: an overview, in *Sediment/Water Interactions,* Sly, P. G. and Hart, B. T., Eds., Kluwer Academic Publishers, Belgium, 1989, 125.
80. **Vanoni, V. A.,** *Sedimentation Engineering,* American Society of Civil Engineers, New York, 1977.
81. **Gibbs, R. J.,** Mechanisms of trace metal transport in rivers, *Science,* 180, 71, 1973.
82. **Drever, J. I.,** *The Geochemistry of Natural Waters,* 2nd ed., Prentice-Hall, Englewood Cliffs, NJ, 1988.
83. **Brady, P. V. and Walther, J. V.,** Control of silicate dissolution rate in neutral and basic pH solutions at 25°C, *Geochim. Cosmochim. Acta,* 53, 2823, 1989.
84. **Stallard, R. F. and Edmonds, J. M.,** Geochemistry of the Amazon. I. Precipitation chemistry and the marine contribution to the dissolved load at the time of peak discharge, *J. Geophys. Res.,* 86, 9844, 1981.
85. **Stallard, R. F. and Edmonds, J. M.,** Geochemistry of the Amazon. II. The influence of geology and weathering environment on the dissolved load, *J. Geophys. Res.,* 88, 9671, 1983.
86. **Stallard, R. F. and Edmonds, J. M.,** Geochemistry of the Amazon. III. Weathering chemistry and limits to dissolved inputs, *J. Geophys. Res.,* 92, 8293, 1987.

87. **Casey, W. H. and Westrich, H. R.,** Control of dissolution rates of orthosilicate minerals by divalent metal-oxygen bonds, *Nature,* 355, 157, 1992.
88. **Charles, D. F.,** *Acidic Deposition and Aquatic Ecosystems: Regional Case Studies,* Springer-Verlag, New York, 1991.
89. **Baker, L. A., Herlihy, A. T., Kaufmann, P. R., and Eilers, J. M.,** Acidic lakes and streams in the United States: the role of acidic deposition, *Science,* 252, 1151, 1991.
90. **Likens, G. E., Keene, W. C., Miller, J. M., and Galloway, J. N.,** Chemistry of precipitation from a remote, terrestrial site in Australia, *J. Geophys. Res.,* 92, 13299, 1987.
91. **Baker, L. A., Herlihy, A. T., Kaufman, P. R., and Eilers, J. M.,** Acid lakes and streams in the United States: the role of acid deposition, *Science,* 252, 1151, 1991.
92. **Driscoll, C. T., Yatsko, C. P., and Unangst, F. J.,** Longitudinal and temporal trends in the water chemistry of the north branch of the Moose River, *Biogeochemistry,* 3, 37, 1987.
93. **Lawrence, G. B., Driscoll, C. T., and Fuller, R. D.,** Hydrologic control of aluminium chemistry in an acidic headwater stream, *Water Resour. Res.,* 24, 659, 1988.
94. **Adriano, D. C.,** *Trace Elements in the Terrestrial Environment,* Springer-Verlag, New York, 1986.
95. **USEPA,** *Quality Criteria for Water — 1986,* U.S. Environmental Protection Agency, Washington, D.C., 1986.
96. CCREM, *Canadian Water Quality Guidelines,* Canadian Council of Resource and Environment Ministers, Inland Waters Directorate, Environment Canada, Ottawa, 1991.
97. **Hart, B. T., Angehrn-Bettinazzi, C., Campbell, I. C., and Jones, M. J.,** Australian water quality guidelines: a new approach for protecting ecosystem "health", *J. Aquat. Ecosystem Health,* 2, 151, 1993.
98. **Hornung, M., Adamson, J. K., Reynolds, B., and Stevens, P. A.,** Influence of mineral weathering and catchment hydrology on drainage water chemistry in three upland sites in England and Wales, *J. Geol. Soc. London,* 143, 627, 1986.
99. **Kennedy, V. C., Kendall, C., Zellweger, G. W., Wyerman, T. A., and Avanzino, R. J.,** Determination of the components of stormflow using water chemistry and environmental isotopes, Mattole river basin, California, *J. Hydrol.,* 84, 107, 1986.
100. **Pearce, A. J.,** Streamflow generation processes: an austral view, *Water Resour. Res.,* 26, 3037, 1990.
101. **Mulholland, P. J., Wilson, G. V., and Jardine, P. M.,** Hydrogeochemical response of a forested watershed to storms: effects of preferential flow along shallow and deep pathways. *Water Resourc. Res.,* 26, 3021, 1990.
102. **Cullen, P., Rosich, R. S., and Bek, P.,** *A Phosphorus Budget for Lake Burley Griffin and Management Implications for Urban Lakes,* AWRC Technical Paper No. 31, Australian Government Publishing Service, Canberra, 1978.
103. **Cosser, P. R.,** Nutrient concentration-flow relationships and loads in the South Pine River, south-eastern Queensland. I. Phosphorus loads, *Aust. J. Mar. Freshwater Res.,* 40, 613, 1989.
104. **Squillace, P. J. and Thurman, E. M.,** Herbicide transport in rivers: importance of hydrology and geochemistry in non-point source contamination, *Environ. Sci. Technol.,* 26, 538, 1992.
105. **Hart, B. T., Day, G., Sharp-Paul, A., and Beer, T.,** Water quality variations during a flood event in the Annan River, north Queensland, *Aust. J. Mar. Freshwater Res.,* 39, 225, 1988.
106. **Sullivan, T. J., Christophersen, N., Muniz, I. P., Seip, H. M., and Sullivan P. D.,** Aqueous aluminium chemistry response to episodic increases in discharge, *Nature,* 323, 324, 1986.
107. **Honeyman, B. D. and Santschi, P. H.,** Metals in aquatic systems; predicting their scavenging residence times from laboratory data remains a challenge, *Environ. Sci. Technol.,* 22, 862, 1988.
108. **Hart, B. T. and Lake, P. S.,** Studies of heavy metal pollution in Australia with particular emphasis on aquatic systems, in *Occurrence and Pathways of Lead, Mercury, Cadmium and Arsenic in the Environment,* Hutchinson, T. C. and Meema, K. M., Eds., SCOPE Series, John Wiley & Sons, Chichester, 1986, 187.
109. **Newbold, J. D., Elwood, J. W., O'Neill, R. V., and Sheldon, A. L.,** Phosphorus dynamics in a woodland stream ecosystem: a study of nutrient spiralling, *Ecology,* 64, 1249, 1983.
110. **Munn, N. L. and Meyer, J. L.,** Habitat-specific solute retention in two small streams: an intersite comparison, *Ecology,* 71, 2069, 1990.
111. **Newbold, J. D., Elwood, J. W., Schulze, M. S., Stark, R. W., and Barmeier, J. C.,** Continuous ammonium enrichment of a woodland stream: uptake kinetics, leaf decomposition and nitrification, *Freshwater Biol.,* 13, 193, 1983.

112. **Triska, F. J., Kennedy, V. C., Avanzino, R. J., Zellweger, G. W., and Bencala, K. E.,** Retention and transport of nutrients in a third-order stream: channel processes, *Ecology,* 70, 1877, 1989.
113. **Triska, F. J., Kennedy, V. C., Avanzino, R. J., Zellweger, G. W., and Bencala, K. E.,** Retention and transport of nutrients in a third-order stream: hyporheic processes, *Ecology,* 70, 1893, 1989.
114. **Grimm, N. B. and Fisher, S. G.,** Nitrogen limitation in a Sonoran Desert stream, *J. N. Am. Benthol. Soc.,* 5, 2, 1986.
115. **Lake, P. S., Barmuta, L. A., Boulton, A. J., Campbell, I. C., and St Clair, R. M.,** Australian streams and northern hemisphere stream ecology: comparisons and problems, *Proc. Ecol. Soc. Aust.,* 14, 61, 1987.
116. **Meyer, J. L.,** A blackwater perspective on riverine ecosystems, *BioScience,* 40, 643, 1990.
117. **Kuwabara, J. S., Leland, H. V., and Bencala, K. E.,** Copper transport along a Sierra Nevada stream, *J. Environ. Eng.,* 110, 646, 1984.
118. **Jackman, A. P., Walters, R. A., and Kennedy, V. C.,** Transport and concentration controls for chloride, strontium, potassium and lead in Uvas Creek, a small cobble-bed stream in Santa Clara County, California, U.S.A. II. Mathematical modeling, *J. Hydrol.,* 75, 111, 1984.
119. **Chapman, B. M., Jones, D. R., and Jung, R. F.,** Processes controlling metal ion attenuation in acid mine drainage streams, *Geochim. Cosmochim. Acta,* 47, 1957, 1983.
120. **Chapman, B. M., Jones, D. R., and Jung, R. F.,** Heavy metal transport in streams — field release experiments, in *Chemistry and Biology of Solid Wastes,* Salomons, W. and Förstner, U., Eds., Springer-Verlag, Heidelberg, 1988, 275.
121. **McKnight, D. M. and Bencala, K. E.,** Diel variations in iron chemistry in an acidic stream in the Colorado Rocky Mountains, *Arctic Alpine Res.,* 20, 492, 1988.
122. **McKnight, D. M. and Bencala, K. E.,** Reactive iron transport in an acidic mountain stream in Summit County, Colorado: a hydrologic perspective, *Geochim. Cosmochim. Acta,* 53, 2225, 1989.
123. **Morgan, J. J. and Stone, A. T.,** Kinetics of chemical processes of importance in lacustrine environments, in *Chemical Processes in Lakes,* Stumm, W., Ed., John Wiley & Sons, New York, 1985, 389.
124. **Honeyman, B. D. and Santschi, P. H.,** A Brownian-pumping model for oceanic trace metal scavenging: evidence from Th isotopes, *J. Mar. Res.,* 47, 951, 1989.
125. **Honeyman, B. D. and Santschi, P. H.,** Coupling adsorption and particle aggregation: laboratory studies of "colloidal pumping" using ^{59}Fe-labelled hematite, *Environ. Sci. Technol.,* 25, 1739, 1991.
126. **Beckett, R., Hotchin, D. M., and Hart, B. T.,** The use of field flow fractionation to study pollutant-colloid interactions, *J. Chromatogr.,* 517, 435, 1990.
127. **Beckett, R. and Hart, B. T.,** Use of field flow fractionation techniques to characterise aquatic particles, colloids and macromolecules, in *Sampling and Characterisation of Aquatic Particles,* Vol 2., Buffle, J. and Van Leeuwen, H. P., Eds., Lewis Publ., London, 1993, 165.
128. **McKnight, D. M., Kimball, B. A., and Bencala, K. E.,** Iron photoreduction in an acidic mountain stream, *Science,* 240, 637, 1988.
129. **Nordstrom, D. K.,** The rate of ferrous iron oxidation in a stream receiving acid mine effluent, *U.S. Geol. Surv. Water-Supply Pap.,* 2270, 113, 1985.
130. **Nordstrom, D. K. and Ball, J. W.,** The geochemical behaviour of aluminium in acidified surface waters, *Science,* 232, 54, 1986.
131. **Filipek, L. H., Nordstrom, D. K., and Ficklin, W. H.,** Interaction of acid mine drainage with waters and sediments of West Squaw Creek in the West Shasta Mining District, California, *Environ. Sci. Technol.,* 21, 388, 1987.
132. **Bencala, K. E. and Walters, R. A.,** Simulation of solute transport in a mountain pool-and-riffle stream: a transient storage model, *Water Resour. Res.,* 19, 718, 1983.
133. **Santschi, P. H.,** Factors controlling the biogeochemical cycles of trace elements in fresh and coastal marine waters as revealed by artificial radioisotopes, *Limnol. Oceanogr.,* 33, 848, 1988.
134. **Fitzgerald, W. F. and Watras, C. J.,** Mercury in surficial waters of rural Wisconsin lakes, *Sci. Tot. Environ.,* 87/88, 223, 1989.

Chapter 10

Trace Metals in Estuaries

Geoffrey E. Millward and Andrew Turner

CONTENTS

I. INTRODUCTION

Estuaries encompass the river-ocean interface, a region which is both chemically and physically dynamic. Pronounced biogeochemical reactivity, including sorption, flocculation, and redox cycling of trace metals, is induced by sharp gradients in the estuarine master variables of salinity, temperature, dissolved O_2, pH, and particle character and concentration that result from the mixing of fresh and saline end members.[1] These processes drastically modify the riverine compositional signal to the oceans and must be fully understood to accurately determine chemical fluxes and to refine geochemical mass balances.[2] Moreover, the commercial and industrial significance of estuaries and their catchments has led to an increasing anthropogenic perturbation of this signal. Consequently, trace metals and their inherent toxicity have formed an integral component of estuarine water quality monitoring programs.[3] This critical review examines the processes controlling the cycling of trace metals in estuaries based on field observations, controlled laboratory experiments, and biogeochemical models.

II. SPACE- AND TIME-DEPENDENT PROCESSES IN ESTUARIES

A. HYDRODYNAMICS

Estuaries have been classified according to physiography and vertical salinity stratification.[4] The latter is controlled by circulation and mixing characteristics, which are dependent on the relative contributions

0-8493-6304-7/95/$0.00+$.50

224 T

Table 1 Physical characteristics of five contrasting estuaries

Characteristic	Amazon	St. Lawrence	Savannah	Scheldt	Tamar
Catchment area, km^2	7.1×10^6	1.0×10^6	2.5×10^4	2.1×10^4	9×10^2
Mean river flow, $m^3\ s^{-1}$	200,000	14,160	300	100	19
Tidal range,[a] m	4.3/3.3	2.6/1.4	2.7/1.9	~4.0	4.7/2.2
Flushing time, days	[b]	~30	~12	60–90	7–14
Stratification	Salt wedge	Partly mixed	Partly mixed	Well mixed	Partly mixed
Sediment discharge, $t\ a^{-1}$	0.5×10^9	5.1×10^6	2.0×10^6	0.1×10^6	0.1×10^5

[a] Spring/neap tidal range near mouth; [b] Not appropriate for salt wedge estuaries.

of river and tidal flows. Thus, river flow dominates in a highly stratified estuary (e.g., Amazon), giving rise to pronounced vertical density gradients, and strong tidal currents induce a vertically well-mixed water column (e.g., Scheldt); intermediate flow regimes typify a partly mixed estuary (e.g., Tamar). The spatial domain of estuarine mixing, which is critical to sediment mobility and the transport of trace metals, is also determined by these hydrodynamic characteristics. For example, the mixing zone of the Amazon extends seaward for several hundred kilometers over the continental shelf,[5] whereas in the Tamar, mixing is essentially complete over 20 kilometers within the shallow upestuary reaches.[6] The characteristics of five estuaries exemplifying contrasts in dimensions and stratification are listed in Table 1.[4–8]

A key hydrodynamic control on estuarine trace metal reactivity is the flushing time, τ, of an estuary or segment thereof:[9]

$$\tau = \frac{1}{Q_w} \int \left(1 - \frac{S}{S_S}\right) dV \tag{1}$$

where Q_W is the river flow, S is the salinity (S_S is the salinity of seawater), and V is the volume. Of particular significance is the relation of τ to the rates of *in situ* chemical reactions.[10] The question arises: are these chemical reactions sufficiently rapid for thermodynamic equilibria to exist within the mixing zone, or are the observed reactant distributions dependent on the interplay between reaction kinetics and flushing time? Such considerations are critical to the approach adopted to estuarine modeling of trace metal transport. Theoretically, for ~99% conversion of reactants to products, equilibrium modeling is only appropriate when the ratio of the first-order reaction half-life to the flushing time is less than 10^{-1}.[10]

B. PARTICLE TRANSPORT

The particulate phase plays a complex yet fundamental role in the cycling of trace metals in estuaries. The solid substrate is a heterogeneous composition of various mineral phases including clays, quartz, feldspars, and carbonates derived from catchment and coastal erosion. In addition, discrete and biogenic phases are produced *in situ* through flocculation and biological activity, respectively. Mineral particles are coated with Fe and Mn oxyhydroxides and organics of both marine and terrestrial origin which essentially isolate the underlying lithogenous material. Adsorbed organic compounds impart a negative surface charge (generally in the range -0.7 to $-2.0 \times 10^{-8}\ m^2\ s^{-1}\ V^{-1}$),[11–13] although a positive surface charge has been observed on particles from the low salinity region of an acidic, Fe-rich environment due to the exposure of oxide surfaces.[14]

Transport of the particulate phase is highly nonconservative, and particles, together with particle-reactive trace metals, may be subjected to many cycles of deposition-resuspension in an estuary, thereby residing considerably longer than water parcels. The residence time of suspendable particles in the Humber Estuary is estimated at 18 years, based on the mass of material held in suspension (3×10^6 t) and the annual supply of riverine sediment ($1.7 \times 10^5\ t\ a^{-1}$).[15] This compares with a freshwater replacement time of up to 40 days[15] and illustrates the potential period required for the decontamination of the estuarine sedimentary regime.[7] Detailed studies of the seasonal cycling and retention of particles in the Tamar Estuary[6,16,17] identify the key role of the turbidity maximum zone (TMZ), a localized region of pronounced suspended particulate matter (SPM) concentration generated by estuarine gravitational circulation.[18] This is maintained by the tidally induced periodic resuspension of local sediment and upestuary transport of resuspendable material by tidal asymmetry. During winter, the TMZ is shifted downestuary by enhanced river flow and augmented by remobilization and downestuary transport of the sediment shoal formed during the previous summer by river currents. The seasonal cycle of sediment transport is depicted in Figure 1[17]; the estimates of sediment fluxes yield a particle residence time in the TMZ of 1.4 years[16] compared with an estuarine flushing time of 7 to 14 days.[6]

The TMZ is a region of effective particle size selection and, in the Tamar Estuary, particle size distributions have been investigated using an *in situ* Fraunhofer laser diffraction technique.[19] Within the TMZ, particles exhibited a unimodal distribution and were characterized by a permanently suspended population of a diameter 20 to 25 μm; elsewhere, particles were larger and exhibited multimodal distributions. Estuarine distributions of specific surface area (SSA, $m^2\ g^{-1}$) of SPM, determined by a gravimetric BET nitrogen adsorption technique, have demonstrated a maximum approximately coincident with the TMZ.[13,20–22] Possible causes are size selection and/or disaggregation of particles,[19] precipitation of fresh Fe-Mn phases,[13,21] and depletion of particulate organic carbon,[23,24] yielding a localized region of high scavenging potential. For example, in the Elbe Estuary the total SSA was found to be 9.9 $m^2\ l^{-1}$ in the TMZ, compared with 1.8 $m^2\ l^{-1}$ and 0.1 $m^2\ l^{-1}$ at the river and marine end members, respectively.[21] Thus, the precondition for metal removal in the TMZ is not only a sufficiently high SPM concentration but also particles with high SSAs. An extension of the nitrogen adsorption technique also allows examination of particle microporosity.[21] Estuarine particles are comprised of slit-shaped pores, with most of the internal volume contained within the size range <5 nm, providing a potential source of variation in the total density of active surface sites for adsorbing metal ions.[25]

III. FIELD SAMPLING AND ANALYSIS FOR TRACE METALS

A. SAMPLING STRATEGY

Although a geographical reference is suitable for sampling relatively immobile deposited sediments and associated pore waters in an estuary, sample collection for dissolved and SPM constituents should be carried out with respect to a conservative index of mixing, which can readily be determined *in situ,* i.e., salinity. Coverage of the required salinity range may be achieved either spatially or at a fixed station over a tidal cycle. Near-surface sampling at intervals of 1 PSU along an axial, midchannel transect provides a reasonable basis for general observations.[26] The sampling density should be increased for the study of *in situ* chemical processes, and multisampling campaigns are necessary to investigate processes driven by cyclic and stochastic events. The specific field strategy adopted is dependent on the objective of the exercise, spatial extent of the mixing zone, and site accessibility (see Table 1).

B. SAMPLE COLLECTION AND STORAGE

1. Deposited Sediment and Pore Waters

Although a surficial scrape of intertidal sediment is conveniently taken manually, subtidal sampling requires the deployment of a coring or grabbing device,[27] taking precautions to avoid contamination from metallic sample housings. Pore waters may be recovered by squeezing core sections between PTFE pistons under *in situ* temperature and redox conditions[28] or by centrifugation under a nitrogen atmosphere.[29] Diffusion-controlled pore-water samplers of varying sophistication have been developed, including an *in situ* dialysis technique to recover pore waters from intertidal sediments.[30–32] A modified suction sampler which gives improved resolution in pore-water sampling through a sediment profile has potential application in estuaries.[33]

2. Water and Suspended Particulate Matter

Recently, the adoption of ultra-clean techniques has been emphasized for the collection and processing of water samples even in relatively contaminated estuarine environments, including rigorous acid cleaning of PTFE sample bottles and filters and isolation from metallic components such as the ship's hull and hydrographic wires.[34–36] To this end, a shipboard clean sampling and laboratory facility originally designed for deep-sea work[37] has since found application in estuarine plume environments of the North Sea.[15]

Water samples are taken by either deployment of suitable containers ranging from basic polyethylene bottles to PTFE-lined samplers which can be triggered remotely whilst submerged, or abstraction from a continuous, on-board pumped supply. Clean filtration is achieved by applying a positive pressure using a purified inert gas, and should be performed immediately in order to preserve *in situ* solid-solution partitioning.[38] The nominal pore size of the filter provides an operational distinction between dissolved and particulate material; this is conventionally 0.45 μm and colloidal-sized particles are, therefore,

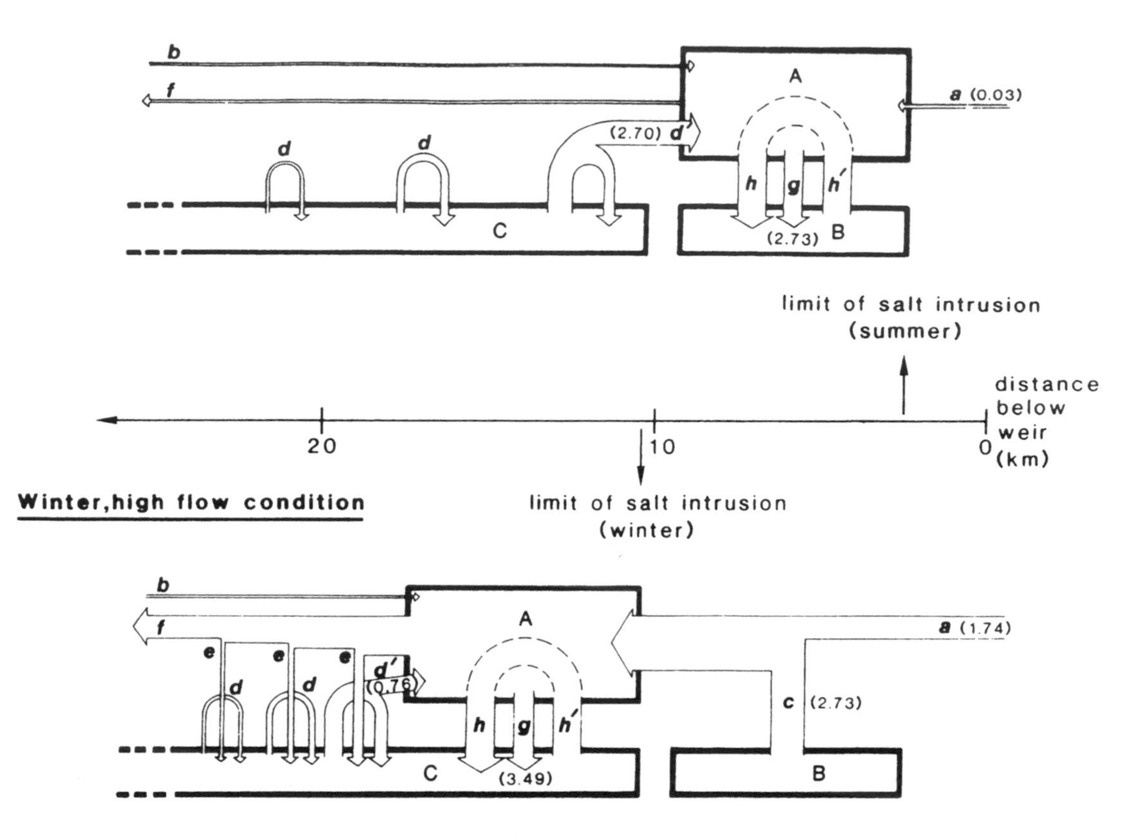

Figure 1 Schematic representation of resuspendable sediment transport in the Tamar Estuary during winter and summer. Reservoirs: A, suspended TMZ; B, summer-accumulating shoal; C, midestuarine sediment shoal. Fluxes: a, riverine particle influx; b, marine particle influx; c, winter remobilization of summer-accumulated sediment shoal; d, asymmetrical tidal pumping (d′ into TMZ); e/f, seaward transport during winter; g, net sedimentation at TMZ; h′, tidal resuspension (h, tidal deposition) with no net transport. Quantified mean fluxes are in units of 10^5 kg d^{-1}. (From Morris, A. W., Bale, A. J., Howland, R. J. M., Millward, G. E., Ackroyd, D. R., Loring, D. H., and Rantala, R. T. T., *Water Sci. Technol.*, 18, 111, 1986. With permission.)

included in the dissolved fraction, although ultrafilters of nominal pore size or molecular weight cutoffs are available for their isolation.[39,40] The size dispersity of typical estuarine species is shown in Figure 2 along with the selectivity of trace metal detection and fractionation procedures. Filtrates are acidified to pH 1 to 2 and stored at a low temperature to stabilize metal ions and minimize microbial action. The SPM retained on filters is washed with ultra-pure water to remove seawater salts, although this may potentially desorb loosely bound metals, and refrigerated or frozen.

Continuous-flow centrifugation provides a more rapid and efficient technique for the recovery of SPM than filtration and is particularly advantageous in low turbidity waters if high mass samples are required. A centrifuge for the collection of estuarine samples is described by van der Sloot and Duinker.[41] The solid material is deposited on PTFE panels lining the interior of the rotor which operates up to 10^5 G and can process about 100 1 in 2 h. A comparison of estuarine filtration and centrifugation techniques and differences in composition of SPM thus recovered has been carried out by Etcheber and Jouanneau.[42]

C. SAMPLE PRETREATMENT AND TRACE METAL DETERMINATION

1. Dissolved Trace Metals

Direct analysis of dissolved trace metals is preferable in order to reduce the potential of contamination via sample manipulation. Differential pulse anodic stripping voltammetry (DPASV) offers a sensitive solution to this problem for the analysis of Cd, Cu, Pb, and Zn.[20,34–36,43,44] Cathodic stripping voltammetry

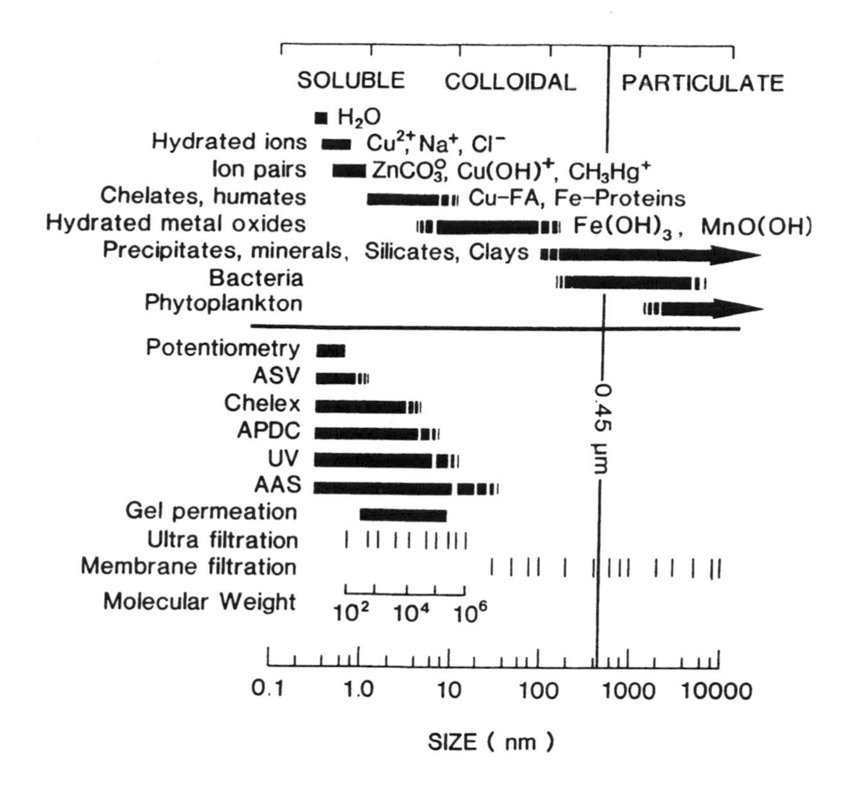

Figure 2 The dimensions of soluble, colloidal, and particulate estuarine species. The size selectivity of various detection and fractionation procedures is also shown. ASV, anodic stripping voltammetry; APDC, solvent extraction with ammonium pyrrolidine dithiocarbamate in methyl iso-butyl ketone; UV, ultra-violet photooxidation; AAS, atomic absorption spectroscopy using graphite furnace. (From Morris, A. W., Ed., *Practical Procedures for Estuarine Studies,* Natural Environment Research Council, Swindon, U.K., 1983. With permission.)

(CSV)[45–47] used in combination with UV-irradiation yields operational concentrations of total and labile dissolved metals. Colorimetry and fluorimetry are in routine use for the determination of dissolved Fe, following reduction and reaction with a complexing agent,[48–51] and dissolved Al, after reaction with Lumogallion,[52,53] respectively.

Inorganic and organic species (including methylated forms) of As, Ge, Sb, Se, and Sn may be determined by atomic absorption spectrophotometry (AAS) following separation by either chromatography (GC or HPLC) or cryogenic trapping and selective volatilization of their hydrides,[54–65] although important As species such as arsenobetaine and arsenocholine cannot be detected.[64] Selenium species[66] and ^{9}Be[67] can be determined after complexation with organic reagents and analysis by GC. Dissolved Hg is detected using cold vapor-AAS, coupled with trapping of the Hg cold vapor by amalgamation on an Au column.[68–70] Combining this technique with oxidation, by either UV irradiation[68] or bromination,[69] provides an analytical fractionation of inorganically and organically bound Hg. A method using head space GC with microwave plasma detection has potential application for methylated Hg species in estuaries.[71]

Conventional AAS is generally not sensitive enough to allow direct determination of dissolved trace metals, with the exception of Mn,[72–74] and preconcentration is required by either ion exchange,[75–80] coprecipitation,[5,81–83] or solvent extraction.[84–86] Preconcentration is also required prior to multielement determinations by either total reflection X-ray fluorescence (TXRF)[22,87] or neutron activation analysis (NAA),[88] the analysis of rare earth elements and Hf and Zr by isotope dilution mass spectrometry,[29,89,90] and radiochemical analysis of U.[91,92]

2. Particulate Trace Metals

Generally, decreases in deposited sediment metal concentrations coincide with increases in grain size, due to a reduction in surface area to weight ratio.[93] Thus, because of regional diversity in sediment grain size distributions, it is recommended that a fine fraction is separated by wet sieving (commonly <63 μm) through nylon mesh prior to analysis. Alternatively, bulk sample analyses should be normalized with respect to the percentage weight of fine fraction[94] or SSA.

Multielement analysis of deposited sediment and SPM may by achieved by TXRF[22,87] or NAA.[88] Metal analysis by AAS, or more recently by inductively coupled plasma-emission spectroscopy (ICP-ES),[29,85] requires the solubilization of metals from solid form. Total dissolution of sediment entails heating the sample in a mixture of HF and aqua regia in sealed PTFE digestion vessels.[95] Milder chemical treatments such as 1 *M* HCl or 25% v/v HOAc are employed to give an operational measure of a nondetrital available fraction of metals.[22,29,96,97] Although the exact mechanisms controlling bioaccumulation of trace metals are poorly understood,[98] a combination of weak acid and reducing agent is thought to provide an indicator of a bioavailable fraction, assuming particle ingestion is the main accumulation route and acidic/reducing conditions occur in the intestinal tract.[99] The application of sequential extractants designed to successively remove metals in various associations, such as exchangeable, carbonate, Fe-Mn oxyhydroxide, organic and detrital, forms the basis of (operational) particulate metal speciation schemes.[99–102] Despite the ability to operationally fractionate metals in the solid phase, a lack in standardization of extraction procedures has prevented systematic interestuarine comparisons of available particulate trace metal concentrations.[8]

IV. APPROACHES TO ESTUARINE METAL STUDIES

A. ANALYSIS OF FIELD DISTRIBUTIONS

The reactivity of dissolved constituents is conventionally interpreted by plotting their concentrations against a (quasi-) conservative index of mixing.[103] Salinity is widely adopted for this purpose, although the oceanographic principles of salinity no longer apply below 1 PSU.[104] Thus, a straight line joining the end members (theoretical dilution line) signifies conservative (noninteractive) behavior; curvilinear relationships signify nonconservative behavior resulting in net input or output of the dissolved constituent (Figure 3). Boyle et al.[105] derived an analytical model to demonstrate the application of constituent-salinity plots in determining reactivity in estuaries. The variation of the flux of the constituent with salinity (i.e. the reactivity) is given by:

$$\frac{dQ_c}{dS} = -Q_w\left(S - S_0\right)\frac{d^2C}{dS^2} \tag{2}$$

where Q_w is the river flow of salinity S_0, and Q_c is the flux of the constituent across an isohaline surface of salinity S and constituent concentration C. For nonconservative behavior, the constituent flux varies with salinity and the second derivative is nonzero. Gain or loss of constituent between two isohaline surfaces is calculated from the difference between the respective effective end-member concentrations (C_*), i.e., the intercept of the tangent (dC/dS) to the constituent-salinity curve at S = 0 PSU,[106] as shown in Figure 3. Thus:

$$Q_1 - Q_2 = Q_w\left(C_{*1} - C_{*2}\right) \tag{3}$$

and for conservative mixing $C_* = C_0$ at all salinities; $C_* < C_0$ signifies upestuary removal and $C_* > C_0$ signifies upestuary addition. For example, in the Ogeechee River during low flow (8 $m^3\ s^{-1}$), Maeda and Windom[92] recorded 0.05 μgU l^{-1}. Concavature in the estuarine distribution of U at low salinities resulted in an effective end-member concentration of –0.82 μgU l^{-1} from which a net removal rate of 7 mgU s^{-1} was derived.

This treatment of observational data assumes that C is a continuous and single-valued function of salinity and that there exist no subsidiary inputs (tributaries, anthropogenic sources). Nonlinearity may be effected by temporal variation in the end-member constituent concentration,[107] the extent of curvature being determined by the ratio of the flushing time (Equation 1) to the period of end-member variation.[9] A further limitation of constituent-salinity plots is their inability to resolve clearly solid-solution exchange processes at very low salinities (<1 PSU).[108] Biogeochemical reactivity is, however, often accentuated in this zone due to the incipient changes of water composition (master variables)[109] coupled with the existence of the estuarine TMZ. Consequently, reactivity at low salinities is often diagnosed from plots of constituent concentration vs. either the logarithm of salinity[108] or axial distance.[79]

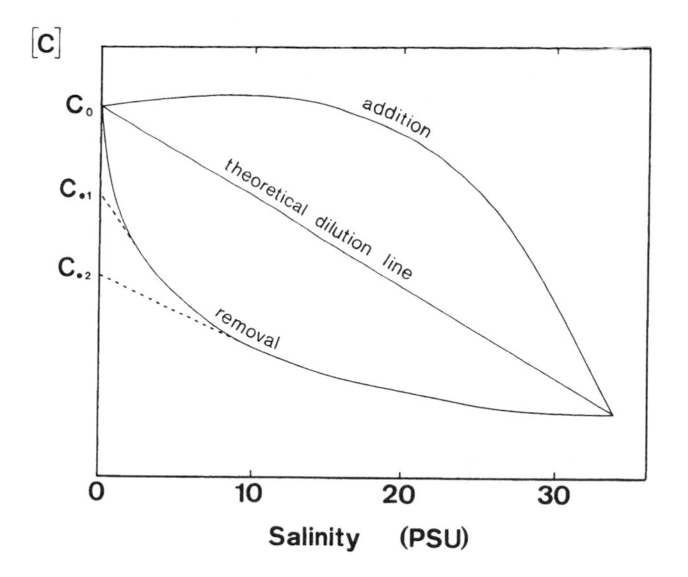

Figure 3 Representation of the estuarine distributions of a dissolved constituent, C, under steady-state conditions.

Knox et al.[31] developed a statistical technique for interpreting the distributions of dissolved constituents A and B within an estuarine segment under steady-state conditions. In the case of conservative mixing of A (or B), data will fit the equation:

$$[A] = a_0 + a_1 S \tag{4}$$

where a_0 and a_1 are constants. If an additional, common source of constituents arises data will fit the equation:

$$[A] = c_0 + c_1 S + k[B] \tag{5}$$

where c_0 and c_1 are constants, and k is the concentration ratio of the source, $[A]_s/[B]_s$. F-ratios are used to establish the significance of additional inputs to the distributions and whether they are common to A and B. Correlations between Mn and NH_4, and between Cu and Zn in the Tamar Estuary according to Equation 5, indicate a common midestuarine source or sources.[31,79] Variable regressions between Mn and Zn, however, suggested additional inputs and/or spatial variability in the relative proportions of source concentrations.[79]

A box model approach for estimating inputs and removal of trace metals within estuaries has also received some attention.[75,84] Bewers and Yeats[84] used axial and vertical observations of salinity and dissolved and particulate trace metals in the St. Lawrence estuary to evaluate the imbalance between influxes from river and deep saline countercurrents and efflux from a surface-mixed layer. Appreciable loss of metals in both phases was attributed to solid-solution reactions in the low salinity region coupled with the selective retention of resuspendable sediment populations. This mass balance has since been extended offshore to assess the net fluxes of estuarine trace metals to the North Atlantic,[110] although the accuracy of this approach needs considerable improvement.[111]

B. LABORATORY EXPERIMENTS

Although environmental chemical reactivity can be inferred from constituent-salinity diagrams, laboratory studies performed under carefully controlled and well-defined conditions are more suited to elucidate the mechanistic and kinetic details of these reactions. Experiments using pure substrates (clays, Fe oxyhydroxides) and synthetic media of adjustable composition (pH, chlorinity, electrolyte concentration) have limited applicability to site-specific and heterogeneous field conditions. Master variable gradients can be simulated by batch mixing of native samples, either turbidized (sorption studies) or nonturbidized (flocculation studies). Spike addition of radioisotopes is commonly performed to replicate their stable analogs; the advantage of this approach lies in the subsequent analysis, which offers rapid, specific, sensitive, and multielement determinations.[112] Bale and Morris[113] suggest that the main drawback of the batch mixing technique is the assumption that the instantaneous promotion of chemical reactions is independent of previous and continuous mixing histories. Accordingly, these authors developed a one-dimensional estuarine mixing profile by pumping, at selected controlled rates, end-member samples from opposite ends through a series of interconnected reactor vessels; such a construction cannot, however, accurately simulate particle mixing.

V. FIELD OBSERVATIONS OF TRACE METALS

A. DISSOLVED TRACE METALS

Representative behaviors of dissolved trace metals in estuaries inferred from constituent-salinity diagrams are presented in Table 2; a more comprehensive compilation is given by Bourg.[104] With the notable exception of Fe, axial distributions of a given trace metal exhibit considerable interestuarine variability, indicating sensitivity to the unique biogeochemical and hydrodynamical conditions of a system. In the absence of complementary laboratory studies, mechanistic interpretations of trace metal behavior rely on a knowledge of these conditions. This may be exemplified by contrasting the chemical behavior of trace metals in two well-characterized but distinctly dissimilar estuaries, viz., the Tamar and Scheldt, despite apparently similar metal distributions. In the relatively dynamic Tamar Estuary, dissolved Cu, Mn, and Zn undergo rapid sorptive removal onto suspended particles of the TMZ in the low salinity reaches (τ of the order hours) and subsequent midestuarine desorption as these particles are advected seawards in association with pore-water infusion from the tidally disturbed bed.[73,79,114] In the Scheldt Estuary, chemical reactivity is governed by a persistent and relatively stagnant (τ several days) upestuary anoxic zone.[115] Thus, the depletion of dissolved Cd, Cu, and Zn at low salinities is due to the precipitation of the insoluble sulfide species; their oxidation and resolubilization occurs downstream as the dissolved O_2 content increases.[85]

The information in Table 2 is largely based on limited survey information from which temporal intraestuarine variability is not always apparent. Intraestuarine variations in trace metal end-member concentrations (principally riverine) and axial distributions may arise through the seasonal cycling of reactions mediated by temperature[76] and biological activity.[5,80,82] The latter is of particular significance to the distributions of methylated As species generated by estuarine organisms.[116] Short-term fluctuations result mainly from tidally induced processes such as infusion from pore waters.[73,79] In a hydrological sense, increasing runoff acts to dilute a steady source.[117,118] Additional chemical consequences are less predictable because the reaction-controlling variables of pH and SPM concentration vary nonlinearly with discharge.[117] Moreover, the discharge-dependent flushing characteristics may impose time constraints on metal-ligand interactions in freshwaters,[119] and the subsequent sorptive behavior of metals in the turbidized mixing zone.[10]

B. PARTICULATE TRACE METALS

Axial distributions of trace metals in surficial sediments[17] and SPM[22,120] generally reflect the mixing of fluvial material of high metal content with marine material of low metal content and a high autochthonous organic component. Although particulate matter is instrumental in biogeochemical phase transformations, these interactions are ineffectual to the bulk solid composition because of the magnitude of the sedimentary metal reservoir.[17] Only large-scale phase redistributions of Fe and Mn have incurred demonstrable alteration to their nondetrital associations,[8,39] as exemplified by midestuarine maxima of acetic acid-leachable particulate Fe and Mn in the Scheldt Estuary caused by precipitation of the respective oxidized species.[8] This simple mixing model is, however, modified by significant anthropogenic effluents, biological production, and particle selection in the TMZ. Short-term variability of SPM composition in the TMZ arises due to the fluctuating relative contributions from resuspendable and permanently suspended particles. The instantaneous concentration of particulate constituent, P (w/w), in the TMZ of suspended particulate matter concentration, SPM (w/v), is given by:[121]

$$P = \frac{SPM_P(P_P - P_T)}{SPM} + P_T \tag{6}$$

where subscripts P and T refer to the permanently and temporarily suspended fractions, respectively. Accordingly, at times of maximum tidal stress, dense, temporarily suspended particles enriched in heavy mineral phases give rise to a population high in detritally bound metals.

Table 2 **Field observations of the behavior of dissolved trace metals in the estuaries described in Table 1**

Estuary	Metals	Trace Metal Behavior	References
Amazon	Cu, Ni, V[a]	Conservative	5, 82
	^{9}Be, Zn	Removal	67, 81
	Cd	Possible addition	5
St. Lawrence	As,[a] Co,[a] Cu, Ni, Mn	Conservative	61, 84
	Fe, Cr	Removal at low salinity	77, 84
	Hg	Removal	69
Savannah	As[a]	Conservative	54, 59
	U[a]	Conservative, removal at low Q_w	92
	Fe	Removal at low salinity	80
	Pb, Sb, Zn	Removal	59, 78, 80
	Cu, Cd, Ni	Addition	76, 80
Scheldt	Mo,[a] Ni, Se	Conservative	74, 88
	As, V, Sb	Removal at low salinity	88
	As(tot.)	Possible addition at low salinity	58
	Cd, Cu, Zn	Removal at low salinity; midestuary addition	74, 85
	Mn	Addition at low salinity, midestuary removal	74
Tamar	U[a]	Conservative	91
	Al, Cd, Cu, Mn, Ni, Zn	Removal at low salinity, midestuary addition	17, 53, 73, 79
	As, Sn	Addition	63

[a] Trace metals displaying a positive correlation with salinity.

VI. MECHANISMS AND MODELING OF ESTUARINE TRACE METAL PROCESSES

A. FLOCCULATION OF COLLOIDAL MATERIAL

The large-scale removal of dissolved Fe from solutions at low salinities[48–51] (see Table 2) is the result of the destabilization and subsequent flocculation of negatively charged Fe-humic colloids as they encounter the major seawater cations. The nature and extent of this process is sensitive to salinity, SPM concentration, bacterial activity, and nominal pore size of filters used to isolate aggregates.[51,113] According to Fox and Wofsy,[50] the reaction follows second-order kinetics with a rate proportional to the square of salinity. Mayer,[49] however, has demonstrated approximate second-order kinetics involving two sequential reactions. Initially, organic-oxide interactions are complete within a few minutes, with the rate being dependent on the frequency of intercolloid collisions but independent of temperature:

$$\frac{dN}{dt} = k_2 N^2 \tag{7}$$

where N is the number of colloids and k_2 is a second-order rate constant. Subsequently, a temperature-dependent aggregation of filterable Fe colloids, with larger aggregates formed in the first reaction, lasts several hours:

$$\frac{d[Fe]_f}{dt} = k_2^{obs}[Fe]_f^2 \tag{8}$$

where $[Fe]_f$ is the filterable Fe concentration and k_2^{obs} is a pseudo second-order rate constant. Gravitational studies of the end-product aggregates suggest characteristic settling velocities of less than 1 m per 15 days,[48] which would preclude their retention in all but the most poorly flushed estuaries.

A number of investigations have achieved appreciable removal of other dissolved trace metals through flocculation of their organically or Fe oxide-bound colloidal forms on mixing filtered river and seawaters.[89,122–124] Sholkovitz[123] demonstrated that this process is salinity dependent (essentially complete at 20 PSU) and that removal follows the order:

$$Fe(95\%) > Cu, Ni(40\%) > Mn(25-45\%) > Al(20\%) > Co(10\%) > Cd(5\%)$$

reflecting the relative affinities of trace metals for seawater anions and colloidal humics/Fe oxides.

B. MANGANESE OXIDATION AND PRECIPITATION

The behavior of Mn in estuarine waters is strongly influenced by its redox chemistry.[72–74,125] Although the oxidation of soluble Mn^{2+} into precipitable MnO_x ($1.5 < x < 2$) in aquatic environments is theoretically a slow process, the reaction is catalyzed by preexisting or freshly generated solid Mn oxide phases[126] and microbial activity.[127,128] Detailed studies of the autocatalytic mechanism have been undertaken using Tamar Estuary waters[73] (see Figure 4). In freshwater, oxidation exhibits zero-order kinetics with respect to dissolved Mn concentration; the rate is reduced with an increase in ionic strength through occupation of catalytic sites by major seawater cations, and the reaction assumes a dependence on the concentration of dissolved Mn (first-order kinetics) as seawater comprises >1 to 2% of the admixture. In addition, the rate is dependent on pH, temperature, and the composition and concentration of SPM.

C. SORPTION

Sorption is of fundamental significance to the geochemical cycling of trace metals in estuaries. The surface adsorption of a metal ion (M^{z+}) is conventionally described in terms of competition with protons for particle surface sites (S–OH):[25,129]

$$S-OH + M^{z+} \underset{k_{-1}}{\overset{k_1}{\rightleftharpoons}} S-OM^{(z-1)^+} + H^+ \qquad \textbf{(9)}$$

where S represents the Al, Fe, and Mn of oxide coatings and exposed mineral phases or the C of carboxyl/hydroxyl functional groups of organic film, and k_1 and k_{-1} are the reaction rate constants. Sorption is a complex function of solution and substrate composition (ionic strength, pH, and concentration and nature of adsorbate/adsorbent and competing ligands)[25] and is demonstrably nonadditive.[130] Not surprisingly, the ability to predict accurately such reactions on estuarine particles suspended in a complex and dynamic medium is lacking.

Laboratory sorption studies have demonstrated a decreasing affinity of trace metals for the solid phase with increasing salinity (or a component thereof; chlorinity, electrolyte concentration).[131–133] Such observations have been attributed to an increasing competition for active particle sorption sites by polyvalent seawater cations, and/or an increasing tendency for the metal to form stable and soluble chlorocomplexes. Using a thermodynamic equilibrium model, Bourg[134] demonstrated that an increase in chlorinity suppressed adsorption in the order:

$$Cd > Ni > Cu, Pb > Zn$$

A reduction in the adsorption of a constituent with salinity is not necessarily indicative of desorption from seaward fluxing particles as this is also determined by the extent of the reaction reversibility. For example, it has been demonstrated that Cd sorption on a variety of synthetic and natural substrates is essentially fully reversible.[135,136] In contrast, by virtue of its critical ionic radius, Cs effectively and irreversibly migrates into interlattice mineral sites.[137] Accordingly, particle-bound Cd undergoes release from particles on estuarine mixing, whereas Cs desorption is restricted to those ions held nonselectively in exchangeable sites on the particle surface. Sanders and Abbe[138] showed that the reversibility of Ag sorption was dependent on the dominant particle population. Sorption was rapid and reversible on lithogenous sediment at low salinities, hence subject to desorption on estuarine mixing due to the formation of soluble chlorocomplexes. However, sorption onto biogenic material (as Ag^+ and AgCl complexes) was largely irreversible, Ag only becoming available on remineralization of lysed cells.

Time-dependent sorption studies have revealed the possibility of two or more sequential reactions.[130,139–141] A proposed mechanism involves initial rapid and reversible adsorption of ions onto particle

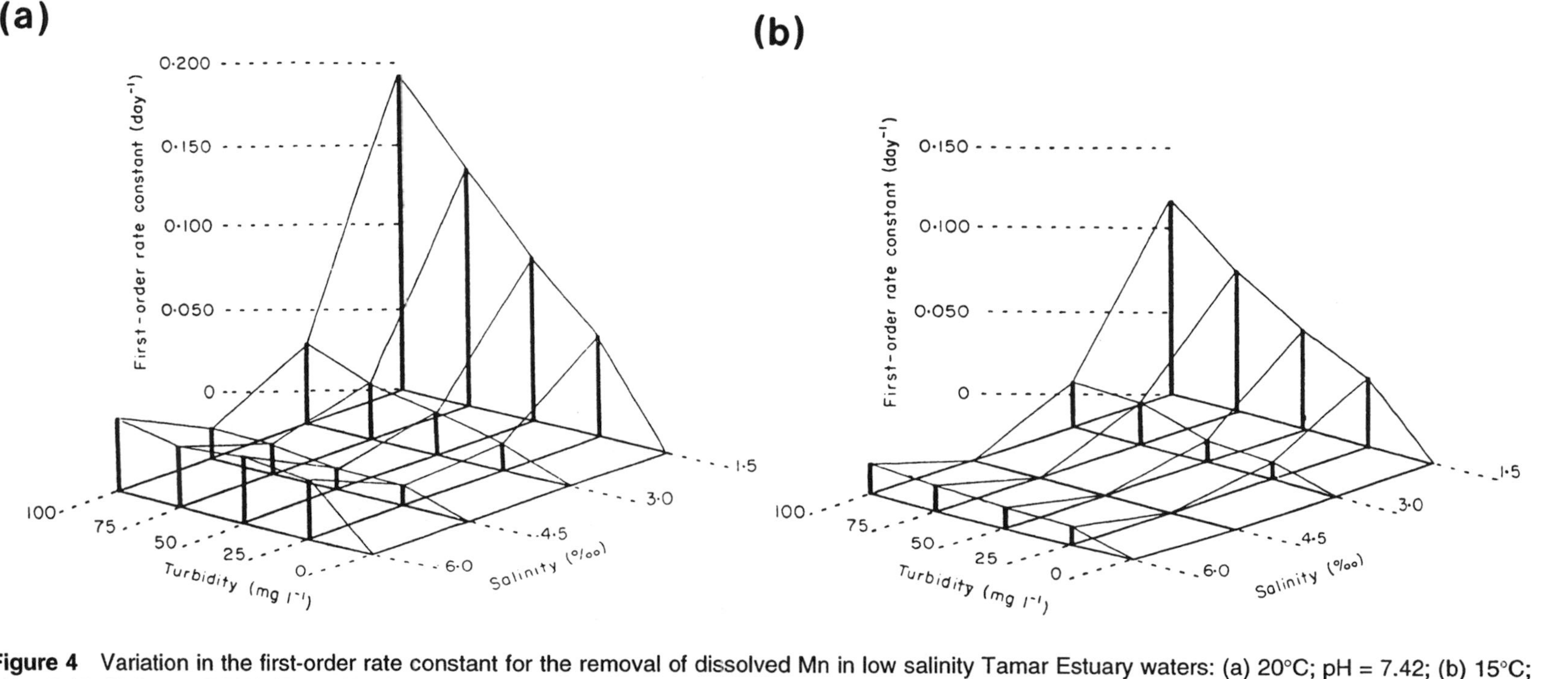

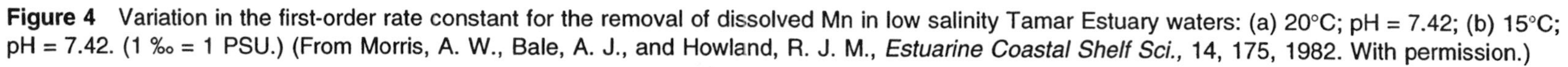
Figure 4 Variation in the first-order rate constant for the removal of dissolved Mn in low salinity Tamar Estuary waters: (a) 20°C; pH = 7.42; (b) 15°C; pH = 7.42. (1 ‰ = 1 PSU.) (From Morris, A. W., Bale, A. J., and Howland, R. J. M., *Estuarine Coastal Shelf Sci.*, 14, 175, 1982. With permission.)

surfaces, followed by a much slower irreversible diffusion into particle matrix sites.[139] The series of consecutive reactions can be described as follows:

$$S + M \underset{k_{-1}}{\overset{k_1}{\rightleftharpoons}} SM \underset{k_{-2}}{\overset{k_2}{\rightleftharpoons}} S'M \underset{k_{-3}}{\overset{k_3}{\rightleftharpoons}} S''M \quad (10)$$

where S denotes sorption sites, SM surface adsorbed metal, S′M matrix-bound metal, and S″M more strongly held matrix-bound metal. Alternatively, Jannasch et al.[141] suggest that the overall sorption process can be described by a series of parallel reactions onto particle surface sites, S, S′, and S″, of increasing binding strength:

$$\begin{aligned} SM &\underset{k_{-1}}{\overset{k_1}{\rightleftharpoons}} M \\ S'M &\underset{k_{-2}}{\overset{k_2}{\rightleftharpoons}} M \\ S''M &\underset{k_{-3}}{\overset{k_3}{\rightleftharpoons}} M \end{aligned} \quad (11)$$

Table 3 gives the first-order rate constants derived from curve fitting of experimental data from estuarine and coastal waters using the model given by Equation 10; k_{-2} (and often k_2) is too small to be established from the duration of the experiments. Sorption half-lives based on the particle concentration normalized first reaction forward rate constant, k_1^*, and an SPM loading of 100 mg l^{-1} range from about 2 hours for Cd to a few seconds for Sn and Hg. An important implication is the significance of sorption even in the most rapidly flushed portions of estuaries.[10]

D. SPECIATION MODELING

Of most importance to the bioavailability, toxicity, and reactivity of trace metals in aquatic environments is not the total concentration but the relative proportions of the component species. Analytical techniques are not generally available for the determination of specific forms of dissolved trace metals, and their prediction, therefore, relies on thermodynamic theory.

The distribution of a metal, M, in solution is given in the form of the mass balance equation:[104,142–144]

$$[M]_T = [M] + \sum_{i,j} \left[M(L_i)_n\right] \quad (12)$$

where $[M]_T$ is the total or analytical concentration of metal, [M] is the concentration of uncomplexed and/or hydrated metal ion, $[M(L_i)_n]$ is the concentration of the nth order complex between M and ligand, L_i, and j is the number of ligand types (e.g., Cl^-, CO_3^{2-}, HCO_3^-, OH^-). The concentration of a complex may be defined in terms of its overall formation constant, $(\beta_i)_n$, the thermodynamic activity coefficient, γ, of the complex, and the thermodynamic activity coefficients and concentrations of uncomplexed metal and uncomplexed ligand:

$$\left[M(L_i)_n\right] = \left((\beta_i)_n [M][L_i]^n \frac{\gamma_M \gamma_{L_i}^n}{\gamma_{M(L_i)_n}}\right) \quad (13)$$

Combining Equations 12 and 13 yields the mass balance:

$$[M]_T = [M] + \sum (\beta_i)_n [M][L_i]^n \frac{\gamma_m \gamma_{L_i}^n}{\gamma_{M(L_i)_n}} \quad (14)$$

Table 3 **Trace metal sorption rate constants for estuarine and coastal waters evaluated according to Equation 10**

Metal	Narragansett Bay[139] Surficial Sediments[a]			Tamar Estuary[140] Suspended Sediments[b]			Puget Sound[141] Suspended Sediments[c]		
	k_1^*	k_{-1}	k_2	k_1^*	k_{-1}	k_2	k_1^*	k_{-1}	k_2
Cd	85	1.0	~0						
Cs	140	1.0	~0						
Cu				7.2×10^3– 4.5×10^4	0.2–3.1				
Hg	3.5×10^4	0.20	~0						
Mn	1.6×10^3	0.04	0.018						
Sn	2.0×10^4	0.10	~0				3.7×10^7	2.3	-0.50
Zn	390	0.32	~0	9.6×10^3– 6.4×10^4	0.5–3.1		7.0×10^4	1.9	0.16

Note: k_1^*in 1 kg^{-1} d^{-1}; k_{-1} in d^{-1}; k_2 in d^{-1}.

[a] S = 30 PSU; pH = 8.1; T = 2°C; SPM = 100 mg l^{-1}; [b] S = 0,1,10 PSU; pH = 7.5,8.2; T = 10°C; SPM = 200,400 mg l^{-1}; [c] S = 28 PSU; pH = 7.9; T = 12°C; SPM = 1 mg l^{-1}.

The effect of ionic strength on the activity coefficients of metal ions can be estimated from an extended Debye-Hückel equation.[142]

For each metal under consideration, Equations 14 and 13 are solved for [M] and $[M(L_i)_n]$, respectively, by computer software, such as MINEQL[83] and CHARON,[143] from the known total metal concentration, ligand composition, and appropriate stability constants. Mantoura et al.[144] determined the stability constants of metals with humic acids isolated from river waters and seawater using a gel filtration chromatographic technique. The sequence of increasing strength of metal-humic complexes generally reflected the Irving-Williams series:

$$Mg < Ca, Cd < Mn < Co < Zn < Ni < Cu < Hg$$

The results of their speciation model showed, with an increase in salinity, a rapid decrease in the proportion of humic-bound metals (except Cu) due to competition for humics from Ca and Mg, an increase in the proportion of chloro-species (>80% for Hg and Cd at salinities > 10 PSU), and a reduction in the proportion of free ion, although for Mn and Zn this was still the dominant species in seawater. For comparison, calculations of trace metal speciation in San Francisco Bay[83] have shown that, at a salinity of 27 PSU and between pH 7.6 and 8.2, >90% of dissolved Cu and <5% dissolved Zn are organically complexed.

Speciation modeling has been extended to include adsorption of free and humic-complexed metals,[129] i.e., SO–M, SO–M–HA, $(SO)_2$–M and $(SO)_2$–M–HA, where SO– are adsorption sites treated as ligands and HA are humic acids. The surface (adsorption) constants of natural sediments were not determined but approximated using silica for a concentration of 30 mg 1^{-1}. The results are shown in Figure 5 and serve to illustrate that metal species are highly nonconservative as a result of the nonlinear relationships of pH, carbonate alkalinity, activity coefficients, and conditional stability constants with salinity.[144] A more sophisticated speciation model incorporating metals in natural multicomponent sediment is precluded by a difficulty in determining precisely relevant parameters such as operational measurements of the relative abundances of these components and their binding capacities, and characteristics of particle surface coatings and aggregation and their effects on metal partitioning.[145]

E. SORPTION MODELING: DISTRIBUTION COEFFICIENTS

The solid-solution partitioning of a constituent can be quantified by the conditional distribution coefficient, K_D, given by:

$$K_D = \frac{P}{C} \quad (15)$$

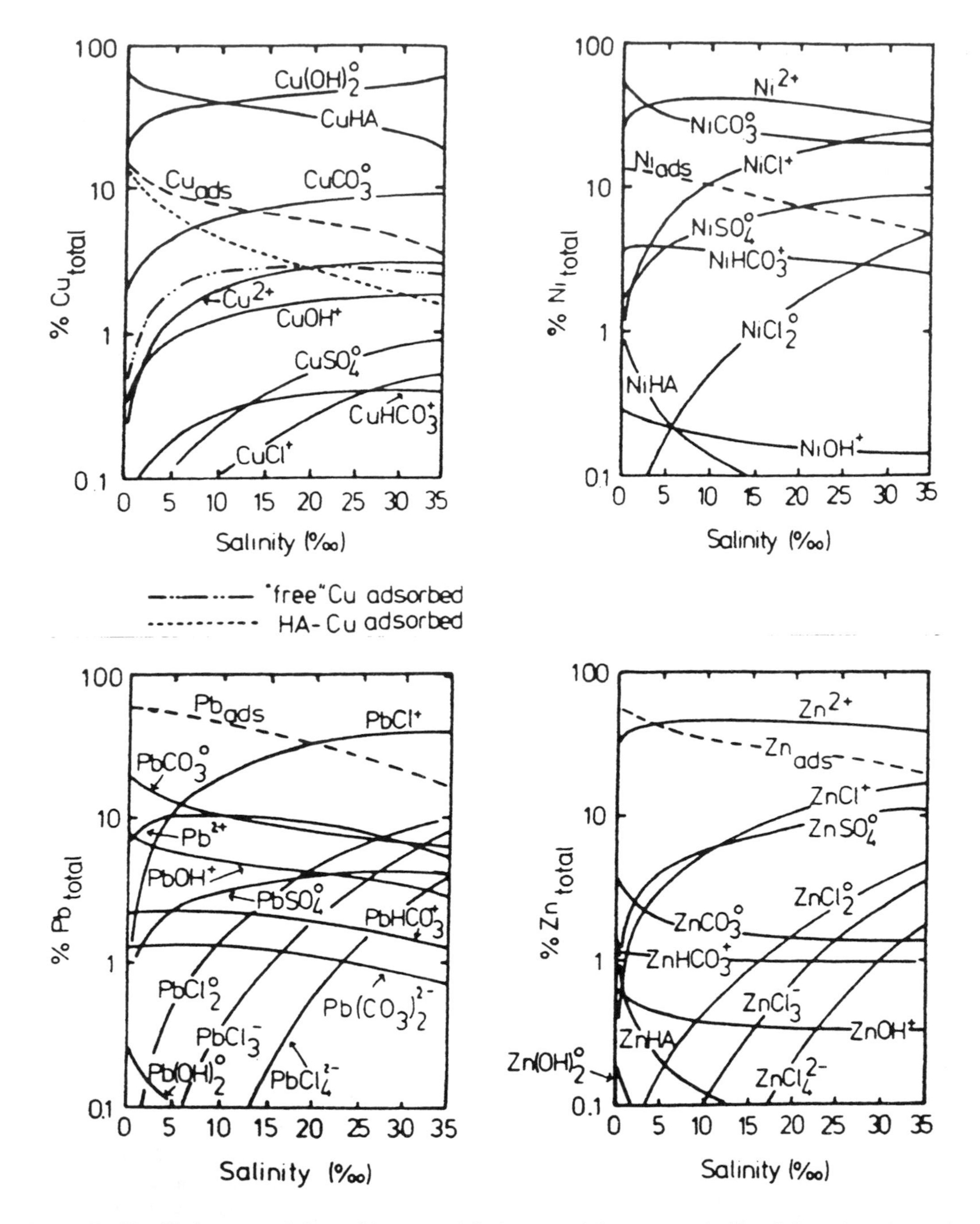

Figure 5 Equilibrium speciation of trace metals in a model estuary of pH = 8.0 and surface site concentration = 1.8×10^5 mol 1^{-1}.[129] 1 ‰ = 1 PSU. (From Bourg, A. C. M., *Trace Metals in Sea Water,* Vol. 9, Nato Conf. Series IV, Wong, C. S., Boyle, E., Bruland, K. W., Burton, J. D., and Goldberg, E. D., Eds., Plenum Press, New York, 1983, 195. With permission.)

where P (w/w) and C (w/v) are the particulate and dissolved concentrations, respectively. This provides a convenient representation of field measurements of trace metals in estuarine and coastal waters[22,43,44,146] and is of potential significance to the assessment of elemental DTIs (dissolved transport indices, the ratio of dissolved to total transport[2]), although application to the interpretation of sorption reactions is limited due to the lack of analytical specificity of the solid phase metal association.[134] K_Ds derived from laboratory studies, however, parameterize those metals inherently involved in sorptive reactivity[132,136,139] and are, therefore, suitable tools for the modeling of estuarine sorption processes.

The change in the equilibrium K_D with salinity as derived from batch mixing experiments can be used to estimate the relative change of concentrations of equilibrated metals in solution, C, and particle bound, X, for a parcel of water advected seawards by solving the mass balance equation:[132]

$$X_{S1} \cdot SPM + C_{S1} = X_{S2} \cdot SPM + C_{S2} \quad \textbf{(16)}$$

where subscripts S1 and S2 signify salinity limits and S2 > S1, and SPM is the conservative concentration of suspended solids. Defining $(K_D)_{S1} = X_{S1}/C_{S1}$ and $(K_D)_{S2} = X_{S2}/C_{S2}$, it follows that:

$$\frac{X_{S2}}{X_{S1}} = \frac{SPM + 1/(K_D)_{S1}}{SPM + 1/(K_D)_{S2}} \quad \textbf{(17)}$$

and:

$$\frac{C_{S2}}{C_{S1}} = \frac{X_{S2} \cdot (K_D)_{S1}}{X_{S1} \cdot (K_D)_{S2}} \quad \textbf{(18)}$$

Li et al.[132] demonstrated that for Bi, Fe, Hg, and Sn $(X_{S2}/X_{S1}) > 1$, and for Ba, Cd, Co, Cs, Mn, Sb, and Zn $(X_{S2}/X_{S1}) < 1$, suggesting adsorption/coagulation and desorption with increasing salinity, respectively.

Equation 16 has been modified by Morris[114] in order to investigate the change in sorptive equilibrium in river water induced by the addition of (re-)suspended particles. Assuming river water is the only source of dissolved metal and salinity-induced sorptive changes are negligible, the mass balance for the turbidized zone is as follows:

$$C_0 + X_0 \cdot SPM_0 + X_a \cdot SPM_a = C_t + X_t(SPM_0 + SPM_t) \quad \textbf{(19)}$$

where superscripts 0, a, and t refer to river water, the added particles, and the resultant turbidized zone, respectively. Given that $K_D = X_0/C_0 = X_t/C_t$ and defining $\alpha = X_a/X_t$ yields:

$$\frac{C_t}{C_0} = \frac{[1 + K_D \cdot SPM_0]}{[1 + K_D \cdot SPM_0 + K_D \cdot SPM_a(1-\alpha)]} \quad \textbf{(20)}$$

For removal to occur, $\alpha < 1$ (i.e., the added particles must be impoverished of particulate equilibrated metal), and the extent of removal increases with increase in both K_D and SPM_a. Figure 6 shows the predicted removal of metals as a function of SPM_a, a potential relative measure of removal in the TMZ of an estuary; values of α have been taken as the ratio of K_D after 1 day to the K_D after 20 days in time-dependent sorption studies conducted in low salinity water (S ~ 0 PSU) by Li et al.[132] (Table 4). Field distributions of Al,[53] and Cu, Ni, and Zn[114] in the upper Tamar Estuary, accord with the general pattern predicted by the model, despite variable environmental conditions (in particular, SPM_0) during sampling. Assuming a value of α of 0.8, and for an SPM_0 concentration of 5 mg l^{-1}, Tamar data yield approximate freshwater K_D values (l Kg^{-1}) as follows:

$$Cu(10^6) > Al(10^5) > Zn(5 \times 10^4) > Ni(10^4)$$

Ackroyd et al.[79] have implemented K_Ds to assess the relative contribution to midestuarine dissolved metal inputs from pore-water infusion and desorption from resuspendable sediment as follows:

$$\lambda = (K_D)_{pw} \rho[(1/v) - 1] \quad \textbf{(21)}$$

where λ is the ratio of the desorption/pore-water inputs, ρ is the density of the particles, v is the fractional pore-water volume, and $(K_D)_{pw}$ is the ratio of exchangeable sorbed metal concentration to dissolved pore-water metal concentration. For typical bed sediments (ρ = 2500 kg m^{-3}; v = 70%), λ approximates to $(K_D)_{pw}$. Mesocosm studies of radiotracers at the sediment-water interface of Narragansett Bay suggest that

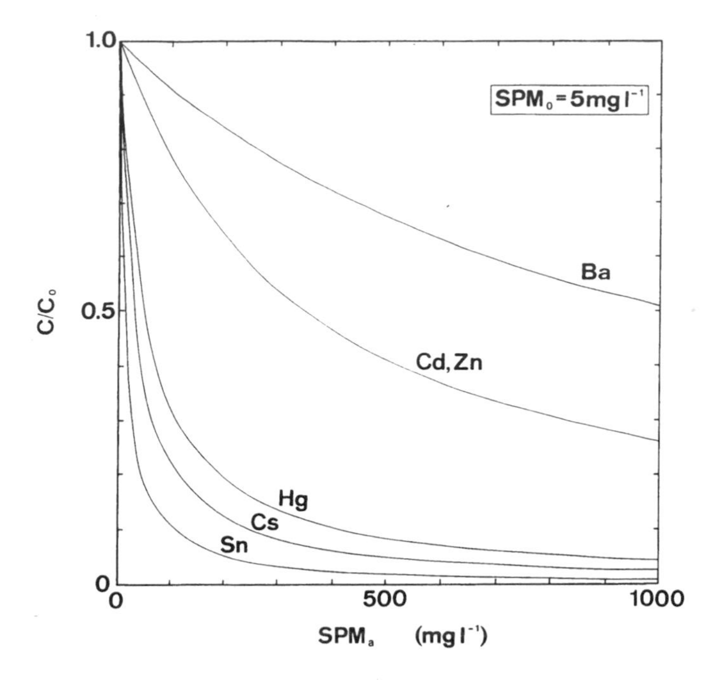

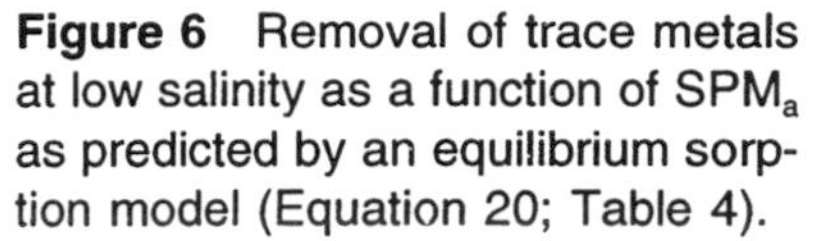

Figure 6 Removal of trace metals at low salinity as a function of SPM_a as predicted by an equilibrium sorption model (Equation 20; Table 4).

$(K_D)_{pw}$s are likely to be considerably lower than K_Ds in the overlying water column, especially for redox sensitive elements;[147] for example, in the top 2 cm of interfacial sediment, $K_D(Cs) = 200$, $K_D(Fe) = 1000$, and $K_D(Mn) = 20$. Nevertheless, an important implication is that the potential contribution to estuarine-dissolved metal inputs from the pore-water reservoir is minor compared with that from desorption of the disturbed sediment itself.

Bale[148] has incorporated K_Ds for Zn derived from batch mixing of end members into a one-dimensional model of solute and particle flux in the Tamar Estuary.[149,150] Solute dispersal coefficients were derived from extensive salinity observations throughout the estuary. The SPM was exchangeable with the bed diurnally and consisted of three populations, riverine, estuarine, and marine. Riverine and marine populations were treated conservatively, the magnitude of the former being related to river flow, and the excess estuarine material comprising the TMZ was related to salinity, tidal range, and river discharge. The exponential relationship between K_D and salinity:

$$K_D = (K_D)_0 \exp(bS) \quad \textbf{(22)}$$

where $(K_D)_0$ quantifies the solid-solution partitioning in river water and b is an empirical curve-fitting constant, and time-dependent end member-dissolved Zn concentrations, were input into the model which was run for one simulated year. Despite the implicit assumption that the particulate Zn inherent in the K_Ds was completely exchangeable, the results were in accord with general observations of dissolved Zn in the estuary,[79] including removal in the TMZ and a midestuarine input under low river discharge.

VII. CONCLUDING COMMENTS: FUTURE WORK

Despite continually improving analytical techniques for the determination of trace metals in environmental samples, there exist limitations in the diagnostic methods available for the interpretation of reactivity from distributional data. Controlled laboratory experiments are, therefore, necessary to define the mechanisms and kinetics of these reactions. In particular, the potential usefulness of basic sorption models merits further investigation of laboratory-derived K_Ds, such as their relation to the composition of medium (master variables) and particles (biogenic, lithogenic, microbial components), and reaction reversibility. To this end, extensive and systematic characterization of estuarine particles (composition, surface area/exchangeable surface sites, surface charge) is also necessary. Thus, it is conceivable that the predictive estuarine chemical models will rely on the integration of more fully parameterized, as opposed to empirical and site-specific, and time-dependent K_Ds with water and particle subpopulation dynamics.

Table 4 **Values for K_D and α used to predict metal removal in turbidized river water**

Metal	K_D at S ~ 0 PSU ($1\ kg^{-1}$)[132] 1 Day	20 Days	α
Ba	4.5×10^3	5.5×10^3	0.82
Cd	1.4×10^4	1.7×10^4	0.82
Cs	8.0×10^4	1.4×10^5	0.57
Hg	3.0×10^4	5.8×10^4	0.52
Sn	1.1×10^5	3.8×10^5	0.29
Zn	0.9×10^4	1.2×10^4	0.75

REFERENCES

1. **Martin, J.-M. and Whitfield, M.,** The significance of the river input of chemical elements to the ocean, in *Trace Metals in Sea Water,* Wong, C. S., Boyle, E., Bruland, K. W., Burton, J. D., and Goldberg, E. D., Eds., NATO Conference Series, Plenum Press, New York, 1983, 265.
2. **Martin, J.-M. and Meybeck, M.,** Elemental mass-balance of material carried by major world rivers, *Mar. Chem.,* 7, 173, 1979.
3. **Michaelis, W., Ed.,** *Estuarine Water Quality Management,* Coastal and Estuarine Studies, Vol. 36, Springer-Verlag, Berlin, 1990, 478.
4. **Dyer, K. R.,** *Estuaries: A Physical Introduction,* John Wiley & Sons, London, 1973.
5. **Boyle, E. A., Huested, S. S., and Grant, B.,** The chemical mass balance of the Amazon plume. II. Copper, nickel, and cadmium, *Deep-Sea Res.,* 29, 1355, 1982.
6. **Uncles, R. J., Elliott, R. C. A., and Weston, S. A.,** Observed fluxes of water, salt and suspended sediment in a partly mixed estuary, *Estuarine Coastal Shelf Sci.,* 20, 147, 1985.
7. **Olausson, E. and Cato, I.,** Eds., *Chemistry and Biogeochemistry of Estuaries,* Wiley-Interscience, Chichester, 1980.
8. **Turner, A., Millward, G. E., and Morris, A. W.,** Particulate metals in five major North Sea estuaries, *Estuarine Coastal Shelf Sci.,* 32, 325, 1991.
9. **Officer, C. B. and Lynch, D. R.,** Dynamics of mixing in estuaries, *Estuarine Coastal Shelf Sci.,* 12, 525, 1981.
10. **Morris, A. W.,** Kinetic and equilibrium approaches to estuarine chemistry, *Sci. Tot. Environ.,* 97/98, 253, 1990.
11. **Hunter, K. A. and Liss, P. S.,** The surface charge of suspended particles in estuarine and coastal waters, *Nature,* 282, 823, 1979.
12. **Loder, T. C. and Liss, P. S.,** Control by organic coatings of the surface charge of estuarine suspended particles, *Limnol. Oceanogr.,* 30, 418, 1985.
13. **Martin, J.-M., Mouchel, J. M., and Nirel, P.,** Some recent developments in the characterisation of estuarine particles, *Water Sci. Technol.,* 18, 83, 1986.
14. **Newton, P. P. and Liss, P. S.,** Positively charged suspended particles: studies in an iron-rich river and its estuary, *Limnol. Oceanogr.,* 32, 1267, 1987.
15. **Turner, A.,** Chemical dynamics in North Sea estuaries and plumes, Ph.D. thesis, Polytechnic South West, Devon, U.K., 1990.
16. **Bale, A. J., Morris, A. W., and Howland, R. J. M.,** Seasonal sediment movement in the Tamar Estuary, *Oceanol. Acta,* 8, 1, 1985.
17. **Morris, A. W., Bale, A. J., Howland, R. J. M., Millward, G. E., Ackroyd, D. R., Loring, D. H., and Rantala, R. T. T.,** Sediment mobility and its contribution to trace metal cycling in a macrotidal estuary, *Water Sci. Technol.,* 18, 111, 1986.
18. **Allen, G. P., Salomon, J. C., Bassolet, P., du Penhoat, Y., and de Granpré, C.,** Effects of tides on mixing and suspended sediment transport in macrotidal estuaries, *Sediment. Geol.,* 26, 69, 1980.
19. **Bale, A. J., Barrett, C. D., West, J. R., and Oduyemi, K. O. K.,** Use of in-situ laser diffraction particle sizing for particle transport studies in estuaries, in *Developments in Estuarine and Coastal Study Techniques,* McManus, J. and Elliott, M., Eds., Olsen and Olsen, Fredensborg, 1989, 133.
20. **Elbaz-Poulichet, F., Holliger, P., Huang, W. W., and Martin, J.-M.,** Lead cycling in estuaries, illustrated by the Gironde, France, *Nature,* 308, 409, 1984.

21. **Millward, G. E., Turner, A., Glasson, D. R., and Glegg, G. A.,** Intra- and inter-estuarine variability of particle microstructure, *Sci. Tot. Environ.,* 97/98, 289, 1990.
22. **Turner, A., Millward, G. E., Schuchardt, B., Schirmer, M., and Prange, A.,** Trace metal distribution coefficients in the Weser Estuary, FRG, *Cont. Shelf Res.,* 12, 1277, 1992.
23. **Morris, A. W., Loring, D. H., Bale, A. J., Howland, R. J. M., and Woodward, E. M. S.,** Particle dynamics, particulate carbon and the oxygen minimum in an estuary, *Oceanol. Acta,* 5, 349, 1982.
24. **Loring, D. H., Rantala, R. T. T., Morris, A. W., Bale, A. J., and Howland, R. J. M.,** Chemical composition of suspended particles in an estuarine turbidity maximum zone, *Can. J. Fish. Aquat. Sci.,* 40 (Suppl. 1), 201, 1983.
25. **Davis, J. A. and Kent, D. B.,** Surface complexation modeling in aqueous geochemistry, in *Mineral-Water Interface Geochemistry,* Reviews in Mineralogy, Vol. 23, Hochella, M. F. and White, A. F., Eds., Mineralogical Society of America, Washington, D.C., 1990, 177.
26. **Morris, A. W.,** Ed., *Practical Procedures for Estuarine Studies,* Natural Environment Research Council, Swindon, U.K., 1983.
27. **Eleftheriou, A. and Holme, N. A.,** Macrofauna techniques, in *Methods for the Study of Marine Benthos,* Holme, N. A. and McIntyre, A. D., Eds., Blackwell Scientific, Oxford, 1984, 140.
28. **Bender, M., Martin, W., Hess, J., Sayles, F., Ball, L., and Lambert, C.,** A whole-core squeezer for interfacial pore-water sampling, *Limnol. Oceanogr.,* 32, 1214, 1987.
29. **Elderfield, H. and Sholkovitz, E. R.,** Rare earth elements in the pore waters of reducing nearshore sediments, *Earth Planet. Sci. Lett.,* 82, 280, 1987.
30. **Schwedhelm, E., Vollmer, M., and Kersten, M.,** Determination of dissolved heavy metal gradients at the sediment-water interface by the use of a diffusion-controlled sampler, *Fresenius Z. Anal. Chem.,* 332, 756, 1988.
31. **Knox, S., Turner, D. R., Dickson, A. G., Liddicoat, M. I., Whitfield, M., and Butler, E. I.,** Statistical analysis of estuarine profiles: application to manganese and ammonium in the Tamar Estuary, *Estuarine Coastal Shelf Sci.,* 13, 357, 1981.
32. **Ebdon, L., Walton, A. P., Millward, G. E., and Whitfield, M.,** Methylated arsenic species in estuarine porewaters, *Appl. Organomet. Chem.,* 1, 427, 1987.
33. **Watson, P. G. and Frickers, T. E.,** A multilevel, in situ pore-water sampler for use in intertidal sediments and laboratory microcosms, *Limnol. Oceanogr.,* 35, 1381, 1990.
34. **Mart, L., Nürnberg, H. W., and Rützel, H.,** Levels of heavy metals in the tidal Elbe and its estuary and the heavy metal input into the sea, *Sci. Tot. Environ.,* 44, 35, 1985.
35. **Balls, P. W. and Topping, G.,** The influence of inputs to the Firth of Forth on the concentrations of trace metals in coastal waters, *Environ. Pollut.,* 45, 159, 1987.
36. **Harper, D. J.,** Dissolved cadmium and lead in the Thames Estuary, *Mar. Pollut. Bull.,* 19, 535, 1988.
37. **Morley, N. H., Fay, C. W., and Statham, P. J.,** Design and use of a clean shipboard handling system for seawater samples, *Adv. Underwater Technol.,* 16, 283, 1988.
38. **Campbell, J. A., Gardner, M. J., and Gunn, A. M.** Short-term stability of trace metals in estuarine water samples, *Anal. Chim. Acta,* 176, 193, 1985.
39. **Moore, R. M., Burton, J. D., Williams, P. J. LeB., and Young, M. L.,** The behaviour of dissolved organic material, iron and manganese in estuarine mixing, *Geochim. Cosmochim. Acta,* 43, 919, 1979.
40. **Whitehouse, B. G.,** Cross-flow filtration of colloids from aquatic environments, *Limnol. Oceanogr.,* 35, 1368, 1990.
41. **van der Sloot, H. A. and Duinker, J. C.,** Isolation of different suspended matter fractions and their trace metal contents, *Environ. Technol. Lett.,* 2, 511, 1981.
42. **Etcheber, H. and Jouanneau, J. M.,** Comparison of the different methods for the recovery of suspended matter from estuarine waters: deposition, filtration and centrifugation: consequences for the determination of some heavy metals, *Estuarine Coastal Shelf Sci.,* 11, 701, 1980.
43. **Valenta, P., Duursma, E. K., Merks, A. G. A., Rützel, H., and Nürnberg, H. W.,** Distribution of Cd, Pb and Cu between the dissolved and particulate phase in the eastern Scheldt and western Scheldt Estuary, *Sci. Tot. Environ.,* 53, 41, 1986.
44. **Golimowski, J., Merks, A. G. A., and Valenta, P.,** Trends in heavy metal levels in the dissolved and particulate phase in the Dutch Rhine-Meuse (Maas) delta, *Sci. Tot. Environ.,* 92, 113, 1990.
45. **van den Berg, C. M. G. and Dharmvanji, S.,** Organic complexation of zinc in estuarine interstitial and surface water samples, *Limnol. Oceanogr.,* 29, 1025, 1984.

46. **van den Berg, C. M. G., Merks, A. G. A., and Duursma, E. K.,** Organic complexation and its control of the dissolved concentrations of copper and zinc in the Scheldt Estuary, *Estuarine Coastal Shelf Sci.,* 24, 785, 1987.
47. **Apte, S. C., Gardner, M. J., and Ravenscroft, J. E.,** An investigation of copper complexation in the Severn Estuary using differential pulse cathodic stripping voltammetry, *Mar. Chem.,* 29, 63, 1990.
48. **Mayer, L. M.,** Retention of riverine iron in estuaries, *Geochim. Cosmochim. Acta,* 46, 1003, 1982.
49. **Mayer, L. M.,** Aggregation of colloidal iron during estuarine mixing: kinetics, mechanisms, and seasonality, *Geochim. Cosmochim. Acta,* 46, 2527, 1982.
50. **Fox, L. E. and Wofsy, S. C.,** Kinetics of removal of iron colloids from estuaries, *Geochim. Cosmochim Acta,* 47, 211, 1983.
51. **Hunter, K. A. and Leonard, M. W.,** Colloidal stability and aggregation in estuaries. I. Aggregation kinetics of riverine dissolved iron after mixing with seawater, *Geochim. Cosmochim. Acta,* 52, 1123, 1988.
52. **Hydes, D. J. and Liss, P. S.,** The behaviour of dissolved aluminium in estuarine and coastal waters, *Estuarine Coastal Mar. Sci.,* 5, 755, 1977.
53. **Morris, A. W., Howland, R. J. M., and Bale, A. J.,** Dissolved aluminium in the Tamar Estuary, southwest England, *Geochim. Cosmochim. Acta,* 50, 189, 1986.
54. **Waslenchuk, D. G. and Windom, H. L.,** Factors controlling the estuarine chemistry of arsenic, *Estuarine Coastal Mar. Sci.,* 7, 455, 1978.
55. **Howard, A. G. and Arbab-Zavar, M. H.,** Determination of inorganic arsenic (III) and arsenic (V), methylarsenic and dimethylarsenic species by selective hydride evolution atomic-absorption spectroscopy, *Analyst,* 106, 213, 1981.
56. **Andreae, M. O., Byrd, J. T., and Froelich, P. N.,** Arsenic, antimony, germanium and tin in the Tejo Estuary, Portugal: modelling a polluted estuary, *Environ. Sci. Technol.,* 17, 731, 1983.
57. **Froelich, P. N., Kaul, L. W., Byrd, J. T., Andreae, M. O., and Roe, K. K.,** Arsenic, barium, germanium, tin, dimethylsulfide and nutrient biogeochemistry in Charlotte Harbor, Florida, a phosphorus-enriched estuary, *Estuarine Coastal Shelf Sci.,* 20, 239, 1985.
58. **Andreae, M. O. and Andreae, T. W.,** Dissolved arsenic species in the Schelde Estuary and watershed, Belgium, *Estuarine Coastal Shelf Sci.,* 29, 421, 1989.
59. **Byrd, J. T.,** Comparative geochemistries of arsenic and antimony in rivers and estuaries, *Sci. Tot. Environ.,* 97/98, 301, 1990.
60. **Seyler, P. and Martin, J.-M.,** Distribution of arsenite and total dissolved arsenic in major French estuaries: dependence on biogeochemical processes and anthropogenic inputs, *Mar. Chem.,* 29, 277, 1990.
61. **Tremblay, G.-H. and Gobeil, C.,** Dissolved arsenic in the St. Lawrence Estuary and the Saguenay Fjord, Canada, *Mar. Pollut. Bull.,* 21, 465, 1990.
62. **Velinsky, D. J. and Cutter, G. A.,** Geochemistry of selenium in a coastal salt marsh, *Geochim. Cosmochim. Acta,* 55, 179, 1991.
63. **Byrd, J. T. and Andreae, M. O.,** Geochemistry of tin in rivers and estuaries, *Geochim. Cosmochim. Acta,* 50, 835, 1986.
64. **Howard, A. G. and Comber, S. D. W.,** The discovery of hidden arsenic species in coastal waters, *Appl. Organomet. Chem.,* 3, 509, 1989.
65. **Ebdon, L., Evans, K., and Hill, S.,** The variation of tributyltin levels with time in selected estuaries prior to the introduction of regulations governing the use of tributyltin-based anti-fouling paints, *Sci. Tot. Environ.,* 68, 207, 1988.
66. **Measures, C. I. and Burton, J. D.,** Behaviour and speciation of dissolved selenium in estuarine waters, *Nature,* 273, 293, 1978.
67. **Measures, C. I. and Edmond, J. M.,** The geochemical cycle of ^{9}Be: a reconnaissance, *Earth Planet. Sci. Lett.,* 66, 101, 1983.
68. **Figueres, G., Martin, J.-M., Meybeck, M., and Seyler, P.,** A comparative study of mercury contamination in the Tagus Estuary (Portugal) and major French estuaries (Gironde, Loire, Rhone), *Estuarine Coastal Shelf Sci.,* 20, 183, 1985.
69. **Cossa, D., Gobeil, C., and Courau, P.,** Dissolved mercury behaviour in the Saint Lawrence Estuary, *Estuarine Coastal Shelf Sci.,* 26, 227, 1988.
70. **Schmidt, D., Freimann, P., and Zehle, H.,** Changes in trace metal levels in the coastal zone of the German Bight, in *Contaminant Fluxes through the Coastal Zone,* Kullenberg, G., Ed., *Rapp. P.v. Réun. Cons. int. Explor. Mer,* 186, 321, 1986.

71. **Lansens, P., Meuleman, Leermakers, M., and Baeyens, W.,** Determination of methylmercury in natural waters by head space gas chromatography with microwave plasma detection on a resin containing dithiocarbamate groups, *Anal. Chim. Acta,* 234, 417, 1990.
72. **Morris, A. W. and Bale, A. J.,** Effect of rapid precipitation of dissolved Mn in river water on estuarine Mn distributions, *Nature,* 279, 318, 1979.
73. **Morris, A. W., Bale, A. J., and Howland, R. J. M.,** The dynamics of estuarine manganese cycling, *Estuarine Coastal Shelf Sci.,* 14, 175, 1982.
74. **Duinker, J. C., Nolting, R. F., and Michel, D.,** Effects of salinity, pH and redox conditions on the behaviour of Cd, Zn, Ni and Mn in the Scheldt Estuary, *Thalassia Jugosl.,* 18, 191, 1982.
75. **Klinkhammer, G. P. and Bender, M. L.,** Trace metal distributions in the Hudson River Estuary, *Estuarine Coastal Shelf Sci.,* 12, 629, 1981.
76. **Windom, H. L., Wallace, G., Smith, R., Dudek, N., Maeda, M., Dulmage, R., and Storti, F.,** Behaviour of copper in southeastern United States estuaries, *Mar. Chem.,* 12, 183, 1983.
77. **Campbell, J. A. and Yeats, P. A.,** Dissolved chromium in the St. Lawrence estuary, *Estuarine Coastal Shelf Sci.,* 19, 513, 1984.
78. **Windom, H. L., Smith, R. G., and Maeda, M.,** The geochemistry of lead in rivers, estuaries and the continental shelf of the southeastern United States, *Mar. Chem.,* 17, 43, 1985.
79. **Ackroyd, D. R., Bale, A. J., Howland, R. J. M., Knox, S., Millward, G. E., and Morris, A. W.,** Distributions and behaviour of dissolved Cu, Zn and Mn in the Tamar Estuary, *Estuarine Coastal Shelf Science,* 23, 621, 1986.
80. **Windom, H. L., Byrd, J., Smith, R., Jr., Hungspreugs, M., Dharmvanij, S., Thumtrakul, W., and Yeats, P.,** Trace metal-nutrient relationships in estuaries, *Mar. Chem.,* 32, 177, 1991.
81. **Shiller, A. M. and Boyle, E. A.,** Dissolved zinc in rivers and estuaries, *Nature,* 317, 49, 1985.
82. **Shiller, A. M. and Boyle, E. A.,** Dissolved vanadium in rivers and estuaries, *Earth Planet. Sci. Lett.,* 86, 214, 1987.
83. **Kuwabara, J. S., Chang, C. C. Y., Cloern, J. E., Fries, T. L., Davis, J. A., and Luoma, S. N.,** Trace metal associations in the water column of south San Francisco Bay, California, *Estuarine Coastal Shelf Sci.,* 28, 307, 1989.
84. **Bewers, J. M. and Yeats, P. A.,** Trace metals in the waters of a partially mixed estuary, *Estuarine Coastal Mar. Sci.,* 7, 147, 1978.
85. **Zwolsman, J. J. G. and van Eck, G. T. M.,** The behaviour of dissolved Cd, Cu and Zn in the Scheldt Estuary, in *Estuarine Water Quality Management,* Coastal and Estuarine Studies, Vol. 36, Michaelis, W., Ed., Springer-Verlag, Berlin, 1990, 413.
86. **Elbaz-Poulichet, F., Guan, D. M., and Martin, J.-M.,** Trace metal behaviour in a highly stratified Mediterranean estuary: the Krka (Yugoslavia), *Mar. Chem.,* 32, 211, 1991.
87. **Prange, A., Niedergesäss, R., and Schnier, C.,** Multielement determination of trace elements in estuaries by TXRF and INAA, in *Estuarine Water Quality Management,* Coastal and Estuarine Studies, Vol. 36, Michaelis, W., Ed., Springer-Verlag, Berlin, 1990, 429.
88. **van der Sloot, H. A., Hoede, D., Wijkstra, J., Duinker, J. C., and Nolting, R. F.,** Anionic species of V, As, Se, Mo, Sb, Te and W in the Scheldt and Rhine Estuaries and the Southern Bight (North Sea), *Estuarine Coastal Shelf Sci.,* 21, 633, 1985.
89. **Hoyle, J., Elderfield, H., Gledhill, A., and Greaves, M.,** The behaviour of rare earth elements during mixing of river and sea waters, *Geochim. Cosmochim. Acta,* 48, 143, 1984.
90. **Boswell, S. M. and Elderfield, H.,** The determination of zirconium and hafnium in natural waters by isotope dilution mass spectrometry, *Mar. Chem.,* 25, 197, 1988.
91. **Toole, J., Baxter, M. S., and Thomson, J.,** The behaviour of uranium isotopes with salinity change in three U.K. estuaries, *Estuarine Coastal Shelf Sci.,* 25, 283, 1987.
92. **Maeda, M. and Windom, H. L.,** Behaviour of uranium in two estuaries of the southeastern United States, *Mar. Chem.,* 11, 427, 1982.
93. **de Groot, A. J., Salomons, W., and Allersma, E.,** Processes affecting heavy metals in estuarine sediments, in *Estuarine Chemistry,* Burton, J. D. and Liss, P. S., Eds., Academic Press, London, 1976, 131.
94. **Ackroyd, D. R., Millward, G. E., and Morris, A. W.,** Periodicity in the trace metal content of estuarine sediments, *Oceanol. Acta,* 10, 161, 1987.
95. **Loring, D. H. and Rantala, R. T. T.,** An intercalibration exercise for trace metals in marine sediments, *Mar. Chem.,* 24, 13, 1988.
96. **Sundby, B., Silverberg, N., and Chesselet, R.,** Pathways of manganese in an open estuarine system, *Geochim. Cosmochim. Acta,* 45, 293, 1981.

97. **Luoma, S. N. and Bryan, G. W.,** A statistical assessment of the form of trace metals in oxidised estuarine sediments employing chemical extractants, *Sci. Tot. Environ.,* 17, 165, 1981.
98. **Luoma, S. N.,** Can we determine the biological availability of sediment-bound trace elements?, *Hydrobiologia,* 176/177, 379, 1989.
99. **Tessier, A. and Campbell, P. G. C.,** Partitioning of trace metals in sediments: relationships with bioavailability, *Hydrobiologia,* 149, 43, 1987.
100. **Lion, L. W., Altmann, R. S., and Leckie, J. O.,** Trace-metal adsorption characteristics of estuarine particulate matter: evaluation of contributions of Fe/Mn oxide and organic surface coatings, *Environ. Sci. Technol.,* 16, 660, 1982.
101. **Kersten, M. and Förstner, U.,** Effect of sample pretreatment on the reliability of solid speciation data on heavy metals — implications for the study of early diagenetic processes, *Mar. Chem.,* 22, 299, 1987.
102. **Martin, J.-M., Nirel, P., and Thomas, A. J.,** Sequential extraction techniques: promises and problems, *Mar. Chem.,* 22, 313, 1987.
103. **Liss, P. S.,** Conservative and non-conservative behaviour of dissolved constituents during estuarine mixing, in *Estuarine Chemistry,* Burton, J. D. and Liss, P. S., Eds., Academic Press, London, 1976, 93.
104. **Bourg, A. C. M.,** Physicochemical speciation of trace elements in oxygenated estuarine waters, in *The Determination of Trace Metals in Natural Waters,* West, T. S. and Nürnberg, H. W., Eds., Blackwell Scientific, Oxford, 1988, 287.
105. **Boyle, E., Collier, R., Dengler, A. T., Edmond, J. M., Ng, A. C., and Stallard, R. F.,** On the chemical mass-balance in estuaries, *Geochim. Cosmochim. Acta,* 38, 1719, 1974.
106. **Officer, C. B.,** Discussion of the behaviour of nonconservative dissolved constituents in estuaries, *Estuarine Coastal Mar. Sci.,* 9, 91, 1979.
107. **Cifuentes, L. A., Schemel, L. E., and Sharp, J. H.,** Qualitative and numerical analyses of the effects of river inflow variations on mixing diagrams in estuaries, *Estuarine Coastal Shelf Sci.,* 30, 411, 1990.
108. **Morris, A. W., Mantoura, R. F. C., Bale, A. J., and Howland, R. J. M.,** Very low salinity regions of estuaries: important sites for chemical and biological reactions, *Nature,* 274, 678, 1978.
109. **Morel, F. M. M., Dzombak, D. A., and Price, N. M.,** Heterogeneous reactions in coastal waters, in *Ocean Margin Processes in Global Change,* Mantoura, R. F. C., Martin, J.-M., and Wollast, R., Eds., John Wiley & Sons, Chichester, 1990, 165.
110. **Bewers, J. M. and Yeats, P. A.,** Transport of river-derived trace metals through the coastal zone, *Neth. J. Sea Res.,* 23, 359, 1989.
111. **GESAMP,** *Land/Sea Boundary Flux of Contaminants: Contributions from Rivers,* Reports and Studies No. 32, UNESCO, Paris, 1987.
112. **Amdurer, M., Adler, D. M., and Santschi, P. H.,** Radiotracers in studies of trace metal behaviour in mesocosms: advantages and limitations, in *Marine Mesocosms Biological and Chemical Research in Experimental Ecosystems,* Grice, G. D. and Reeve, M. R., Eds., Springer-Verlag, Berlin, 1982, 81.
113. **Bale, A. J. and Morris, A. W.,** Laboratory simulation of chemical processes induced by estuarine mixing: the behaviour of iron and phosphate in estuaries, *Estuarine Coastal Shelf Sci.,* 13, 1, 1981.
114. **Morris, A. W.,** Removal of trace metals in the very low salinity region of the Tamar Estuary, England, *Sci. Tot. Environ.,* 49, 297, 1986.
115. **Wollast, R.,** The Scheldt Estuary, in *Pollution of the North Sea: An Assessment,* Salomons, W., Bayne, B. L., Duursma, E. K., and Förstner, U., Eds., Springer-Verlag, Berlin, 1988, 184.
116. **Howard, A. G., Apte, S. C., Comber, S. D. W., and Morris, R. J.,** Biogeochemical control of the summer distribution and speciation of arsenic in the Tamar Estuary, *Estuarine Coastal Shelf Sci.,* 27, 427, 1988.
117. **Shiller, A. M. and Boyle, E. A.,** Variability of dissolved trace metals in the Mississippi River, *Geochim. Cosmochim. Acta,* 51, 3273, 1987.
118. **Cossa, D., Tremblay, G. H., and Gobeil, C.,** Seasonality in iron and manganese concentrations of the St. Lawrence River, *Sci. Tot. Environ.,* 97/98, 185, 1990.
119. **Hering, J. G. and Morel, F. M. M.,** The kinetics of trace metal complexation: implications for metal reactivity in natural waters, in *Aquatic Chemical Kinetics,* Stumm, W., Ed., John Wiley & Sons, New York, 1990, 145.
120. **Jouanneau, J. M., Etcheber, H., and Latouche, C.,** Impoverishment and decrease of metallic elements associated with suspended matter in the Gironde Estuary, in *Trace Metals in Sea Water,* Wong, C. S., Boyle, E., Bruland, K. W., Burton, J. D., and Goldberg, E. D., Eds., NATO Conference Series, Plenum Press, New York, 1983, 245.

121. **Morris, A. W., Bale, A. J., Howland, R. J. M., Loring, D. H., and Rantala, R. T. T.,** Controls of the chemical composition of particle populations in a macrotidal estuary (Tamar Estuary, U.K.), *Cont. Shelf Res.,* 7, 1351, 1987.
122. **Sholkovitz, E. R.,** Flocculation of dissolved organic and inorganic matter during the mixing of river water and seawater, *Geochim. Cosmochim. Acta,* 40, 831, 1976.
123. **Sholkovitz, E. R.,** The flocculation of dissolved Fe, Mn, Al, Cu, Ni, Co and Cd during estuarine mixing, *Earth Planet. Sci. Lett.,* 41, 77 1978.
124. **Sholkovitz, E. R. and Copland, E. R.,** The coagulation, solubility and adsorption properties of Fe, Mn, Cu, Ni, Cd, Co and humic acids in a river water, *Geochim Cosmochim. Acta,* 45, 181, 1981.
125. **Yeats, P. A. and Strain, P. M.,** The oxidation of manganese in seawater: rate constants based on field data, *Estuarine Coastal Shelf Sci.,* 31, 11, 1990.
126. **Diem, D. and Stumm, W.,** Is dissolved Mn^{2+} being oxidised by O_2 in absence of Mn-bacteria or surface catalysts?, *Geochim. Cosmochim. Acta,* 48, 1571, 1984.
127. **Vojak, P. W. L., Edwards, C., and Jones, M. V.,** Evidence for microbiological manganese oxidation in the River Tamar Estuary, south west England, *Estuarine Coastal Shelf Sci.,* 20, 661, 1985.
128. **Sunda, W. G. and Huntsman, S. A.,** Microbial oxidation of manganese in a North Carolina Estuary, *Limnol. Oceanogr.,* 32, 552–564.
129. **Bourg, A. C. M.,** Role of fresh water/sea water mixing on trace metal adsorption phenomena, in *Trace Metals in Sea Water,* Wong, C. S., Boyle, E., Bruland, K. W., Burton, J. D., and Goldberg, E. D., Eds., Plenum Press, New York, 1983, 195.
130. **Honeyman, B. D. and Santschi, P. H.,** Metals in aquatic systems, *Environ. Sci. Technol.,* 22, 862, 1988.
131. **Millward, G. E. and Moore, R. M.,** The adsorption of Cu, Mn and Zn by iron oxyhydroxide in model estuarine solutions, *Water Res.,* 16, 981, 1982.
132. **Li, Y.-H., Burkhardt, L., and Teraoka, H.,** Desorption and coagulation of trace elements during estuarine mixing, *Geochim. Cosmochim. Acta,* 48, 1879, 1984.
133. **Johnson, C. A.,** The regulation of trace element concentrations in river and estuarine waters contaminated with acid mine drainage: the adsorption of Cu and Zn on amorphous Fe oxyhydroxides, *Geochim. Cosmochim. Acta,* 50, 2433, 1986.
134. **Bourg, A. C. M.,** Trace metal adsorption modelling and particle-water interactions in estuarine environments, *Cont. Shelf Res.,* 7, 1319, 1987.
135. **Comans, R. N. J. and van Dijk, C. P. J.,** Role of complexation processes in cadmium mobilization during estuarine mixing, *Nature,* 336, 151, 1988.
136. **Turner, A., Millward, G. E., Bale, A. J., and Morris, A. W.,** The solid-solution partitioning of radiotracers in the southern North Sea — in situ radiochemical experiments, *Cont. Shelf Res.,* 12, 1311, 1992.
137. **Evans, D. W., Alberts, J. J., and Clark, R. A., III,** Reversible ion-exchange fixation of cesium-137 leading to mobilization from reservoir sediments, *Geochim. Cosmochim Acta,* 47, 1041, 1983.
138. **Sanders, J. G. and Abbe, G. R.,** The role of suspended sediments and phytoplankton in the partitioning and transport of silver in estuaries, *Cont. Shelf Res.,* 7, 1357, 1987.
139. **Nyffeler, U. P., Li, Y.-H., and Santschi, P. H.,** A kinetic approach to describe trace-element distribution between particles and solution in natural aquatic systems, *Geochim. Cosmochim. Acta,* 48, 1513, 1984.
140. **Millward, G. E., Glegg, G. A., and Morris, A. W.,** Cu and Zn removal kinetics in estuarine waters, *Estuarine Coastal Shelf Sci.,* 35, 37, 1992.
141. **Jannasch, H. W., Honeyman, B. D., Balistrieri, L. S., and Murray, J. W.,** Kinetics of trace element uptake by marine particles, *Geochim. Cosmochim. Acta,* 52, 567, 1988.
142. **Turner, D. R., Whitfield, M., and Dickson, A. G.,** The equilibrium speciation of dissolved components in freshwater and seawater at 25°C and 1 atm pressure, *Geochim. Cosmochim. Acta,* 45, 855, 1981.
143. **van Eck, G. T. M. and De Rooij, N. M.,** Development of a water quality and bio-accumulation model for the Scheldt Estuary, in *Estuarine Water Quality Management,* Coastal and Estuarine Studies, Vol. 36, Michaelis, W., Ed., Springer-Verlag, Berlin, 1990, 95.
144. **Mantoura, R. F. C., Dickson, A., and Riley, J. P.,** The complexation of metals with humic materials in natural waters, *Estuarine Coastal Mar. Sci.,* 6, 387, 1978.
145. **Luoma, S. N. and Davis, J. A.,** Requirements for modeling trace metal partitioning in oxidised estuarine sediments, *Mar. Chem.,* 12, 159, 1983.

146. **Olsen, C. R., Cutshall, N. H., and Larsen, I. L.,** Pollutant-particle associations and dynamics in coastal marine environments: a review, *Mar. Chem.,* 11, 501, 1982.
147. **Santschi, P. H., Amdurer, M., Adler, D., O'Hara, P., Li, Y.-H., and Doering, P.,** Relative mobility of radioactive trace elements across the sediment-water interface in the MERL model ecosystems of Narragansett Bay, *J. Mar. Res.,* 45, 1007, 1987.
148. **Bale, A. J.,** The Characteristics, Behaviour and Heterogeneous Chemical Reactivity of Estuarine Suspended Particles, Ph.D. thesis, Plymouth Polytechnic, Devon, U.K. 1987.
149. **Harris, J. R. W., Bale, A. J., Bayne, B. L., Mantoura, R. F. C., Morris, A. W., Nelson, L. A., Radford, P. J., Uncles, R. J., Weston, S. A., and Widdows, J.,** A preliminary model of the dispersal and biological effect of toxins in the Tamar Estuary, England, *Ecol. Model.,* 22, 253, 1984.
150. **Uncles, R. J., Woodrow, T. Y., and Stephens, J. A.,** Influence of long-term sediment transport on contaminant dispersal in a turbid estuary, *Cont. Shelf Res.,* 7, 1489, 1987.

Chapter 11

Trace Elements in the Oceans

John R. Donat and Kenneth W. Bruland

CONTENTS

I. INTRODUCTION

Over the past 15 years a revolution has occurred in marine and environmental chemistry regarding our knowledge of the distributions and chemical behavior of trace elements in natural waters, particularly seawater. Important factors initiating this revolution were the development and adoption of noncontaminating or "clean" techniques for collection, preservation, storage, and analysis of seawater samples, and major advances in modern analytical methods and instruments. As a result, seawater concentrations of many trace elements have been shown to be factors of 10 to 1000 lower than those previously accepted. Vertical concentration profiles have been found to be consistent with known biological, physical, and/or geochemical processes operating within the ocean. As the oceanic concentrations of most of the trace elements have been determined, the oxidation states and chemical forms or speciation of certain trace elements in the oceans have been identified. Complexation of certain trace

0-8493-6304-7/95/$0.00+$.50

metals with inorganic and organic ligands has been demonstrated. Organometallic forms of certain trace elements have been identified and their concentrations determined. Thus, a new picture of the trace element composition of seawater has developed.

The progress made in advancing our understanding of trace element chemistry in seawater up through the early 1980s was demonstrated in the major collection of papers edited by Wong et al.[1] Thorough reviews of this field have been prepared by Burton and Statham,[2–4] Bruland,[5] and Whitfield and Turner.[6] We refer the reader to these sources and the references cited within them for detailed information. In this chapter, we do not attempt a comprehensive survey of the elements for which reliable data are now available. Rather, we concentrate on presenting recent highlights regarding some of the more interesting aspects of oceanic trace element chemistry, and we summarize background information on input and removal processes, and sampling, preanalysis, and analytical considerations.

II. CONTRIBUTING AND REMOVAL PROCESSES

Trace elements in seawater have two major external sources: (1) atmospheric or riverine inputs of weathering products of the exposed continents; and (2) inputs resulting from the interactions of seawater with newly formed oceanic crustal basalt at ridge-crest spreading centers via both high temperature hydrothermal activity and low temperature interactions with newly formed oceanic crust.

While the riverine inputs of the major elements to the oceans are known fairly accurately, those of the majority of the trace metals are not, primarily because the concentrations of many of the trace elements in rivers are not known reliably. This situation results most likely from insufficient control of sample contamination and lack of validation of the data obtained. Examples of reliable studies that have been performed include those of Shiller and Boyle[7] and Trefrey et al.[8] who, using sampling and analytical methods proven to eliminate or reduce trace element contamination of natural water samples, determined concentrations of Cu, Ni, Cd, Fe, and Zn in the Mississippi River that were much lower than those for the same region published annually in the *U.S. Geological Survey Water Data Reports.* Shiller and Boyle[9] demonstrated that dissolved Zn concentrations commonly quoted as typical in rivers are as much as a factor of 100 too high. Table 1 presents the most recent reliable data on world average riverine trace element concentrations and fluxes to ocean margins.[10] Martin and Windom[10] state that a need still exists for additional reliable data from river systems draining geologically, climatologically, and physiologically diverse watersheds.

Chester and Murphy[11] and Buat-Ménard[12] have recently reviewed the literature regarding the geochemical significance of atmospheric inputs of trace elements to the oceans. Arimoto et al.[13] have provided a similar review specific to the Pacific Ocean. Chester and Murphy[11] compiled reports showing that the atmospheric flux of some metals, especially Pb, Zn, and Cd, to surface coastal waters can equal or exceed their riverine flux. They also attempted to illustrate the relative importance of atmospheric and riverine inputs of various trace metals to individual ocean basins such as the North and South Pacific and the North Atlantic (Table 2). Their estimates indicate that atmospheric deposition has the most influence on the surface waters of the North Atlantic and the least on those of the South Pacific. In addition, they estimate that, averaged over an entire ocean basin, the atmospheric dissolved fluxes for Mn, Ni, Co, Cr, V, and Cu are an order of magnitude less than their riverine dissolved fluxes; the two fluxes are the same order of magnitude for Al, Fe, and Cd, but the atmospheric fluxes are dominant for Zn and Pb. If, however, just open ocean regimes are examined, atmospheric inputs overwhelmingly predominate for particle reactive elements such as Fe and Al because the riverine inputs of these elements are largely removed by intense particle scavenging in the coastal ocean.

The estimates made by Chester and Murphy[11] required estimates of the solubility of aerosol-associated trace metals in seawater. Such solubilities are experimentally difficult to obtain and the values are the subject of much discussion. For example, the wide variability in the estimates of the solubility of Fe(III) in surface seawater, and their dependence upon the pore size of the filter used, provide a striking example of our lack of definition of this process. In the first study to examine aerosol dissolution using sufficiently high dilution factors such that the solubility of Fe was not exceeded, Zhuang et al.[14] found the "saturated concentration" of dissolved atmospheric Fe passing through a 0.4-μm filter to be 10 to 17 n*M*, while that passing through a 0.05-μm filter was 5 to 8 n*M*. Sunda[15] estimated the solubility of truly dissolved Fe(III) to be 1.5 n*M* at pH 8.2 and 20 to 25°C. In a review of Fe biogeochemistry in seawater, Wells[16] argues for a maximum solubility of 0.1 n*M*. Wells[16] suggests that the discrepancy in the different values is due to the existence of colloidal Fe. Reliable estimates of the soluble fractions of trace metals such as Fe in seawater and data on colloidal metals in the open ocean are still needed.

Table 1 **Atmospheric dissolved vs. fluvial dissolved inputs to regions of the world ocean (μg/cm²/year)**

	Oceanic Region					
	North Atlantic		North Pacific		South Pacific	
Element	Fluvial Dissolved Flux[a]	Atmospheric Dissolved Flux	Fluvial Dissolved Flux	Atmospheric Dissolved Flux	Fluvial Dissolved Flux	Atmospheric Dissolved Flux
Al	0.56	0.25	0.24	0.09	0.12	0.0066
Fe	0.18	0.24	0.076	0.062	0.038	0.0035
Mn	0.18	0.025	0.076	0.004	0.038	0.0013
Ni	0.017	0.008	0.0074	—	0.0037	—
Co	0.0044	0.00061	0.0019	0.00004	0.0009	0.000006
Cr	0.011	0.0014	0.0047	—	0.0023	—
V	0.022	0.0043	0.0094	0.0017	0.0047	—
Cu	0.037	0.0075	0.016	0.0021	0.0079	0.0013
Pb	0.0022	0.093	0.0009	0.0022	0.0005	0.0004
Zn	0.016	0.059	0.0071	0.029	0.0035	0.0026
Cd	0.0018	0.0016	0.0008	0.00023	0.0004	—

[a] No attempt has been made to adjust the North Atlantic fluvial flux for European and North American anthropogenic inputs.

From Chester, R. and Murphy, K. J. T., Metals in the marine atmosphere, in *Heavy Metals in the Marine Environment,* Rainbow, P. S. and Furness, R. W., Eds., CRC Press, Boca Raton, Florida, 1990, 27. With permission.

Hydrothermal activity associated with the formation of a new oceanic crust exerts a major influence on the composition of seawater. High temperature systems, such as the "black smokers" on the East Pacific Rise in which hot (~350°C), acidic (pH ~3.5), sulfide-, and metal-rich solutions emerge directly into surrounding seawater, are major oceanic sources of the trace elements Fe and Mn, and the minor elements Li and Rb, and the major oceanic sinks for Mg and SO_4^{2-}. More detailed discussions of the trace element solution chemistry of hydrothermal systems can be found in the literature reviewed by Thompson[17] and Edmond,[18] in the NATO Advanced Research Institute volume edited by Rona et al.,[19] and in papers by Von Damm et al.[20,21]

Most dissolved trace elements are ultimately removed from seawater to marine sediments primarily by adsorption onto sinking particles ("scavenging") or by incorporation into biological phases via active uptake by phytoplankton followed by sinking of biological detritus. However, prior to their ultimate removal, trace elements may undergo various degrees of recycling which can involve chemical desorption reactions or redissolution from particles ("regeneration") as the particle carrier phases oxidize and/or dissolve. Regeneration may occur within the water column, and within surficial sediments followed by diffusion of the dissolved trace elements back into the water column, allowing these elements to participate again in their internal cycling within the oceans.

III. SAMPLING AND PREANALYSIS HANDLING

The recent revolution in our knowledge of the distributions and forms of trace elements in seawater results primarily from major advances in the development of: (1) "clean techniques" to eliminate or control contamination during sampling, storage, and analysis, (2) preconcentration and separation techniques, and (3) extremely sensitive analytical instruments.[5] These major advances in sampling and preanalysis handling techniques are reviewed below; the advances in analytical techniques and instruments are presented in the subsequent section.

A. GENERAL CONTAMINATION CONTROL CONSIDERATIONS

The importance of using clean, noncontaminating methods in all stages of trace element determinations was emphasized by Moody and Beary:[22]

Table 2 **Riverine trace element concentrations and fluxes to ocean margins**

	Concentrations		Fluxes	
	Dissolved µg/l	Particulate µg/g	Dissolved kg/yr	Particulate kg/yr
Metalloids				
As	1.7	5	65	75
Nutrient-type				
Cd	0.01	1.2	0.4	18
Cu	1.5	100	58	1,500
Ni	0.5	90	19	1,350
Zn	0.6	250	23	3,750
Geochemically controlled				
Al	50	94,000	1,925	1,410,000
Co	0.1	20	3.8	300
Fe	40	48,000	1,540	720,000
Mn	8.2	1,050	320	15,750
Pb	0.03	35	1.1	525
U	0.24	3	9.2	45
Rare earth elements				
Ce	0.08	80	3.1	1,200
Eu	0.001	1.5	0.04	23
La	0.05	45	1.9	675
Sm	0.008	7	0.31	105

From Martin, J.-M. and Windom, H. L., Present and future roles of ocean margins in regulating marine biogeochemical cycles of trace elements, in *Ocean Margin Processes in Global Change,* Mantoura, R. F. C., Martin, J.-M., and Wollast, R., Eds., John Wiley & Sons, New York, 1991, 45. With permission.

> "In the present-day real world of trace element analyses, the theoretical or even practical sensitivity limit of the instrument is very often not the limiting factor in the accuracy of an analysis. Instead the size and variability of the analytical blank is the principal limitation."

Trace element analysts must be concerned with the reduction and control of this blank through all steps of the analysis, from choosing containers to making the final quantitative measurement. Sample contamination can be caused by the method of handling or containment, by apparatus, by the use of analytical reagents, by the environment in which the sampling or analysis is performed, and even by the presence and influence of the analyst.

The specific means used to reduce the analytical blank may vary from element to element, but use of Class 100, clean-air work spaces, and purified reagents has become common for present day trace element analysis of natural waters. For most laboratories involved with the analysis of trace elements in seawater, the cost of a complete Class 100 laminar-flow laboratory is prohibitive. Instead, a more pragmatic and less stringent approach is adopted whereby the laboratory is flushed with a nonlaminar flow of filtered air, and within this "clean" area, several Class 100, laminar-flow work benches provide dust-free conditions for all critical manipulations.[23] Additional precautions include elimination of corroding metal components (or those with the potential to corrode) and unnecessary instrumentation, and covering walls and ceilings with nonshedding paint or plastic.

Reduction of blanks can be extremely difficult during work at sea aboard a research vessel because the ambient laboratory air and work area can be severely contaminated with high trace element levels. However, clean-air work areas can be arranged by mounting a portable clean-room container or van onto the research vessel's deck or by constructing a clean room within the vessel's laboratory space from heavyweight polyethylene sheeting. These options have worked very effectively for a number of different research groups.

High purity water, acids, and reagents are essential for a trace element analytical scheme. Although Milli-Q systems (Millipore) appear adequate to produce high purity water, the method of subboiling distillation has been the primary means to prepare purified acids and many other reagents (see Moody and Beary[22]).

Bottles for sample storage should, in general, be Teflon or polyethylene because of their inherent cleanliness as well as their ability to be cleaned further of any trace element content by acid leaching. Cleaning procedures can vary from lab to lab, but generally involve the initial use of detergents and/or solvents (e.g., ethanol, acetone), extended strong-acid cleaning with or without heating and/or ultrasonification, with repeated rinses with successively cleaner water between soak solutions, and final storage in ultraclean weak acids and/or water. The final stages of cleaning and handling of the sample bottles are performed within a Class 100 clean bench. The cleaned bottles are sealed in multiple polyethylene bags to protect them from dust during storage and transport.

B. SAMPLING

The requirements for contamination control during sampling vary tremendously from one trace element to another, dependent primarily upon the concentration of the element or species in seawater and its ubiquity. For example, a research vessel can be a significant source of contamination (e.g., from stack dust, steel hydrowire, paint, brass fittings, galvanized materials, sacrificial Zn anodes, bilge and holding tank discharges, etc.) for many trace elements during collection of the water samples. For some trace elements (e.g., Ba, V, As, and Se) contamination is relatively easy to control, whereas for other elements (e.g., Pb, Zn, Fe, Hg, and Sn) more painstaking procedures are required to collect uncontaminated samples.

The requirement for clean procedures can best be appreciated by reviewing C.C. Patterson's zealous and successful efforts to determine common Pb accurately in seawater. The decrease of three orders of magnitude in the accepted concentration of Pb in seawater over the last four decades is not a real oceanographic effect, but rather an artifact of successive improvements in reduction and control of the level of contamination introduced during sampling, storage, and analysis.[24] Although Patterson had at his disposal an elegant analytical technique, isotope dilution mass spectrometry, the problems of field contamination in collecting a deep seawater sample with a Pb concentration of only 5 p*M* (one part per trillion) had to be overcome.

Schaule and Patterson[24] developed their own exotic deep water sampler in order to collect samples uncontaminated with Pb. Subsequently, Martin and co-workers[25] showed that commercially available, Teflon-coated GO-Flo (General Oceanics) samplers mounted on nonmetallic (Kevlar) hydroline[26] enabled them to produce a large set of accurate and precise Pb concentration data. The use of Teflon-coated GO-Flo sampling bottles mounted on a nonmetallic hydroline was developed and first reported by Bruland and co-workers.[26] This sampling system is currently the most commonly used for obtaining deep water profiles of trace metals. Recently, researchers at Moss Landing Marine Laboratories have designed and assembled a Nylon-II coated, stainless steel rosette system which holds eight 30-liter, Teflon-coated and modified GO-Flo bottles. This system is deployed on a polyurethane-coated Keylar conducting cable, and was used to collect uncontaminated water samples during the recent IronEx experiments near the Galapagos Islands.[26a]

Consistent data for open-ocean surface waters for elements such as Pb, Hg, Cd, Cu, and Zn have been obtained from samples either collected by hand from rafts rowed well away from the research vessel,[26,27] collected by extending a pole over the side of the ship while the ship is steaming at approximately 2 knots,[28] or pumped while underway from a towed "fish".[29]

C. FILTRATION AND STORAGE

The definition of "dissolved" trace metals is operational and is dependent upon the filter pore size used. Nuclepore polycarbonate filters of either 0.3- or 0.4-μm pore size have been utilized for the vast majority of trace metal studies, although some researchers have recently switched to using 0.2 μm pore size filters to ensure removal of all microorganisms. After capturing the sample, the GO-Flo sampler is usually pressurized with filtered, high-purity nitrogen through a Swagelok fitting at the top of the sampler. The water sample is then pushed out of a Teflon stopcock at the bottom of the sampler, through Teflon tubing to the filtration system. A commonly used filtration system consists of precleaned, 142-mm, 0.4-μm pore size Nuclepore polycarbonate filters contained in Teflon filter holders. Water samples for dissolved trace metal assays are then collected from the filter effluent line within a clean environment.

There is considerable variety among investigators on the method chosen to preserve samples for later analysis. Ideally, the analysis would occur *in situ.* However, the bulk of the studies that have been performed to date have involved sample analysis back in the home laboratory of the investigator or, at best, aboard the research vessel. Depending upon the analysis to be performed, samples have been preserved by acidification to pH values between 1 and 2 with either ultrapure nitric or hydrochloric acid, by storing the unacidified sample in the dark at temperatures from 1 to 4°C, or by storing the sample frozen. Certain analyses preclude some of the above preservation methods. For example, samples for metal speciation determinations should not be acidified.

D. PRECONCENTRATION STEPS

Because most of the instrumental analytical methods currently available do not have the requisite selectivity, sensitivity, or freedom from matrix interferences despite great advances in sensitivity made recently, trace elements in natural waters, particularly seawater, must be concentrated (and separated from the matrix for some techniques) prior to assay. The majority of the oceanographically consistent trace element results have been obtained using a wide variety of selective preconcentration steps prior to the analysis: hydride generation techniques for the determination of the trace metalloids, electrochemical preconcentration in anodic stripping voltammetry, adsorptive electrochemical preconcentration in cathodic stripping voltammetry, liquid-liquid extraction with chelating agents such as dithiocarbamates, chelating cation exchange with resins such as Chelex-100, and coprecipitation methods using, for example, Co pyrrolidine dithiocarbamate as a carrier.

These preconcentration techniques also serve to minimize matrix effects which can cause difficulties when nanomolar and picomolar levels of trace elements are being determined in a solution that is roughly 0.5 *M* NaCl. In general, the use of various chemical methods of concentration, matrix alteration, and species isolation is an essential supplement to the powerful instrumental techniques available. Many of the concentration steps would provide a sample amenable to analysis by a variety of detection techniques. Radioactive isotopes of the elements of interest have proven to be invaluable as yield tracers in developing these various preconcentration methods.

IV. ANALYTICAL CONSIDERATIONS

The sensitivity of a number of instrumental techniques has improved dramatically over the last 15 years. For example, there has been the development of graphite-furnace atomic absorption spectrometry and inductively coupled plasma emission spectrometry, as well as major advances in the areas of anodic stripping voltammetry, cathodic stripping voltammetry (preceded by a metal-chelate adsorption step), gas and liquid chromatography, mass spectrometry, etc. In the following sections, selected applications of these various approaches are presented.

A. GRAPHITE-FURNACE ATOMIC ABSORPTION SPECTROMETRIC METHODS

With regard to the transition metals, graphite-furnace atomic absorption spectrometry has been the single most important instrumental technique used in advancing our knowledge of the dissolved metal distributions in seawater. It has primarily been used on samples that have been preconcentrated by solvent-solvent extractions using chelating agents such as dithiocarbamates,[26,30–33] 8-hydroxyquinoline,[34,35] and dithizone;[36,37] by chelating ion-exchange resins such as Chelex-100[26,31,38] and immobilized 8-hydroxyquinoline resins;[39,40] and by coprecipitations of metal chelates such as Co dithiocarbamates.[41] Van Geen and Boyle[42] developed an automated preconcentration method based on adsorption of the metal complexes of sodium bis(2-hydroxyethyl) dithiocarbamate onto a hydrophobic resin. Graphite-furnace atomization has the advantage of being highly sensitive and using only a small sample size (e.g., 10 to 100 μl). This makes the method amenable to sequential multielement analysis. For example, Boyle and Edmond,[41] Kingston et al.,[38] Bruland et al.,[26] Smith and Windom,[37] Sturgeon et al.,[33] and van Geen and Boyle[42] used various preconcentration techniques and GFAAS to determine Co, Cu, Cd, Fe, Mn, Ni, Pb, and Zn in seawater. The recent use of atomization platforms within the graphite furnace has been a substantial advance to minimize matrix effects and ensure uniform atomization from sample to sample.

B. VOLTAMMETRIC STRIPPING METHODS

The attributes of voltammetric methods and their applications to trace element analysis of seawater have been reviewed by Nürnberg[43] and more recently by van den Berg.[44] Advantages of voltammetric methods include their sensitivity, small sample volume requirement, amenability towards real-time shipboard analysis, and speciation determination capability.

Differential pulse anodic stripping voltammetry (DPASV) using a rotating, glassy carbon, mercury film electrode under conditions developed to minimize contamination sources and to enhance sensitivity for seawater matrices[31,45] has been effectively utilized for determining both the concentrations and speciation of Cu, Pb, Cd, and Zn in seawater. In DPASV, preconcentration is achieved electrochemically and as an integral (i.e., not separate) part of the determination, typically for deposition periods of 5 to 30 min.

Recently, sensitive techniques based on cathodic stripping voltammetry (CSV) have been developed by van den Berg and co-workers (see review by van den Berg[44] and references cited therein) and others[46-48] that preconcentrate the trace metal-of-interest by adsorption of surface-active complexes onto a hanging mercury drop electrode (HMDE) followed by a reductive stripping step producing a cathodic current. The advantage of CSV is the wide applicability of the technique. Unlike DPASV, amalgamation of the trace metal-of-interest with the mercury electrode is not required for preconcentration, thus essentially opening up the periodic table for analysis by this method. Trace metals determined in seawater by this approach to date include Al, Cd, Co, Cr, Cu, Fe, Ni, Pb, Pt, Sb, Se, Sn, Ti, V, and Zn. Detection limits by this approach are typically in the subnanomolar and picomolar range.

C. GAS CHROMATOGRAPHIC METHODS

Measures and co-workers have developed electron capture/gas chromatographic methods for the determination of Se(IV),[49] Be,[50] and Al[51] in seawater. The methods are based upon forming a volatile organic chelate or derivative of the trace element in the sample, extraction into a small volume of organic solvent, and determination by electron capture gas chromatography. These techniques are extremely sensitive, and detection limits in the picomolar range can be achieved. In addition, they use small sample volumes, the instrumentation costs are low, and the methods have been used at sea to obtain large sets of near-real-time data. Surprisingly, these methods have not caught on in popularity amongst other trace element labs. A liability of the methods is that many metal chelates are unstable at the high temperatures required for sufficient volatility.

D. MASS SPECTROMETRIC METHODS

Historically, isotope dilution mass spectrometry (IDMS) has played an important role in trace element determinations in natural waters. Patterson and co-workers[24,52,53] used this elegant instrumental technique to achieve the first accurate Pb concentrations in ocean waters. Stukas and Wong[54] used thermal source IDMS to determine Cu, Cd, Pb, Zn, Ni, and Fe concentrations simultaneously in seawater in a single experiment. Berman, Sturgeon, and co-workers at the National Research Council of Canada have used it to put confidence limits on their reported values for their seawater standard reference materials.[33,55,56] It has been used for the bulk of the studies on rare earth elements in seawater.[57-61] However, the liabilities of this approach are the initial cost of instrumentation and relatively slow sample throughput. On the other hand, the power of the technique lies in measurement of the isotopic ratios of those elements with geochemically interesting signatures or fingerprints, such as Pb. Currently, there is a great deal of interest on the potential of inductively coupled plasma emission mass spectrometry (ICPMS). This technique is proving particularly useful for refractory elements such as Ti[62] and Au.[63,64]

E. CHEMILUMINESCENCE DETECTION METHODS

Recently, chemiluminescence methods for the determination of Co, Cu, Mn, and Fe with detection limits in the picomolar and nanomolar ranges have been introduced. Boyle et al.[65] developed a method for the analysis of Co in natural waters by cation-exchange liquid chromatography using luminol chemiluminescence detection (LC/CL). Johnson and co-workers have developed methods for the determination of picomolar levels of Co[66] and nanomolar levels of Cu,[67] Mn,[68] and Fe(II)[69] in seawater by flow injection analysis with chemiluminescence detection (FIA/CL). Obata et al.[69a] developed an automated shipboard method for determining Fe(III) in seawater which couples chelating resin column preconcentration with chemiluminescence detection. These techniques are sensitive, rapid, require small sample volumes, and are particularly amenable to shipboard and even *in situ* analyses.

F. MISCELLANEOUS METHODS

In addition to the methods discussed above, a variety of other methods have been developed for various elements as we show in the following examples. Cutter[70,71] developed a selective hydride generation method for the determination of Se in seawater involving cryogenic trapping of the hydride, followed by thermal desorption and determination by atomic absorption spectrometry using a quartz cuvette burner with an air-hydrogen flame (quartz burner/AAS). Andreae[72] described the development and application of hydride generation techniques for determination of As, Sb, Ge, and Sn species in seawater. The hydrides of these elements were quantified by a variety of detection systems: quartz burner/AAS for As and Sb species, electron capture detection for some methylarsines, flame photometric detection for Sn species, and graphite-furnace atomic absorption detection for Ge and Sn species. Gill and Bruland[73] measured Hg concentrations and speciation in natural waters using a selective sample reduction method and detection via two-stage Au amalgamation to introduce gas-phase Hg° into the cell of a nondispersive atomic fluorescence system (AFS). The AFS system is markedly more sensitive than similar systems using atomic absorption spectrometric detection.[74] Cutter et al.[75] developed a selective hydride generation, gas chromatographic/photoionization detection method for the simultaneous determination of inorganic As and Sb species (As (III) + Sb(III) and As (III + V) + Sb(III + V)) in natural waters.

V. CONCENTRATION LEVELS

The advances in analytical chemistry and instrumentation and control of potential contamination have enabled corresponding advances in our knowledge of the true oceanic concentrations of dissolved metals and their distributions. Since the review by Bruland,[5] the seawater concentrations of Ti, Ga, the Pt group metals, Re, Te, Au, Zr, and Hf have been determined. Thus, we now know reliably the concentrations and distributions of most of the elements in the periodic table. The surface and deepwater concentrations of the trace elements in open North Atlantic and North Pacific Ocean waters are presented in Table 3. Whitfield and Turner[6] have grouped the trace elements into three principal categories reflecting their biogeochemical interactions with particles: conservative, recycled, and scavenged.

A. CONSERVATIVE ELEMENTS

The conservative trace elements interact weakly with particles, have long residence times (> 10^5 yr) relative to the mixing time of the oceans, and have concentrations that maintain a constant ratio to one another, varying only as a result of water mass mixing. Trace elements in this category include the monovalent cations Li^+, Rb^+, and Cs^+, Tl^+, the negatively charged carbonato complex ($UO_2(CO_3)_3^{4-}$) of the oxy-cation UO_2^{2+}, the oxy-anion MoO_4^{2-}, and, interestingly, the methylated forms of Ge monomethylgermanic acid ($CH_3Ge(OH)_3^{\circ}$) and dimethylgermanic acid ($(CH_3)_2Ge(OH)_2^{\circ}$).

B. RECYCLED ELEMENTS

Recycled (or nutrient-type) elements appear to be involved with the internal cycles of biologically derived particulate material. Consequently, their concentrations are depleted in surface waters by direct uptake by phytoplankton and/or adsorption onto biogenic particles, and increase with depth as sinking detrital particles undergo microbial decomposition of their organic phases, and/or dissolution of their inorganic mineral phases (e.g., opal, calcium carbonate). In addition, the concentrations of the recycled elements in deep waters increase along the direction of the main advective flow of deep water in the world ocean. Consequently, their concentrations in deep North Pacific waters are higher than in deep North Atlantic waters. The residence times of the recycled elements are intermediate (10^3 to 10^5 yr) to those of the conservative and scavenged elements.

Two of the most striking examples of recycled elements are Cd[26,76–79] and Zn.[78–80] The vertical distribution of Cd correlates closely with that of phosphate, suggesting that Cd is cycled with the formation and decomposition of organic soft tissues, while the vertical distribution of Zn correlates most closely with that of silicate (Figure 1). Depletion of these elements in surface waters and their regeneration at depth causes deep waters to be enriched relative to surface waters by up to 1000-fold for Cd and 180-fold for Zn. Deepwater enrichment of these metals and the deepwater flow pattern results in substantial fractionation between ocean basins; deep North Pacific Cd and Zn concentrations are 3 and 5 times greater, respectively, than the concentrations at comparable depths in the North Atlantic.

C. SCAVENGED ELEMENTS

Because of their strong interactions with particles, the scavenged elements have very short oceanic residence times (<10^3 yr). Their concentrations are maximum near, and decrease with distance from, their sources which include rivers, atmospheric dust, hydrothermal sources, and bottom sediments. In general,

Table 3 **Concentrations and distribution types of the trace elements in ocean waters**

Element	Concentration Units	North Pacific		North Atlantic		Distribution Type
		Surface	Deep	Surface	Deep	
Be	p*M*	4–6	28–32	10	20	R + S
Al	n*M*	0.3–5	0.5–2	30–43	15–35	S
Sc	p*M*	8	18	14	20	R
Ti	p*M*	4–8	200–300	30–60	200	R + S
V	n*M*	32	36	23		R
Cr	n*M*	3	5	3.5	4.5	R
Mn	n*M*	0.5–3	0.08–0.5	1–3	0.25–0.5	S
Fe	n*M*	0.02–0.5	0.5–1	0.05–1	0.6–1	R + S
Co	p*M*	4–50	10–20	18–300	20–30	S
Ni	n*M*	2	11–12	2	6	R
Cu	n*M*	0.5–1.3	4.5	1.0–1.3	2	R + S
Zn	n*M*	0.1–0.2	8.2	0.1–0.2	1.6	R
Ga	p*M*	12	30	25–30	?	S
Ge	p*M*	5	100	1	20	R
As	n*M*	20	24	20	21	S
Se	n*M*	0.5	2.3	0.5	1.5	R
Y	n*M*				(0.15)	?
Zr	p*M*	12–95	275–325	100	(150)[c]	R + S
Nb	p*M*		(<50)			?
Ru	f*M*		(<50)			?
Rh			(?)			?
Pd	p*M*	0.18	0.66			
Ag	p*M*	1–5	23			R
Cd	p*M*	1–10	1000	1–10	350	R
In	p*M*	<0.5–1.8		2.7	0.9	S?
Sn	p*M*	4		10–20	8	?
Sb	n*M*	(0.74–1.13)		1.7		?
Te	p*M*	1.2	1	1–1.5	0.4–1	S
Cs	n*M*	2.3	—	—	—	C
La	p*M*	20	50–70	12–15	80–84	S + R
Ce	p*M*	10	3	80	40–60	S
Pr	p*M*	3–4	7–9	3–4	10	S + R
Nd	p*M*	13–16	40–50	12.8	24.9	S + R
Sm	p*M*	2.6–2.8	8–9	3–4	7.6–8.0	S + R
Eu	p*M*	0.73	2.3	0.6–0.8	1.6–1.8	S + R
Gd	p*M*	3.8	12–13			S + R
Tb	p*M*	0.56	1.8–2.1	0.73	1.4–1.6	S + R
Dy	p*M*		6.1	4.8	5.1	?
Ho	p*M*	0.7–1.0	4–5	1.5–1.8	2.5–2.7	S + R
Er	p*M*		5.8	4.1	5.1	?
Tm	p*M*	0.3–0.5	2.0–2.5	0.7–1.0	1.1–1.3	S + R
Yb	p*M*	1.9–2.8	13–17	3.8–5.1	7.0–7.4	S + R
Lu	p*M*	0.3–0.4	2.3–3.1	0.7–0.8	1.5–1.6	S + R
Hf	p*M*	0.2–0.4	1–2	0.4	(0.8)[c]	?
Ta	p*M*		(<14)			?
W	n*M*		(0.6)			?
Re	p*M*	28–82		32–43		?
Os				(?)		?
Ir	f*M*	(5–30)[a]				?
Pt	p*M*	0.4	0.3,1.2	0.2–0.4[b]	0.2–0.4[b]	?[d]
Au	f*M*	50–150		50–150		?

Table 3 (Continued) **Concentrations and distribution types of the trace elements in ocean waters**

Element	Concentration Units	North Pacific Surface	North Pacific Deep	North Atlantic Surface	North Atlantic Deep	Distribution Type
Hg	p*M*	0.5–10	2–10	1–7	1	S
Tl	p*M*	60–80	80	60–70	60	C
Pb	p*M*	14–50	3–6	100–150	20	S
Bi	p*M*	0.2	0.02	0.25		S

Note: R, Recycled type; S, Scavenged type; C, Conservative type.

[a] SIO Pier sample; [b] Indian Ocean concentrations; [c] depth = 1250 m; [d] R, S, and C distributions have been reported.

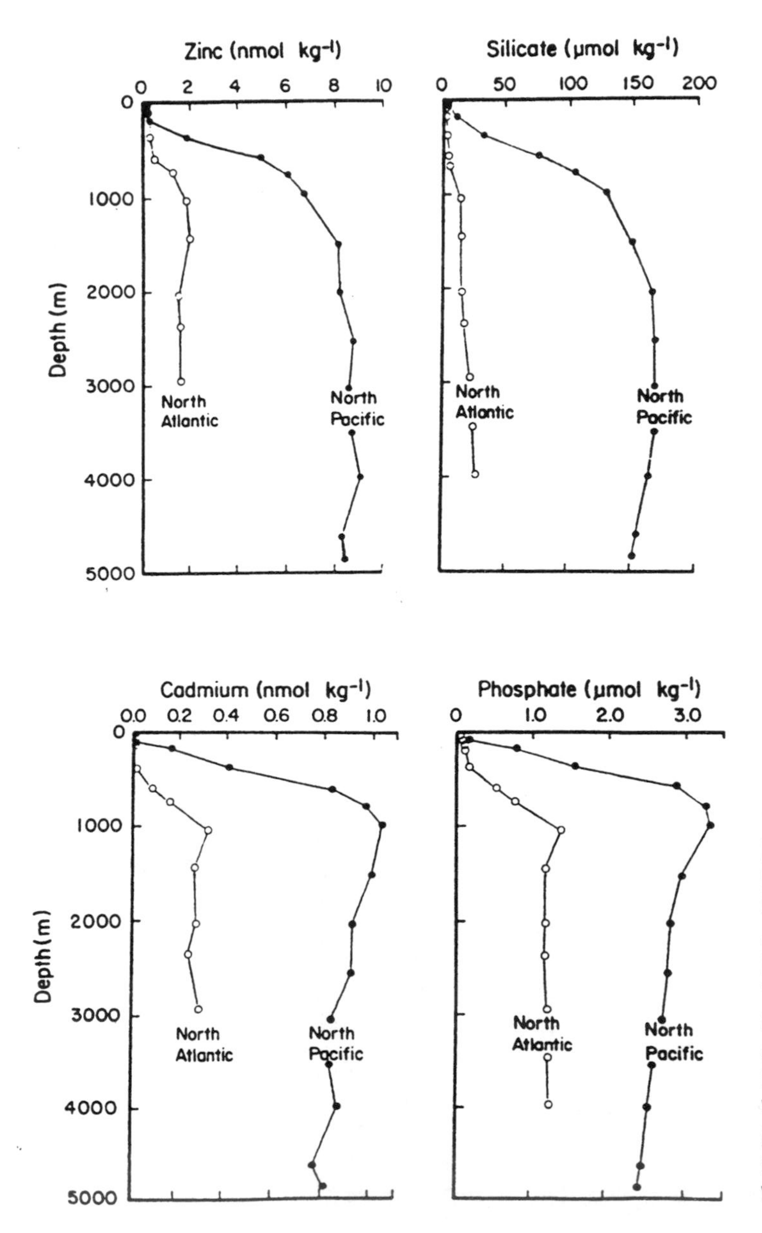

Figure 1 Vertical concentration profiles of dissolved cadmium, phosphate, zinc, and silicic acid in the North Pacific and North Atlantic Oceans. (From Bruland, K.W., *Chemical Oceanography,* Vol. 8, Riley, J.P. and Chester, R., Eds., Academic Press, 1983; and Bruland, K.W. and Franks, R.P., *Trace Metals in Seawater,* Wong, C.S., Boyle, E., Bruland, K.W., Burton, J.D., and Goldberg, E.D., Eds., Plenum Press, 1983. With permission.)

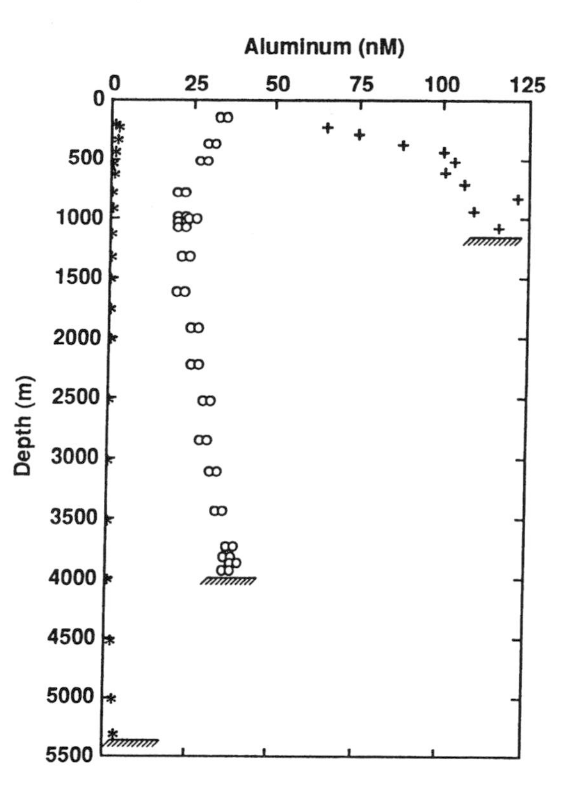

Figure 2 Vertical concentration profiles of dissolved aluminum: (*) North Pacific;[86,87] (o) North Atlantic;[81] (+) Mediterranean (From Orians, K.J. and Bruland, K.W., *Nature,* 316, 427, 1985; Orians, K.J. and Bruland, K.W., *Earth. Planet. Sci. Letts.,* 78, 402, 1986; Hydes, D.J., *Science,* 205, 1261, 1979; and Measures, C.I. and Edmond, J.M., *J. Geophys. Res.,* 93, 591, 1988. With permission.)

concentrations of the scavenged elements decrease along the direction of deepwater flow due to continuing particle scavenging. Thus, concentrations in the North Pacific are lower than in the North Atlantic.

The seawater distribution of Al best illustrates the general characteristics of the scavenged elements (Figure 2). Seawater concentrations of dissolved Al are maximal in surface waters due to atmospheric input, minimal at middepths due to particle scavenging, and increase toward the seafloor.[81–87] North Pacific deepwater Al concentrations[86,87] are 8- to 40-fold lower than those in the North Atlantic[81,82,85,88,89] and about 100-fold lower than those in the Mediterranean.[90–93] This marked interocean fractionation is the reverse of that shown by the recycled elements and the greatest yet observed for any element in seawater.

D. ELEMENTS SHOWING HYBRID DISTRIBUTIONS

Some trace elements, such as Cu and Fe, have distributions that are influenced by both recycling and scavenging processes. Like recycled elements, dissolved Cu[78,94,95] and Fe[96–98] are often depleted in surface waters in areas of high productivity and regenerated at depth. In less productive waters or in areas with higher external aeolian inputs, these elements can exhibit surface maxima more indicative of scavenged elements.

Instead of the rapid increase in concentration with depth shown by recycled elements, dissolved Cu concentrations increase only gradually with depth, due to the combined effects of regeneration and scavenging in deep waters. Also consistent with the recycled nature of its behavior, dissolved Cu concentrations in Atlantic deep waters are one half to two thirds of those in the deep Pacific.[79,99] The oceanic residence time of Cu is about 1000 years.

Seawater constituents showing recycled behavior like nitrate, phosphate, and Cd are assimilated primarily in surface waters, are regenerated with depth, and have long oceanic residence times. Unlike these elements, regenerated Fe is scavenged relatively intensely throughout the oceans, including deep waters. The short oceanic residence time of Fe (thought to be less than 100 years) reflects this scavenging. Thus, the oceanic concentrations of Fe result from a balance between regeneration from the "rain" of particulate matter and scavenging.

E. RECENT ADVANCES

The recent advances in our knowledge of the trace elements in the oceans can be placed into two general categories: (1) concentrations and distributions and (2) biogeochemical processes.

1. Concentrations and Distributions

Through the use of the newer analytical techniques mentioned in previous sections, especially those amenable to shipboard analyses, several researchers have gone beyond producing single vertical or horizontal profiles. These researchers have made large numbers of measurements allowing construction of 2- and even 3-dimensional ocean basin-scale distribution maps for certain elements (e.g., Al and Fe).

Using electron capture gas chromatography, Measures and Edmond[100] determined dissolved Al concentrations in water samples from 18 vertical profiles and 35 surface samples aboard ship during the South Atlantic Ventilation Experiment. The data coverage provided by the shipboard analyses allowed vertical sections of Al concentrations in the South Atlantic to be constructed (Figure 3) which were useful in examining the input and distribution mechanisms of Al in a region of the Atlantic for which no previous information existed. Subsurface Al distributions were used to examine the evolution of the enriched Al signal of recently descended North Atlantic Deep Water propagating through the South Atlantic.

Martin et al.[101] reported a dissolved Fe section in the Northeast Pacific from 25°N, 137°W north to Alaska (Figure 4), obtained by APDC/DDDC-chloroform organic extraction preconcentration and GFAAS detection. This section shows general surface depletion and enrichment with depth above 500 m for all stations. Inputs of Fe along the continental margin in the near-surface waters at Stn T-9 and in the oxygen-minimum zones (800 to 1000 m) at offshore Stns T-2 to T-6 and nearshore Stn T-8 are also evident. The small maximum at T-4 was suggested to result from water flowing offshore from California 1500 km to the east, whereas the relatively Fe-poor waters at Stn T-6 are thought to be flowing towards California from the central Pacific.

In addition to obtaining basin-wide distributions of the trace elements in the oceans, researchers continue to develop and apply new analytical techniques to obtain zero- and first-order understandings of the oceanic concentrations and distributions of the remaining elements in the periodic table (e.g., Ti, Ga, Ru, Pd, Ir, Pt, Au, Re, Te, Zr, and Hf).

Dissolved Ti have been determined concentrations in North Pacific[62] and North Atlantic[62,101a] waters. Although the inorganic speciation of Ti is predicted to be dominated by particle reactive hydrolysis species (predominantly $TiO(OH)_2^o$), dissolved Ti shows a recycled-type profile with surface water depletion and deep water enrichment (Figure 5). However, indications of scavenging behavior also exist. North Pacific surface water concentrations range from 4 to 8 p*M* and increase linearly with depth to maximum concentrations of 200 to 300 p*M* near the bottom. North Atlantic surface concentrations are ~60 p*M* and increase linearly with depth to ~200 p*M* at 2000 m. In the northwest Atlantic, Skrabal[101a] reported that dissolved Ti concentrations decrease from 390 p*M* off the mouth of Delaware Bay to ~30 to 60 p*M* in the outer shelf and slope waters. Locally-high Ti concentrations of 90 to 140 p*M* were determined at the dynamic shelf/slope boundary, due perhaps to shallow upwelling. Orians et al.[62] estimated residence times of Ti in the deep waters of the relatively highly productive North Pacific Sub-Arctic gyre of 100 to 200 years, and they speculated that the residence time of Ti in the deep waters of the North Pacific central gyre (lower productivity) might be longer, about 800 to 900 years.

Orians and Bruland[102] reported the vertical distribution of dissolved Ga in the North Pacific and down to 1000 m in the Gulf Stream. They determined Ga concentrations after preconcentrating their filtered samples using the chelating ion-exchange resin Chelex-100 and evaporation (North Pacific) or using an 8-hydroxyquinoline resin (Gulf Stream), followed by detection by GFAAS. North Pacific concentrations increase from surface values of 12 to 18 p*M* near 300 m, decrease to 7 to 10 p*M* between 500 to 1000 m, and increase to 30 p*M* from 1000 m to the bottom. Gulf Stream concentrations are approximately three times higher than those at comparable depths in the North Pacific. The vertical distribution of dissolved Ga reflects surface, subsurface, and deep water sources combined with particle scavenging throughout the water column. Thus, Ga appears to show a hybrid oceanic distribution; its moderately short deep water residence time of 750 years and interocean fractionation (concentrations in North Atlantic > North Pacific) reflect scavenging behavior, while its vertical distribution reflects recycling behavior. Shiller[103] observed that the Ga/Al ratios of seawater are distinctly greater than that in crustal rocks and suggested that the Ga enrichment could be due to anthropogenic input or, more likely, due to preferential dissolution of solid phase Ga and preferential removal of dissolved Al.

Results of the first analyses of the trace metals Ru,[104] Ir,[105] Pd,[106] and Re[107,107a] in Pacific Ocean waters have been reported. Pacific Ocean seawater appears to have Ru[104] and Ir[105] concentrations in the low femtomolar range. Pd concentrations in Pacific surface waters are 0.18 p*M* and deep water concentrations are 0.66 p*M*.[106] The vertical distribution of Re in Pacific Ocean waters shows no consistent trend; dissolved concentrations range from 28 to 82 n*M*.[107]

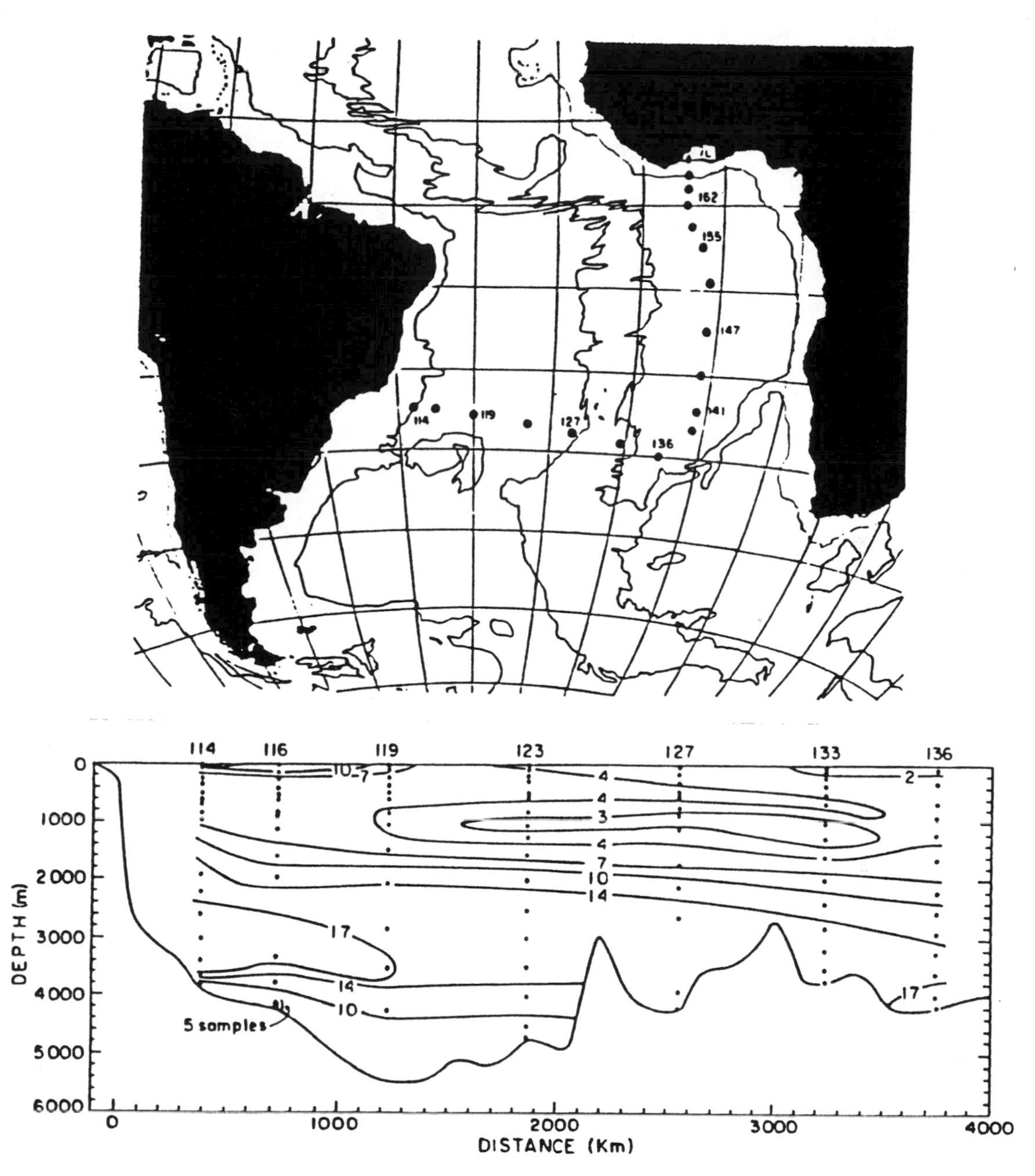

Figure 3 (Upper) Station positions for leg III of the South Atlantic Ventilation Experiment (SAVE) cruise. (Lower) Vertical cross-section of dissolved aluminum concentrations (n*M*) in the western basin of the South Atlantic, SAVE Stns 114 to 136. (From Measures, C.I. and Edmond, J.M., *J. Geophys. Res.*, 95, C4, 5331, 1990. With permission.)

Falkner and Edmond[64] reported vertical profiles of Au from the Atlantic, Pacific, and Mediterranean determined using anion exchange preconcentration and ICPMS detection.[63] Atlantic and Pacific concentrations were similar (~50 f*M* to 150 f*M*) with Au concentrations appearing fairly uniform with depth. Mediterranean deep water concentrations are higher, ranging from 100 to 150 f*M*.

Lee and Edmond[108] have recently determined the seawater concentrations of Te for the first time. Te exists in two oxidation states in seawater, (IV) and (VI). The distributions of both oxidation states reflect surface enrichment and strong scavenging at depth. Total Te concentrations appear to average about 1.2 to 1.3 p*M* in surface waters and ~0.5 to 0.6 p*M* at depths below 2500 m.

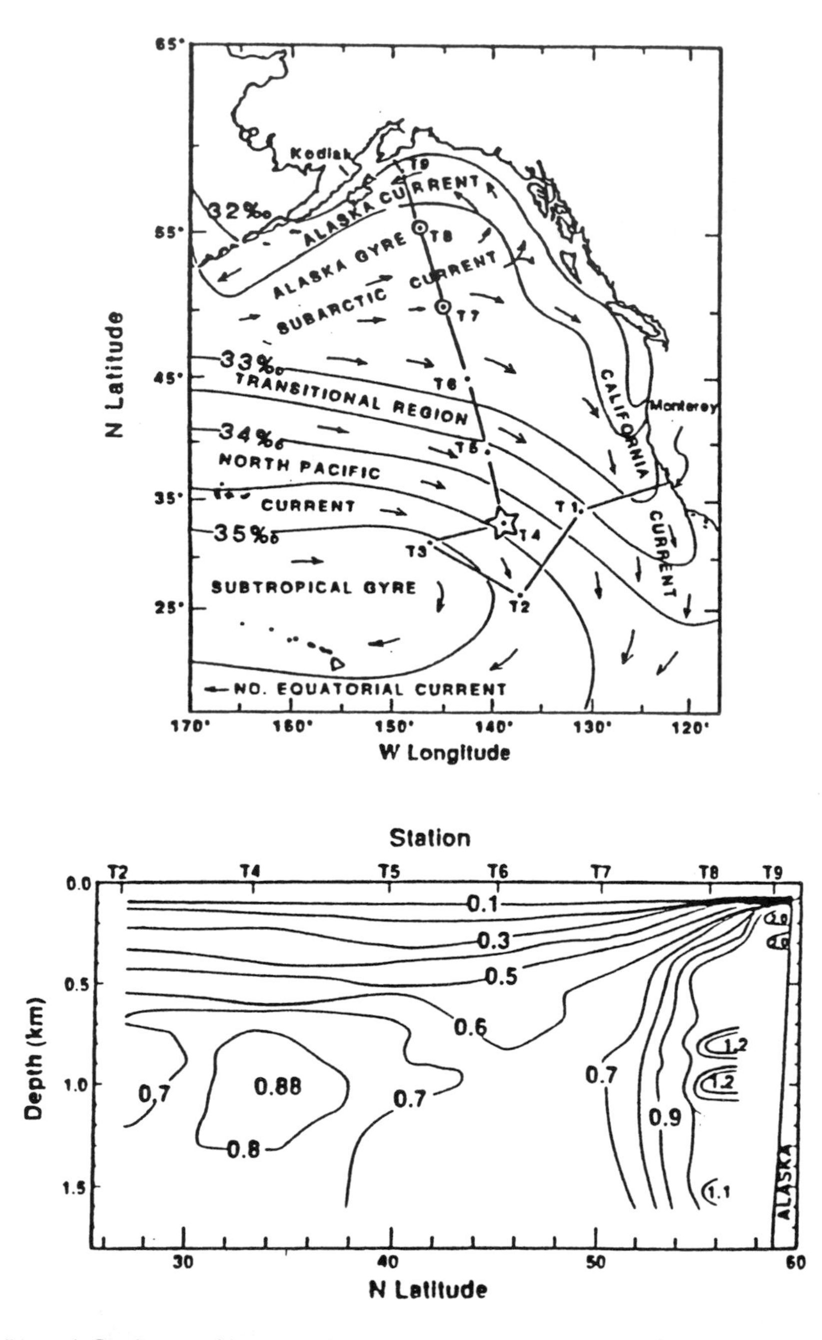

Figure 4 (Upper) Station positions, surface current systems, and surface isohalines for Vertical Transport and Exchange (VERTEX) seasonal and Alaska transect cruises in the Northeast Pacific. (Lower) Vertical cross-section of dissolved iron concentrations in the Northeast Pacific, VERTEX Stns T2 to T9. (From Martin, J.H., Gordon, R.M., Fitzwater, S.E., and Broenkow, W.W., *Deep-Sea Res.*, 36, 649, 1989. With permission.)

Concentrations of Pt have been determined in Pacific,[107a,109,110] Atlantic,[107a] and Indian Ocean waters.[112] However, the distributions reported in these studies sharply contrast each other. In the Pacific, Hodge et al.[109,110] and Goldberg[111] reported a recycled-type profile for Pt, while in Indian Ocean waters Jacinto and van den Berg[112] reported a scavenged-type profile (Figure 6). Most recently, Colodner's[107a] observations agree with neither of the previous reports; she measured Pt concentrations in both Pacific and Atlantic waters that were invariant with depth! Although the vertical profiles of dissolved Pt show smooth variation with depth (one criterion for "oceanographic consistency"), the data presently available

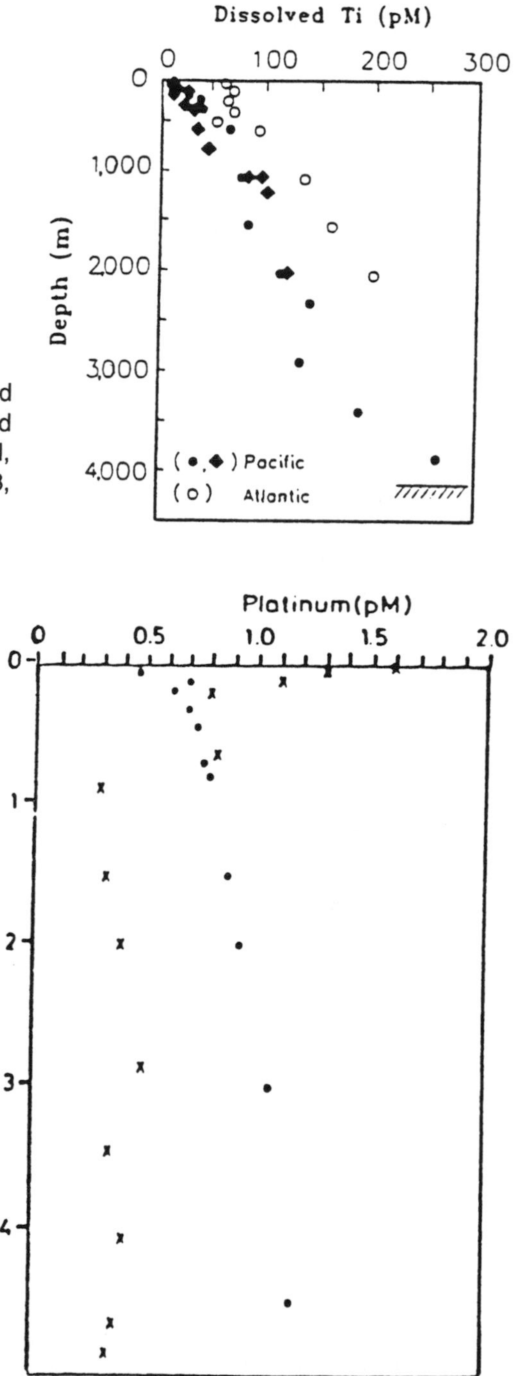

Figure 5 Vertical concentration profiles of dissolved titanium in the North Pacific: 50°N, 145°W (o), and 33°N, 139°W (♦); and in the North Atlantic: 32°N, 64°W (o). (From Orians, K.J., et al., *Nature,* 348, 323, 1990. With permission.)

Figure 6 Vertical concentration profiles of dissolved platinum: (o) North Pacific;[110] (x) western Indian Ocean. (From Jacinto, G.S. and van den Berg, C.M.G., *Nature,* 338, 333, 1989. With permission.)

for Pt are not "oceanographically consistent" with respect to their basin-to-basin variation. More information is required to settle this apparent discrepancy.

McKelvey and Orians have measured the first vertical profiles of Zr[112a,b] and Hf[112b] in the North Pacific and North Atlantic. Both elements show surface depletion and deep water enrichment, and evidence of coastal and bottom sources. North Pacific Zr and Hf concentrations range, respectively, from 12 to 95 p*M* and 0.2 to 0.4 p*M* in surface waters, to 275 to 325 p*M* and 1 to 2 p*M* in deep waters. A single North Atlantic profile showed surface Zr and Hf concentrations, respectively, of ~100 p*M* and ~0.4 p*M*, and concentrations of ~150 p*M* and ~0.8 p*M* at 1250 m depth.

2. Trace Element Biogeochemical Processes

Recent highlights regarding biogeochemical processes involving trace elements include photomediated processes such as photochemical control of Mn, Fe, and Cu redox cycling, and the formation of, and interactions of dissolved trace elements with, colloids.

Calculations for oxygenated seawater based on the redox couple between MnO_2(s) or MnOOH(s) and Mn^{2+} suggest that the equilibrium concentration of dissolved Mn should be much less than 1 n*M*, possibly less than 1 p*M*, depending on the oxide mineral phase with which it is in equilibrium. The observed values for dissolved Mn in open ocean waters are normally between 0.1 and 3 n*M*,[97,98,101] and a maximum in the dissolved Mn concentration is a prominent feature in surface waters of the oceans. Sunda and Huntsman[113,114] have presented evidence that the higher-than-expected concentrations and the surface maximum of dissolved Mn may be maintained by photoreduction of Mn oxides and by photoinhibition of Mn oxidizing microorganisms. Sunda and Huntsman[114] observed pronounced diel cycles in the formation rates of particulate Mn in coastal surface waters of the Bahama Islands which, they suggest, is caused by nightly removal of dissolved Mn^{2+} by Mn-oxidizing bacteria, followed by its daily regeneration and maintenance by reductive photodissolution processes and photoinhibition of the Mn-oxidizing bacteria. Reductive photodissolution of Mn oxides may have important implications for organic matter in seawater because this process may involve light-activated ligand-to-metal charge transfer reactions between Mn oxides and bacterial extracellular organic polymers associated with them or from photochemical production of reductants such as O_2^- and H_2O_2 in seawater or within bacterial MnO_x aggregates.[115]

Fe may be the most important of all the trace elements essential to phytoplankton, and evidence exists that Fe may limit primary production in nutrient-rich, oligotrophic open ocean regimes having low atmospheric Fe inputs.[98,101,116,117] Photoreductive dissolution of Fe(III)-oxides and Fe colloids, and photoreduction of organically chelated Fe may be a major mechanism influencing the availability of Fe to phytoplankton. Waite and Morel[118] and Rich and Morel[119] demonstrated the importance of photoreductive dissolution of particulate and colloidal Fe in seawater. The photoreaction does not appear to proceed in the absence of organics in synthetic seawater[119] or in UV-irradiated coastal seawater.[118] Although no evidence exists for complexation of Fe by organic ligands in oceanic waters, the photoreduction of particulate and colloidal Fe may have potentially important implications for the concomitant photooxidation of organic matter associated with Fe in seawater.

Thermodynamic equilibrium considerations predict that nearly all of the dissolved Cu in seawater should exist as Cu(II). However, Cu(II) may be reduced to Cu(I) in seawater by a number of potentially important reactions, many of which are photochemically induced.[120] For example, Cu(I) may be produced by direct photoreduction of Cu(II) organic complexes (with simultaneous oxidation of the organic ligand via Cu catalysis) and/or by reduction of Cu(II) by photochemically produced reductants. Moffett and Zika[121] and Moffett et al.[122] have provided evidence for these mechanisms operating in surface seawaters. Vertical profiles of Cu(I) in surface seawaters are consistent with a photochemical production mechanism: concentrations show a surface maximum and a decrease with depth.[121] The vertical concentration profile of L_1, the stronger of two Cu-complexing organic ligands whose complexes dominate the speciation of dissolved Cu(II) in oceanic surface waters, in the Sargasso Sea shows concentrations decreasing toward the surface, possibly due to photochemical decomposition potentially involving Cu-catalyzed photooxidation.[122]

As mentioned previously, removal of most trace elements from seawater is controlled by the oceanic cycle of particulate matter, through reactions in the water column and at the sediment interface. For particle reactive metals that have no biological function, partitioning onto marine particulate material is the dominant removal process. Partitioning onto particles controls the dissolved concentration of many trace elements in the oceans. Several recent studies have focused on the role of colloids (particles in the 0.5 nm to 0.4μm range) in particle sedimentation[123] and in particle scavenging of trace elements in seawater (e.g., Honeyman and Santschi[124] and references cited therein; Moran and Moore[125]).

Farley and Morel[123] studied the dynamics of coagulation and settling of particles in well-mixed systems for wastewater discharge. In analytical, numerical, and laboratory studies, they observed that during the initial period of sedimentation a characteristic particle size distribution develops as smaller particles are aggregated into larger particles through coagulation, resulting in an increased rate of solids removal but relatively constant particle mass concentrations. After this initial period, the rate of particle removal (via sedimentation) becomes characteristic of the particle mass concentration, regardless of the initial particle concentration. Farley and Morel[123] determined that the characteristic rate of solids removal as a function of particle mass concentration appears to be described by the summation of three power laws corresponding to particular coagulation mechanisms including differential settling, shear, and Brownian motion.

Honeyman and Santschi[124] used the same form of Farley and Morel's[123] equation describing the particle-concentration-dependent particle removal rate in establishing their conceptual and mathematical "Brownian-pumping" model. The "Brownian-pumping" model describes the transfer of truly dissolved metal species to larger, filterable particles through a colloidal intermediate emphasizing Brownian motion as the coagulation mechanism. In this model, the transfer consists of two rate steps: (1) rapid formation of metal/colloid surface site complexes, and (2) slow coagulation of colloids with filterable particles. This model indicates that trace element behavior in natural waters results from the tight coupling of chemical equilibrium and physico-chemical interactions between particles. It also provides a framework for interpreting the behavior of the particle reactive trace element Th and perhaps other trace elements prone to adsorptive particulate scavenging.

Moran and Moore[125] used cross-flow filtration to study the size distribution of Al and organic carbon in an operationally defined colloidal size range (1 nm to 0.45μm) in coastal and open ocean waters off Nova Scotia. They determined that colloidal Al was <5 to 15% of the dissolved Al. Colloidal organic carbon was < 10 to 15% of the dissolved organic carbon (measured using UV-photooxidation).

VI. SPECIATION

A. INTRODUCTION AND BACKGROUND

Although our knowledge of the oceanic concentrations and distributions of the trace metals has advanced dramatically, it has become increasingly clear that this information alone is insufficient for providing a complete understanding of a trace metal's biological and geochemical interactions. Trace metals dissolved in seawater can exist in different oxidation states and chemical forms (species) including free solvated ions, inorganic complexes (e.g., with Cl^-, OH^-, CO_3^{2-}, SO_4^{2-}, etc.), organometallic compounds, and organic complexes (e.g., with phytoplankton metabolites, proteins, humic substances).

Knowledge of the distribution of a trace metal's total dissolved concentration amongst its various forms (speciation) is extremely important because the different oxidation states and chemical forms undergo very different biological and geochemical interactions. For example, Fe(III) and Mn(IV) are much less soluble than their reduced forms (Fe(II) and Mn(II)). The toxicity and nutrient availability of several transition metals to phytoplankton have been shown to decrease as a result of complexation with ligands such as EDTA, indicating that the toxicity and availability of these metals are proportional to their free metal ion activities (see Sunda[15] for a recent review). Organic complexation may greatly decrease[126,127] or, in some cases, even increase[127–129] adsorption of metals onto metal oxide particles. Thus, speciation information is necessary to attempt to fully understand a trace metal's marine biogeochemical cycle.

B. INORGANIC SPECIATION

Inorganic forms of the trace elements in seawater include hydrated metal ions and complexes with inorganic ligands. These inorganic forms can also include trace elements in different oxidation states because the potential required for the elements to change valence states falls within the range of the oxidizing/reducing potentials developed in the elements' immediate environment. Important examples of trace elements exhibiting redox changes in seawater include Fe (Fe(II)/Fe(III)),[130–132] Mn (Mn(II)/Mn(IV)),[130–132] Se (Se(IV)/Se(VI)),[49,133–136] Cr (Cr(III)/Cr(VI)),[137] and Cu (Cu(I)/Cu(II)).[120–122] The higher of the two oxidation states mentioned are the thermodynamically stable forms of these elements in oxidizing seawater. However, most natural waters, especially surface waters, never attain complete chemical equilibrium due to the relatively steady input of solar energy, some of which can cause production, either biochemically (e.g., photosynthesis) or chemically (e.g., photochemistry), of species that are out of thermodynamic equilibrium with their environment.

The theoretical inorganic speciation of the trace metals in seawater has been summarized in two landmark papers by Turner et al.[138] and Byrne et al.[139] Turner et al.[138] used a database of stability constants for more than 500 metal complexes to calculate the inorganic speciation for 58 trace elements in model seawater at pH 8.2, 25°C, and 1 atm. Byrne et al.[139] extended the models of Turner et al.[138] by considering the results of recent metal-ligand equilibrium studies and by considering the influence of temperature and pH on speciation. In this section, we only briefly summarize the inorganic speciation of the trace metals in seawater described in these two papers.

The extent to which a trace element is complexed by inorganic ligands is given by α, the element's inorganic side reaction coefficient:[140]

$$\alpha = 1 + \Sigma\beta_{MXi}\left[X_i'\right] \quad (1)$$

where β is the overall conditional stability constant for the inorganic complex MX_i of the trace element M with the inorganic ligand X_i, and $[X_i']$ is the concentration of uncomplexed X_i. The inorganic side reaction coefficient α is also equal to the ratio of the sum of the concentrations of all inorganic species of the trace element M ([M′]) to the concentration of its free hydrated cation $[M^{n+}]$:

$$\alpha = \left[M'\right]/\left[M^{n+}\right] \quad (2)$$

The free hydrated divalent cation form dominates the dissolved inorganic speciation of Zn and the first transition series metals Mn, Fe, Co, and Ni. Strongly hydrolyzed trace metals include the trivalent trace metal cations Al^{3+}, Ga^{3+}, Tl^{3+}, Fe^{3+}, and Bi^{3+}. The inorganic side reaction coefficients of this group of strongly hydrolyzed trace elements with respect to complexation by OH^- range from $10^{5.76}$ for Al^{3+} to $10^{20.47}$ for Tl^{3+}. The inorganic speciation of the hydrolyzed trace metals is strongly influenced by pH and temperature. For example, at a constant temperature of 5°C, the side reaction coefficient of Al^{3+} varies from $10^{5.76}$ at pH 7.6 to $10^{9.39}$ at pH 8.2, and at a constant pH of 7.6, it ranges from $10^{5.76}$ at 5°C to $10^{7.23}$ at 25°C.[139]

The inorganic speciation of Fe is complex and still not adequately understood. The literature reviews by Turner et al.[138] and Byrne et al.[139] indicate that the hydrolysis species $Fe(OH)_3^\circ$ is the dominant species of dissolved Fe(III) in surface seawater. In contrast, Zafiriou and True[141] and Hudson and Morel[142] discounted the importance of the neutral $Fe(OH)_3^\circ$ hydrolysis species and argued that the $Fe(OH)_2^+$ species is dominant. The different assumptions used by these authors concerning which thermodynamic data are correct, and thus, which hydrolysis species is predominant, result in nearly a two order-of-magnitude difference in the estimates of the inorganic side reaction coefficient for Fe(III). This discrepancy needs to be resolved particularly because Fe may be the most important biologically essential trace metal, and potential differences in the reactivity and behavior of hydrolysis species such as $Fe(OH)_2^+$ and $Fe(OH)_3^\circ$ may influence the bioavailability of Fe.

Trace metals whose dissolved speciation is dominated by chloride complexation include the noble metal cations Ag^+ (predominantly as $AgCl_3^{2-}$), Au^+ (as $AuCl_2^-$), Pd^{2+} (as $PdCl_4^{2-}$), and Pt^{2+} (as $PtCl_4^{2-}$), and Cu^+ (as $CuCl_3^{2-}$), Cd^{2+} (as $CdCl_2^\circ$ and $CdCl_3^+$), and Hg^{2+} (as $HgCl_4^{2-}$). The dissolved speciation of these metals is only moderately affected by temperature and pH. Of this group, Hg^{2+} is complexed by chloride to the greatest extent. The side reaction coefficient of Hg^{2+} with respect to chloride complexation is $10^{15.10}$ at 5°C, which, for a typical Hg concentration of 5 p*M*, represents an average of only 1 free Hg^{2+} ion for every 400 l of seawater!

The dissolved speciation of most of the lanthanides and some of the actinides is dominated by carbonate complexes, and is influenced considerably by temperature and pH, although less than the strongly hydrolyzed metal cations.

C. ORGANIC SPECIATION

Organic forms of the trace elements in seawater can include organometallic compounds in which the trace element is covalently bound to carbon (e.g., methyl forms of As, Ge, Hg, Sb, Se, Sn, and Te; ethyl-Pb forms; butyl-Sn forms), and complexes with organic ligands (e.g., with phytoplankton metabolites and proteins, humic substances). The concentrations of naturally occurring methylated forms of As,[143,144] Ge,[145,146] Hg,[147] Sb,[148] and Sn[72,149,150] have been determined in seawater. A most astonishing discovery is that 90% of the oceanic Ge exists as methylated forms ($CH_3Ge(OH)_3^\circ$ and $(CH_3)_2Ge(OH)_2^\circ$) that are so stable to degradation that they have been called the "Teflon of the sea". The remarkable stability of these species is reflected in their conservative vertical profiles, which markedly contrast those of other methylated metal species and with the recycled behavior of inorganic H_4GeO_4.

Even though the existence of organic trace metal complexes in seawater had been postulated as much as 60 years ago,[151,152] and subsequent attempts were made to demonstrate the presence and characterize the nature of these complexes, the organically complexed fraction of the total concentration of certain trace metals in seawater has been reliably estimated only recently. Previous literature reviews (e.g., Mantoura[153] and van den Berg et al.[154]) of metal complexation determinations in seawater indicated little

Table 4 **Determinations of the fraction of organically complexed copper in seawater**

Location	Percent Organic Cu	Technique	Reference
San Francisco Bay	80–92	CLE/DPCSV DPASV CRCP/GFAAS	Donat et al.[161b]
Indian Ocean	>99.7	CLE/DPCSV	Donat & van den Berg[48]
North Sea	>99.9	CLE/DPCSV	Donat & van den Berg[48]
Sargasso Sea	98.8	CLE/LP/GFAAS	Moffett et al.[122]
Sargasso Sea	93	CLE/DPCSV DPASV	Donat & Bruland[161a]
North Pacific	99.4–99.8	DPASV	Coale & Bruland[160,161]
New York coast	99.8	FPA	Hering et al.[208]
Biscayne Bay	99.6	CLE/LP/GFAAS	Moffett & Zika[159]
Narragansett Bay	99.9	CLE/SPE/GFAAS	Sunda & Hanson[158]
Coastal Peru	98	CLE/SPE/GFAAS	Sunda & Hanson[158]
North Atlantic	89–99.8	MnO_2 ads.	Buckley & van den Berg[157]
North Atlantic	98.8–99.4	CLE/DPCSV	Buckley & van den Berg[157]
South Atlantic	99.9	CLE/DPCSV	van den Berg[156]
Coastal Florida	98.7	Bioassay	Sunda & Ferguson[155]
Mississippi Plume	99.1	Bioassay	Sunda & Ferguson[155]
New York Bight	>95	DPASV	Huizenga & Kester[209]
Irish Sea	94–98	MnO_2 ads.	Van den Berg[126]

Note: CLE/DPCSV = Competitive ligand equilibration/differential pulse cathodic stripping voltammetry; CRCP/GFAAS = Chelating resin column partitioning/graphite furnace atomic absorption spectrometry; CLE/LP/GFAAS = Competitive ligand equilibration/liquid partitioning/graphite-furnace atomic absorption spectrometry; DPASV = Differential pulse anodic stripping voltammetry; FPA = Fixed potential amperometry; CLE/SPE/GFAAS = Competitive ligand equilibration/solid phase extraction/graphite-furnace atomic absorption spectrometry; MnO_2 ads. = Manganese dioxide adsorption.

agreement between values for ligand concentrations, conditional stability constants, and the overall extent of organic complexation for Cu — in fact, previous estimates of the organically complexed fraction for Cu ranged from 0 to 100%. However, recent work shows that this fraction dominates the dissolved speciation of Cu[48,122,155–161a,b] (Table 4) and Zn[162–164] in oceanic surface water and is, therefore, of utmost importance in calculating the free metal ion concentrations of these two trace metals. These observations differ markedly from frequently cited model predictions by Mantoura et al.[165] that, of all the transition metals, only Cu should be complexed significantly (~10%) with organic ligands, and that dissolved humic substances (as well as dissolved organic compounds in general) should not substantially influence the speciation of Zn in seawater.

Recent vertical profiles obtained in the central northeast Pacific for the speciation of at least two trace elements, Cu and Zn, demonstrate progress made in determining that the chemical speciation of these two trace elements is dominated by organic complexation, although the chemical nature of the complexing ligands remains unknown.

1. Copper

Recent anodic stripping voltammetric measurements reported by Coale and Bruland[160,161] indicate that greater than 99.7% of total dissolved Cu(II) in central northeast Pacific surface waters shallower than 200 m is bound in strong organic complexes, primarily with the stronger (L_1) of two Cu-complexing ligands which has an average concentration of ~1.8 n*M* in the upper 100 m (Figure 7a). The concentration of this ligand exceeds the concentration of dissolved Cu from the surface down to ~200 m which, in concert with the strength of its Cu complexes, causes the high degree of organic complexation observed in the upper 200 m (Figure 7b). Because the concentration of dissolved Cu in these surface waters is only ~0.5 to 1.0 n*M*, the high extent of organic complexation reduces the fraction of inorganic Cu species to

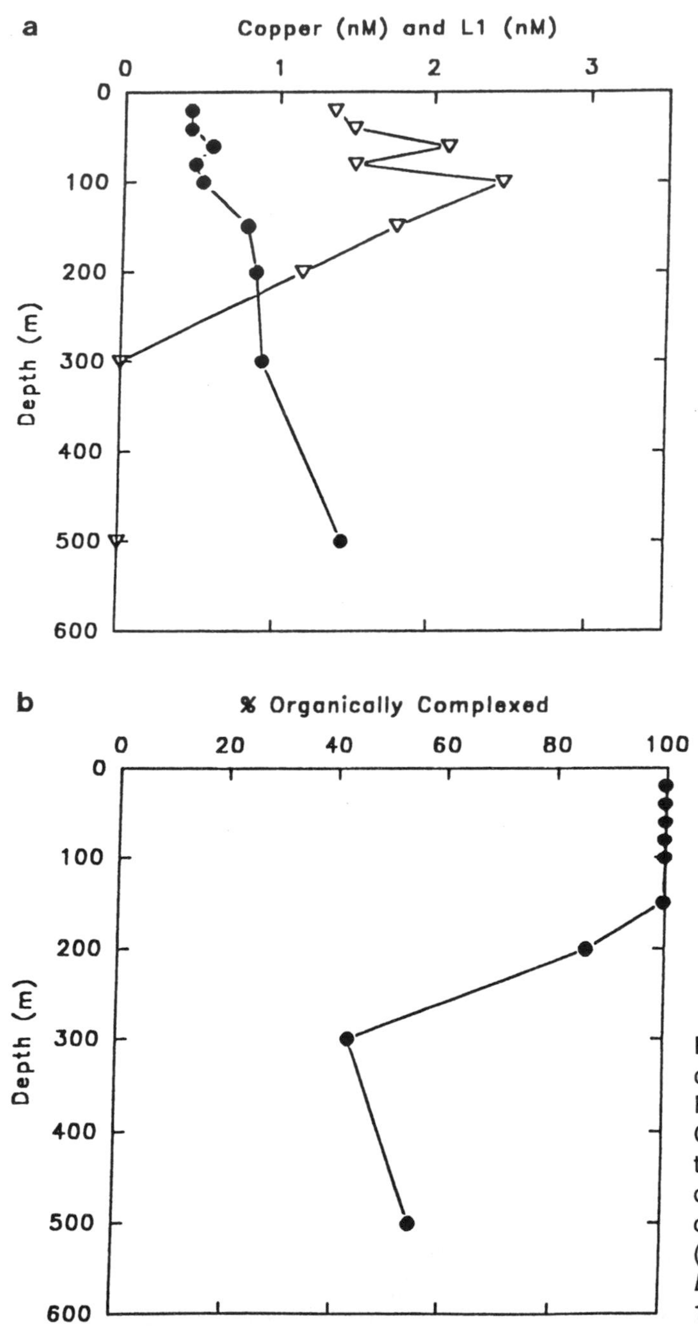

Figure 7 Dissolved copper (II) speciation vs. depth in central Northeast Pacific surface waters: (a) Dissolved Cu(II) concentrations; (b) concentrations of L_1, the stronger copper-complexing organic ligand; (c) log concentration values of free Cu^{2+} ion. (From Coale, K.H. and Bruland, K.W., *Deep-Sea Research,* 47(2), 317, 1990. With permission.)

less than 0.3% of the total dissolved Cu, and free hydrated Cu^{2+} amounts to only ~4% of this inorganic fraction. Whereas total dissolved Cu concentrations in the central northeast Pacific increase by approximately 3-fold from the surface to middepths, complexation of Cu by strong organic ligands causes the free hydrated Cu^{2+} concentration to vary from approximately $10^{-13.1}$ *M* at the surface to approximately $10^{-9.9}$ *M* at 300 m (Figure 7c), approximately a *2000-fold* increase!

In surface waters of the Sargasso sea[122,161a] and south San Francisco Bay,[161b] the concentrations of the strongest Cu-complexing ligand have been determined to be equal to or less than the dissolved Cu concentration. The weaker ligand can then dominate Cu speciation, slightly decreasing the overall extent of organic complexation, and increasing the inorganic Cu fraction and free Cu^{2+} concentrations. Some evidence exists that the ligand concentrations and extent of organic complexation can vary seasonally.

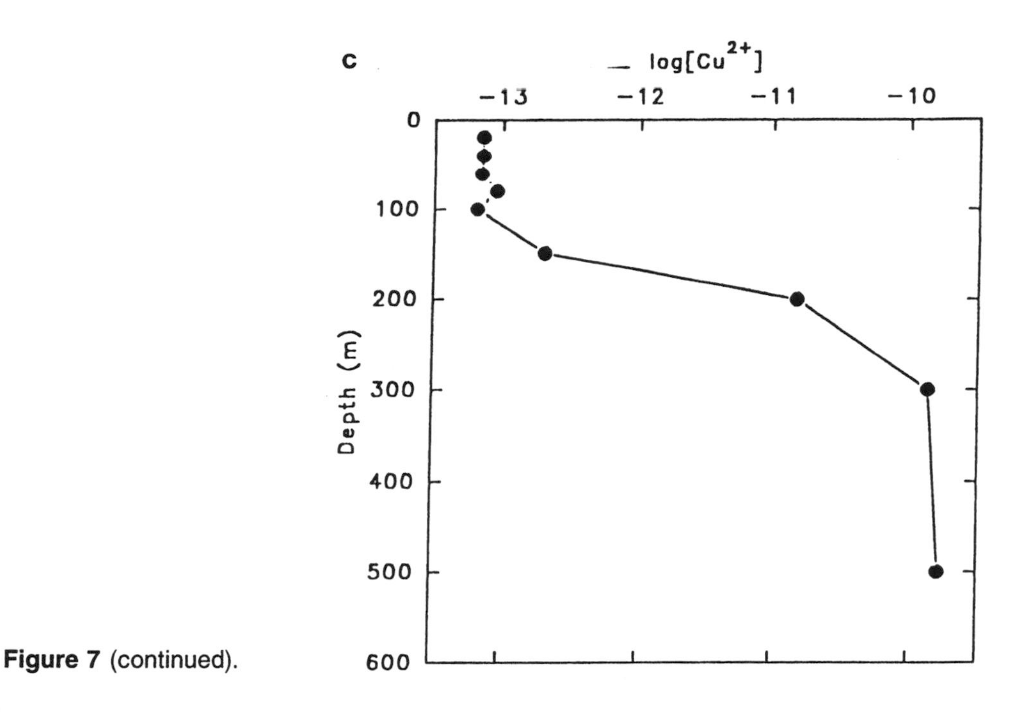

Figure 7 (continued).

2. Zinc

Recent anodic stripping voltammetry measurements in central North Pacific waters reported by Bruland[163] indicate that an average of 98.7% of the Zn in surface waters shallower than 200 m is complexed with a relatively Zn-specific organic ligand (or ligand class). Using two independent voltammetric techniques, Donat and Bruland[164] confirmed that strong organic complexes dominate the dissolved speciation of Zn in Northeast Pacific surface waters. Bruland[163] determined that the concentration of the Zn-complexing organic ligand or ligand class averages ~1.2 n*M* in the upper 200 m and exceeds the concentration of dissolved Zn from the surface down to ~350 m (Figure 8a). This excess in the ligand's concentration relative to that of dissolved Zn, and the strength of its Zn complexes, causes the high degree of organic complexation in the upper 300 m (Figure 8b). The free Zn^{2+} concentration varies from ~$10^{-11.8}$*M* at depths shallower than 200 m, to ~$10^{-8.6}$ *M* at 600 m (Figure 8c) — approximately a *1400-fold* increase! The high degree of organic complexation for Zn reduces the concentration of inorganic Zn species to ~2 p*M* (2×10^{-12} *M*) and the free Zn^{2+} ion concentration to values as low as 1 p*M* (1×10^{-12} *M*; pZn = 12).

3. Other Metals

Only a few initial reports exist on the importance of organic complexation to the dissolved speciation of other metals in the open ocean, including Cd, Pb, Co, and Ni. Bruland[166] reported results of anodic stripping voltammetric measurements indicating that 70% of the dissolved Cd in central North Pacific surface waters was bound in strong complexes by relatively Cd-specific organic ligands existing at low concentrations (~0.1 n*M*). Bruland[166] found this ligand class only within the surface 175 m, and it had a concentration maximum at depths between 40 and 100 m. Bruland[166] determined that the concentration of inorganic forms of Cd varied from ~0.7 p*M* in surface waters to 800 p*M* at 600 m. Considering both organic and inorganic complexation, the concentration of free Cd^{2+} ranged from 20 f*M* in surface waters to 22 p*M* at 600 m — a 1000-fold variation! Chlorocomplexes appear to dominate the inorganic speciation of dissolved Cd in intermediate and deep waters, while organic complexation is important in influencing Cd speciation within oceanic surface waters.

Capodaglio et al.[167] determined the organically complexed fraction of dissolved Pb in surface waters of the eastern North Pacific by anodic stripping voltammetry. Approximately 50% of the dissolved Pb appeared to complexed by one class of strong organic ligands existing at concentrations between 0.2 and 0.5 n*M*. Inorganic and organic complexation of Pb in eastern North Pacific surface waters results in a free Pb^{2+} concentration of ~0.4 p*M*.

Organic complexation of dissolved Co and Ni in oceanic waters has not been reported; however, a few reports exist on the concentrations of organically complexed Co and Ni in estuarine and coastal samples.

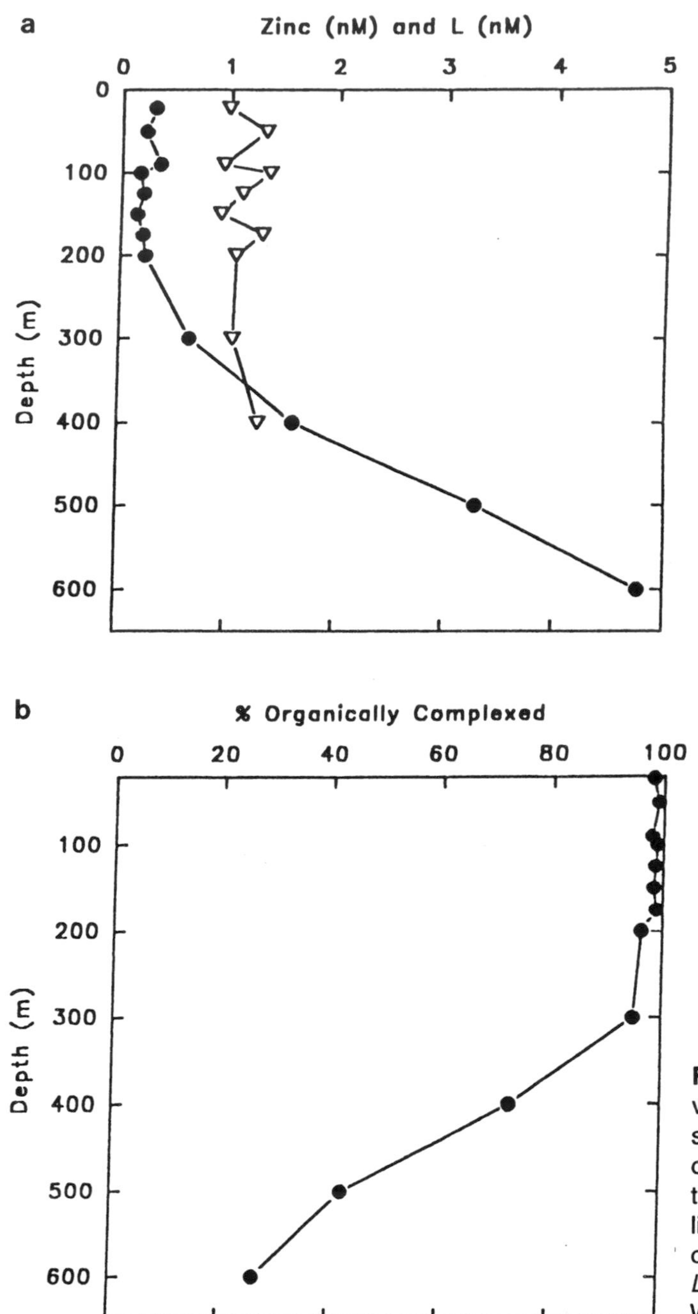

Figure 8 Dissolved zinc speciation vs. depth in central Northeast Pacific surface waters: (a) Dissolved Zn concentrations; (b) concentrations of the strong zinc-complexing organic ligand; (c) log concentration values of free Zn^{2+} ion. (After Bruland, K.W., *Limnol. Oceanogr.*, 34, 267, 1989. With permission.)

Zhang et al.[168] determined that a variable fraction (45 to 100%; average 73%) of dissolved Co was very strongly complexed in the Scheldt River Estuary and the Irish Sea. Donat and Bruland[47] found evidence that ~50% of dissolved Co in coastal and open ocean seawater reference materials (CASS-1 and NASS-1) was bound in strong organic complexes, after adjustment of the pH from 1.6 to 7.6. Van den Berg and Nimmo[169] and Nimmo et al.[170] found that ~30 to 50% of dissolved Ni in the U.K. coastal waters was bound in extremely strong organic complexes. Donat et al.[161b] determined the same extent of organic complexation for Ni in south San Francisco Bay.

Fe^{3+} has a great tendency to form complexes with natural organic ligands (i.e., dissolved humic substances; see Sholkovitz[171] and references cited therein), and organic complexation is probably responsible for elevated concentrations of dissolved Fe in many estuarine and coastal waters.[171–173] Recent reviews by Sunda[15] and Wells[16] point out that most researchers have either ignored the possibility of organic complexation of Fe in the open ocean, or have argued that it is not significant. However, Gledhill

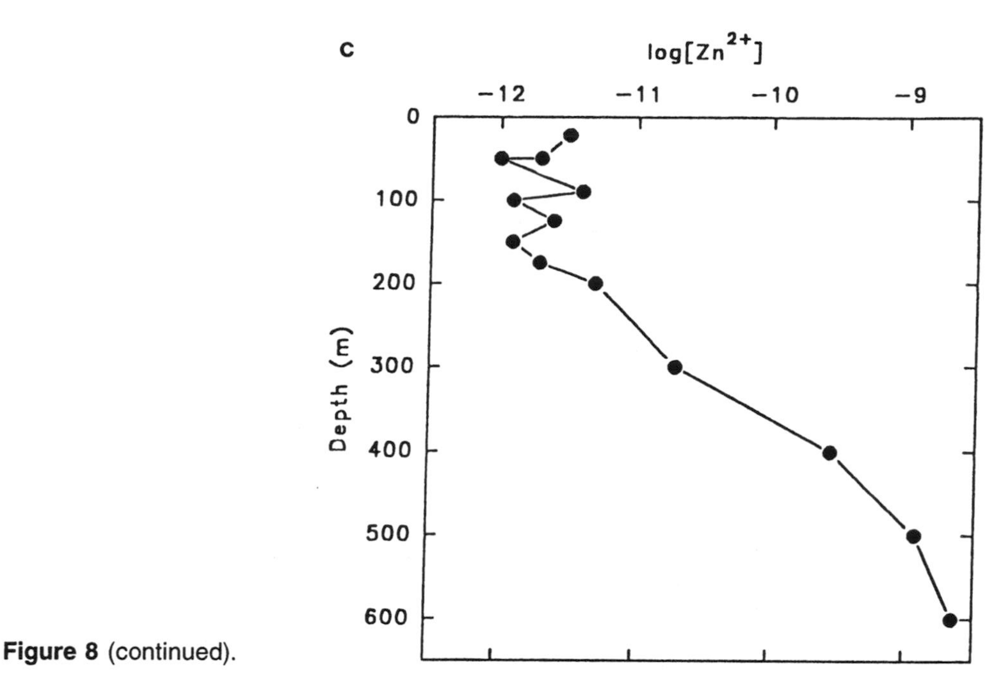

Figure 8 (continued).

and van den Berg[173a] (North Sea) and Rue and Bruland[173b] (North Pacific) have recently determined that >99% of dissolved Fe(III) is bound in strong complexes by a ligand class having a concentration of 1 to 5 n*M* in excess of the ambient dissolved Fe concentration. Although siderophore complexes are probably important to Fe speciation in some freshwater systems, no convincing evidence yet exists for their existence in the oceans.

Although the actual chemical structures of the trace metal complexing organic ligands are still unknown, the data presented in the preceding sections demonstrate the importance of complexation of certain trace metals with organic ligands which exist at low concentrations yet form very strong complexes (i.e., have high conditional stability constants). The recent accumulation of evidence indicating complexation of trace metals by naturally occurring organic ligands in the open ocean represents revolutionary progress beyond the previous status of this research field.

4. Biological Implications of Organic Complexation and Speciation of Trace Elements in Seawater

Almost three decades ago, Johnston[174,175] proposed that certain organic compounds present in seawater in trace quantities may exert an important influence on the primary production of marine communities. Barber and Ryther[176] suggested:

> "in recently upwelled water phytoplankton growth is initially limited not by inorganic nutrients, trace metals, nor vitamin deficiencies, but rather by the absence of certain chelating substances. As the upwelling water ages, the organisms apparently gradually enrich the water with organic compounds some of which may be effective chelators."

Barber and Ryther[176] proposed that these chelators might be effective at reducing to nontoxic levels, the toxic free Cu^{2+} concentration thought to exist in freshly upwelled waters.

Compelling indications of biological sources for the Cu complexing ligands are provided by field data indicating that organic complexation of Cu is most pronounced in the surface euphotic zone, in general being highest at depths near the productivity maximum, and decreasing by one to three orders of magnitude below the vernal mixed layer in both the central northeast Pacific[160,161] and in the Sargasso Sea.[122]

The domination of the speciation of dissolved Zn by organic complexes may suggest a biological influence, as discussed above for Cu. However, the biological advantage of organic complexation of this essential and potentially biolimiting element is not readily apparent. Bruland[163] made the following speculations: (1) perhaps organic complexation may help keep Zn in solution, thereby preventing or retarding its adsorption onto particles; (2) perhaps some phytoplankton may be able to assimilate organically complexed Zn to the exclusion of other phytoplankton; and/or (3) perhaps complexation of Zn first occurs inside a cell with intracellular ligands such as phytochelatins, the cytoplasmic detoxifying metal chelators found in higher plants and at least some algae,[177–179] followed by release of the chelators through phytoplankton DOC leakage and/or zooplankton grazing.

Moffett et al.[122] provided direct laboratory evidence for production of a strong Cu-binding ligand by a marine photoautotroph. This ligand has Cu-complexing strength identical to that of the stronger ligand observed by these same researchers in the Sargasso Sea and by Coale and Bruland[161] in the central northeast Pacific. The laboratory results Moffett and co-workers[122] obtained from a preliminary survey of four marine phytoplankton species (three species of eukaryotes and one prokaryote) indicated that only the cyanobacterial species *Synechococcus* produced a chelator forming Cu complexes of the same strength as those observed in the Sargasso Sea. These observations are consistent with earlier work in freshwater systems by McKnight and Morel,[180] who reported that among freshwater phytoplankton taxa, only the cyanobacteria appear to produce strong Cu-binding organic ligands. Because *Synechococcus* appear to be widespread in the surface waters of the open ocean,[181] and can account for as much as 50% of the primary productivity in some regions,[182,183] chelators produced by this genus might have a major influence on Cu speciation in many oceanic regions.

Production of a strong Cu-complexing ligand like L_1 may reflect strong selective pressure to detoxify Cu by lowering its free ion concentration. Evidence for this includes the work of Brand et al.[184] who examined the reproduction rates of clone cultures of 38 marine phytoplankton species in media containing different free Cu^{2+} activities. The Cu toxicity responses among the various species showed a phylogenetic trend: cyanobacteria were the most sensitive, diatoms were the least sensitive, and coccolithophores and dinoflagellates showed intermediate sensitivity. The reproduction rates of most of the cyanobacteria were reduced at Cu^{2+} concentrations above 10^{-12} *M*, while most eukaryotic algae maintained maximum reproduction rates at Cu^{2+} activities as high as 10^{-11} *M*. If cyanobacteria produced an organic ligand of the same concentration and binding strength as the stronger ligand L_1, it would complex Cu and lower the free Cu^{2+} concentration in oceanic surface waters to levels noninhibitory to their growth. However, intense upwelling could bring water having Cu^{2+} concentrations approaching 10^{-11} *M* into the surface photic zone — concentrations potentially high enough to limit reproduction of some phytoplankton species, especially cyanobacteria.[185] Cu^{2+} toxicity due to high Cu^{2+} concentrations in upwellings was speculated to cause a decline in cyanobacterial abundance after a vertical mixing event in the North Pacific Central Gyre.[186] Thus, free Cu^{2+} may influence species composition and seasonal species succession within local phytoplankton communities, and enough evidence exists to suggest that Cu^{2+} concentrations may influence picoplankton (cyanobacteria) production in high nutrient-low chlorophyll areas.

Brand et al.[187] examined the potential limitation of marine phytoplankton reproductive rates by Zn, Mn, and Fe, and suggested that differences among species in their abilities to reproduce in the presence of low free ion activities of these nutrient trace metals could result in species shifts in phytoplankton communities subjected to changes in trace metal or organic complexation regimes. For example, the reproductive rates of neritic species were generally limited by free Zn^{2+} concentrations below $10^{-11.5}$ *M*, while those of oceanic species were either not limited or only slightly limited at the lowest Zn^{2+} concentration set in the experiment, about 10^{-13} *M*. The habitat-related patterns in Zn^{2+} requirements of oceanic and neritic species are consistent with the oceanic-neritic distributions of free Zn^{2+} concentration.[163] The low surface water free Zn^{2+} concentration of $10^{-12.5}$ *M* may be low enough to limit the growth rate of many neritic phytoplankton species, but generally not that of oceanic species. However, intense upwelling could bring water into the surface layer whose free Zn^{2+} concentration would be high enough to support reproduction of more neritic-type species. Thus, the similarity between the nutritional requirement for Zn^{2+} and its distributional patterns has important implications for phytoplankton community species composition and species succession.

Fe may be the most important trace element nutrient to phytoplankton, and its deficiency may limit primary production in certain oceanic areas characterized by high nutrient and low chlorophyll concentrations such as the subarctic and equatorial Pacific and the Southern Ocean.[101,116,117] Fe also influences algal nitrogen metabolism in the ocean because it is required for nitrate reduction and nitrogen fixation. Price et al.[188] determined that additions of Fe caused the indigenous phytoplankton in the equatorial Pacific to switch from using primarily NH_4^+ for growth to using NO_3^-, and that this switch in nitrogenous nutrition was accompanied by a shift in the phytoplankton community composition. Although the speciation of dissolved Fe is extremely complex and presently unknown with any reliability, it has very important biological implications. For example, photoreduction of organic Fe(III) complexes and Fe(III) oxides, and the presence of Fe(III)-siderophore complexes, could enhance Fe's biological availability.

Brand et al.[184] examined the sensitivity of 20 species of marine phytoplankton to free Cd^{2+}. Cd toxicity for even the most sensitive species in their experiments occurred only at free Cd^{2+} exceeding 10^{-10} *M*, well outside the range of Cd^{2+} concentrations determined by Bruland[166] in the central North Pacific. This comparison led Bruland[166] to suggest that Cd toxicity is probably not an important selective factor to phytoplankton growth or ecology in oceanic waters. However, Brand et al.[184] stated that Cd toxicity could still be a selective factor in estuaries due to increases in Cd concentrations from riverine and anthropogenic sources.

Price and Morel[189] recently hypothesized that Cd might promote growth of Zn-limited oceanic phytoplankton. Phytoplankton can exhibit substantially reduced growth rates due to Zn limitation at a free Zn^{2+} concentration of 10^{-12} *M*, a concentration identical to that determined by Bruland[163] in oceanic surface waters. This low free Zn^{2+} concentration results from both the low concentration of total dissolved Zn and its high degree of organic complexation in the surface ocean. Price and Morel[189] reported that, in laboratory experiments using seawater with low free Zn^{2+} concentrations mimicking oceanic surface water conditions, Cd stimulates the growth of the marine diatom *Thalassiosira weissflogii* by substituting for Zn in certain macromolecules. The substitution appears to be highly effective, allowing Zn-deficient cells to grow at 90% of their maximum rate when supplied with Cd. Price and Morel[189] suggest that this biochemical metal substitution of Cd for Zn by phytoplankton could account for the pronounced nutrient-type oceanographic distribution of Cd.

Dissolved Co exists at concentrations of only 4 to 50 p*M* in the North Pacific.[98,101] These low values suggest a potential role of Co as a biolimiting element in seawater. Co is the central metal atom in the corrin ring core of vitamin B_{12} (cobalamin), and is one of the two metals for which organic complexation is known to enhance uptake by phytoplankton; siderophore-bound Fe is the other metal. Vitamin B_{12} is a required growth factor for many algal species, particularly some diatoms, chrysophytes, and dinoflagellates.[190] Some phytoplankton have been observed to produce a glycoprotein that binds to vitamin B_{12} making it unavailable to other species,[191] supporting the idea that this form of organically complexed Co may be important in interspecies competitive interactions.

Dissolved Ni exhibits a nutrient-type distribution in the oceans and its concentration ranges from 2 to 12 n*M*.[78,192] Price and Morel[193] have provided evidence indicating a potential biochemical role for Ni in urea assimilation by marine phytoplankton. They observed that phytoplankton growing on urea as a sole nitrogen source are limited by low free Ni^{2+} concentrations, but those growing on nitrate or ammonium are not. Ni is known to be an essential cofactor in the enzyme urease. Price and Morel's[193] results indicate that the concentration of dissolved Ni present in oceanic surface waters would not limit urea assimilation if all of the dissolved Ni was biologically available. However, if a substantial amount of the Ni present in oceanic surface waters is rendered unavailable by complexation with organic ligands, the resulting low free Ni^{2+} concentrations could potentially limit urea assimilation. This may be important in areas in which production is supported largely through regeneration, since urea is a common waste product of zooplankton.

Most field investigations of trace metal/phytoplankton interactions have focused on the effects of a single metal, largely ignoring the effects of metal ion ratios on algal growth.[185] However, laboratory culture studies clearly show that many metals may act synergistically and antagonistically to influence growth limitation or toxicity. Thus, it is necessary to quantify and experimentally control the availability of all combinations of potentially antagonistic metals in order to accurately describe metal/biota interactions. Brand et al.[187] have argued that simultaneous limitation by Zn, Mn, and Fe may be more severe than limitation by any one of these metals alone (i.e., synergism). Even more significantly, culture studies have repeatedly demonstrated that the uptake and metabolism of nutrient metal ions is a function of the concentration of other metals, which often act in a competitive or antagonistic manner. In phytoplankton, such metal antagonisms have been reported for Cu and Mn;[115,194–196] Cu and Fe;[195] Cu and Zn;[197] Cd and Fe;[198,199] Mn and Fe;[195,200] and Mn and Zn.[196]

Because antagonistic effects have been reported between Fe and toxic metals,[195,198,199] low $[Fe^{3+}]$ to $[Cu^{2+}]$ ratios in seawater, for instance, could plausibly explain the growth stimulation observed in Fe addition experiments, such as those reported by Martin et al.[101] Some cases of primary production limitation, apparently due to Fe alone, could actually be due to antagonistic effects of Cu and Fe. This explanation is especially difficult to dismiss in incubation experiments in which relatively high Fe concentrations (>10n*M*) have been added (i.e., Menzel et al.;[201] Barber;[202] deBaar et al.;[203] Buma et al.[204]). In these cases, hydrous ferric oxides may form and adsorb Cu^{2+} ions, resulting in substantially lowered free Cu^{2+} ion concentrations.[205–207] Any observed growth stimulation could be due to alleviation of Cu toxic effects by purely abiotic, chemical adsorption processes occurring in the incubation bottles.

Although this is not a true case of biological metal antagonism, it again emphasizes that metal ratios are critical and must be closely monitored in this type of experiment. Even in cases where the soluble concentration of Fe has apparently not been exceeded (e.g., Martin et al.[101,117]), relatively low Fe additions (1 n*M*) could cause a substantial favorable shift in the [Fe^{3+}] to [Cu^{2+}] ratio. Proper interpretation of the results of metal addition experiments demands that all potentially interacting trace metals be considered and quantified.

More field data on metal antagonisms are necessary to support or negate hypotheses which have been formed primarily from laboratory studies. The artifacts in this type of study can be fiendishly cunning, and correct interpretation of their results requires a thorough understanding of both the oceanic chemistry of trace metals and their potential interactions with the biota. Thus, the advancement of our understanding of trace metal influences on oceanic productivity will require a unified collaboration between clever biologists and clever chemists.

VII. SUMMARY

As a result of major advances made since 1975 in analytical techniques and instrumentation and in the elimination or control of contamination during sample collection, storage, and analysis, our knowledge of the concentrations, distributions, speciation, and, therefore, our understanding of the biogeochemical cycling of the trace elements in the oceans has progressed dramatically. We now have a first-order understanding of the distributions of most of the trace elements in the oceans. The development of extremely sensitive analytical techniques amenable to obtaining measurements at sea have provided basin-wide cross sections of the distributions of some of the trace elements (e.g., Al, Mn, and Fe), allowing their use as sensitive and important oceanographic tracers. The biogeochemical cycling of certain trace elements like Cu, Fe, and Mn have been found to involve mediation by light and microorganisms. Our knowledge of the inorganic speciation of the trace elements in seawater has progressed to include estimations of the influences of temperature and pH. Convincing evidence has accrued demonstrating not just the presence of organically complexed metals in seawater, but the overwhelmingly dominant role of organic complexation in the speciation of dissolved Cu and Zn in surface seawater. Organic complexation of certain trace elements seems to have important biological implications to phytoplankton in the sea, as it may dominate the speciation of these elements and control their availability as nutrients and toxicants.

REFERENCES

1. **Wong, C. S., Boyle, E., Bruland, K. W., Burton, J. D., and Goldberg, E. D., Eds.,** *Trace Metals in Seawater,* Plenum Press, New York, 1983.
2. **Burton, J. D. and Statham, P. J.,** Occurrence, distribution, and chemical speciation of some minor dissolved constituents in ocean waters, in *Environmental Chemistry,* Vol. 2, Specialist Periodical Report, Royal Society of Chemistry, London, 1982, 234.
3. **Burton, J. D. and Statham, P. J.,** Trace metals as tracers in the ocean, *Philos. Trans. R. Soc. London Ser. A,* 325, 127, 1988.
4. **Burton, J. D. and Statham, P. J.,** Trace metals in seawater, in *Heavy Metals in the Marine Environment,* Rainbow, P. S. and Furness, R. W., Eds., CRC Press, Boca Raton, FL, 1990, 5.
5. **Bruland, K. W.,** Trace elements in sea-water, in *Chemical Oceanography,* Vol. 8, Riley, J. P. and Chester, R., Eds., Academic Press, London, 1983, 157.
6. **Whitfield, M. and Turner, D. R.,** The role of particles in regulating the composition of seawater, in *Aquatic Surface Chemistry,* Stumm, W., Ed., John Wiley & Sons, New York, 1987, 457.
7. **Shiller, A. M. and Boyle, E. A.,** Variability of dissolved trace metals in the Mississippi River, *Geochim. Cosmochim. Acta,* 51, 3273, 1987.
8. **Trefrey, J. H., Nelson, T. A., Trocine, R. P., Metz, S., and Vetter, T.,** Trace metal fluxes through the Mississippi River delta system, in *Contaminant Fluxes Through the Coastal Zone,* Kullenberg, G., Ed., Rapp. P.-v. Reun. Const. Int. Explor. Mer., 1986, 277.
9. **Shiller, A. M. and Boyle, E.,** Dissolved zinc in rivers, *Nature,* 317, 49, 1985.
10. **Martin, J.-M. and Windom, H. L.,** Present and future roles of ocean margins in regulating marine biogeochemical cycles of trace elements, in *Ocean Margin Processes in Global Change,* Mantoura, R. F. C., Martin, J.-M., and Wollast, R., Eds., John Wiley & Sons, New York, 1991, 45.

11. **Chester, R. and Murphy, K. J. T.,** Metals in the marine atmosphere, in *Heavy Metals in the Marine Environment,* Rainbow, P. S. and Furness, R. W., Eds., CRC Press, Boca Raton, FL, 1990, 27.
12. **Buat-Ménard, P.,** Air to sea transfer of anthropogenic trace metals, in *The Role of Air-Sea Exchange in Geochemical Cycling,* Buat-Ménard, P., Ed., NATO ASI Series, D. Reidel Pub. Co., 1985, 477.
13. **Arimoto, R., Duce, R. A., and Ray, B. J.,** Concentrations, sources and air-sea exchange of trace elements in the atmosphere over the Pacific Ocean, in *Chemical Oceanography,* Vol. 10, SEAREX: The Sea/Air Exchange Program, Riley, J. P., Chester, R., and Duce, R. A., Eds., Academic Press, London, 1989, 27.
14. **Zhuang, G., Duce, R. A., and Kester, D. R.,** The dississolution of atmospheric iron in surface seawater of the open ocean, *J. Geophys. Res.-Oceans,* 95(C9), 6207, 1990.
15. **Sunda, W. G.,** Trace metal interactions with marine phytoplankton, *Biol. Oceanogr.,* 6, 411, 1991.
16. **Wells, M. L.,** The availability of iron in seawater: a perspective, *Biol. Oceanogr.,* 6, 463, 1991.
17. **Thompson, G.,** Hydrothermal fluxes in the ocean, in *Chemical Oceanography,* Vol. 8, Riley, J. P. and Chester, R., Eds., Academic Press, London, 1983, 272.
18. **Edmond, J. M.,** U.S. research on oceanic hydrothermal chemistry: 1987–1990, in *Rev. Geophysics, Supp.,* 645, 1991.
19. **Rona, P. A., Boström, K., Laubier, L., and Smith, K. L., Jr., Eds.,** *Hydrothermal Processes at Seafloor Spreading Centers,* Plenum Press, New York, 1983.
20. **von Damm, K. L., Edmond, J. M., Grant, B., Measures, C. I., Walden, B., and Weiss, R. F.,** Chemistry of submarine hydrothermal solutions at 21°N, East Pacific Rise, *Geochim. Cosmochim. Acta,* 49, 2197, 1985.
21. **von Damm, K. L., Edmond, J. M., Measures, C. I., and Grant, B.,** Chemistry of submarine hydrothermal solutions at Guaymas Basin, Gulf of California, *Geochim. Cosmochim. Acta,* 49, 2221, 1985.
22. **Moody, J. R. and Beary, E. S.,** Purified reagents for trace metal analysis, *Talanta,* 29, 1003, 1982.
23. **Mart, L.,** Minimization of accuracy risks in voltammetric ultratrace determination of heavy metals in natural waters, *Talanta,* 29, 1035, 1982.
24. **Schaule, B. K. and Patterson, C. C.,** Lead concentration in the Northeast Pacific: evidence for global anthropogenic perturbations, *Earth Planet. Sci. Lett.,* 54, 97, 1981.
25. **Martin, J. H., Knauer, G. A., and Broenkow, W. W.,** VERTEX: the lateral transport of manganese in the northeast Pacific, *Deep-Sea Res.,* 32, 1405, 1986.
26. **Bruland, K. W., Franks, R. P., Knauer, G. A., and Martin, J. H.,** Sampling and analytical methods for the determination of Cu, Cd, Zn, and Ni at the nanogram per liter level in seawater, *Anal. Chim. Acta,* 105, 233, 1979.

26a. **Hunter, C., Gordon, M., Fitzwater, S., and Johnson, K.,** A rosette/winch system for the collection of trace metal clean, large volume, discrete seawater samples, *EOS,* 75(3), 150, 1994.

27. **Mart, L.,** Prevention of contamination and other accuracy risks in voltammetric trace metal analysis of natural waters. II. Collection of surface water samples, *Fresenius Z. Anal. Chem.,* 299, 97, 1979.
28. **Boyle, E. A., Huested, S. S., and Jones, S. P.,** On the distribution of Cu, Ni, and Cd in the surface waters of the North Atlantic and North Pacific Ocean, *J. Geophys. Res.,* 86, 8048, 1981.
29. **Kremling, K.,** The distribution of Cd, Cu, Ni, Mn, and Al in surface waters of the open Atlantic and European shelf areas, *Deep-Sea Res.,* 32, 531, 1985.
30. **Danielsson, L. G., Magnusson, B., and Westerlund, S.,** An improved metal extraction procedure for the determination of trace metals in seawater by atomic absorption spectrometry with electrothermal atomization, *Anal. Chim. Acta,* 98, 47, 1978.
31. **Bruland, K. W., Coale, K. H., and Mart, L.,** Analysis of seawater for dissolved Cd, Cu, and Pb: an intercomparison of voltammetric and atomic absorption methods, *Mar. Chem.,* 17, 285, 1985.
32. **Sturgeon, R. E., Berman, S. S., Desaulniers, A., and D. S. Russell,** Pre-concentration of trace metals from sea-water for determination by graphite-furnace atomic-absorption spectrometry, *Talanta,* 27, 85, 1980.
33. **Sturgeon, R. E., Berman, S. S., Desaulniers, J. A. H., Mykytiuk, A. P., McLaren, J. W., and Russell, D. S.,** Comparison of methods for the determination of trace elements in seawater, *Anal. Chem.,* 52, 1585, 1980.
34. **Klinkhammer, G.,** Determination of manganese in seawater by flameless atomic absorption spectrometry after pre-concentration with 8-hydroxyquinoline in chloroform, *Anal. Chem.,* 52, 117, 1980.
35. **Landing, W. M. and Bruland, K. W.,** Manganese in the North Pacific, *Earth Planet. Sci. Lett.,* 49, 45, 1980.

36. **Armannsson, H.,** Dithizone extraction and flame atomic absorption spectrometry for the determination of cadmium, zinc, lead, copper, nickel, cobalt, and silver in sea water and biological tissues, *Anal. Chim. Acta,* 110, 21, 1979.
37. **Smith, R. G., Jr. and Windom, H. L.,** A solvent extraction technique for determining nanogram per liter concentrations of cadmium, copper, nickel, and zinc in sea water, *Anal. Chim. Acta,* 113, 39, 1980.
38. **Kingston, H. M., Barnes, I. L., Brady, T. J., and Rains, T. C.,** Separation of eight transition elements from alkali and alkaline earth elements in estuarine and sea water with chelating resin and their determination by graphite furnace atomic absorption spectrometry, *Anal. Chem.,* 50, 2064, 1978.
39. **Sturgeon, R. E., Berman, S. S., Willie, S. N., and Desaulniers, J. A. H.,** Pre-concentration of trace metals from seawater with silica-immobilized 8-hydroxyquinoline, *Anal. Chem.,* 53, 2337, 1981.
40. **Landing, W. M., Haraldsson, C., and Paxéus, N.,** Vinyl polymer agglomerate based transition metal cation chelating ion-exchange resin containing the 8-hydroxyquinoline functional group, *Anal. Chem.,* 58, 3031, 1986.
41. **Boyle, E. A. and Edmond, J. M.,** Determination of copper, nickel, and cadmium in sea water by APDC chelate coprecipitation and flameless atomic absorption spectrometry, *Anal. Chim. Acta,* 91, 189, 1977.
42. **van Geen, A. and Boyle, E.,** Automated preconcentration of trace metals from seawater and freshwater, *Anal. Chem.,* 62, 1705, 1990.
43. **Nürnberg, H. W.,** Trace analytical procedures with modern voltammetric determination methods for the investigation and monitoring of ecotoxic heavy metals in natural waters and atmospheric precipitates, *Sci. Tot. Environ.,* 37, 9, 1984.
44. **van den Berg, C. M. G.,** The electroanalytical chemistry of sea water, in *Chemical Oceanography,* Vol. 9, Riley, J. P. and Chester, Eds., Academic Press, London, 1988, 197.
45. **Mart, L. Nürnberg, H. W., and Rutzel, H.,** Comparative studies on cadmium levels in the North Sea, Norwegian Sea, Barents Sea, and the Eastern Arctic Ocean. *Fresenius Z. Anal. Chem.,* 317, 201, 1984.
46. **Pihlar, B., Valenta, P., and Nürnberg, H. W.,** New high-performance analytical procedure for the voltammetric determination of nickel in routine analysis of waters, biological materials and food, *Fresenius Z. Anal. Chem.,* 307, 337, 1981.
47. **Donat, J. R. and Bruland, K. W.,** Direct determination of dissolved cobalt and nickel in seawater by cathodic stripping voltammetry preceeded by collection of cyclohexane-1,2-dione dioxime complexes, *Anal. Chem.,* 60, 240, 1988.
48. **Donat, J. R. and van den Berg, C. M. G.,** A new cathodic stripping voltammetric method for determining organic copper complexation in seawater, *Mar. Chem.,* 38(1-2), 69, 1992.
49. **Measures, C. I. and Burton, J. D.,** Gas chromatographic method for the determination of selenite and total selenium in sea water, *Anal. Chim. Acta.,* 120, 177, 1980.
50. **Measures, C. I. and Edmond, J. M.,** Determination of beryllium in natural waters in real time using electron capture detection gas chromatography, *Anal. Chem.,* 58, 2065, 1986.
51. **Measures, C. I. and Edmond, J. M.,** Shipboard determination of aluminum in seawater at the nanomolar level by electron capture detection gas chromatography, *Anal. Chem.,* 61, 544, 1989.
52. **Schaule, B. K. and Patterson, C. C.,** Perturbations of the natural lead depth profile in the Sargasso Sea by industrial lead, in *Trace Metals in Seawater,* Wong, C. S., Boyle, E., Bruland, K. W., Burton, J. D., and Goldberg, E. D., Eds., Plenum Press, New York, 1983, 487.
53. **Flegal, A. R. and Patterson, C. C.,** Vertical concentration profiles of lead in the central Pacific at 15°N and 20°S, *Earth Planet. Sci. Lett.,* 64, 19, 1983.
54. **Stukas, V. J. and Wong, C. S.,** Accurate and precise analysis of trace levels of Cu, Cd, Pb, Zn, Fe, and Ni in sea water by isotope dilution mass spectrometry, in *Trace Metals in Seawater,* Wong, C. S., Boyle, E., Bruland, K. W., Burton, J. D., and Goldberg, E. D., Eds., Plenum Press, New York, 1983, 513.
55. **Berman, S. S., Sturgeon, R. E., Desaulniers, J. A. H., and Mykytiuk, A. P.,** Preparation of the sea water reference material for trace metals, NASS-1, *J. Mar. Pollut. Bull.,* 14, 69, 1983.
56. **Beauchemin, D. and Berman, S. S.,** Determination of trace metals in reference water standards by inductively coupled plasma mass spectrometry with on-line preconcentration, *Anal. Chem.,* 61, 1857, 1989.
57. **Piepgras, D. J. and Wasserburg, G. J.,** Neodymium isotopic variations in sea water, *Earth Planet. Sci. Lett.,* 50, 128, 1980.
58. **Elderfield, H. and Greaves, M. J.,** The rare earth elements in seawater, *Nature,* 296, 214, 1982.

59. **Elderfield, H. and Greaves, M. J.,** Determination of the rare earth elements in sea water, in *Trace Metals in Seawater,* Wong, C. S., Boyle, E., Bruland, K. W., Burton, J. D., and Goldberg, E. D., Eds., Plenum Press, New York, 1983, 427.
60. **German, C. R. and Elderfield, H.,** Rare earth elements in Saanich Inlet, British Columbia, a seasonally anoxic basin, *Geochim. Cosmochim. Acta,* 53, 2561, 1989.
61. **German, C. R. and Elderfield, H.,** Rare earth elements in the NW Indian Ocean, *Geochim. Cosmochim. Acta,* 54, 1929, 1990.
62. **Orians, K. J., Boyle, E. A., and Bruland, K. W.,** Dissolved titanium in the open ocean, *Nature,* 348, 322, 1990.
63. **Falkner, K. K. and Edmond, J. M.,** Determination of gold at femtomolar levels in natural waters by flow injection inductively coupled plasma quadrupole mass spectrometry, *Anal. Chem.,* 62, 1477, 1990.
64. **Falkner, K. K. and Edmond, J. M.,** Gold in seawater, *Earth Planet. Sci. Lett.,* 98, 208, 1990.
65. **Boyle, E. A., Handy, B., and van Geen, A.,** Cobalt determination in natural waters using cation-exchange liquid chromatography with luminol chemiluminescence detection, *Anal. Chem.,* 59, 1499, 1987.
66. **Sakamoto-Arnold, C. M. and Johnson, K. S.,** Determination of picomolar levels of cobalt in seawater by flow injection analysis with chemiluminescence detection, *Anal. Chem.,* 59, 1789, 1987.
67. **Coale, K. H., Stout, P. M., Johnson, K. S., and Sakamoto, C. M.,** Determination of copper in seawater using a flow injection method with chemiluminescence detection, *Anal. Chim. Acta,* 266, 345, 1992.
68. **Chapin, T. P., Johnson, K. S., and Coale, K. H.,** Rapid determination of manganese in sea water by flow injection analysis with chemiluminescence detection, *Anal. Chim. Acta,* 249, 469, 1991.
69. **Elrod, V. A., Johnson, K. S., and Coale, K. H.,** Determination of subnanomolar levels of iron(II) and total dissolved iron in seawater by flow injection analysis with chemiluminescence detection, *Anal. Chem.,* 63, 893, 1991.

69a. **Obata, H., Karatani, H., and Nakayama, E.,** Automated determin of iron in seawater by chelating resin concentration and chemiluminescence detection, *Anal. Chem.,* 65, 1524, 1993.

70. **Cutter, G. A.,** Species determination of selenium in natural waters, *Anal. Chim. Acta,* 98, 59, 1978.
71. **Cutter, G. A.,** Elimination of nitrite interference in the determination of Se by hydride generation, *Anal. Chim. Acta,* 149, 391, 1983.
72. **Andreae, M. O.,** The determination of the chemical species of some of the "hydride elements" (arsenic, antimony, tin and germanium) in seawater: methodology and results, in *Trace Metals in Seawater,* Wong, C. S., Boyle, E., Bruland, K. W., Burton, J. D., and Goldberg, E. D., Eds., Plenum Press, New York, 1983, 1.
73. **Gill, G. A. and Bruland, K. W.,** Mercury speciation in surface freshwater systems in California and other areas, *Environ. Sci. Technol.,* 24, 1392, 1990.
74. **Gill, G. A. and Fitzgerald, W.,** Picomolar mercury measurements in seawater and other materials using stannous chloride reduction and two-stage gold amalgamation with gas phase detection, *Mar. Chem.,* 20, 227, 1987.
75. **Cutter, L. S., Cutter, G. A., and San Diego-McGlone, M. L. C.,** Simultaneous determination of inorganic arsenic and antimony species in natural waters using selective hydride generation with gas chromatography/photoionization detection, *Anal. Chem.,* 63, 1138, 1991.
76. **Boyle, E. A., Sclater, F., and Edmond, J. M.,** On the marine chemistry of cadmium, *Nature,* 263, 42, 1976.
77. **Martin, J. H., Bruland, K. W., and Broenkow, W. W.,** Cadmium transport in the California Current, in *Marine Pollutant Transfer,* Windom, H. and Duce, R., Eds., D.C. Heath, Lexington, MA, 1976, 159.
78. **Bruland, K. W.,** Oceanographic distributions of cadmium, zinc, nickel, and copper in the North Pacific, *Earth Planet. Sci. Lett.,* 47, 176, 1980.
79. **Bruland, K. W. and Franks, R. P.,** Mn, Ni, Cu, Zn and Cd in the western north Atlantic, in *Trace Metals in Seawater,* Wong, C. S., Boyle, E., Bruland, K. W., Burton, J. D., and Goldberg, E. D., Eds., Plenum Press, New York, 1983, 395.
80. **Bruland, K. W., Knauer, G. A., and Martin, J. H.,** Zinc in north-east Pacific water, *Nature,* 271, 741, 1978.
81. **Hydes, D. J.,** Aluminium in seawater: control by inorganic processes, *Science,* 205, 1260, 1979.
82. **Hydes, D. J.,** Distribution of aluminium in waters of the North East Atlantic 25°N to 35°N, *Geochim. Cosmochim. Acta,* 47, 967, 1983.

83. **Moore, R. M.,** Oceanographic distributions of zinc, cadmium, copper and aluminum in waters of the central Arctic, *Geochim. Cosmochim. Acta,* 45, 2475, 1981.
84. **Olafsson, J.,** Mercury concentrations in the North Atlantic in relation to cadmium, aluminium, and oceanographic parameters, in *Trace Metals in Seawater,* Wong, C. S., Boyle, E., Bruland, K. W., Burton, J. D., and Goldberg, E. D., Eds., Plenum Press, New York, 1983, 475.
85. **Measures, C. I., Grant, B., Khadem, M., Lee, D. S., and Edmond, J. M.,** Distribution of Be, Al, Se, and Bi in the surface waters of the western North Atlantic and Caribbean, *Earth. Planet. Sci. Lett.,* 71, 1, 1984.
86. **Orians, K. J. and Bruland, K. W.,** Dissolved aluminum in the central North Pacific, *Nature,* 316, 427, 1985.
87. **Orians, K. J. and Bruland, K. W.,** The biogeochemistry of aluminum in the Pacific Ocean, *Earth Planet. Sci. Lett.,* 78, 397, 1986.
88. **Moore, R. M., and Millward, G. E.,** Dissolved-particulate interactions of aluminium in ocean waters, *Geochim. Cosmochim. Acta,* 48, 235, 1984.
89. **Measures, C. I., Edmond, J. M., and Jickells, T. D.,** Aluminium in the northwest Atlantic, *Geochim. Cosmochim. Acta,* 50, 1423, 1986.
90. **MacKenzie, F. T., Stoffyn, M., and Wollast, R.,** Aluminum in seawater: control by biological activity, *Science,* 199, 680, 1978.
91. **Caschetto, S. and Wollast, R.,** Vertical distribution of dissolved aluminum in the Mediterranean Sea, *Mar. Chem.,* 7, 141, 1979.
92. **Hydes, D. J., De Lange, G. J., and De Baar, H. J. W.,** Dissolved aluminium in the Mediterranean, *Geochim. Cosmochim. Acta,* 52, 2107, 1988.
93. **Measures, C. I. and Edmond, J. M.,** Aluminium as a tracer of the deep outflow from the Mediterranean, *J. Geophys. Res.,* 93, 591, 1988.
94. **Boyle, E. A. and Edmond, J. M.,** Copper in surface waters south of New Zealand, *Nature,* 253, 107, 1975.
95. **Boyle, E. A., Sclater, F., and Edmond, J. M.,** The distribution of dissolved copper in the Pacific, *Earth Planet. Sci. Lett.,* 37, 38, 1977.
96. **Gordon, R. M., Martin, J. H., and Knauer, G. A.,** Iron in northeast Pacific waters, *Nature,* 299, 611, 1982.
97. **Landing, W. M. and Bruland, K. W.,** The contrasting biogeochemistry of iron and manganese in the Pacific Ocean, *Geochim. Cosmochim. Acta,* 51, 29, 1987.
98. **Martin, J. H. and Gordon, R. M.,** Northeast Pacific iron distributions in relation to phytoplankton productivity. *Deep-Sea Res.,* 35, 177, 1988.
99. **Moore, R. M.,** The distribution of dissolved copper in the eastern Atlantic Ocean, *Earth Planet. Sci. Lett.,* 41, 461, 1978.
100. **Measures, C. I. and Edmond, J. M.,** Aluminium in the South Atlantic: steady state distribution of a short residence time element, *J. Geophys. Res.,* 95 C4, 5331, 1990.
101. **Martin, J. H., Gordon, R. M., Fitzwater, S., and Broenkow, W. W.,** VERTEX: phytoplankton/iron studies in the Gulf of Alaska, *Deep-Sea Res.,* 36, 649, 1989.

101a. **Skrabad, S. A.,** The estuarine and marine geochemistry of titanium, Ph.D. thesis, University of Delaware, Newark, 1993.

102. **Orians, K. J. and Bruland, K. W.,** Dissolved gallium in the open ocean, *Nature,* 332, 717, 1988.
103. **Shiller, A. M.,** Enrichment of dissolved gallium relative to aluminum in natural waters, *Geochim. Cosmochim. Acta,* 52, 1879, 1988.
104. **Koide, M., Stallard, M., Hodge, V., and Goldberg, E. D.,** Preliminary studies on the marine chemistry of ruthenium, *Neth. J. Sea. Res.,* 1986.
105. **Goldberg, E. D., Hodge, V. F., Kay, P., Stallard, M., and Koide, M.,** Some comparative marine chemistries of platinum and iridium, *Appl. Geochem.,* 1, 227, 1986.
106. **Lee, D. S.,** Palladium and nickel in northeast Pacific waters, *Nature,* 305, 5929, 1983.
107. **Koide M., Hodge, V., Yang, J. S., and Goldberg, E. D.,** Determination of rhenium in marine waters and sediments by graphite furnace atomic absorption spectrophotometry, *Anal. Chem.,* 59, 1802, 1987.

107a. **Colodner, D. C., Boyle, E. A., and Edmond, J. M.,** Determination of rhenium and platinum in natural waters and sediments, and iridium in sediments by flow injection isotope dilution inductively coupled plasma mass spectrometry, *Anal. Chem.,* 65, 1419, 1993.

108. **Lee, D. S. and Edmond, J. M.,** Tellurium species in seawater, *Nature,* 313, 782, 1985.

109. **Hodge, V. F., Stallard, M., Koide, M., and Goldberg, E. D.,** Platinum and the platinum anomaly in the marine environment, *Earth Planet. Sci. Lett.,* 72, 158, 1985.
110. **Hodge, V. F., Stallard, M., Koide, M., and Goldberg, E. D.,** Determination of platinum and iridium in marine waters, sediments, and organisms, *Anal. Chem.,* 58, 616, 1986.
111. **Goldberg, E. D.,** Comparative chemistry of the platinum and other heavy metals in the marine environment, *Pure Appl. Chem.,* 59, 565, 1987.
112. **Jacinto, G. S. and van den Berg, C. M. G.,** Different behavior of platinum in the Indian and Pacific Oceans, *Nature,* 338, 332, 1989.
112a. **McKelvey, B. A. and Orians, K. J.,** Dissolved zirconium in the North Pacific Ocean, *Geochim. Cosmochim. Acta,* 57, 3801, 1993.
112b. **McKelvey, B. A. and Orians, K. J.,** The marine geochemistry of zirconium and hafnium, *EOS,* 75(3), 78, 1994.
113. **Sunda, W. G. and Huntsman, S. A.** Effect of sunlight on redox cycles of manganese in the southwestern Sargasso Sea, *Deep-Sea Res.,* 35, 1297, 1988.
114. **Sunda, W. G. and Huntsman, S. A.,** Diel cycles in microbial manganese oxidation and manganese redox speciation in coastal waters of the Bahama Islands, *Limnol. Oceanogr.,* 35, 325, 1990.
115. **Sunda, W. G. and Huntsman, S. A.,** Effect of competitive interactions between manganese and copper on cellular manganese and growth in estuarine and oceanic species of the diatom *Thalassiosira, Limnol. Oceanogr.,* 28, 924, 1983.
116. **Martin, J. H. and Fitzwater, S. E.,** Iron deficiency limits phytoplankton growth in the northeast Pacific subarctic, *Nature,* 331, 341, 1988.
117. **Martin, J. H., Gordon, R. M., and Fitzwater, S.,** Iron in Antarctic waters, *Nature,* 345, 156, 1990.
118. **Waite, T. D. and Morel, F. M. M.,** Photoreductive dissolution of colloidal iron oxides in natural waters, *Environ. Sci. Technol.,* 18, 860, 1984.
119. **Rich, H. W. and Morel, F. M. M.,** Availability of well-defined iron colloids to the marine diatom *Thalassiosira weissflogii, Limnol. Oceanogr.,* 35, 652, 1990.
120. **Moffett, J. W. and Zika, R. G.,** Oxidation kinetics of Cu(I) in seawater: implications for its existence in the marine environment, *Mar. Chem.,* 13, 239, 1983.
121. **Moffett, J. W. and Zika, R. G.,** Measurement of copper(I) in surface waters of the subtropical Atlantic and Gulf of Mexico, *Geochim. Cosmochim. Acta,* 52, 1849, 1988.
122. **Moffett, J. W., Zika, R. G., and Brand, L. E.,** Distribution and potential sources and sinks of copper chelators in the Sargasso Sea, *Deep-Sea Res.,* 37, 27, 1990.
123. **Farley, K. J. and Morel, F. M. M.,** Role of coagulation in the kinetics of sedimentation, *Environ. Sci. Technol.,* 20, 187, 1986.
124. **Honeyman, B. D. and Santschi, P. H.,** A Brownian-pumping model for oceanic trace metal scavenging: evidence from Th Isotopes, *J. Mar. Res.,* 47, 951, 1989.
125. **Moran, S. B. and Moore, R. M.,** The distribution of colloidal aluminum and organic carbon in coastal and open ocean waters off Nova Scotia, *Geochim. Cosmochim. Acta,* 53, 2519, 1989.
126. **van den Berg, C. M. G.,** Determination of copper complexation with natural organic ligands in seawater by equilibration with MnO_2. II. Experimental procedures and application to surface seawater, *Mar. Chem.,* 11, 323, 1982.
127. **Davis, J. A.,** Complexation of trace metals by adsorbed natural organic matter, *Geochim. Cosmochim. Acta,* 48, 679, 1984.
128. **Davis, J. A. and Leckie, J. O.,** Effect of adsorbed complexing ligands on trace metal uptake by hydrous oxides, *Environ. Sci. Technol.,* 12, 1309, 1978.
129. **Bourg, A. C. M. and Schindler, P. W.,** Ternary surface complexes. I. Complex formation in the system silica-Cu(II)-ethylenediamine, *Chimia,* 32, 166, 1978.
130. **Spencer, D. W. and Brewer, P. G.,** Vertical advection diffusion and redox potentials as controls on the distribution of manganese and other trace metals dissolved in waters of the Black Sea, *J. Geophys. Res.,* 76, 5877, 1971.
131. **Emerson, S., Cranston, R. E., and Liss, P. S.,** Redox species in a reducing fjord: equilibrium and kinetic considerations, *Deep-Sea Res.,* 26A, 859, 1979.
132. **Bacon, M. P., Brewer, P. G., Spencer, D. W., Murray, J. W., and Goddard, J.,** Lead-210, polonium-210, manganese, and iron in the Cariaco Trench, *Deep-Sea Res.,* 27A, 119, 1980.
133. **Measures, C. I., McDuff, R. E., and Edmond, J. M.,** Selenium redox chemistry at GEOSECS I reoccupation, *Earth Planet. Sci. Lett.,* 49, 102, 1980.

134. **Measures, C. I., Grant, B. C., Mangum, B. J., and Edmond, J. M.,** The relationship of the distribution of dissolved selenium IV and VI in three oceans to physical and biological processes, in *Trace Metals in Seawater,* Wong, C. S., Boyle, E., Bruland, K. W., Burton, J. D., and Goldberg, E. D., Eds., Plenum Press, New York, 1983, 73.
135. **Cutter, G. A.,** Selenium in reducing waters, *Science,* 217, 829, 1982.
136. **Cutter, G. A. and Bruland, K. W.,** The marine biogeochemistry of selenium: a re-evaluation, *Limnol. Oceanogr.,* 29, 1179, 1984.
137. **Murray, J. W., Spell, B., and Paul, B.,** The contrasting geochemistry of manganese and chromium in the eastern tropical Pacific Ocean, in *Trace Metals in Seawater,* Wong, C. S., Boyle, E., Bruland, K. W., Burton, J. D., and Goldberg, E. D., Eds., Plenum Press, New York, 1983, 643.
138. **Turner, D. R., Whitfield, M., and Dickson, A. G.** The equilibrium speciation of dissolved components in freshwater and seawater at 25°C and 1 atm pressure, *Geochim. Cosmochim. Acta,* 45, 855, 1981.
139. **Byrne, R. H., Kump, L. R., and Cantrell, K. J.,** The influence of temperature and pH on trace metal speciation in seawater, *Mar. Chem.,* 25, 163, 1988.
140. **Ringbom, A.,** *Complexation in Analytical Chemistry,* Wiley-Interscience, New York, 1963.
141. **Zafiriou, O. C. and True, M. B.,** Interconversion of Fe(III) hydroxy complexes in seawater, *Mar. Chem.,* 8, 281, 1980.
142. **Hudson, R. J. M. and Morel, F. M. M.,** Iron transport in marine phytoplankton: kinetics of cellular and medium coordination reactions, *Limnol. Oceanogr.,* 35, 1002, 1990.
143. **Andreae, M. O.,** Determination of arsenic species in natural waters, *Anal. Chem.,* 49, 820, 1977.
144. **Andreae, M. O.,** Arsenic speciation in seawater and interstitial waters: the influence of biological-chemical interactions on the chemistry of a trace element, *Limnol. Oceanogr.,* 24, 440, 1979.
145. **Hambrick, G. A., III, Froelich, P. N., Jr., Andreae, M. O., and Lewis, B. L.,** Determination of methylgermanium species in natural waters by graphite furnace atomic absorption spectrometry with hydride generation, *Anal. Chem.,* 56, 421, 1984.
146. **Lewis, B. L., Froelich, P. N., and Andreae, M. O.,** Methylgermanium in natural waters, *Nature,* 313, 303, 1985.
147. **Mason, R. P. and Fitzgerald, W. F.,** Alkylmercury species in the equatorial Pacific, *Nature,* 347, 457, 1990
148. **Andreae, M. O., Asmodé, J.-F., Foster, P., and Van't dack, L.,** Determination of antimony (III), antimony (V), and methylantimony species in natural waters by graphite furnace atomic absorption spectrometry with hydride generation, *Anal. Chem.,* 53, 287, 1981.
149. **Byrd, J. T. and Andreae, M. T.,** Tin and methyltin species in seawater: concentrations and fluxes, *Science,* 218, 565, 1982.
150. **Andreae, M. O. and Byrd, J. T.,** Determination of tin and methyltin species by hydride generation and detection with graphite-furnace atomic absorption or flame emission spectrometry, *Anal. Chim. Acta,* 156, 147, 1984.
151. **Harvey, H. W.,** *Biological Chemistry and Physics of Sea Water,* Cambridge University Press, Cambridge, 1928.
152. **Cooper, L. H. N.,** Some conditions governing the solubility of iron, *Proc. R. Soc. London,* B124, 299, 1937.
153. **Mantoura, R. F. C.,** Organo-metallic interactions in natural waters, in *Marine Organic Chemistry,* Duursma, E. K. and Dawson, R., Eds., Elsevier Science Publ. Co., New York, 1981, 179.
154. **van den Berg, C. M. G., Buckley, P. J. M., and Dharmvanij, S.,** Determination of ligand concentrations and conditional stability constants in seawater. Comparison of the DPASV and MnO_2 adsorption techniques, in *Complexation of Trace Metals in Natural Waters,* Kramer, C. J. M. and Duinker, J. C., Eds., Nijhoff/Junk Publishers, 1984, 213.
155. **Sunda, W. G. and Ferguson, R. L.** Sensitivity of natural bacterial communities to additions of copper and to cupric ion activity: a bioassay of copper complexation in seawater, in *Trace Metals in Seawater,* Wong, C. S., Boyle, E., Bruland, K. W., Burton, J. D., and Goldberg, E. D., Eds., Plenum Press, New York, 1983, 871.
156. **van den Berg, C. M. G.,** Determination of the complexing capacity and conditional stability constants of complexes of copper (II) with natural organic ligands in seawater by cathodic stripping voltammetry of copper-catechol complex ions, *Mar. Chem.,* 15, 1, 1984.
157. **Buckley, P. J. M. and van den Berg, C. M. G.,** Copper complexation profiles in the Atlantic Ocean, *Mar. Chem.,* 19, 281, 1986.

158. **Sunda, W. G. and Hanson, A. K., Jr.,** Measurement of free cupric ion concentration in seawater by a ligand competition technique involving copper sorption onto C_{18} Sep Pak cartridges, *Limnol. Oceanogr.*, 32, 537, 1987.
159. **Moffett, J. W. and Zika, R. G.,** Solvent extraction of copper acetylacetonate in studies of copper (II) speciation in seawater, *Mar. Chem.*, 21, 301, 1987.
160. **Coale, K. H. and Bruland, K. W.,** Copper complexation in the Northeast Pacific, *Limnol. Oceanogr.*, 33, 1084, 1988.
161. **Coale, K. H. and Bruland, K. W.,** Spatial and temporal variability in copper complexation in the North Pacific, *Deep-Sea Res.*, 47, 317, 1990.
161a. **Donat, J. R. and Bruland, K. W.,** Organic copper complexation in Sargasso Sea surface waters: comparison to the North Pacific, *EOS*, 72(51), 44, 1992.
161b. **Donat, J. R., Lao, K. A., and Bruland, K. W.,** Speciation of dissolved copper and nickel in south San Francisco Bay: a multi-method approach, *Anal. Chim. Acta*, 284(3), 547, 1994.
162. **van den Berg, C. M. G.,** Determination of the zinc complexing capacity in seawater by cathodic stripping voltammetry of zinc-APDC complex ions, *Mar. Chem.*, 16, 121, 1985.
163. **Bruland, K. W.,** Oceanic zinc speciation: complexation of zinc by natural organic ligands in the central North Pacific, *Limnol. Oceanogr.*, 34, 267, 1989.
164. **Donat, J. R. and Bruland, K. W.,** A comparison of two voltammetric techniques for determining zinc speciation in Northeast Pacific Ocean waters, *Mar. Chem.*, 28, 301, 1990.
165. **Mantoura, R. F. C., Dickson, A., and Riley, J. P.,** The complexation of metals with humic materials in natural waters, *Estuarine Coastal Mar. Sci.*, 6, 387, 1978.
166. **Bruland, K. W.,** Complexation of cadmium by natural organic ligands in the central North Pacific, *Limnol. Oceanogr.*, 37(5), 1008, 1992.
167. **Capodaglio, G., Coale, K. H., and Bruland, K. W.,** Lead speciation in surface waters of the Eastern North Pacific, *Mar. Chem.*, 29, 221, 1990.
168. **Zhang, H., van den Berg, C. M. G., and Wollast, R.,** The determination of interactions of cobalt (II) with organic compounds in seawater using cathodic stripping voltammetry, *Mar. Chem.*, 28, 285, 1990.
169. **van den Berg, C. M. G. and Nimmo, M.,** Determination of interactions of nickel with dissolved organic material in seawater using cathodic stripping voltammetry, *Sci. Tot. Environ.*, 60, 185, 1987.
170. **Nimmo, M., van den Berg, C. M. G., and Brown, J.,** The chemical speciation of dissolved nickel, copper, vanadium, and iron in Liverpool Bay, Irish Sea, *Estuarine Coastal Shelf Sci.*, 29, 57, 1989.
171. **Sholkovitz, E. R., Boyle, E. A., and Price, N. B.,** The removal of dissolved humic acids and iron during estuarine mixing, *Earth Planet. Sci. Lett.*, 40, 130, 1978.
172. **Boyle, E. A., Edmond, J. M., and Sholkovitz, E. R.,** The mechanism of iron removal in estuaries, *Geochim. Cosmochim. Acta*, 41, 1313, 1977.
173. **Sholkovitz, E. R.,** The flocculation of dissolved Fe, Mn, Al, Cu, Ni, Co, and Cd during estuarine mixing, *Earth Planet. Sci. Lett.*, 41, 77, 1978.
173a. **Gledhill, M. and van den Berg, C. M. G.,** Determination of complexation of Fe(III) with natural organic complexing ligands using cathodic stripping voltammetry, *Mar. Chem.*
173b. **Rue, E. L. and Bruland, K. W.,** Complexation of Fe(III) by natural organic ligands in the central North Pacific as determined by competitive ligand equilibration/adsorptive cathodic stripping voltammetry, *Mar. Chem.*, submitted.
174. **Johnston, R.,** Seawater, the natural medium of phytoplankton. I. General features, *J. Mar. Biol. Assoc. UK*, 43, 427, 1963.
175. **Johnston, R.,** Seawater, the natural medium of phytoplankton. II. Trace metals and chelation, and general discussion, *J. Mar. Biol. Assoc. UK*, 44, 87, 1964.
176. **Barber, R. T. and Ryther, J. H.,** Organic chelators: factors affecting primary production in the Cromwell Current upwelling, *J. Exp. Mar. Biol. Ecol.*, 3, 191, 1969.
177. **Grill, E., Winnacker, E. L., and Zenk, M. H.,** Phytochelatins: the principal heavy metal-complexing peptides of higher plants, *Science*, 230, 674, 1985.
178. **Grill, E., Winnacker, E. L., and Zenk, M. H.,** Phytochelatins, a class of heavy metal-binding peptides from plants, are functionally analogous to metallothioneins, *Proc. Natl. Acad. Sci. U.S.A.*, 84, 439, 1987.
179. **Gekeler, W., Grill, E., Winnacker, E. L., and Zenk, M. H.,** Algae sequester heavy metals via synthesis of phytochelatin complexes, *Arch. Microbiol.*, 150, 197, 1988.
180. **McKnight, D. M. and Morel, F. M. M.,** Release of weak and strong copper-complexing agents by algae, *Limnol. Oceanogr.*, 24, 823, 1979.

181. **Waterbury, J. B., Watson, S. W., Guillard, R. R. L., and Brand, L. E.,** Widespread occurrence of a unicellular marine planktonic cyanobacterium, *Nature,* 277, 293, 1979.
182. **Glover, H. E., Keller, M. D., and Guillard, R. R. L.** Light quality and oceanic ultraphytoplankters, *Nature,* 319, 142, 1986.
183. **Iturriaga, R. and Marra, J.,** Temporal and spatial variability of chroococcoid cyanobacteria *Synechococcus* spp. specific growth rates and their contribution to primary production in the Sargasso Sea, *Mar. Ecol. Prog. Ser.,* 44, 175, 1988.
184. **Brand, L. E., Sunda, W. G., and Guillard, R. R. L.,** Reduction of marine phytoplankton reproduction rates by copper and cadmium, *J. Exp. Mar. Biol. Ecol.,* 96, 225, 1986.
185. **Bruland, K. W., Donat, J. R., and Hutchins, D. A.,** Interactive influences of bioactive trace metals on biological production in oceanic waters, *Limnol. Oceanogr.,* 36(8), 1555, 1991.
186. **DiTullio, G. R. and Laws, E. A.,** Impact of an atmospheric-oceanic disturbance on phytoplankton community dynamics in the North Pacific Central Gyre, *Deep-Sea Res.,* 38, 1305, 1991.
187. **Brand, L. E., Sunda, W. G., and Guillard, R. R. L.,** Limitation of marine phytoplankton reproductive rates by zinc, manganese, and iron, *Limnol. Oceanogr.,* 28, 1182, 1983.
188. **Price, N. M., Andersen, L. F., and Morel, F. M. M.,** Iron and nitrogen nutrition of equatorial Pacific plankton, *Deep-Sea Res.,* 38, 1361, 1991.
189. **Price, N. M. and Morel, F. M. M.,** Cadmium and cobalt substitution for zinc in a zinc-deficient marine diatom, *Nature,* 344, 658, 1990.
190. **Swift, D. G.,** Vitamins and phytoplankton growth, in *The Physiological Ecology of Phytoplankton,* Morris, I., Ed., Blackwell Scientific Publications, Oxford, 1980, 329.
191. **Pitner, I. J. and Altmeyer, V. L.,** Vitamin B_{12} binder and other algal inhibitors, *J. Phycol.,* 15, 391, 1979.
192. **Sclater, F. F., Boyle, E. A., and Edmond, J. M.,** On the marine geochemistry of nickel, *Earth Planet. Sci. Lett.,* 31, 119, 1976.
193. **Price, N. and Morel, F. M. M.,** Co-limitation of phytoplankton growth by nickel and nitrogen, *Limnol. Oceanogr.,* 1991.
194. **Sunda, W. G., Barber, R. T., and Huntsman, S. A.,** Phytoplankton growth in nutrient rich seawater; importance of copper-manganese cellular interactions, *J. Mar. Res.,* 39, 567, 1981.
195. **Murphy, L. S., Guillard, R. R. L., and Brown, J. F.,** The effects of iron and manganese on copper sensitivity in diatoms: differences in the responses of closely related neritic oceanic species, *Biol. Oceanogr.,* 3, 187, 1984.
196. **Sunda, W. G.,** Neritic-oceanic trends in trace-metal toxicity to phytoplankton communities, in Proceedings of the Fourth International Ocean Disposal Symposium, Plymouth, England, April 1983, 1986, 19.
197. **Rueter, J. G. and Morel, F. M. M.,** The interaction between zinc deficiency and copper toxicity as it affects the silicic acid uptake mechanisms of *Thalassiosira pseudonana, Limnol. Oceanogr.,* 26, 67, 1981.
198. **Foster, P. L. and Morel, F. M. M.,** Reversal of cadmium toxicity in a diatom: an interaction between cadmium activity and iron, *Limnol. Oceanogr.,* 27, 745, 1982.
199. **Harrison, G. I. and Morel, F. M. M.,** Antagonism between cadmium and iron in the marine diatom *Thalassiosira weisflogii, J. Phycol.,* 19, 495, 1983.
200. **Harrison, G. I. and Morel, F. M. M.,** Response of the marine diatom *Thalassiosira weissflogii* to iron stress, *Limnol. Oceanogr.,* 31, 989, 1986.
201. **Menzel, D. W., Hulburt, E. M., and Ryther, J. H.,** The effects of enriching Sargasso Sea water on the production and species composition of phytoplankton, *Deep-Sea Res.,* 10, 209, 1963.
202. **Barber, R. T.,** Organic ligands and phytoplankton growth in nutrient-rich seawater, in *Trace Metals and Metal-Organic Interactions in Natural Waters,* Singer, P. C., Ed., Ann Arbor Science, Ann Arbor, MI, 1973, 321.
203. **deBaar, H. J. W., Buma, A. G. J., Nolting, R. F., Cadee, G. C., Jacques, G., and Treguer, P. J.,** On iron limitation of the Southern Ocean: experimental observations in the Weddell and Scotia Seas, *Mar. Ecol. Prog. Ser.,* 65, 105, 1990.
204. **Buma, A. G. J., deBaar, H. J. W., Nolting, R. F., and van Bennekom, A. J.,** Metal enrichment experiments in the Weddell-Scotia Seas: effects of Fe and Mn on various plankton communities, *Limnol. Oceanogr.,* 36(8), 1865, 1991.

205. **Steemann-Nielsen, E. and Wium-Anderson, S.,** Copper ions as poison in the sea and in fresh water, *Mar. Biol.,* 6, 93, 1970.

206. **Huntsman, S. A. and Sunda, W. G.,** The role of trace metals in regulating phytoplankton growth, in *The Physiological Ecology of Phytoplankton,* Morris, I., Ed., Studies in Ecology, Vol. 7, Blackwell Scientific, Oxford, 1980, 285.

207. **Stauben, J. L. and Florence, T. M.,** Interactions of copper and manganese: a mechanism by which manganese alleviates copper toxicity to the marine diatom, *Nitzschia dostenium* (Ehrenberg) W. Smith, *Aquat. Toxicol.,* 7, 241, 1985.

208. **Hering, J. G., Sunda, W. G., Ferguson, R. L., and Morel, F. M. M.,** A field comparison of two methods for the determination of copper complexation: bacterial bioassay and fixed-potential amperometry, *Mar. Chem.,* 20, 299, 1987.

209. **Huizenga, D. L. and Kester, D. R.,** The distribution of total and electrochemically available copper in the northwestern Atlantic Ocean, *Mar. Chem.,* 13, 281, 1983.

Chapter 12

State of the Art and Future Trends

Brit Salbu and Eiliv Steinnes

CONTENTS

I. INTRODUCTION

As discussed in detail in previous chapters, trace element studies in natural waters are of great importance in order to increase our knowledge on natural processes in general and the chemical behavior and biological significance of the elements in particular. Another major goal for the study of trace elements in natural systems is to understand the relationship between anthropogenic releases (source terms) and future consequences for man and the environment. Such assessments can form the basis for authorized industrial releases, decisions on measures to be taken, and international consensus on critical loads with respect to trace metals and long-lived radionuclides. In order to estimate the transfer from various sources to critical response systems, information on trace elements in natural waters is needed.

In different natural water systems, the concentrations of trace elements vary by orders of magnitude reflecting variable influences from anthropogenic sources, geological and biological cycles, as well as key processes taking place in the water column. However, information on total concentrations is of limited value for assessment of future consequences, as the mobility and the bioavailability of trace elements will depend on the presence of different physico-chemical forms, interactions, and transformation processes as well as the kinetics involved.

In order to reduce the uncertainties of predictive models for trace elements and radionuclides in different aquatic systems several issues are of major concern:

- To ascertain contamination-free sampling and preanalysis handling
- To ascertain representative sampling (flowing waters) and sufficient volume of samples (interstitial waters)
- To establish concentration levels (except for oceans) for most of the trace elements and long-lived artificially produced radionuclides, and to identify sources of release
- To obtain information on physico-chemical forms, related in particular to mobility and bioavailability
- To identify factors and key processes influencing mobility and bioavailability for modeling purposes

For trace elements in natural water systems, investigations related to concentration levels, physico-chemical forms, and especially the key processes and mechanisms influencing mobility and bioavailability are required (see Chapters 1 and 2). The major challenges within analytical chemistry are to provide sufficiently sensitive and specific means to allow such studies to be undertaken (see Chapter 3). For multicomponent and multispecies systems, the use of data analysis and statistical methods is needed (see Chapter 4).

0-8493-6304-7/95/$0.00+$.50

II. PRIORITIES FOR FUTURE WORK

Among the chemical elements occurring on earth, 12 to 14 are present in most natural waters in concentrations that place them in the major element category. If we disregard the noble gas elements, about 65 elements occur in stable form at trace concentrations in all natural waters. In addition, 6 elements forming part of the U and Th decay series are present as relatively short-lived isotopes ($t_{1/2} < 2 \cdot 10^3$ yr). However, measurable amounts of man-made long-lived radioisotopes of transuranic elements (e.g., Pu and Am), Tc and I have also been introduced into the hydrological cycle.

If the present knowledge on trace elements in natural waters as summarized in the preceding chapters is compared with the above facts, it is clear that most of our experience lies with a fairly small number of naturally occurring trace elements, at least as far as fresh waters are concerned. A majority of the reported studies deal with the major crustal elements Al, Fe, Mn, and the eight elements Cr, Ni, Cu, Zn, As, Cd, Hg, and Pb, which are often referred to as heavy metals. In most cases, these elements are selected for study because of potential toxic effects towards biota. The elements V, Co, and Se are also fairly well represented, while the remaining 50 elements have been studied only sporadically in fresh waters. Some of these studies are relatively old, and the quality of data could be questioned.

Marine scientists, on the other hand, can present a much more complete list of elements studied in marine waters, as is evident from Table 3, Chapter 11. A great number of the studies leading to these data employed preconcentration procedures prior to the analytical determination of trace elements. It is reasonable to assume that more extensive use of such techniques in freshwater studies will have a positive impact on our knowledge about trace element occurrence and behavior in natural waters, particularly if combined with sensitive multielement analytical techniques such as ICP-MS.

Regarding the speciation of trace elements in natural waters, the current knowledge is very limited. Investigations on trace elements in fresh waters deal almost invariably with "total" or "filtrate" concentrations. Filtrates usually refer to samples from which suspended particles larger than 0.4 or 0.45 μm are removed. As different physico-chemical forms, especially of multivalent trace metals, may be present in the filtrate, e.g., colloids, caution should be taken when data is interpreted. The combination of size and charge fractionation techniques (ultrafiltration, ion-exchange chromatography, liquid-liquid extraction) as utilized for determination of Al species (see Chapter 3) demonstrates, however, the potential for development within this field.

In ocean waters, the feasibility of voltammetric methods has provided substantial information on labile species of certain trace metals. In general, however, most of the literature data on trace element speciation in natural waters is still based on thermodynamical calculations. As natural waters are dynamic systems and certain processes (e.g., solid-water interactions) may be rather slow, the assumption of equilibria may be questioned. Especially during episodic events, or in zones where waters of different qualities mix, these models are of little relevance. Experimental studies on trace element speciation with special reference to mobility and bioavailability is therefore highly needed, not the least because of the potential for improvement in the uncertainties of predicting models.

In the following discussion, the state of the art in relation to the specific water systems addressed in previous chapters is summarized and the need for future work is focused.

A. PRECIPITATION

Although considerable data on trace elements in precipitation have been collected over the last 15 to 20 years in many countries, the data produced until recently is of limited value because sufficient care was not taken to avoid significant contamination problems during collection and preanalysis handling of precipitation samples (see Chapter 5). Therefore, there remains much to be learned about the wet deposition of trace elements in most areas of the world.

Most precipitation monitoring networks are restricted to a limited number of "anthropogenic" trace metals. In order to extend the knowledge about other elements and to quantify their contribution to atmospheric cycles, a large amount of work is needed, in particular to identify the relative contributions from various natural and anthropogenic sources. Multielement studies, e.g., by ICP-MS combined with multivariate statistics, would be highly beneficial. In a recent study based on ICP-MS, the contributing source categories for about 25 elements at Norwegian background stations were identified.[1]

However, information on the physico-chemical forms of trace elements in precipitation is rather scarce. The importance of speciation was demonstrated after the Chernobyl accident, where radioactive particles, being a major component in fallout close to the site and a minor component in far distant areas, influenced the mobility and bioavailability of associated radionuclides.[2,3] As the size distribution pattern is essential for predicting models, the source term should be improved if, for instance, cascade impactors were utilized for sampling of air or fractionation techniques were used for precipitation.

B. ICE AND SNOW SHEETS

Although a substantial fraction of the freshwater reservoir on earth is stored in snow and ice caps in the polar regions, information on the significance of this huge reservoir in global cycling of trace element is limited due to the slow turnover rate of this "fossil" water. Trace elements in polar ice have received considerable attention, however, as dated ice cores may reveal the historical development of the global atmospheric transport of trace elements with regard to the contribution from neutral sources as well as anthropogenic input.

The first convincing evidence of recent atmospheric pollution of Pb in polar regions was reported already in 1969.[4] However, the most reliable trace element data from polar core studies are from the last 10 years. Substantial reduction in contamination during sampling and preanalysis handling has facilitated a quite accurate record of not only the recent anthropogenic contribution of elements such as Pb, Cd, and Zn, but also variations in the natural atmospheric fluxes of these elements over the last 150,000 years.[5] It appears that virtually 100% of the preindustrial Pb can be explained by the crustal component. As far as recent trace element deposition in Antarctica is concerned, convincing evidence for human contribution of Pb is reported, while recent concentrations of Zn, Cd, and Cu in Antarctic ice are close to the levels observed in ice several thousand years old. In Antarctica, the 1940 Pb level appeared to be 10 times the preindustrial one, while a similar situation was evident in Greenland already in 1750.

The work done so far has revealed only a small fraction of the total information on long-term variations in global fluxes of trace elements that is still hidden in the polar ice caps. Prehistoric levels of most trace elements are frequently in the order of a few pg g^{-1} or less, which means that direct determination is in most cases very difficult, and chemical preconcentration techniques must be employed.[6] The low concentrations involved and the potential risk of preanalysis contamination illustrate the analytical challenge. Thus, speciation work is left for the future.

C. PORE WATERS

1. Sediment Waters

Trace element studies in sediment interstitial waters have most commonly included the crustal major elements Al, Mn, Fe, and the additional trace metals V, Cr, Ni, Cu, Zn, As, Cd, and Pb. Investigations of other trace elements in interstitial waters are fairly scarce and further work is awaited, in particular for elements that undergo redox reactions within the Eh range normally encountered in surface sediments. One specific area where more work is needed (see Chapter 6) is studies to understand early diagnosis of trace elements. Another important area for future work is investigations on key processes and binding mechanisms governing solid-water interactions as well as resuspension. Especially for sediments previously contaminated by trace metals or radionuclides from anthropogenic sources, future releases may represent a hazard. Furthermore, sediment quality criteria based on sediment-water studies should be established, as the bioavailability of trace elements in sediments is related to their chemical activity in the interstitial water phase.

2. Soil Waters

The collection of sufficient volumes of interstitial waters from soils to enable investigation of trace elements is often a problem. This may be a main reason why the information on trace element distributions in soil interstitial waters seems to be rather limited (see Chapter 6). It is also obvious that contamination problems associated with sampling often represent a serious limitation for such studies.

Soil water trace element studies appear to be very important for at least two reasons. First, it is conceivable that soil waters would reflect the plant-available part of trace elements better than total soil samples or various extracts commonly used, in particular if the speciation in the interstitial phase could be revealed. Second, the mobility of pollutants within the surface layers of soil may be conveniently studied. Even though information on the speciation is essential, it seems that speciation studies are more difficult in soil waters than in most other natural water systems, partly due to representative sampling problems.

Besides more specific process-oriented studies, there is a need for more general knowledge of the occurrence and behavior of trace elements in the soil solution system, e.g., concentration levels and forms in different soil types, seasonal variations, dependence of pH and Eh, interactions with natural occurring components, influence of microorganisms, etc. Soil interstitial water trace element studies, being essential for understanding vertical transport and plant uptake, appear to be a vast area for future work.

D. GROUNDWATERS

Very few published studies of groundwater systems include analyses of trace elements, and, consequently, accurate trace element concentration data that is representative of uncontaminated groundwater systems is scarce (Chapter 7). Moreover, existing data are almost exclusively total concentrations, which may vary by two to three orders of magnitude between groundwaters in different geological settings. Only eight elements (i.e., Cr, Ni, Cu, Zn, As, Cd, Hg, Pb) have been selected for discussion in Chapter 7, because the knowledge on other trace elements in groundwaters at natural concentration levels is rather limited.

In principle, the concentrations and physico-chemical forms of trace elements in groundwaters are largely determined by pH and redox conditions, as well as the presence and concentration of various complexing agents. Based on such data, models are used for calculation of the distribution of trace elements species. Microchemical phenomena that occur for elements at trace concentrations (see Chapter 2), e.g., adsorption and coprecipitation, however, severely limit the value of model calculations. The lack of stability constants for complexes with natural organic substances, and the fact that equilibrium is rarely achieved in natural waters, further limit the predicting power of these models. Development of procedures allowing experimental determination of speciation sufficiently rapidly and without altering the original hydrochemical conditions, e.g., on line systems, is, therefore, a high priority task.

E. LAKE WATERS

Reported studies of trace elements in lake waters largely confine themselves to the crustal major elements Al, Fe, Mn, and some trace metals that are toxic to aquatic organisms, i.e., Cr, Co, Ni, Cu, Zn, As, Cd, Hg, and Pb. For other trace elements, e.g., the essential elements V, Se, Mo, and I, the available information is limited. Reliable data for some elements that occur in relatively high proportions in crustal rocks such as Ti, Rb, Sr, Zr, Ba, and rare earths are also rather scarce, and the same applies to natural and anthropogenic radionuclides.

The literature on trace elements in lakes is quite scattered, and much information is likely to be found in publications with limited distribution. Thus, Chapter 8 is not complete with regard to available data, but presumably provides a representative survey. Most available data are either "total" or "filterable" concentrations. A relatively limited number of studies have proceeded further in attempting to define the "truly soluble" (low molecular weight) fraction of trace elements using techniques such as *in situ* dialysis and hollow fiber filtration (see Chapter 3). However, an attempt is made in Chapter 8 to classify the 20 most commonly studied trace elements into 5 groups according to their speciation behavior. It is clear that factors such as ionic strength, humic substances, pH, and, not least, biological activity affect the behavior of trace elements in lakes. Furthermore, lakes cover an extreme range of systems, e.g., from small oligotrophic humic lakes in Fennoscandia to large saline lakes in East Africa.

There is still a great need for further information on the occurrence and behavior of trace elements in lakes, their speciation, bioavailability, and turnover times. It is also essential to identify natural sources and levels, since most of the available data are on lakes where anthropogenic sources have affected the system to a varying extent.

F. RIVER WATERS

The behavior of trace elements in flowing water (see Chapter 9) is additionally complicated because of the dynamics of the system. Trace elements are distributed between three major compartments: water, seston, and benthic, with transfers occurring between these compartments depending upon the physical, chemical, and biological conditions at the time. Many of the budgets for trace element transport quoted in the literature may also be seriously underestimated, as high flow events are not adequately sampled. On the other hand, baseline trace element concentrations in uncontaminated rivers may be significantly lower than presently accepted values because of contamination problems during sampling and preanalysis handling. Additionally, the majority of speciation schemes in use at the present time fail to account for trace elements associated with the colloidal fraction, which is an important transporting agent especially for trace metals. Furthermore, episodic events (pH, flow, temperature, etc.) will influence the distribution of species. Consequently, there is still a great need for further work on trace elements in rivers.

Conceptually, the key processes likely to influence the behavior of trace elements in rivers are well known from many laboratory studies. However, laboratory-based information is not easily transferable to explain quantitatively and qualitatively the actual behavior in nonequilibrium natural water systems. Future success in explaining the behavior of trace elements in rivers is likely to arise from approaches that more completely couple the physical, chemical, and biological components occurring in the system.

G. ESTUARINE WATERS

Existing knowledge on trace element behavior in estuaries is largely based on field studies in a limited number of rivers, where the behavior of a given element is classified according to its distribution between a particulate and a filterable fraction. In most of these studies, colloids are included in the assumed dissolved fraction and, therefore, may give rise to significant misinterpretation.

Limitations exist in the diagnostic methods available for the interpretation of trace element reactivity from distributional data (see Chapter 10). Controlled laboratory experiments are, therefore, needed in order to identify the mechanisms and kinetics of the reactions involved. In particular, further investigations are recommended on laboratory-derived distribution coefficients, such as their relations to the composition of medium (key variables), particles and colloids (biogenic, lithogenic, microbial components), and reaction reversibility. Extensive and systematic characterization of estuarine particles and colloids (composition, surface area/exchangeable surface sites, surface charge) is also needed. In addition, information of processes and kinetics occurring in nonequilibrium mixing zone systems is needed for improving the uncertainties of predictive models for describing the behavior of trace elements in estuaries and the transfer to oceans.

H. OCEAN WATERS

As opposed to the situation for fresh waters, most elements in the periodic table have been subject to investigation in marine waters, and the total concentration and the distribution in the water column are fairly well known for over 40 trace elements (see Chapter 11, Table 3), even though the concentrations involved are usually very low.

As far as speciation is concerned, experimental studies have, to a large extent, been concentrated on Cu and Zn. A limited number of studies have been reported for other trace metals such as Co, Ni, Cd, and Pb. Some elements are known to occur in the ocean as organometallic compounds, e.g., methyl forms of Ge, As, Se, Sn, Sb, and Ag, which, in part, define their speciation. In general, however, available information on trace element speciation in ocean water is still largely dependent on theoretical calculations using thermodynamical data. Much work remains to be carried out in order to verify experimentally the extent to which the calculated speciation is representative for the various trace elements. In particular, further studies on the complexation of metals with naturally occurring organic ligands, as well as their association with colloidal material, are expected in the years to come.

REFERENCES

1. **Berg, T., Røyset, O., and Steinnes, E.,** 28 trace elements in atmospheric precipitation at six background stations in Norway 1989-1990, *Atmos. Environ.*, in press.
2. **Loshchilov, N. A., Kashparov, V. A., Yudin, Ye. B., Protsak, V. P., Zhurba, M. A., and Pashakov, A. E.,** The radiobiological impact of hot beta-particles from the Chernobyl fallout. *Risk Assessment, IAEA, Vienna,* Part I, 34-39, 1992.
3. **Oughton, D. H., Salbu, B., Riise, G., Lien, H., Østby, G., and Nøren, A.,** Radionuclide Mobility and bioavailability in Norwegian- and Soviet soils, *Analyst,* 117, 481, 1992.
4. **Murozumi, M., Chow, T. J., and Patterson, C. C.,** Chemical concentrations of pollutant lead aerosols, terrestrial dusts and seasalts in Greenland and Antarctic snow strata, *Geochim. Cosmochim. Acta.,* 33, 1242, 1969.
5. **Boutron, C. F., Candelone, J. -P., and Görlach, U. I.,** Time variations in natural and anthropogenic heavy metals in Antarctic and Greenland ice and snow, *Sci. Tot. Environ.,* submitted.
6. **Suttie, E. D. and Wolff, E. W.,** Preconcentration method for electrothermal atomic absorption spectrometric analysis for heavy metals in Antarctic snow at sub ng kg^{-1} levels, *Anal. Chim. Acta,* 258, 229, 1992.

INDEX

A

B

C

D

E

F

G

H

I

K

L

M

N

O

P

Q

R

S

T

U

V

W

Y

Z